A Bridge Between Lie Theory and Frame Theory

A Bridge Between Lie Theory and Frame Theory

Applications of Lie Theory to Harmonic Analysis

Vignon Oussa
Bridgewater State University
Bridgewater, MA, USA

WILEY

Published by John Wiley & Sons, Inc., Hoboken, New Jersey.
Published simultaneously in Canada.

Library of Congress Cataloging-in-Publication Data

Names: Oussa, Vignon, author.
Title: A bridge between Lie theory and frame theory : applications of Lie theory to harmonic analysis / Vignon Oussa.
Description: Hoboken, New Jersey : Wiley, [2025] | Includes bibliographical references and index.
Identifiers: LCCN 2024057768 (print) | LCCN 2024057769 (ebook) | ISBN 9781119712138 (hardback) | ISBN 9781119712145 (adobe pdf) | ISBN 9781119712152 (epub)
Subjects: LCSH: Frames (Vector analysis) | Lie groups. | Geometry, Differential. | Harmonic analysis.
Classification: LCC QA433 .O87 2023 (print) | LCC QA433 (ebook) | DDC 512/.482–dc23/eng/20241226
LC record available at https://lccn.loc.gov/2024057768
LC ebook record available at https://lccn.loc.gov/2024057769

Cover Design: Wiley
Cover Image: © Flavio Coelho/Getty Images

Set in 9.5/12.5pt STIXTwoText by Straive, Chennai, India

Contents

Preface

Duffin and Schaeffer developed frame theory in the 1950s to solve problems in nonharmonic Fourier series. The quest for redundant and flexible, basis-like reproducing systems for signal analysis led to the rediscovery of frames in the early 1980s. The seminal work of Daubechies, Meyer, Grossman, and others underscored the influential role of frames in the study of signal analysis via wavelet theory and time-frequency analysis. Frame theory, a branch of harmonic analysis, has now matured into a dynamic and active field, drawing its strengths from diverse areas like representation theory and Lie theory. The proposed book centers around the discretization problem of representations of Lie groups, which can be articulated as follows: Given a representation of a Lie group, under what conditions is it possible to sample one of its orbits for constructing frames with prescribed properties?

The book aims to offer a differential geometric treatment of the mathematics involved in finding a satisfactory solution to the discretization problem for a class of representations relevant to current research in wavelet, Gabor wavelet, shearlet theories, and their various generalizations.

The first part of the book delivers a self-contained exposition of differential geometry, Lie theory, representation theory, and frame theory. This material will arm the reader with the technical tools necessary to probe the discretization problem of a class of representations at a deep level. The second part of the book applies the theories laid down in the initial section to characterize a type of representations of Lie groups that can be discretized to construct frames and other basis-like systems. For example, Lie groups with frames of translates, sampling, and interpolation spaces on Lie groups are examined. Inspired by applications in signal analysis, the book introduces novel constructions of frames possessing additional sought-after features like boundedness, compact support, continuity, fast decay, and smoothness. The book features an extensive list of meticulously composed and concrete examples. Moreover, it draws relevant connections with ongoing dynamic research problems in frame theory, wavelet theory, time-frequency analysis, and other related branches of harmonic analysis.

Massachusetts 2023 *Vignon Oussa*

Acknowledgments

The evolution of novel approaches to solving problems in mathematics is often a nonlinear process. Various individuals significantly shaped my growth throughout the journey of writing this book. I want to specifically acknowledge the profound influence Hartmut Führ's book, "Abstract Harmonic Analysis of Continuous Wavelet Transforms," has had on my mathematical development. As a Ph.D. student, Hartmut's book was not just a gateway to my own mathematical journey but also exemplified excellent mathematical writing to my novice self. Additionally, in his monograph, Hartmut introduced original results that sparked my interest in further expanding these ideas.

Later, at the onset of my career as an Assistant Professor of Mathematics at Bridgewater State University, I was fortunate enough to visit Hartmut in Aachen, Germany. During these three weeks, I had the opportunity to work, learn, and collaborate with him directly. I distinctly remember presenting him with a rough draft containing ideas related to the central theme of the present book (the discretization problem of representations for constructing frames), and his encouragement and insightful feedback were invaluable in helping me refine and expand these concepts. His input significantly boosted my confidence and further fueled my drive for mathematical research. These experiences underscore the profound and lasting impact effective mentorship can have on a young mathematician's career trajectory.

Writing a book is a daunting task, necessitating patience, dedication, and perseverance. When I was first approached to draft a proposal for this book, I received unwavering support from many individuals, including Hartmut Führ, Karlheinz Gröchenig, Akram Aldroubi, and Jens Christensen. As I stepped back from research to devote myself to this book, I encountered several challenges that tested my resolve to complete this project. However, the sustained support of friends, colleagues, and my children: Sena, Kemi, and Donah, kept me steadfast. I am particularly indebted to Hartmut once again, who generously proofread significant portions of the book, providing invaluable feedback that has been integrated into subsequent revisions.

The completion of this book was supported by the National Science Foundation under Grant No. 2205852. I gratefully acknowledge the NSF's support, which made this work possible. I would also like to thank Professor Kasso Okoudjou for his invaluable support and collaboration throughout this project.

Vignon Oussa

1

Introduction

As a doctoral candidate at Saint-Louis University, I have had the opportunity to study Frame Theory, taught by Professor Darrin Speegle, Representation Theory, taught by Professor Bradley Currey, and Differential Geometry, a course conducted by Professor James Hebda. In my pursuit of a research direction, I found myself captivated by the field of Abstract Harmonic Analysis, specifically continuous wavelets on noncommutative domains [3]. Through my explorations, I have come to appreciate the immense utility of the tools provided by Abstract Harmonic Analysis across other subjects within Harmonic Analysis, including Frame Theory, Wavelet Theory, and Shearlet Theory [1, 4, 6].

Recognizing the broad significance of Frame Theory within the Harmonic Analysis community, my aim was to enhance the accessibility and clarity of the connections between Abstract Harmonic Analysis and Frame Theory in academic literature. However, my previous approach, which heavily leaned on Representation Theory, potentially obscured these connections. This was compounded by the fact that only a minority of researchers in Harmonic Analysis possess expertise in Representation Theory.

In response, I endeavored to minimize reliance on Representation Theory in framing constructions of frames, wavelets, and their broader generalizations. Instead, I introduced a series of constructions characterized by a more pronounced emphasis on Lie Theory and differential geometry.

Despite dispensing with stringent Representation-theoretic assumptions, the Lie theoretic/differential geometric approach still necessitated a decent understanding of Differential Geometry and Lie Theory. Thus, I decided to pen this book in order to clarify these connections and make them more comprehensible to researchers and graduate students interested in studying Frame Theory within the context of wavelet, time-frequency, and shearlet theories.

A pertinent question arises for any vector space, whether finite- or infinite-dimensional, structured as a Hilbert space: how can we systematically construct basis-like objects such that any vector within the given vector space can be

A Bridge Between Lie Theory and Frame Theory: Applications of Lie Theory to Harmonic Analysis, First Edition. Vignon Oussa.

expressed as a series expansion by appropriately scaling the elements of the chosen building block? The term "basis-like" underscores that we do not necessarily aim for an outcome where every vector has a unique series expansion in terms of the provided building blocks. Randomly selecting vectors is not an efficient approach for the systematic construction of these building blocks, particularly if we are seeking blocks with prescribed properties. One approach, established throughout the literature, begins with a unitary representation. A unitary representation is a continuous map between a group and a group of unitary operators acting on the given Hilbert space. For example, if the group is connected, a particular orbit associated with the representation may be excessively redundant to be of use. Thus, we might consider sampling an orbit of a representation along a countable set, with the hope that the countable set of vectors retained can serve as building blocks for the space. However, the following question remains: how can we achieve this?

In this book, I delve into the fundamental question of what the primary building blocks of any data collection are and how one can systematically generate such a collection. Any data, whether it be a recorded piece of music, a movie, or a signal from distant galaxies, can be regarded as a function defined over a set. This set, generally has the structure of a (smooth) manifold and can be equipped with a measure with respect to which a satisfactory theory of integration can be established. This allows us to view the space of all possible data on a fixed manifold as a Hilbert space. The study of frames, wavelets and shearlets in this context, will serve as the central theme of this book.

As a first example, let us consider the picture below:

Color intensity map represented on a 2×2 grid.

This data could be naturally viewed as a function defined over a finite 2×2 grid (a zero-dimensional manifold). Different real numbers represent varying intensities of the same color in each cell, and we may for instance code this image as the following square matrix of order two:

$$M = \begin{bmatrix} 1 & 2 \\ 2 & 1 \end{bmatrix}.$$

Thus, there is a one-to-one correspondence between the set of all pictures described in the format described above and the set of all real square matrices of order two. Indeed, let

$$V = \left\{ \begin{bmatrix} x & y \\ z & w \end{bmatrix} : x, y, z, w \in \mathbb{R} \right\}$$

be the vector space of objects representing pictures modeled as square matrices of order two, with basis $\{E_1, E_2, E_3, E_4\}$, where

$$E_1 = \begin{bmatrix} 1 & 0 \\ 0 & 0 \end{bmatrix}, \quad E_2 = \begin{bmatrix} 0 & 1 \\ 0 & 0 \end{bmatrix}, \quad E_3 = \begin{bmatrix} 0 & 0 \\ 1 & 0 \end{bmatrix}, \quad E_4 = \begin{bmatrix} 0 & 0 \\ 0 & 1 \end{bmatrix}.$$

Every picture is represented by the linear combination

$$p = \sum_{k=1}^{4} c_k E_k = \begin{bmatrix} c_1 & c_2 \\ c_3 & c_4 \end{bmatrix}$$

for some unique sequence of real numbers $(c_k)_{k=1}^{4}$. Moreover, the collection $\{E_1, E_2, E_3, E_4\}$ is a set of basic building blocks for V. Lastly, consider the isomorphism ϕ that transforms every square matrix of order two to its flatten version. Applying ϕ to a matrix with elements c_1, c_2, c_3, and c_4, results in a vertical vector with elements c_1, c_2, c_3, and c_4 from top to bottom.

$$\phi\left(\begin{bmatrix} c_1 & c_2 \\ c_3 & c_4 \end{bmatrix} \right) = \begin{bmatrix} c_1 \\ c_2 \\ c_3 \\ c_4 \end{bmatrix},$$

and letting $\Gamma = \langle T \rangle$ be the cyclic group of order four generated by matrix

$$T = \begin{bmatrix} 0 & 0 & 0 & 1 \\ 1 & 0 & 0 & 0 \\ 0 & 1 & 0 & 0 \\ 0 & 0 & 1 & 0 \end{bmatrix}.$$

That is

$$\Gamma = \left\{ \begin{bmatrix} 1 & 0 & 0 & 0 \\ 0 & 1 & 0 & 0 \\ 0 & 0 & 1 & 0 \\ 0 & 0 & 0 & 1 \end{bmatrix}, \begin{bmatrix} 0 & 0 & 0 & 1 \\ 1 & 0 & 0 & 0 \\ 0 & 1 & 0 & 0 \\ 0 & 0 & 1 & 0 \end{bmatrix}, \begin{bmatrix} 0 & 0 & 1 & 0 \\ 0 & 0 & 0 & 1 \\ 1 & 0 & 0 & 0 \\ 0 & 1 & 0 & 0 \end{bmatrix}, \begin{bmatrix} 0 & 1 & 0 & 0 \\ 0 & 0 & 1 & 0 \\ 0 & 0 & 0 & 1 \\ 1 & 0 & 0 & 0 \end{bmatrix} \right\}.$$

The first matrix is an identity matrix of order four. The second matrix has its first element in the fourth position, its second element in the first position, its third element in the second position, and its fourth element in the third position, with all other elements as zeros. The third matrix has its first element in the third position, its second element in the fourth position, its third element in the first position, and its fourth element in the second position. The fourth matrix has its first element in the second position, its second element in the third position, its third element in the fourth position, and its fourth element in the first position. Letting $E = \text{col}_4(T)$ (the vector obtained by extracting the 4th-column of T) it is easy to verify that

$$\{E_1, E_2, E_3, E_4\} = \{\phi^{-1}(T^k E) : 0 \le k \le 3\}.$$

In other words, the set of basic building blocks $\{E_1, E_2, E_3, E_4\}$ of V is generated by allowing a finite cyclic group of order 4 to act on a single element of this set. This process provides us with a systematic method for constructing a basis for the vector space of pictures made of 2×2 pixels with different intensities of the same color.

The interested reader is invited to interact with a graphical user interface available through Mathematica for further exploration of these concepts. [Click here]

More generally, let U be an invertible matrix of order 4 commuting with T and therefore every element of Γ. To this end, it suffices to allow U to be an invertible circulant matrix of the type

$$U = \sum_{k=0}^{3} u_k T^k = \begin{bmatrix} u_0 & u_3 & u_2 & u_1 \\ u_1 & u_0 & u_3 & u_2 \\ u_2 & u_1 & u_0 & u_3 \\ u_3 & u_2 & u_1 & u_0 \end{bmatrix}$$

such that

$$\det(U) = \left((u_0 - u_2)^2 + (u_1 - u_3)^2 \right) (u_0 - u_1 + u_2 - u_3)(u_0 + u_1 + u_2 + u_3) \neq 0.$$

In other words, U is a polynomial function of T, and for any integer k, $UT^k E = T^k UE$. As a result,

$$\mathcal{B} = \left\{\phi^{-1}\left(T^k UE\right) : 0 \le k \le 3\right\} = \left\{\phi^{-1}\left(\begin{bmatrix} 0 & 0 & 0 & 1 \\ 1 & 0 & 0 & 0 \\ 0 & 1 & 0 & 0 \\ 0 & 0 & 1 & 0 \end{bmatrix}^k \begin{bmatrix} u_0 \\ u_1 \\ u_2 \\ u_3 \end{bmatrix}\right) : 0 \le k \le 3\right\}$$

is a basis for V as well. This procedure describes a very convenient method for constructing in a systematic fashion, a large class of bases for V.

For a second example, we consider a thin rod modeled as the open unit interval $(0, 1)$ (a bounded one-dimensional smooth manifold). Let us imagine that the temperature of the rod is naturally regarded as a function defined over the open set $(0, 1)$. Assume furthermore that the space of all possible temperatures on this rod is modeled as the Hilbert space $L^2(0, 1)$. Thanks to Fourier analysis [7], it is known that the collection

$$\mathcal{E} = \left\{e_k : t \mapsto e^{2\pi ikt} : k \in \mathbb{Z}\right\} \tag{1.1}$$

of complex exponentials parametrized by the integers forms an orthonormal basis for $L^2(0, 1)$. Moreover, any square-integrable function (in this case, temperature) on a unit interval can be written as a series expansion of complex exponentials suitably scaled by some complex coefficients. Precisely, for any vector $v \in \mathcal{H}$, there exists a sequence $\left(c_k\right)_{k\in\mathbb{Z}}$ such that

$$v = \sum_{k\in\mathbb{Z}} c_k e_k \tag{1.2}$$

with convergence in the norm of $L^2(0, 1)$. Next, the sequence $\left(c_k\right)_{k\in\mathbb{Z}}$ of coefficients obtained in such a series expansion are scalars given by projecting the fixed vector v on each element of this building block. That is, each complex number c_k is uniquely determined as follows:

$$c_k = \left\langle v, e_k\right\rangle = \int_0^1 v(t)\, e^{-2\pi ikt}\, dt.$$

Moreover, the series obtained from this procedure (1.2) is stable, since convergence is unconditional when one is concerned with orthonormal bases. To construct this orthonormal basis from a group acting on a single vector, one may proceed as follows. First, let $\mathbb{R}$ be the additive group acting unitarily in $L^2(0, 1)$ by multiplication by complex exponentials as follows. Given real numbers x, t and a vector $f \in L^2(0, 1)$,

$$\left[x \star f\right](t) = e^{2\pi itx} f(t)$$

and it turns out that $\mathcal{E} = \mathbb{Z} \star 1_{(0,1)}$ is an orthonormal basis obtained by sampling the orbit $\mathbb{R} \star 1_{(0,1)}$ along the integers [7]. Furthermore, let U be a bounded invertible operator such that for every integer k, $U\left(k \star 1_{(0,1)}\right) = k \star \left(U1_{(0,1)}\right)$. Then $\mathbb{Z} \star$

$\left(U1_{(0,1)}\right)$ is a Riesz basis [1] for $L^2(0,1)$. For instance, letting ϕ be a uniformly continuous function on $(0,1)$ with an empty zero set and letting $U = U_\phi$ be a bounded invertible multiplicative operator acting as follows: $[Uf](t) = \phi(t) f(t)$ then one obtains that $\mathbb{Z} \star U1_{(0,1)} = \mathbb{Z} \star \phi(t)$ is a Riesz basis for $L^2(0,1)$.

If we were to replace the rod discussed in the previous example by a rod of infinite length modeled as the real line (an unbounded one-dimensional manifold), we would be interested in constructing an orthonormal basis for $L^2(\mathbb{R})$, which is obtained by sampling the group orbit of some vector. A viable strategy would be to view the real line as an infinite union of finite intervals. By constructing an orthonormal basis for each finite interval, the union of such bases would therefore form a basis for the entire space itself. To accomplish this, let us consider the Heisenberg group

$$H = \left\{ M_{(x,y,z)} = \begin{bmatrix} 1 & x & z \\ 0 & 1 & y \\ 0 & 0 & 1 \end{bmatrix} : x, y, z \in \mathbb{R} \right\}$$

acting unitarily on $L^2(\mathbb{R})$ by translation, modulation (frequency-translation) and multiplication by complex exponentials as follows. Given real numbers x, y, z,

$$\begin{aligned} \left[M_{(0,y,0)} \star f\right](t) &= e^{-2\pi i y t} f(t) && \text{(frequency-translation)}, \\ \left[M_{(x,0,0)} \star f\right](t) &= f(t-x) && \text{(time-translation action)}, \\ \left[M_{(0,0,z)} \star f\right](t) &= e^{2\pi i z} f(t) && \text{(phase translation)}. \end{aligned}$$

In addition, it is worth pointing out that the Heisenberg group is also isomorphic to a semidirect product group of the type $\mathbb{R}^2 \rtimes \mathbb{R}$ endowed with the noncommutative group operation

$$(v,t)(w,s) = \left(v + e^{tX} w, t+s\right),$$

where

$$X = \begin{bmatrix} 0 & 1 \\ 0 & 0 \end{bmatrix} \quad \text{and} \quad e^{tX} = \begin{bmatrix} 1 & t \\ 0 & 1 \end{bmatrix}.$$

In any case, one searches for a pair (f, Γ) such that $f \in L^2(\mathbb{R})$ and Γ is a countable subset of the Heisenberg group such that $\Gamma \star f$ is a basis-like set for $L^2(\mathbb{R})$. For instance, it is easy to verify that

$$\left\{ M_{(j,0,0)} M_{(0,k,0)} \star 1_{\left[-\frac{1}{2},\frac{1}{2}\right]} : k, j \in \mathbb{Z} \right\} = \left\{ M_{(j,k,jk)} \star 1_{\left[-\frac{1}{2},\frac{1}{2}\right]} : k, j \in \mathbb{Z} \right\}$$

is an orthonormal basis for $L^2(\mathbb{R})$. This construction also gives rise to a large class of orthonormal bases. Indeed, let U be a unitary operator which leaves the set

$$\Lambda = \left\{ M_{(j,0,0)} M_{(0,k,0)} : k, j \in \mathbb{Z} \right\} = \left\{ M_{(j,k,jk)} : k, j \in \mathbb{Z} \right\}$$

invariant. Then

$$UM_{(j,k,jk)}\star 1_{\left[-\frac{1}{2},\frac{1}{2}\right]} = M_{(j,k,jk)}\star\left(U1_{\left[-\frac{1}{2},\frac{1}{2}\right]}\right)$$

and $\left\{M_{(j,k,jk)}\star\left(U1_{\left[-\frac{1}{2},\frac{1}{2}\right]}\right) : k,j\in\mathbb{Z}\right\}$ is an orthonormal basis for $L^2(\mathbb{R})$. For instance, if $\mathcal{F}$ is the Fourier transform acting on $L^2(\mathbb{R})$ then, $\mathcal{F}\Lambda = \Lambda\mathcal{F}$,

$$\left[\mathcal{F}1_{\left[-\frac{1}{2},\frac{1}{2}\right]}\right](\xi) = s(\xi) = \frac{\sin(\pi\xi)}{\pi\xi}$$

and as a result, $\{M_{(j,k,jk)}\star s : k,j\in\mathbb{Z}\}$ must also be an orthonormal basis for $L^2(\mathbb{R})$.

For a fourth example, let $\mathcal{H}$ be the space of square-integrable vectors on the real line. Next, let $G = \mathbb{R}\rtimes\mathbb{R}$ be the semidirect product group equipped with the following operation: $(v,t)(w,s) = \left(v+e^t w, t+s\right)$ and let G acts unitarily in $L^2(\mathbb{R})$ as follows:

$$(v,t)\star f(s) = e^{2\pi i e^{-s} v} f(s-t).$$

The interested reader is invited to verify that the above is indeed a group action, which is also norm-preserving. Part of the group acts by translation, while another part of the group acts with a modulation-like action. Next, let h, f be compactly supported functions on the real line, and let us assume that f is supported on some closed interval of the form $[-\epsilon,\epsilon]$ for some positive real number ϵ. Then the change of variable $\xi = e^{-s}$ in the second integral below gives

$$\begin{aligned}\langle h,(v,0)\star f\rangle_{L^2(\mathbb{R})} &= \int_{[-\epsilon,\epsilon]} h(s)\,e^{-2\pi i e^{-s} v}\overline{f(s)}ds\\ &= \int_{[e^{-\epsilon},e^{\epsilon}]} \frac{h\left(\ln\left(\xi^{-1}\right)\right)\overline{f\left(\ln\left(\xi^{-1}\right)\right)}}{\xi} e^{-2\pi i\xi v}\,d\xi\end{aligned}$$

and as such,

$$\sum_{k\in\mathbb{Z}}\left|\langle h,(k,0)\star f\rangle_{L^2(\mathbb{R})}\right|^2 = \sum_{k\in\mathbb{Z}}\left|\int_{[e^{-\epsilon},e^{\epsilon}]} \frac{h\left(\ln\left(\xi^{-1}\right)\right)\overline{f\left(\ln\left(\xi^{-1}\right)\right)}}{\xi} e^{-2\pi i\xi k}\,d\xi\right|^2. \tag{1.3}$$

Observing that the length of the interval $[e^{-\epsilon},e^{\epsilon}]$ is equal to $|[e^{-\epsilon},e^{\epsilon}]| = 2\sinh(\epsilon)$, if $\epsilon = \operatorname{arcsinh}(1/2)$ then $[e^{-\epsilon},e^{\epsilon}]$ is a unit interval, and as such, the trigonometric system $\left\{\xi\mapsto e^{-2\pi i\xi k} : k\in\mathbb{Z}\right\}$ must be an orthonormal basis for $L^2([e^{-\epsilon},e^{\epsilon}])$. Observing additionally that the scalar

$$\left|\int_{[e^{-\epsilon},e^{\epsilon}]} \frac{h\left(\ln\left(\xi^{-1}\right)\right)\overline{f\left(\ln\left(\xi^{-1}\right)\right)}}{\xi} e^{-2\pi i\xi k}\,d\xi\right|$$

is the length of the projection of the vector $\xi \mapsto \frac{h(\ln(\xi^{-1}))\overline{f(\ln(\xi^{-1}))}}{\xi}$ onto $\xi \mapsto e^{-2\pi i \xi k}$, thanks to Parsevalś equality, the following is immediate:

$$\left\| \left(\int_{[e^{-\epsilon}, e^{\epsilon}]} \frac{h\left(\ln\left(\xi^{-1}\right)\right)\overline{f\left(\ln\left(\xi^{-1}\right)\right)}}{\xi} e^{-2\pi i \xi k}\, d\xi \right)_{k \in \mathbb{Z}} \right\|_{l^2(\mathbb{Z})}$$
$$= \left\| \xi \mapsto \frac{h\left(\ln\left(\xi^{-1}\right)\right)\overline{f\left(\ln\left(\xi^{-1}\right)\right)}}{\xi} \right\|_{L^2(\mathbb{R})}.$$

Returning to (1.3), since

$$\left| \frac{h\left(\ln\left(\xi^{-1}\right)\right)\overline{f\left(\ln\left(\xi^{-1}\right)\right)}}{\xi} \right|^2 = \left| \frac{h\left(\ln\left(\xi^{-1}\right)\right) f\left(\ln\left(\xi^{-1}\right)\right)}{\xi^{1/2}} \right|^2 \frac{1}{\xi},$$

it follows that

$$\sum_{k \in \mathbb{Z}} \left| \langle h, (k, 0) \star f \rangle_{L^2(\mathbb{R})} \right|^2 = \int_{[e^{-\epsilon}, e^{\epsilon}]} \left| \frac{h\left(\ln\left(\xi^{-1}\right)\right) f\left(\ln\left(\xi^{-1}\right)\right)}{\xi^{1/2}} \right|^2 \frac{d\xi}{\xi}.$$

Suppose additionally that $\xi \mapsto \xi^{-1/2} f\left(\ln\left(\xi^{-1}\right)\right)$ is the indicator function of the unit interval $[e^{-\epsilon}, e^{\epsilon}]$. In other words, $f(s) = 1_{[-\varepsilon, \varepsilon]}(s) \cdot e^{-\frac{s}{2}}$ and

$$\sum_{k \in \mathbb{Z}} \left| \langle h, (k, 0) \star f \rangle_{L^2(\mathbb{R})} \right|^2 = \int_{[e^{-\epsilon}, e^{\epsilon}]} \left| h\left(\ln\left(\xi^{-1}\right)\right) \right|^2 \frac{d\xi}{\xi}$$
$$= \left\| (h \circ (-\ln)) \cdot 1_{e[-\epsilon, \epsilon]} \right\|^2_{L^2\left(\mathbb{R}, \frac{d\xi}{\xi}\right)}.$$

Changing variable again by allowing ξ to be equal to e^{-s}, we obtain

$$\sum_{k \in \mathbb{Z}} \left| \langle h, (k, 0) \star f \rangle_{L^2(\mathbb{R})} \right|^2 = \int_{[-\epsilon, \epsilon]} |h(s)|^2 ds = \left\| h \cdot 1_{[-\epsilon, \epsilon]} \right\|^2_{L^2(\mathbb{R}, ds)}.$$

Furthermore, since $|[-\epsilon, \epsilon]| = 2\epsilon = 2\text{arcsinh}\,(1/2) = \alpha$ and since the collection of sets $\left\{ \left[-\frac{\alpha}{2}, \frac{\alpha}{2}\right) + \ell : \ell \in \alpha\mathbb{Z} \right\}$ forms a perfect tiling of the real line, appealing to the fact that $\star$ is a unitary action, we derive the following:

$$\sum_{\ell \in \alpha\mathbb{Z}} \sum_{k \in \mathbb{Z}} \left| \langle h, (0, \ell)(k, 0) \star f \rangle_{L^2(\mathbb{R})} \right|^2 = \sum_{\ell \in \alpha\mathbb{Z}} \int_{\left[-\frac{\alpha}{2}, \frac{\alpha}{2}\right]} |(0, -\ell) \star h(s)|^2 ds.$$

Furthermore, since $\left[(0, -\ell) \star h\right](s) = h(s + \ell)$,

$$\sum_{\ell \in \alpha\mathbb{Z}} \sum_{k \in \mathbb{Z}} \left| \langle h, (0, \ell)(k, 0) \star f \rangle_{L^2(\mathbb{R})} \right|^2 = \sum_{\ell \in \alpha\mathbb{Z}} \int_{\left[-\frac{\alpha}{2}, \frac{\alpha}{2}\right]} |h(s + \ell)|^2 ds$$
$$= \int_{\mathbb{R}} |h(s)|^2 ds$$
$$= \|h\|^2_{L^2(\mathbb{R})}.$$

Finally, appealing to the fact that the set of all compactly supported functions on the real line is dense in $L^2(\mathbb{R})$, we conclude that the mapping

$$h \mapsto \left(\left\langle h, (0,\ell)(k,0) \star \left(s \mapsto 1_{[-\varepsilon,\varepsilon]}(s) \cdot e^{-\frac{s}{2}} \right) \right\rangle_{L^2(\mathbb{R})} \right)_{(k,\ell)\in\mathbb{Z}\times\alpha\mathbb{Z}}$$

is an isometric embedding of $L^2(\mathbb{R})$ into the sequence Hilbert space $l^2(\mathbb{Z} \times \alpha\mathbb{Z})$. In other words, whatever information is encoded in any square-integrable vector h can also be compressed into the countable set

$$\beta = \left(\left\langle h, \left(e^{\ell} k, \ell\right) \star \left(s \mapsto 1_{[-\varepsilon,\varepsilon]}(s) \cdot e^{-\frac{s}{2}} \right) \right\rangle_{L^2(\mathbb{R})} \right)_{(k,\ell)\in\mathbb{Z}\times\alpha\mathbb{Z}}.$$

Additionally, it is possible to fully recover h from β. As such, we may view the countable set β as a building block for $L^2(\mathbb{R})$.

This approach for constructing building blocks of function spaces by sampling the group orbit of a fixed vector is a method that has successfully been tackled in a variety of subjects: wavelet, generalized wavelet, shearlet, and generalized shearlet theory. Generally, this approach is described as follows. Fix a group acting unitarily on a Hilbert space. Given any fixed vector in the space, the collection of vectors produced via this group action is called an orbit. Moreover, the collection of all possible orbits forms a partition of the space. If for instance, the given action is faithful (that is, only the identity element fixes a vector), the given group parametrizes every orbit, and one aims to sample a fixed orbit to construct a countable building block for the given space.

Similarly to the previous example, with wavelet theory [1, 4], the underlying group is also the affine group (that is, the group of affine transformations on the real line) acting by translation and dilation in $L^2(\mathbb{R})$ (a difference choice of action). This group can also be realized as a matrix group of the type

$$A = \left\{ M_{(a,x)} = \begin{bmatrix} a & x \\ 0 & 1 \end{bmatrix} : a, x \in \mathbb{R} \quad \text{and} \quad a \neq 0 \right\}.$$

Note that A is isomorphic to a semidirect product group of the form $\mathbb{R} \rtimes (\mathbb{R}\setminus\{0\})$ with group operation given by

$$(x, a)(y, b) = (x + ay, ab).$$

Furthermore A acts unitarily on $L^2(\mathbb{R})$ as follows:

$$\left[M_{(a,0)} \star f\right](t) = a^{-1/2} f\left(a^{-1}t\right) \quad \text{(dilation action)}$$
$$\left[M_{(1,x)} \star f\right](t) = f(t - x) \quad \text{(translation action)}$$

and it is a useful exercise for the reader to verify that $(a, x) \mapsto M_{(a,x)} = M_{(1,x)} \circ M_{(a,0)}$ is an injective homomorphism (a monomorphism). Additionally, one seeks building blocks of the form $\Gamma \star f$ for some countable set $\Gamma \subset A$. To this end, recall that the Fourier transform, formally defined as:

$$\mathcal{F} : f \mapsto \left(\xi \mapsto \int_{\mathbb{R}} f(t)\, e^{-2\pi i t\xi}\, dt \right)$$

is a unitary operator which intertwines translation with pointwise multiplication by complex exponentials. In this context, it follows that

$$\begin{aligned}\mathcal{F}\left[M_{(1,x)} \star f\right](\xi) &= \int_{\mathbb{R}} f(t-x)\, e^{-2\pi i t\xi}\, dt \\ &= e^{-2\pi i x\xi} \int_{\mathbb{R}} f(t)\, e^{-2\pi i t\xi}\, dt \\ &= e^{-2\pi i x\xi} \mathcal{F} f(\xi).\end{aligned}$$

Next, let f be the Fourier inverse of the indicator function of $E = \left[-1, -\frac{1}{2}\right] \cup \left[\frac{1}{2}, 1\right]$. That is,

$$f(t) = \int_{-1}^{-1/2} e^{2\pi i t\xi}\, d\xi + \int_{1/2}^{1} e^{2\pi i t\xi}\, d\xi = \frac{\sin(2\pi t) - \sin(\pi t)}{\pi t}.$$

As a subset of the real line, since the collection of sets

$$\left\{ \left(\left[-1, -\frac{1}{2}\right) \cup \left[\frac{1}{2}, 1\right) \right) + k : k \in \mathbb{Z} \right\}$$

forms a tiling of the real line, it follows that $\left\{ \xi \mapsto e^{-2\pi i k\xi} \cdot 1_E(\xi) : k \in \mathbb{Z} \right\}$ is an orthonormal basis for $L^2(E)$. Appealing to the fact that unitary operators map orthonormal bases to orthonormal bases, one verifies that the collection $\left\{ M_{(1,k)} \star f : k \in \mathbb{Z} \right\}$ is an orthonormal basis for the Hilbert subspace $\mathcal{F}^{-1}\left(L^2(E)\right)$. Next, for any nonzero real number a,

$$\begin{aligned}\mathcal{F}\left[M_{(a,0)} \star f\right](\xi) &= a^{-1/2} \int_{\mathbb{R}} f\left(a^{-1} t\right) e^{-2\pi i t\xi}\, dt \\ &= a^{1/2} \int_{\mathbb{R}} f(t)\, e^{-2\pi i t(a\xi)}\, dt \\ &= a^{1/2} \mathcal{F} f(a\xi)\end{aligned}$$

and since $\left\{ 2^j E : j \in \mathbb{Z} \right\}$ tiles the real line as well,

$$L^2(\mathbb{R}) = \bigoplus_{j \in \mathbb{Z}} \mathcal{F}^{-1}\left(L^2\left(2^j E\right)\right).$$

Noting additionally that

$$M_{(2^j,0)} M_{(1,k)} = \begin{bmatrix} 2^j & 0 \\ 0 & 1 \end{bmatrix} \begin{bmatrix} 1 & k \\ 0 & 1 \end{bmatrix} = \begin{bmatrix} 2^j & 2^j k \\ 0 & 1 \end{bmatrix},$$

we derive that the system of vectors

$$\left\{ M_{(2^j,0)} M_{(1,k)} \star f : k, j \in \mathbb{Z} \right\} = \left\{ M_{(2^j, 2^j k)} \star f : k, j \in \mathbb{Z} \right\}$$

is an orthonormal basis for $L^2(\mathbb{R})$.

Shearlet theory on the other hand, uses a group obtained by extending the affine group by some shearing action. Precisely, let S be a semidirect product group of the form $S = \mathbb{R}^3 \rtimes \mathbb{R}^3$ with group operation given by $(v,t)(w,s) = (v + t \cdot w, t + s)$, where

$$t \cdot w = \exp\left(\begin{bmatrix} t_1 & t_2 & t_3 \\ 0 & t_1 & t_2 \\ 0 & 0 & t_1 \end{bmatrix}\right)\begin{bmatrix} w_1 \\ w_2 \\ w_3 \end{bmatrix} = \begin{bmatrix} e^{t_1} & t_2 e^{t_1} & \frac{1}{2} e^{t_1}\left(t_2^2 + 2t_3\right) \\ 0 & e^{t_1} & t_2 e^{t_1} \\ 0 & 0 & e^{t_1} \end{bmatrix}\begin{bmatrix} w_1 \\ w_2 \\ w_3 \end{bmatrix}.$$

Then S acts unitarily on $L^2\left(\mathbb{R}^3\right)$ as follows:

$$(v,t) \star f(s) = \exp\left(2\pi i \left\langle \begin{bmatrix} e^{-s_1} \\ -s_2 e^{-s_1} \\ \frac{1}{2} s_2^2 e^{-s_1} - s_3 e^{-s_1} \end{bmatrix}, \begin{bmatrix} v_1 \\ v_2 \\ v_3 \end{bmatrix} \right\rangle\right) \cdot f(s - t).$$

This group operation includes shearing transformations, which are a kind of linear transformation that "slants" the space, besides translations and dilations. Moreover, one seeks a basis-like set for $L^2\left(\mathbb{R}^3\right)$ of the form $\Gamma \star f$ for some countable set $\Gamma \subset S$ and some vector $f \in L^2\left(\mathbb{R}^3\right)$. To this end, we may proceed as follows. Let E be a compact subset of $\mathbb{R}^3$ with a nonempty interior. Next, let h be a compactly supported and smooth function defined on $\mathbb{R}^3$ (a test function). Also, let $b : \mathbb{R}^3 \to \mathbb{R}^3$ be a smooth vector-valued function defined as follows:

$$b(s) = \begin{bmatrix} e^{-s_1} \\ -s_2 e^{-s_1} \\ \frac{1}{2} s_2^2 e^{-s_1} - s_3 e^{-s_1} \end{bmatrix} = \exp\left(\begin{bmatrix} -s_1 & 0 & 0 \\ -s_2 & -s_1 & 0 \\ -s_3 & -s_2 & -s_1 \end{bmatrix}\right)\begin{bmatrix} 1 \\ 0 \\ 0 \end{bmatrix}.$$

Then letting ω be a continuous function defined on $\mathbb{R}^3$, one verifies that

$$\left\langle h, (\kappa, 0) \star \left(\omega \cdot 1_E\right)\right\rangle = \int_E h(s)\, \omega(s)\, e^{-2\pi i \langle b(s), \kappa \rangle}\, ds.$$

We will next prove that up to a change of variable, the formula above is the Fourier transform of a compactly supported function. Indeed, since b is everywhere injective, letting $\xi = b(s)$, we obtain that

$$\begin{aligned} \left\langle h, (\kappa, 0) \star \left(\omega \cdot 1_E\right)\right\rangle &= \int_{b(E)} h\left(b^{-1}(\xi)\right) \omega\left(b^{-1}(\xi)\right) e^{-2\pi i \langle \xi, \kappa \rangle} d\left(b^{-1}(\xi)\right) \\ &= \int_{b(E)} \frac{h\left(b^{-1}(\xi)\right) \omega\left(b^{-1}(\xi)\right)}{\left|\xi_1^3\right|} e^{-2\pi i \langle \xi, \kappa \rangle}\, d\xi. \end{aligned}$$

The factor $\left|\xi_1^3\right|^{-1}$ in the integrand arises from the change of variables. This factor compensates for the way in which the transformation by $b(s)$ stretches or compresses different regions of the space. Suppose next that E is small enough to ensure that $b(E)$ is contained in a fundamental domain of the integer lattice $\mathbb{Z}^3$.

Then summation over $\mathbb{Z}^3$ gives

$$\sum_{\kappa\in\mathbb{Z}^3}\left|\left\langle h,(\kappa,0)\star\left(\omega\cdot 1_E\right)\right\rangle\right|^2=\sum_{\kappa\in\mathbb{Z}^3}\left|\int_{b(E)}\frac{h\left(b^{-1}(\xi)\right)\omega\left(b^{-1}(\xi)\right)}{\left|\xi_1^3\right|}e^{-2\pi i\langle\xi,\kappa\rangle}d\xi\right|^2.$$

Noting that the sequence $\left(\int_{b(E)}\frac{h(b^{-1}(\xi))\omega(b^{-1}(\xi))}{|\xi_1^3|}e^{-2\pi i\langle\xi,\kappa\rangle}\,d\xi\right)_{\kappa\in\mathbb{Z}^3}$ represents the Fourier coefficients of $\xi\mapsto\frac{h(b^{-1}(\xi))\omega(b^{-1}(\xi))}{|\xi_1^3|}1_{b(E)}(\xi)$, it follows that

$$\sum_{k\in\mathbb{Z}^3}\left|\left\langle h,(\kappa,0)\star\left(\omega\cdot 1_E\right)\right\rangle\right|^2=\int_{b(E)}\left|\frac{h\left(b^{-1}(\xi)\right)\omega\left(b^{-1}(\xi)\right)}{\left|\xi_1^3\right|^{1/2}}\right|^2\frac{d\xi}{\left|\xi_1^3\right|}.$$

Suppose now that ω is a continuous function satisfying $\omega\left(b^{-1}(\xi)\right)=\left|\xi_1^3\right|^{1/2}$. Then

$$\begin{aligned}\sum_{k\in\mathbb{Z}^3}\left|\left\langle h,(\kappa,0)\star\left(\omega\cdot 1_E\right)\right\rangle\right|^2&=\int_{b(E)}\left|\frac{h\left(b^{-1}(\xi)\right)\left|\xi_1^3\right|^{1/2}}{\left|\xi_1^3\right|^{1/2}}\right|^2\frac{d\xi}{\left|\xi_1^3\right|}\\&=\int_{b(E)}\left|h\left(b^{-1}(\xi)\right)\right|^2\frac{d\xi}{\left|\xi_1^3\right|}.\end{aligned}$$

Furthermore, the change of variable $\xi=b(s)$ allows us to proceed as follows:

$$\sum_{k\in\mathbb{Z}^3}\left|\left\langle h,(\kappa,0)\star\left(\omega\cdot 1_E\right)\right\rangle\right|^2=\int_E|h(s)|^2ds.$$

Finally, suppose that E is a fundamental domain for a full-rank lattice L. Then

$$\sum_{\ell\in L}\sum_{k\in\mathbb{Z}^3}\left|\left\langle h,((0,\ell)(\kappa,0))\star\left(\omega\cdot 1_E\right)\right\rangle\right|^2=\sum_{\ell\in L}\int_E|h(s+\ell)|^2ds=\|h\|^2_{L^2(\mathbb{R}^3)}$$

and since the set of all compactly supported functions on the real line is dense in $L^2\left(\mathbb{R}^3\right)$, we may conclude that

$$h\mapsto\left(\left\langle h,((0,\ell)(\kappa,0))\star\left(\omega\cdot 1_E\right)\right\rangle_{L^2(\mathbb{R}^3)}\right)_{(k,\ell)\in\mathbb{Z}^3\times L}$$

is an isometric embedding of $L^2\left(\mathbb{R}^3\right)$ into the sequence space $l^2\left(\mathbb{Z}^3\times L\right)$.

1.1 Organization of the Book

The underlying set (affine, Heisenberg, and shearlet groups) in each of the construction above has the structure of a semidirect product Lie group. Moreover, the operators arising from each group action are contained in the range of some unitary representation (called induced characters) of the groups at hand [2, 5]. Thus, to generalize the methods of wavelet, Gabor, and shearlet theories, a certain degree of fluency in the subjects of Lie theory seems useful.

The main aim of the present book is to (a) offer an introduction to both Lie and frame (basis-like object) theories and (b) provide new constructions of bases-like objects by only exploiting differential geometric properties of the group action of interest. Mainly motivated by the classical construction of wavelets, Gabor wavelets, shearlets, and their various generalizations, we will be particularly interested in a type of group action obtained from a class of representations known as induced characters. When concerned with the construction of basis-like objects associated with induced characters, we will establish a novel theory based only on differential geometric properties of the group action of interest. We will aim to describe a general theory in which all examples described above fit. To this end, we will consider semidirect product groups of the type $G = N \rtimes H$, where N is a simply connected commutative Lie group and H is a closed subgroup of the automorphism group of N. For any fixed character χ (a one-dimensional representation of N), there is an associated action of G on the space of square-integrable vectors on H (with respect to an H-invariant measure $d\mu$) such that given $f \in L^2(H, d\mu)$, $(n,h) \in N \times H$,

$$\left[(n,h) \star f\right](s) = \chi\left(s^{-1}(n)\right) \cdot f\left(h^{-1}s\right) = \left[s \bullet \chi\right](n) f\left(h^{-1}s\right),$$

where $s(n)$ is determined by the action of $s \in H$ as an automorphism map on N. By only considering specific differential geometric properties of the smooth map $s \mapsto s \bullet \chi$, we will provide explicit construction of frames generated by a fixed vector exhibiting a number of desired prescribed properties (compact support and various level of smoothness). The theory developed in this book stands in stark contrast with the representation-theoretic approach, which is far more common in the literature. Additionally, we hope that the interested reader will find the (differential geometric) approach proposed in this text relevant to ongoing research topics on the subjects of wavelet theory, time-frequency analysis, and shearlet theory.

Chapter 2 provides a comprehensive introduction to differentiable manifolds, drawing on Dr. James Hebda's Differentiable Geometry course at Saint Louis University, textbooks by Spivak and John Lee, and discussions on math .stackexchange.com. It revisits Euclidean calculus concepts like the Implicit Function Theorem before introducing topological manifolds as locally linear, Hausdorff, and second countable spaces. By layering a differentiable structure on topological manifolds, the chapter extends familiar notions from Euclidean calculus, such as diffeomorphism and derivatives, to these broader structures. It explores different classes of submanifolds, introduces the concept of a derivation in the context of smooth functions on a manifold, and discusses how derivations form a Lie algebra. Furthermore, it defines the pushforward of a derivation in the context of a differentiable function between two manifolds, establishes the concept of tangent vectors, and concludes by introducing forms and tensor fields to facilitate integration on a manifold.

Chapter 3 introduces the basic concepts and results of Lie theory, which studies the properties of differentiable manifolds with group structures. Key topics covered include vector fields, Lie derivatives, Lie brackets, Lie groups, Lie algebras, exponential maps, invariant measures, homogeneous spaces, matrix Lie groups, and solvable Lie groups. The chapter draws from various sources, including lecture notes and textbooks, with the goal of preparing the groundwork for further developments in frame theory on manifolds.

Chapter 4 explores the action of Lie groups on vector spaces, particularly function spaces over manifolds, from the perspective of representation theory. This branch of mathematics represents abstract groups as linear transformations on vector spaces. Initiated by Sophus Lie and later expanded by Hermann Weyl and Eugene Wigner, the chapter discusses concepts like unitary representations of Lie groups and induced representations, which allow for the creation of new representations from existing ones. Key topics such as invariant spaces, equivalent representations, irreducible representations, Schur's lemma, direct integral, and induced representation are introduced, with examples and applications to the integrability theory of representations. The chapter recommends texts by Folland and Kaniuth and Taylor [2, 5] for a more comprehensive understanding of these topics.

Chapter 5 explores the concept of frames, which are sets of vectors that can be used to represent any vector in a Hilbert space. Frames generalize the notion of bases and orthonormality, essential tools in linear algebra and functional analysis. The chapter also reviews the concepts of orthonormal bases, Riesz bases, and Parseval's equality, which are related to frames [1, 4].

Chapter 6 investigates frames in the Hilbert space of square-integrable functions on a Euclidean manifold. The chapter focuses on the discretization of two types of induced representations of the affine Lie group and the Heisenberg Lie group. These examples pave the way for a more general class of frames in later chapters.

Chapter 7 develops a general and unified theory of induced representations for linear Lie groups, which can be used to construct frames for function spaces over any such group. The chapter also provides an algorithm for generating frames that have desirable properties for various applications in analysis and operator theory. The chapter builds on the frames obtained from the affine and Heisenberg Lie groups in Chapter 6.

Chapter 8 introduces a novel method for constructing frames on homogeneous spaces, which are spaces obtained by quotienting a Lie group by a closed subgroup. The chapter focuses on the unit sphere, which is a homogeneous space of the special orthogonal group, and the Klein bottle, viewed as a homogeneous space.

Chapter 9 addresses the discretization problem for both left and right regular representations of a Lie group acting on square-integrable vectors in a given space. This is because the two representations are unitarily equivalent. Specifically, this chapter demonstrates that a Lie group can admit frames of translates if it contains a closed subgroup, which is isomorphic to the two-dimensional affine group, which was introduced earlier.

Chapter 10 focuses on the search for analogs of the Whittaker–Kotel'nikov–Shannon sampling series for noncommutative groups. The chapter revisits the concept of admissible representations and presents a methodology based on oscillation estimates by Gröchenig and Führ for discretizing an admissible representation to construct frames. It broadens the concept of bandlimitation on the real line, represented by Paley–Wiener-type spaces, to a wider class of noncommutative Lie groups. The chapter provides sufficient conditions that offer a favorable resolution to the discretization problem.

Finally, Chapter 11 distinguishes itself from previous chapters by focusing on a specific type of finite frames. These frames are induced by the action of a Lie group and possess an additional characteristic known as "full spark." As elaborated in Chapter 11, a "full spark" in a finite-dimensional vector space is a spanning set for the vector space that has the added feature of any subset, equal in size to the dimension of the space, being a basis. This means that any information encoded by a full spark frame is highly robust to erasures, a concept that is explained in greater detail within the chapter.

1.2 Proficiency Expectations

This textbook is primarily written for readers at the graduate student level or above. A fundamental proficiency in real analysis, including differential calculus, integration theory, and topology, is assumed. Our ideal reader has a good understanding of these foundational principles and is ready to delve deeper into more complex concepts. For those who wish to brush up on these fundamentals or explore them in greater detail, a list of basic references will be provided. These references, while not exhaustive, represent a starting point for the topics covered in this book. The aim of this book is not just to impart knowledge, but to guide you on a journey of discovery. Our starting point is a solid grounding in differential geometry. From there, we delve into the world of Lie groups, representation theory, and frames.

1.3 Aims

The primary goal of this book, as suggested by the title, is to present a unique approach to the construction of frames through the action of Lie groups The structure of the book is largely dictated by the necessity to furnish readers with the necessary technical tools required to undertake various steps in this construction process. At its core, this book provides a problem-oriented introduction to the use of Lie group methods in frame construction. Its main objective is to develop and present a cohesive treatment of this material, a significant portion of which I have personally developed. The early sections of the book are specifically designed to help readers understand, appreciate, and potentially contribute to this scientific pursuit. By the time readers reach the conclusion of this book, they should have acquired a comprehensive understanding of the principles and techniques involved in the construction of frames using Lie group methods. This understanding will provide the foundation for their further exploration of the subject.

1.4 Scope and Material Selection

Recognizing the diverse audience this book caters to, it is important to note that the material chosen has been purposefully selected to align with the book's objectives, rather than aiming to provide a comprehensive review of the current state of the art.

Frame theory is a significant field in its own right, with comprehensive works penned by authors like Ole Christensen and Chris Heil. This book does not seek to compete with such exhaustive works. Instead, the selection of general frame theory in this book is motivated by its practical relevance to the material that follows.

Similarly, the sections on Lie theory represent a specific choice of material designed to serve the book's broader objectives, rather than attempting to provide a comprehensive and self-contained introduction. Certain aspects of Lie theory, such as the adjoint and coadjoint actions of Lie groups and Lie algebras, their interplay, coadjoint orbits, differentials of homomorphisms, and so on, are touched upon lightly. While these areas could constitute an entire book in themselves, they are not the primary focus of this work.

However, certain foundational results, such as those of the inverse function theorem and the implicit function theorem, are included in this book. These are considered basic, indispensable knowledge, essential for understanding the advanced topics that this book delves into.

1.5 Catering to Diverse Learning Approaches and Expertise Levels

This book serves as a problem-driven introduction to the use of Lie-theoretic methods in frame construction. Rather than offering an exhaustive account of all theoretical aspects, it centers on those which are instrumental to the content of subsequent sections. The extent of exploration into specific areas, such as Lie theory and frames, is primarily determined by their necessity for further interaction within the content. The book occupies a unique niche in the literature, catering to the varied preferences of potential readers. Some readers prefer a thorough, ground-up approach to learning theory, taking the time to explore fundamentals in depth before turning to concrete examples or applications. In contrast, others thrive on having a clear motivation guiding them through the material. This book is designed with the latter type of reader in mind, leveraging its strengths, particularly its wealth of concrete examples.

In serving as an entry point to both frame theory and, more importantly, Lie theory, the book presents fundamental definitions and results in tandem with concrete questions and constructions. This pairing not only motivates the introduction of these notions but also provides a solid foundation for the reader. To complement the introductory material on Lie theory or frame theory, systematic suggestions for further reading are provided, ensuring readers understand that the material presented in the book is not exhaustive. This approach aligns well with the decision to provide references at the end of each chapter rather than a full list at the end of the book.

Finally, the book's structure takes into consideration the needs of readers who may be mathematicians with strong expertise in one of the main areas of the book and less expertise in others. These readers may wish to skip certain parts of the book and revisit others on multiple occasions, possibly after or while reading later parts of the book. Enhancing the book's accessibility to enable this kind of nonlinear reading can significantly boost its utility, making it an effective resource for a wide range of readers.

For instance, for a reader with a strong background in frame theory but less understanding of differential geometry, Lie theory, and representation theory, I would recommend the following pathway through the book:

1. **Chapter 5 "Frame Theory":** Start here since you are already comfortable with frame theory. This will give you a sense of the book's approach and style.
2. **Chapter 2 "Differentiable Manifolds":** Move to this chapter next. This chapter provides foundational knowledge in differential geometry, which is essential for understanding the later chapters on Lie theory and representation theory. Spend time on the sections that are new to you, and do not hesitate to revisit this chapter as you work through the book.
3. **Chapter 3 "Lie Theory":** This chapter is a good next step. While it may be challenging due to your lack of background in the area, it is crucial for understanding the applications of Lie theory to frame construction in later chapters. Take your time to ensure you understand the key concepts.

4. **Chapter 4 "Representation Theory":** After you have grasped the basics of Lie theory, move on to this chapter. The concepts introduced here are essential for understanding the later chapters that delve into more complex applications.
5. **Chapters 6–10:** These chapters apply the theories discussed in the earlier chapters to different settings. You should be able to follow the frame theory sections comfortably, and your newly acquired knowledge in differential geometry, Lie theory, and representation theory will help you understand the rest.
6. **Chapter 11 "Finite Frames Maximally Robust to Erasures":** Finally, this chapter provides a nice capstone to the book, pulling together many of the threads from earlier chapters.

For a reader with a strong background in Lie theory and differential geometry, but limited knowledge of frame theory, I would recommend the following approach to navigate through the book:

1. **Chapter 5 "Frame Theory"**: Start with this chapter to build a foundation in frame theory. Although the concepts may be new, your strong background in Lie theory and differential geometry will likely make the material more accessible.
2. **Chapter 1 "Introduction"**: After gaining some understanding of frame theory, visit this chapter to get a general overview of the book's organization and aims.
3. **Chapters 2 "Differentiable Manifolds" and 3 "Lie Theory"**: Skim through these chapters. Your expertise in differential geometry and Lie theory should allow you to go through this material quickly. However, pay attention to how these concepts are applied in the context of frame theory.
4. **Chapter 4 "Representation Theory"**: Although not directly related to frame theory, the concepts in this chapter will be useful in later chapters that deal with more complex applications.
5. **Chapters 6–10**: These chapters apply the theories discussed in the earlier chapters to different settings. As you read, try to focus on how the frame theory concepts you learned in Chapter 5 are applied in these settings.
6. **Chapter 11 "Finite Frames Maximally Robust to Erasures"**: Lastly, this chapter synthesizes many of the concepts from the earlier chapters and applies them in a complex setting. The material in this chapter should be easier to understand now that you have a foundation in frame theory and have seen how it interacts with Lie theory and differential geometry.

References

1 Ole Christensen. *An introduction to frames and Riesz bases*. Applied and Numerical Harmonic Analysis. Birkhäuser, Boston, MA, 2003.

2 Gerald B. Folland. *A course in abstract harmonic analysis*, volume 29. CRC Press, 2016.

3 Hartmut Führ. *Abstract harmonic analysis of continuous wavelet transforms*, volume 1863 of *Lecture Notes in Mathematics*. Springer-Verlag, Berlin, 2005.

4 Christopher Heil. *A basis theory primer: expanded edition*. Springer Science & Business Media, 2010.

5 Eberhard Kaniuth and Keith F. Taylor. *Induced representations of locally compact groups*, volume 197. Cambridge University Press, 2013.

6 Gitta Kutyniok and Demetrio Labate. Introduction to shearlets. In *Shearlets*, pages 1–38. Springer, 2012.

7 Elias M. Stein and Rami Shakarchi. *Fourier analysis: an introduction*, volume 1. Princeton University Press, 2011.

2

Differentiable Manifolds

The main aim of the ensuing chapter is to provide the reader with an introduction to the concept of differentiable manifolds. The structure of this chapter has been inspired by the curriculum of a year-long course in differentiable geometry offered by Dr. James Hebda at Saint-Louis University in the academic year 2011–2012.

We begin by revisiting established results from Euclidean calculus, notably the Implicit Function Theorem and related concepts. These will later form the foundation for developing critical constructs like frames. Topological manifolds, as locally linear, Hausdorff, and second countable topological spaces, extend the familiar notions of curves and surfaces in Euclidean spaces. We introduce these initially, before overlaying a differentiable structure onto the topological structure to delineate differentiable manifolds. This layered approach allows us to expand Euclidean calculus concepts like diffeomorphism and derivatives to the broader context of differentiable manifolds. We delve into diverse submanifold classes, including immersed and closed embedded submanifolds.

The next portion of the chapter introduces derivations on the algebra of smooth functions on a manifold. These are linear operators that satisfy a product rule, echoing what we see in single-variable calculus. The set of derivations gives rise to a vector space, closed under Lie brackets – a skew-symmetric bilinear form obeying Jacobi's identity, which forms a Lie algebra. This concept will resurface in later chapters.

The pushforward of a derivation with respect to a given differentiable function between two manifolds is defined and we proceed to describe tangent vectors on a manifold at a fixed point as a particular class of derivations with specified properties. Finally, we present forms and tensor fields on manifolds, laying the groundwork for a comprehensive notion of integration on a manifold.

A Bridge Between Lie Theory and Frame Theory: Applications of Lie Theory to Harmonic Analysis, First Edition. Vignon Oussa.

2.1 Calculus on Euclidean Space

This section reviews some important calculus results for Euclidean spaces. These results are well known and widely available in the literature, so we will only state them without proofs. We assume that the reader has some background in multivariable calculus, and we refer to Spivak's book [2] for a more detailed and rigorous treatment of these concepts.

Recall that a Euclidean n-space $\mathbb{R}^n$ is the set of all ordered n-tuples of real numbers equipped with its standard norm $\| \, \| : \mathbb{R}^n \to [0, \infty)$, where

$$\|x\| = \left\|(x_1, x_2, \ldots, x_n)\right\| = \left(\sum_{k=1}^{n} x_k^2\right)^{1/2}.$$

The norm $\|x\|$ of a vector x returns the Euclidean distance of the segment line joining the origin of space to the location occupied by x. With this norm, one is able to measure the distance between two points (or two vectors).

Define a distance function $\text{dist} : \mathbb{R}^n \times \mathbb{R}^n \to [0, \infty)$ by $\text{dist}(x, y) = \|x - y\|$. This function also induces a topology known as the standard topology on the Euclidean n-space. A basis for this topology is the set of all open balls with rational radii centered at all possible points with rational coordinates. The nonnegative integer n is called the dimension of $\mathbb{R}^n$ and $\mathbb{R}^n$ together with norm $x \mapsto \|x\|$ defines a metric space denoted by the pair $(\mathbb{R}^n, \| \, \|)$.

Let U be an open subset of the set of real numbers and let f be a real-valued function defined on U. Recall that the derivative of f at a point p is defined as a limit of a difference quotient. Precisely, we write this derivative as:

$$L = \lim_{x \to p} \frac{f(x) - f(p)}{x - p} \tag{2.1}$$

whenever this limit exists. The quantity whose limit is being evaluated is just the ratio of the shorter legs of the right triangle determined by $(p, f(p))$, $(x, f(p))$, and $(x, f(x))$. This corresponds to measuring the slope (the tangent value of the angle between the tangent line and the horizontal axis or an average rate of change) of the secant line through $(p, f(p))$ and $(x, f(x))$. As the quantity $x - p$ approaches zero, a derivative exists if and only if the associated family of secant lines converges to a tangent line to the graph of f through the point $(p, f(p))$. Put another way, (2.1) can be understood as follows. The quantity

$$\left| \frac{f(x) - f(p) - L \cdot (x - p)}{x - p} \right|$$

can be made as small as desired provided that the gap between x and p is sufficiently narrow. Regarding, L as a linear transformation acting on $\mathbb{R}$ such that

$L : x \mapsto Lx$, it follows that f is differentiable at a point p whenever

$$\lim_{x \to p} \frac{f(x) - f(p) - L(x - p)}{x - p} = 0.$$

Inspired by this definition, we will now extend this concept into higher dimensions. This new definition replaces absolute value by vector-norm and scalar-by-scalar multiplication with operator-vector multiplication. Let U be a nonempty open subset of $\mathbb{R}^n$ and let p be a fixed point in U. We say that a function $f : U \to \mathbb{R}^m$ is differentiable at $p \in U$ if there exists a linear map $L : \mathbb{R}^n \to \mathbb{R}^m$ such that

$$\lim_{x \to p} \frac{\|f(x) - f(p) - L(x - p)\|}{\|x - p\|} = 0. \tag{2.2}$$

In other words, for any prescribed positive value ϵ, there exists a positive real number δ such that if $x \in U$ and $0 < \|x - p\| < \delta$ then

$$\frac{\|f(x) - f(p) - L(x - p)\|}{\|x - p\|} < \epsilon.$$

Clearly, the definition above also implies that for any fixed $\epsilon > 0$, there exists $\delta > 0$ such that

$$\|f(x) - f(p) - L(x - p)\| < \epsilon \|x - p\|$$

whenever $0 < \|x - p\| < \delta$. A closer look at the definition of a derivative leads to the following interpretation: "if f is differentiable at p then f can be approximated near p by a linear operator." Moreover, this approximation is quite satisfactory in the sense that $\|f(x) - f(p) - L(x - p)\|$ approaches zero faster than linearly when $\|x - p\|$ goes to zero. Colloquially, we say that f is locally linear, because for x near point p, the difference between $f(x)$ and $f(p)$ is very close to the value $L(x - p)$.

Lemma 2.1 The linear operator L defined in (2.2) is unique and is denoted by $Df(p)$.

For functions defined as composition in terms of other functions, we need a rule for computing derivatives in chain (see Theorem 2.1).

Theorem 2.1 ***(Chain Rule)*** Let U, V, W be open subsets of $\mathbb{R}^n, \mathbb{R}^m, \mathbb{R}^q$, respectively. If $f : U \to V$ is differentiable at $p \in U$ and $g : V \to W$is differentiable at $f(p)$ then $g \circ f : U \to W$ is differentiable at p and additionally, $D(g \circ f)(p) =$
$Dg(f(p)) \circ Df(p)$.

The following completely characterizes a differentiable function on an open set as a function whose components (composition of f with component-wise projections) are all differentiable as well. This of course requires an application of the Chain Rule. Additionally, the derivative associated with f at a fixed point, has for a matrix representation, the matrix obtained by computing the partial derivatives of each component function of f.

Proposition 2.1 Let U be an open subset of $\mathbb{R}^n$. Then $f : U \to \mathbb{R}^m$ is differentiable at $p \in U$ if and only if for $f = (f_1, \dots, f_m)$, each component function f_k of f is differentiable at p. Moreover, a matrix representation of $Df(p)$ takes the form

$$Jf(p) = \begin{bmatrix} \frac{\partial f_1(p)}{\partial x_1} & \cdots & \frac{\partial f_1(p)}{\partial x_n} \\ \vdots & \ddots & \vdots \\ \frac{\partial f_m(p)}{\partial x_1} & \cdots & \frac{\partial f_m(p)}{\partial x_n} \end{bmatrix}. \tag{2.3}$$

Remark 2.1 It is important to be aware that the existence of partial derivatives of f at point p does not automatically imply that f is differentiable at p. For instance, let $f : \mathbb{R}^2 \to \mathbb{R}$ such that

$$f(x, y) = \begin{cases} \frac{xy}{x^2+y^2}, & \text{if } (x, y) \neq (0, 0), \\ 0, & \text{if } (x, y) = (0, 0). \end{cases}$$

One easily verifies that f is not even continuous at $(0, 0)$. To see this, it suffices to convert (x, y) in polar coordinates and computes its limit as (x, y) tends to the zero vector. Indeed,

$$\lim_{(x,y)\to(0,0)} \frac{xy}{x^2 + y^2} = \lim_{r\to 0} \sin\theta\cos\theta \text{ does not exist.}$$

On the other hand, however, both partial derivatives of f exist and are equal to zero at $(0, 0)$. To ensure differentiability, its suffices to also ensure that the partial derivatives of f are also continuous (see Proposition 2.5).

Next, we show that a linear operator and its derivative must coincide.

Lemma 2.2 If $T : \mathbb{R}^n \to \mathbb{R}^m$ is a linear operator then $DT(p) = T$ for all $p \in \mathbb{R}^n$. Moreover, if $\sigma : \mathbb{R}^2 \to \mathbb{R}$ such that $\sigma(x, y) = x + y$ then $D\sigma(p, q) = \sigma(p, q)$ for all $(p, q) \in \mathbb{R}^2$.

Proof: For arbitrary $p \in \mathbb{R}^n$, assuming that T is a linear operator, we obtain

$$\lim_{\|h\|\to 0} \frac{\|T(p+h) - T(p) - T(h)\|}{\|h\|} = \lim_{\|h\|\to 0} \frac{\|T(p) - T(p) + T(h) - T(h)\|}{\|h\|} = 0.$$

Therefore, $DT(p) = T(p)$ for all $p \in \mathbb{R}^n$. Next, assuming that $\sigma : (x, y) \mapsto x + y$, since σ is a linear operator, it follows that $D\sigma(p, q) = \sigma(p, q)$ for all $(p, q) \in \mathbb{R}^2$.

Proposition 2.2 Let $\kappa : \mathbb{R}^2 \to \mathbb{R}$ be a function such that $\kappa(x,y) = xy$. Then $D\kappa(p,q)$ is the linear operator which maps (x,y) to $qx + py$.

Proposition 2.3 Let U be an open subset of $\mathbb{R}^n$ and let us assume that $f : U \to \mathbb{R}^m$ is differentiable at $p \in U$. Suppose additionally that $g_1, \dots, g_r : U \to \mathbb{R}$ are real-valued functions which are differentiable at p and let $\Phi : U \to \mathbb{R}^m$ such that $\Phi(x) = f\left(g_1(x), \dots, g_r(x)\right)$. Then Φ is differentiable at p and additionally,

$$J\Phi(p) = \begin{bmatrix} \frac{\partial f_1(\Psi(p))}{\partial x_1} & \cdots & \frac{\partial f_1(\Psi(p))}{\partial x_n} \\ \vdots & \ddots & \vdots \\ \frac{\partial f_m(\Psi(p))}{\partial x_1} & \cdots & \frac{\partial f_m(\Psi(p))}{\partial x_n} \end{bmatrix} \begin{bmatrix} \frac{\partial g_1(p)}{\partial x_1} & \cdots & \frac{\partial g_1(p)}{\partial x_n} \\ \vdots & \ddots & \vdots \\ \frac{\partial g_r(p)}{\partial x_1} & \cdots & \frac{\partial g_r(p)}{\partial x_n} \end{bmatrix}. \tag{2.4}$$

Proposition 2.4 Let $f : \mathbb{R}^n \to \mathbb{R}$ and let $g : \mathbb{R}^n \to \mathbb{R}$. Then the following holds true:

1. (Sum rule) $D(f+g)(p) = Df(p) + Dg(p)$.
2. (Product rule) $D(f \cdot g)(p) = f(p)Dg(p) + g(p)Df(p)$.

The following gives a simple sufficient condition under which differentiability is guaranteed.

Proposition 2.5 Let U be an open subset of $\mathbb{R}^n$ and let us assume that $f = \left(f_1, \dots, f_m\right) : U \to \mathbb{R}^m$ and that for all $(j,k) \in \{1, \dots, m\} \times \{1, \dots, n\}$, the partial derivatives $\frac{\partial f_j(p)}{\partial x_k}$ exist and are continuous at a point p in U (in other words, f is continuously differentiable on p.) Then f is differentiable.

Remark 2.2 It is also worth noting that continuity of partial derivatives is not a necessary condition for differentiability. For instance, let

$$f(x) = \begin{cases} x^2 \sin\left(\frac{1}{x}\right), & \text{if } x \neq 0, \\ 0, & \text{if } x = 0. \end{cases}$$

Since $\lim_{h\to 0} \frac{h^2 \sin\left(\frac{1}{h}\right)}{h} = 0$, it follows that

$$f'(x) = \begin{cases} 2x \sin\left(\frac{1}{x}\right) - \cos\left(\frac{1}{x}\right), & \text{if } x \neq 0 \\ 0, & \text{if } x = 0 \end{cases}$$

and clearly, f is differentiable everywhere but its derivative is not continuous at zero.

2.1.1 The Inverse Function Theorem and Its Applications

We have seen earlier that a differentiable function is locally linear in the sense that if a function is differentiable at some point, then near that point, it can be approximated by some linear function. This suggests that Linear Algebra could, in principle, be a valuable tool for studying nonlinear processes near points where differentiability holds. Naturally, one may then ask the following: which local properties of a differentiable function are preserved by linear approximation by its derivative? The Inverse Function Theorem states that the invertibility of the derivative of a function at some point implies that the function is locally invertible.

Theorem 2.2 ***(Inverse Function Theorem)*** Let U be an open subset of $\mathbb{R}^n$ and let us assume that $f = \left(f_1, \dots, f_n\right) : U \to \mathbb{R}^n$ is a function which is continuously differentiable at $p \in U$ such that $\det\,(Jf\,(p)) \neq 0$. Then there exist open sets $O_p \subseteq U$ and $O_{f(p)} \subseteq \mathbb{R}^n$ such that the function $\varphi = f|_{O_p} : O_p \to O_{f(p)}$ has an inverse $\varphi^{-1} : O_{f(p)} \to O_p$ which is differentiable at every point in $O_{f(p)}$. Moreover, the Jacobian of the inverse of φ evaluated at p is the Jacobian of φ evaluated at the inverse of p. That is,

$$J\left(\varphi^{-1}\right)(y) = \left[(J\varphi)\left(\varphi^{-1}(y)\right)\right]^{-1}.$$

2.1.1.1 The Implicit Function and Constant Rank Theorems

Given a continuously differentiable function f, taking $V \times W$ into V where V, W are some real vector spaces, if $f(p, q) = 0$, what does the Jacobian written in block-form as shown below

$$Jf\,(p, q) = \begin{bmatrix} J_1 & J_2 \\ J_3 & \boxed{J_4} \end{bmatrix}$$

of f tell us about its zero set?

The Implicit Function Theorem says that knowing additionally that J_4 is invertible is quite useful in the following sense. One can be sure that there is differentiable curve passing through the zero set of f. To further illustrate this point, let us contemplate for a second solving an equation of the type $x^2 + y^2 = 4$. Naively, one may ask, which values of y should we stay away from, if one is considering solving this equation for y. The Implicit Function Theorem suggests to first compute the Jacobian of $f\,(x, y) = x^2 + y^2 - 4$

$$Jf\,(x, y) = \left(2x, \boxed{2y}\right).$$

Then as long as y is not zero, say $y > 0$, then $y = \sqrt{4 - x^2}$ is certainly a curve contained in the zero set of f.

Theorem 2.3 ***(The Implicit Function Theorem)*** Let U be an open subset of $\mathbb{R}^n \times \mathbb{R}^m$ containing a point (p, q) and let $f : U \to \mathbb{R}^m$ be a continuously differentiable function such that $f(p, q) = 0$. Next, let

$$A = \begin{bmatrix} \frac{\partial f_1(p,q)}{\partial x_{n+1}} & \cdots & \frac{\partial f_1(p,q)}{\partial x_{n+m}} \\ \vdots & \ddots & \vdots \\ \frac{\partial f_m((p,q))}{\partial x_{n+1}} & \cdots & \frac{\partial f_m(p,q)}{\partial x_{n+m}} \end{bmatrix}.$$

If A is invertible then there exist an open set $O_p \subseteq \mathbb{R}^n$ containing p and an open set $O_q \subseteq \mathbb{R}^m$ containing q such that for every $x \in O_p$ there exists a unique $y = \alpha(x) \in O_q$ such that $x \mapsto f(x, \alpha(x)) = 0$ is a differentiable function.

The Constant Rank Theorem gives a local characterization of all continuously differentiable functions whose rank is constant on a neighborhood of some fixed point. In fact, the result suggests that any differentiable function whose Jacobian has a constant rank around a point behaves locally like a linear projection.

Theorem 2.4 ***(The Constant Rank Theorem)*** Let $U \subseteq \mathbb{R}^n$ and $V \subseteq \mathbb{R}^m$ $(n \geq m)$ be open sets and let $f : U \to V$ be a continuously differentiable function such that the function $x \mapsto \text{rank}(Jf(x))$ is constantly equal to m on U. Then there exists an open set O contained in the domain of f and a differentiable function h mapping O onto $\mathbb{R}^m$ such that $f(h(x)) = \left(x_{n-m+1}, \ldots, x_n\right)$.

2.2 Topological Manifolds

The standard definition of a manifold aims to capture our intuition for curves and surfaces. Informally, a manifold is a mathematical object that locally resembles an n-dimensional Euclidean space. To set the stage up, for a precise definition, let us begin by taking a careful look at the unit circle which is a prototype of a manifold which is not in a global sense, Euclidean.

Let $\mathbb{S}^1 = \left\{(x, y) \in \mathbb{R}^2 : x^2 + y^2 = 1\right\}$ be a unit circle,equipped with the subspace topology of $\mathbb{R}^2$ induced by the standard norm $\| \, \|$. We claim that for any point $(x, y) \in \mathbb{S}^1$, there exists an open set O contained in $\mathbb{S}^1$ which contains (x, y) as an element, and additionally, there exists a map $\omega : O \to (-1, 1) \subset \mathbb{R}$ such that ω is a homeomorphism. Suppose (x, y) is assumed to be the north pole of this unit circle. Then the upper hemisphere of a unit circle can be continuously deformed to an open interval of length 2. This process can also be reversed continuously, and in fact, smoothly to recover the original hemisphere. This creates a one-to-one association between the northern hemisphere of the unit circle and the open interval $(-1, 1)$. Repeating this procedure for each four hemispheres of the unit circle, let $\{O_1, O_2, U_1, U_2\}$ be a finite open cover of the unit circle, where

$$O_1 = \{(x, y) \in \mathbb{S}^1 : y > 0\} = \left\{\left(x, \sqrt{1 - x^2}\right) \in \mathbb{R}^2 : x \in (-1, 1)\right\},$$
$$O_2 = \{(x, y) \in \mathbb{S}^1 : y < 0\} = \left\{\left(x, -\sqrt{1 - x^2}\right) \in \mathbb{R}^2 : x \in (-1, 1)\right\},$$

and

$$U_1 = \{(x, y) \in \mathbb{S}^1 : x > 0\} = \left\{\left(\sqrt{1 - y^2}, y\right) \in \mathbb{R}^2 : y \in (-1, 1)\right\},$$
$$U_2 = \{(x, y) \in \mathbb{S}^1 : x < 0\} = \left\{\left(-\sqrt{1 - y^2}, y\right) \in \mathbb{R}^2 : y \in (-1, 1)\right\}.$$

Next, let

$$\omega_1 : O_1 \to (-1, 1), \quad \text{where} \quad \omega_1(x, y) = x,$$
$$\omega_2 : O_2 \to (-1, 1), \quad \text{where} \quad \omega_2(x, y) = x,$$

and

$$\mu_1 : U_1 \to (-1, 1), \quad \text{where} \quad \mu_1(x, y) = y,$$
$$\mu_2 : U_2 \to (-1, 1), \quad \text{where} \quad \mu_2(x, y) = y.$$

Each function $\omega_1, \omega_2, \mu_1, \mu_2$ defines a homeomorphism between O_1, O_2, U_1, U_2, respectively, and the open interval $(-1, 1)$. Geometrically, this means that for every point $p \in \mathbb{S}^1$, there exists an open subset of the unit circle containing p which is naturally identified with an open subset of the real line. In other words, the unit circle is locally Euclidean. This characterization of the unit circle $\mathbb{S}^1$ captures one of the central properties of a topological manifold (see Definition 2.1, Part 1).

Given a function $f : \mathbb{S}^1 \to \mathbb{R}$ defined on the unit circle, let

$$(E, \varphi) \in \left\{\left(O_1, \omega_1\right), \left(O_2, \omega_1\right), \left(U_1, \mu_1\right), \left(U_2, \mu_2\right)\right\}.$$

To analyze the behavior of f near a point $(x, y) \in E$, we will first need to restrict f to the open set E. Secondly, assuming that one has full access to the tools of Euclidean Differential Geometry, it is convenient also to compose f with the inverse of φ. Precisely, let $f_E : E \to \mathbb{R}$ be the restriction of f onto E. Then the function

$$f_E \circ \varphi^{-1} : (-1, 1) \to \mathbb{R}$$

is a real-valued function defined on a subset of a Euclidean space. Furthermore, $f_E \circ \varphi^{-1}$ could also be viewed as a "local" representation of the original function f. Regarding the unit circle as a collection of charts-sets of Euclidean nature, "glued" together, any differentiable function defined on the unit circle can be analyzed (locally) by first composing it with a suitable fixed smooth function $\varphi \in \left\{\omega_1, \omega_1, \mu_1, \mu_2\right\}$. Remark however that, since any finite open covering of the unit circle must contain pairs of overlapping sets, any emergent differentiable property of f_E, deemed useful, must also be independent of a specific choice of an open set around $(x, y) \in E$.

For the time being, we will refrain from further addressing the more subtle aspects of the structure of differentiable manifold, which we have yet to define. A proper definition of a differentiable manifold requires several steps. First, we define a topological manifold as follows.

Definition 2.1 A topological manifold of dimension n is a topological space M satisfying the following conditions:

1. M is locally Euclidean. In other words, for any $x \in M$, there exists an open neighborhood $U \subset M$ containing x together with a homeomorphism $\delta : U \to V$ such that V is an open subset of $\mathbb{R}^n$.
2. M is Hausdorff, in the sense that for arbitrary distinct points x, y in the manifold M, there exist open subsets X and Y of M such that $x \in X, y \in Y$, and $X \cap Y$ is an empty set.
3. M is second countable in the sense that the topology which we endow M with, has a countable basis.

When M is a topological manifold of dimension n, we say that M is a topological n-manifold and we write $n = \dim M$. The following are standard examples of topological manifolds. We leave it to the reader to verify that all axioms in Definition 2.1 are satisfied.

Example 2.1 ***(Examples Of Manifolds)***

1. $\mathbb{R}^n$ endowed with the standard Euclidean metric

$$(x, y) \mapsto \left(\sum_{k=1}^{n} \left(x_k - y_k\right)^2 \right)^{1/2}.$$

2. $\mathbb{S}^n = \left\{x \in \mathbb{R}^{n+1} \ : \ \|x\| = 1\right\}$equipped with the subspace topology of $\mathbb{R}^{n+1}$.

Definition 2.2 Let M be a topological n-manifold. A coordinate chart on M is defined as a pair of the type (O, ω), where O is an open subset of M and $\omega : O \to U \subseteq \mathbb{R}^n$ is a homeomorphism from O onto $U = \omega(O)$.

Note that since every manifold is locally Euclidean, then it is necessarily the case that every point is contained in the domain of some chart. If additionally, $\omega(m)$ is the zero element in the Euclidean space $\mathbb{R}^n$, we say that (O, ω) is a chart centered at m. The set O is known as the coordinate domain of its points and the map ω is called a local coordinate map. Moreover, letting $\omega(m) = \left(\omega_1(m), \ldots, \omega_n(m)\right)$, we say that the n-tuple $(\omega_1, \dots, \omega_n)$ are the local coordinates or local components on O.

Let M be a topological n-manifold. By definition, M is covered by a family of open sets $\{(O_\alpha, \omega_\alpha) : \alpha \in I\}$ where I is a parametrizing set and each domain O_α is homeomorphic to $\mathbb{R}^n$.

An atlas for the manifold M is a family of charts on a topological n-manifold M such that the union of all sets O_α for each α in I equals M, where I is a parametrizing set.

Definition 2.3 A family of charts $\{(O_\alpha, \omega_\alpha) : \alpha \in I\}$ on a topological n-manifold M such that

$$M = \bigcup_{\alpha \in I} O_\alpha$$

is called an atlas for the manifold M.

2.2.1 Differentiable Structures

To develop the tools needed to extend Calculus from Euclidean realms to manifolds, we need some additional conditions.

Let M be a topological n-manifold. Given two charts $(O_\alpha, \omega_\alpha)$ and (O_β, ω_β) defined on M, we say that $(O_\alpha, \omega_\alpha)$ and (O_β, ω_β) are C^∞-related or smoothly compatible if either (a) $O_\beta \cap O_\alpha$ is empty or (b) both change of charts $\omega_\alpha \circ \omega_\beta^{-1} : \omega_\beta (O_\alpha \cap O_\beta) \to \omega_\alpha (O_\alpha \cap O_\beta)$ and $\omega_\beta \circ \omega_\alpha^{-1} : \omega_\alpha (O_\alpha \cap O_\beta) \to \omega_\beta (O_\alpha \cap O_\beta)$ are C^∞-functions and therefore diffeomorphisms as well.

Definition 2.4 A differentiable (or C^∞) atlas on a topological n-manifold M is a collection of charts $\mathcal{A} = \{(O_\alpha, \omega_\alpha) : \alpha \in I\}$ such that

1. $\bigcup_{\alpha \in I} O_\alpha$ is an open cover of M.
2. For any $\alpha, \beta \in I$, $(O_\alpha, \omega_\alpha)$ and (O_β, ω_β) are C^∞-related.

Theorem 2.5 shows that any fixed atlas of a manifold can be extended in a unique fashion.

Theorem 2.5 Let M be a topological n-manifold. Every differentiable atlas of M is contained in a unique maximal differentiable atlas on M.

Proof: Let $\mathcal{A} = \{(O_\alpha, \omega_\alpha) : \alpha \in I\}$ be a differentiable atlas for M and let

$$\mathcal{B} = \{(U, \upsilon) : (U, \upsilon) \text{ is } C^\infty \text{ -related to } (O_\alpha, \omega_\alpha) \text{ for every } \alpha \in I\}.$$

We aim to show that $\mathcal{B}$ is a maximal atlas containing $\mathcal{A}$. First, since every element in $\mathcal{A}$ is related to every element in $\mathcal{A}$, it is clear that $\mathcal{A}$ is a subset of $\mathcal{B}$. Next, we will show that $\mathcal{B}$ is a differentiable atlas as well. Put $\mathcal{B} = \left\{\left(U_\beta, v_\beta\right) : \beta \in J\right\}$. Since $\mathcal{A}$ is a subset of $\mathcal{B}$, $\bigcup_{\beta \in J} U_\beta = M$. This shows that the first part of Definition 2.4 is met. Next, let $(U, \varphi), (V, \psi)$ be two arbitrary elements of $\mathcal{B}$ such that $U \cap V$ is not empty. To prove the stated result, it suffices to show that $(U, \varphi), (V, \psi)$ are C^∞-related (see Definition 2.4, Part 2). By assumption, it is clear that $\psi \circ \varphi^{-1} : \varphi(U \cap V) \to \psi(U \cap V)$ is a homeomorphism (with respect to the usual topology). To establish that $\psi \circ \varphi^{-1}$ is C^∞, we will need to show that $\psi \circ \varphi^{-1}$ is C^∞ in any neighborhood of an arbitrary point in the open set $\varphi(U \cap V)$. To this end, let $u \in U \cap V$. Since $\mathcal{A}$ is a differentiable atlas, there exists $\alpha \in I$ such that $u \in O_\alpha$. Additionally, $\left(O_\alpha, \omega_\alpha\right)$ is C^∞-related to (U, φ) and (V, ψ), respectively. Note that u belongs to the open set $U \cap V \cap O_\alpha$ and additionally, we may factor $\psi \circ \varphi^{-1}$ so that $\psi \circ \varphi^{-1} = \left(\psi \circ \omega_\alpha^{-1}\right) \circ \left(\omega_\alpha \circ \varphi^{-1}\right)$. As a result, $\psi \circ \varphi^{-1}$ is differentiable on a neighborhood of $\varphi(u)$ since it is a composition of two differentiable maps. Similarly, the inverse of $\psi \circ \varphi^{-1}$ is differentiable and this shows that $(U, \varphi), (V, \psi)$ are C^∞-related. Finally, to show that $\mathcal{B}$ is maximal, let us suppose that there exists a differentiable atlas C on M such that $\mathcal{B} \subseteq C$. Since any given pair of charts in C are smoothly related, then every chart in C is related to every chart in $\mathcal{A}$ and $C \subseteq \mathcal{B}$.

Definition 2.5 A maximal differentiable atlas on a topological n-manifold is called a differentiable structure on the manifold.

It is worth noting that the notion of smoothly compatible charts can be relaxed as follows. Two charts $\left(O_\alpha, \omega_\alpha\right)$ and $\left(O_\beta, \omega_\beta\right)$ on a topological manifold M are smoothly compatible if either (1) $O_\beta \cap O_\alpha$ is empty or (2) the change of coordinate map $\omega_\alpha \circ \omega_\beta^{-1} : \omega_\beta\left(O_\alpha \cap O_\beta\right) \to \omega_\alpha\left(O_\alpha \cap O_\beta\right)$ and its inverse $\omega_\beta \circ \omega_\alpha^{-1} : \omega_\alpha\left(O_\alpha \cap O_\beta\right) \to \omega_\beta\left(O_\alpha \cap O_\beta\right)$ are of differentiable class C^k for some natural number k. In this case, we will say that M has a C^k-differentiable structure. Analogously, we may also require that the change of coordinate map and its inverse are real-analytic functions. When this is the case, the corresponding structure is commonly referred to as a C^ω-structure.

Throughout this book, we will mainly be concerned with manifolds endowed with an infinitely smooth differentiable structure. As such, by differentiable, we will always mean differentiable of class C^∞. Moreover, we will often refer to a differentiable manifold simply as a manifold. In other words, unless we specify otherwise, a manifold should always be implicitly understood as a triple consisting of a set, a topology which is Hausdorff, second countable, locally Euclidean, and a differentiable structure.

Example 2.2 ***(The Standard Differentiable Structure on the Euclidean n-space)*** Let Id be the identity map on $\mathbb{R}^n$ which sends every point to itself, and let $\mathcal{A}$ be a maximal differentiable atlas containing the chart $(\mathbb{R}^n, \text{Id})$. The unique structure induced by $\mathcal{A}$ is called the standard differentiable structure on $\mathbb{R}^n$.

Remark 2.3 ***(Differentiable Product Manifold)*** Let M_1 and M_2 be two differentiable manifolds of dimensions n_1 and n_2, respectively. Then we can turn the topological manifold $M_1 \times M_2$ into a differentiable manifold of dimension $n_1 + n_2$ as follows. First, suppose that $\mathcal{A}_1$ and $\mathcal{A}_2$ are two maximal differentiable atlases inducing a smooth structure on M_1 and M_2, respectively. Next, let

$$\mathcal{A} = \left\{(U \times V, \varphi \times \psi) : (U, \varphi) \in \mathcal{A}_1, (V, \psi) \in \mathcal{A}_2\right\}.$$

In the above, $\varphi \times \psi : U \times V \to \varphi(U) \times \psi(V)$ such that $[\varphi \times \psi](x, y) = (\varphi(x), \varphi(y)) \in \mathbb{R}^{n_1+n_2}$ and $\mathcal{A}$ is a differentiable maximal atlas inducing a differentiable structure on $M_1 \times M_2$ called a differentiable product manifold.

Example 2.3 shows that it is possible for a manifold to admit several differentiable structures that are not compatible with each other.

Example 2.3 Let $\varphi : \mathbb{R} \to \mathbb{R}$ such that $\varphi(x) = x^5$. Clearly, φ is a homeomorphism with inverse $y \mapsto \varphi^{-1}(y) = \sqrt[5]{y}$. Moreover, $\varphi \circ \varphi^{-1}$ is the identity map defined on the real line and it follows from Definition 2.4 that $\{(\mathbb{R}, \varphi)\}$ is a differentiable atlas on the real line. By Theorem 2.5, even though $\{(\mathbb{R}, \varphi)\}$ is contained in a maximal atlas which we shall denote by $\mathcal{A}_\varphi$, the differentiable structure induced by this atlas is not the same as the standard smooth structure on the real line. To see this, it suffices to show that $(\mathbb{R}, \varphi)$ and $(\mathbb{R}, \text{id})$ are not smoothly compatible. Indeed,

$$\varphi \circ \text{id}^{-1} = \varphi \circ \text{id} = \varphi \quad \text{and} \quad \text{id} \circ \varphi^{-1} = \varphi^{-1},$$

and since $y \mapsto \varphi^{-1}(y) = \sqrt[5]{y}$ is not differentiable at zero, $(\mathbb{R}, \varphi)$ and $(\mathbb{R}, \text{id})$ are not elements of the same maximal atlas.

We will now turn to an issue of central concern in this section: the derivative of a function defined between two manifolds. Given a manifold M which is a subset of $\mathbb{R}^m$, let $f : M \to \mathbb{R}$. We have defined the derivative of f at a point p to be the unique linear operator $L : \mathbb{R}^m \to \mathbb{R}$ satisfying the following condition:

$$\lim_{x \to p} \frac{|f(x) - f(p) - L(x - p)|}{\|x - p\|_{\mathbb{R}^m}} = 0. \tag{2.5}$$

Note that this definition is generally problematic if M is not assumed to be a subset of some Euclidean space $\mathbb{R}^m$. In fact, the manifold M may not even have a group structure with respect to which the operation $x - p$ is meaningful. To develop a satisfactory definition of a derivative on a manifold that is compatible with the derivative of functions between subsets of Euclidean spaces, it seems reasonable to find a way to exploit the assumption that every manifold is locally Euclidean. As such, naturally, it makes sense to define $f : M \to \mathbb{R}$ as differentiable at $p \in M$ if for some chart (U, φ) such that $p \in U$, the map $f \circ \varphi^{-1} : \varphi(U) \to \mathbb{R}$ is a differentiable function as defined in (2.2). Note that this definition is independent of the choice of a chart. Indeed, given any other chart (U', φ') belonging to a maximal atlas containing (U, φ) such that $p \in U'$, since

$$f \circ [\varphi']^{-1} = (f \circ \varphi^{-1}) \circ \left(\varphi \circ [\varphi']^{-1}\right)$$

is differentiable on $U \cap U'$ as well, since (U, φ) and (U', φ') are smoothly compatible. This observation leads to a more useful definition of the concept of differentiable map between manifolds.

Definition 2.6 Let M, N be two differentiable manifolds. A function $f : M \to N$ is said to be differentiable (or smooth or C^∞) if the following two conditions are satisfied:

1. $f : M \to N$ is continuous.
2. For every pair of charts (U, φ) on M and (V, ψ) on N such that $f(U)$ is a subset of V, the function $\psi \circ f \circ \varphi^{-1}$ is of class C^∞.

The function $\psi \circ f \circ \varphi^{-1}$ in the definition above will be called a local representative of f. Moreover to decide if a function f defined between two manifolds is smooth at a point, it is not necessary to check if every local representative of f is differentiable. In fact, Proposition 2.6 suggests that the existence of a local chart with respect to which, a local representative of f of class C^∞ is both a necessary and a sufficient condition for the differentiability of f.

Proposition 2.6 Let M, N be two differentiable manifolds. The following statements are equivalent:

1. $f : M \to N$ is differentiable.
2. For every point p in the manifold M, there exist smooth charts (U, φ) on M and (V, ψ) on N such that p belongs to U, $f(U)$ is contained in V and $\psi \circ f \circ \varphi^{-1} : \varphi(U) \to \psi(V)$ is of class C^∞.

Proof: Let us assume that $f : M \to N$ is differentiable. Next, let (V, ψ) be a chart on N such that $f(p)$ belongs to its domain V. By virtue of continuity of f, the inverse image $f^{-1}(V)$ is an open subset of the manifold M containing p as a point.

Next, let (O, ω) be a smooth chart on M such that p belongs to O. Let U be the open set obtained by intersecting open sets O and $f^{-1}(V)$ and let φ be the restriction of ω to the set $f^{-1}(V) \cap O = U$. Clearly, (U, φ) is an element of the maximal differentiable atlas containing (O, ω) and since f is assumed to be differentiable, it follows that $\psi \circ f \circ \varphi^{-1} : \varphi(U) \to \psi(V)$ is of class C^{∞}.

Let us now assume that for every point p in the manifold M, there exist smooth charts (U, φ) on M and (V, ψ) on N such that p belongs to U, $f(U) \subseteq V$ and $\psi \circ f \circ \varphi^{-1} : \varphi(U) \to \psi(V)$ is of class C^{∞}. To prove that $f : M \to N$ is differentiable (see Definition 2.6), it suffices to establish that (1) $f : M \to N$ is continuous and (2) for every pair of charts (U', φ') on M and (V', ψ') on N such that $f(U')$ is a subset of V', the function $\psi' \circ f \circ [\varphi']^{-1}$ is of class C^{∞}. To show continuity, let p be an arbitrary point on M. By assumption, $\psi \circ f \circ \varphi^{-1} : \varphi(U) \to \psi(V)$ is of class C^{∞}. The continuity of f at p is therefore a direct consequence of the fact that both ψ and φ are homeomorphisms. Next, let (U', φ') be a chart on M and let (V', ψ') be a chart on N such that $f(U')$ is a subset of V'. To show that $\psi' \circ f \circ [\varphi']^{-1}$ is of class C^{∞}, it is sufficient to prove that $\psi' \circ f \circ [\varphi']^{-1}$ is differentiable in some neighborhood of any point $z \in \varphi'(U')$. Let p be the unique point in U' such that $\varphi'(p) = z$. By assumption, there exist smooth charts (U, φ) on M and (V, ψ) on N such that p belongs to U, $f(U)$ is contained in V and $\psi \circ f \circ \varphi^{-1} : \varphi(U) \to \psi(V)$ is of class C^{∞}. Moreover,

$$\psi' \circ f \circ [\varphi']^{-1} = (\psi' \circ \psi^{-1}) \circ (\psi \circ f \circ \varphi^{-1}) \circ \left(\varphi \circ [\varphi']^{-1}\right) \tag{2.6}$$

defined on $\varphi'(U \cap U')$ is differentiable as the composition of three differentiable functions. This completes the proof.

Proposition 2.7 Let M be a differentiable manifold. Then the identity map on M is differentiable.

Proof: Our proof strategy consists of showing that any local representative of the identity map taking a smooth manifold to itself coincides with an identity function mapping an open subset of $\mathbb{R}^n$ to itself. On the one hand, it is clear that the identity map is a continuous function. On the other, given a fixed point p in M, let (O, ω) be a chart whose domain contains p as a point. Letting id_M be the identity map on M and letting $\mathrm{id}_{\omega(O)}$ be the identity map on $\omega(O) \subseteq \mathbb{R}^n$, it follows that $\omega \circ \mathrm{id}_M \circ \omega^{-1} = \mathrm{id}_{\omega(O)}$. Since $\mathrm{id}_{\omega(O)}$ is differentiable, id_M must be differentiable as well.

Proposition 2.8 The composition of two differentiable functions is differentiable.

Proof: Let M_1, M_2, M_3 be three differentiable manifolds and let $f : M_1 \to M_2, g : M_2 \to M_3$. For arbitrary p in the manifold M_1, select a chart (U_1, ϕ_1) on M_1 and another chart (U_2, ϕ_2) on M_2 such that $f(U_1)$ is contained in U_2. By assumption, $\phi_2 \circ f \circ \phi_1^{-1} : \phi_1(U_1) \to \phi_2(U_2)$ is differentiable. Similarly, we may select charts (U_2', ϕ_2') on M_2 and (U_3, ϕ_3) on M_3 such that $f(p) \in U_2'$ and $g(U_2')$ is contained in U_3 and $\phi_3 \circ g \circ [\phi_2']^{-1}$ is differentiable. Next, let us consider $U = f^{-1}(U_2 \cap U_2') \cap U_1$ and let ω be the restriction of ϕ_1 onto U. Then

$$\phi_3 \circ (g \circ f) \circ \omega^{-1} = \left(\phi_3 \circ g \circ [\phi_2']^{-1}\right) \circ \left(\phi_2' \circ \phi_2^{-1}\right) \circ \left(\phi_2 \circ f \circ \phi_1^{-1}\right)$$

and it follows that $g \circ f : M_1 \to M_3$ is a differentiable function.

Definition 2.7 Let M, N be two differentiable manifolds. We say that $f : M \to N$ is a diffeomorphism if f is a differentiable function which has a differentiable inverse. In other words, $f : M \to N$ is a diffeomorphism if there exists a function $g : N \to M$ such that $f \circ g$ is the identity map on N and $g \circ f$ is the identity map on M. Additionally, we say that two differentiable manifolds are diffeomorphic if there exists a diffeomorphism mapping one manifold onto the other.

A differentiable manifold equipped with an atlas allows us to use Euclidean-like techniques to study the local properties of the given manifold. However, one cannot generally guarantee that the local features described in the domains of the charts can be patched together without violating global smoothness. The tools that will allow us to glue together local features into global ones on a manifold are provided by a set of mathematical objects known as partitions of unity. A proof strategy for the existence of a partition of unity is mainly taken from the Textbook of Warner [3].

Definition 2.8 Let M be a topological space

1. A collection of sets $C = \{U_\alpha : \alpha \in A\}$ is an open cover for a subset $X \subseteq M$ if $X \subset \bigcup_{\alpha \in A} U_\alpha$.
2. Any subcollection of C which an open cover for X is called a subcover.
3. A refinement $C' = \{U_\alpha' : \alpha \in A'\}$ of $C = \{U_\alpha : \alpha \in A\}$ is a cover such that every element in C' is a subset of some element of C.
4. Let X be a subset of M and let $C = \{U_\alpha : \alpha \in A\}$ be an open cover of X. Then C is called locally finite if for every point $x \in X$, there exists an open subset O_x of x such that $C_x = \{U \in C : U \cap O_x \text{ is not empty}\}$ is a finite subset of C.
5. A topological space is called paracompact if every open cover of the space has a locally finite refinement.
6. A topological space is called locally compact if every point admits a neighborhood whose closure is compact.[1]

1 According to our definition, manifolds are locally compact.

Definition 2.9 Let X be a topological space and let $C = \{U_\alpha : \alpha \in A\}$ be an open cover of X. A partition of unity subordinate to the cover C is a collection of C^∞ functions $\mathcal{P} = \{f_\alpha : \alpha \in A\}$ satisfying the following conditions:

1. For every $f_\alpha \in \mathcal{P}$, $0 \le f_\alpha(x) \le 1$, for all $x \in X$.
2. For every $\alpha \in A$, the function $f_\alpha \in \mathcal{P}$ is supported on U_α.
3. Each point in X admits a neighborhood intersecting only finitely many elements in the collection of sets given by $\{\text{support of} f_\alpha : \alpha \in A\}$.
4. $\sum_{\alpha\in A} f_\alpha(x) = 1$, for all $x \in X$.

It is worth highlighting that in the definition of a partition of unity, each point in the space X admits a neighborhood intersecting only finitely many elements in the collection of sets $\{\text{support of} f_\alpha : \alpha \in A\}$. Moreover, although the indexing set A may very well be uncountable, the summation $\sum_{\alpha\in A} f_\alpha$ is in practice just a finite sum.

In the special case where X has a C^k-differentiable structure, if $\mathcal{P} = \{f_\alpha : \alpha \in A\}$ is a partition of unity such that each f_α is of class C^k then we say that $\mathcal{P}$ is a C^k-partition of unity.

We shall prove that any open cover of a manifold admits a C^∞-partition of unity subordinate to the given cover (see Theorem 2.6). The proof of Theorem 2.6 requires Lemmas 2.3 and 2.4.

Lemma 2.3 guarantees that any smooth manifold is a paracompact topological space.

Lemma 2.3 If X is a topological space which is locally compact, Hausdorff and second countable then X is paracompact.

Proof: First, we establish the existence of a countable collection of open sets $\{O_j : j \in \mathbb{N}\}$ such that for every natural number j, the following holds true: (1) $\overline{O_j}$ is compact, (2) $\overline{O_j} \subset O_{j+1}$, and (3) $X = \bigcup_{k=1}^{\infty} O_k$. To this end, appealing to the fact that X is Hausdorff and locally compact, there exists a countable basis $\{V_k : k \in \mathbb{N}\}$ of X such that each $\overline{V_k}$ $(k \in \mathbb{N})$ is compact. Next, let $O_1 = V_1$ and define O_k inductively as follows: $O_k = \bigcup_{j=1}^{j_k} V_j$ and j_{k+1} is the smallest natural number greater than j_k such that the topological closure of O_k is contained in $\bigcup_{j=1}^{j_{k+1}} V_j$. Proceeding inductively, we obtain $\{O_j : j \in \mathbb{N}\}$ as desired.

Let $C = \{U_\alpha : \alpha \in A\}$ be an open cover of X. Conventionally, let O_0 be the empty set. Then, each set $\overline{O_k} \backslash O_{k-1}$ is compact and contained in $\overline{O_{k+1}} \backslash O_{k-2}$. Appealing to the topological definition of compactness, for each $k \ge 3$, choose a finite subcover of the open cover $\left\{U_\alpha \cap \left(\overline{O_{k+1}} \backslash O_{k-2}\right) : \alpha \in A\right\}$ of the set $\overline{O_k} \backslash O_{k-1}$. Additionally, select a finite subcover $\left\{U_\alpha \cap O_3 : \alpha \in A\right\}$ of the closure of O_2. Then the collection of sets described above is a locally finite refinement of C containing open sets whose closures are compact.

Lemma 2.4 Given a positive real number r, let $B_{\infty,r} = \left\{x \in \mathbb{R}^n : \|x\|_\infty < r\right\}$ and let $C_{\infty,r} = \left\{x \in \mathbb{R}^n : \|x\|_\infty \leq r\right\}$. Then there exists a function $f : \mathbb{R}^n \to [0, \infty)$ of class C^∞ such that $f \equiv 1$ on the closed set $C_{\infty,1}$ and vanishes outside of the open $B_{\infty,2}$ and additionally, f is smooth.

Proof: Let

$$f_1(x) = \begin{cases} e^{-\frac{1}{x}}, & \text{if } x > 0 \\ 0, & \text{if } x \leq 0 \end{cases}, \quad f_2(x) = \frac{f_1(2-x)}{f_1(2-x) + f_1(x-1)}$$

and $f_3(x) = f_2(|x|)$. Next, define $f : \mathbb{R}^n \to \mathbb{R}$ such that

$$\left(x_1, \ldots, x_n\right) \mapsto f\left(x_1, \ldots, x_n\right) = \prod_{k=1}^{n} f_3\left(x_k\right)$$

is constantly equal to 1 on $C_{\infty,1}$ and vanishes outside of $B_{\infty,2}$. Then clearly, f is smooth as well, and this completes the proof.

Theorem 2.6 ***(Partition of Unity)*** Let M be a differentiable manifold of dimension n and let $C = \left\{U_\alpha : \alpha \in A\right\}$ be an open cover of M. Then there exists a partition of unity $\mathcal{P} = \left\{f_\alpha : \alpha \in A\right\}$ which is subordinate to the cover C.

Proof: The proof of this result is twofold.

Step 1: First, let O_0 be the empty set and let $\left\{O_j : j \in \mathbb{N}\right\}$ be a countable collection of sets such that for every natural number j, $\overline{O_j}$ is compact, $\overline{O_j} \subset O_{j+1}$ and $M = \bigcup_{k=1}^{\infty} O_k$ (see Lemma 2.3). For every point p in M, we let $\ell(p)$ be the largest integer such that p belongs to the set $M \backslash \overline{O_{\ell(p)}}$. Next, let $U_{\alpha(p)}$ be an element of C such that $p \in U_{\alpha(p)}$. Fix a chart $\left(U^{(p)}, \varphi^{(p)}\right)$ centered at p such that $U^{(p)}$ is contained in $U_{\alpha(p)} \cap \left(O_{\ell(p)+2} \backslash \overline{O_{\ell(p)}}\right)$ and additionally, the range of $\varphi^{(p)}$ contains $C_{\infty,2}$. Next, let $f_p = 1_{U^{(p)}} \cdot \left(\varphi^{(p)} \circ f\right)$ where f is as described in Lemma 2.4. Then it is clear that f_p is a smooth function which is constantly equal to 1 on $\left[\varphi^{(p)}\right]^{-1}\left(C_{\infty,1}\right)$. Additionally, the support of f_p is compact and contained in $U^{(p)} \subset U_{\alpha(p)} \cap \left(O_{\ell(p)+2} \backslash \overline{O_{\ell(p)}}\right)$.

Step 2: Secondly, for every natural number n, we may select a finite set of points

$$\{q_n \in M : q_n \in F_n\}$$

such that $\left\{ \left[\varphi^{(q_n)}\right]^{-1} (C_{\infty,1}) : q_n \in F_n \right\}$ is an open cover of the set $\overline{O_n} \backslash O_{n-1}$. Since the family of functions $F = \left\{ f_{q_n} : q_n \in F_n \text{ and } n \in \mathbb{N} \right\}$ is a countable set, we may reorder F so that $F = \{g_k : k \in \mathbb{N}\}$. Let $D(g_k)$ be the domain of each g_k. Then, the collection of sets $\{D(g_k) : k \in \mathbb{N}\}$ is locally finite. Next, observe the map $\Psi : p \mapsto \sum_{j \in \mathbb{N}} g_j(p)$ is well defined on M. Letting $\widetilde{g}_k = \frac{g_k}{\Psi}$, it follows that $\{\widetilde{g}_k : k \in \mathbb{N}\}$ is a partition of unity subordinate to the open cover C. Moreover, the support of each function $\widetilde{g}_k$ is compact.

Definition 2.10 Let M, N be two differentiable manifolds and let $f : M \to N$ be a differentiable (or smooth or C^∞) map. The rank of the function f at a point p in M is defined as the rank of the linear transformation (dimension of the range space) obtained by computing the Jacobian of the map $\psi \circ f \circ \varphi^{-1}$ at the point $\varphi(p)$, where (1) (U, φ) is a chart on M such that $p \in U$, (2) (V, ψ) is a chart on N such that $f(p) \in V$ and (c) $f(U)$ is a subset of V.

Note that the axioms used in defining a smooth manifold ensure that the notion rank of smooth functions, as described above, is well defined since the rank of the derivative of a local representative, is invariant under change of charts.

Proposition 2.9 Let M, N be two differentiable manifolds such that $\dim M = \dim N = n$. If $f : M \to N$ is a differentiable map which is injective and surjective and if the rank of f is equal to n at every point in the manifold M then f is a diffeomorphism.

Proof: Let us suppose that $f : M \to N$ is a differentiable map which is injective and surjective. Suppose additionally that the rank of f is equal to n at every point in the manifold M. As a bijection of sets, f admits an inverse function f^{-1}. Fix a point $y \in N$. Next, let (V, ψ) is a chart on N such that $y \in V$ and set $x = f^{-1}(y)$. Appealing to the differentiability of f, we may select a chart (U, φ) on the manifold M such that $x \in U$, the image of U through f is contained in V, and additionally, $\psi \circ f \circ \varphi^{-1}$ is a differentiable function, mapping $\varphi(U)$ onto $\psi(V)$. By assumption, the derivative of $\psi \circ f \circ \varphi^{-1}$ at $\varphi(x)$ has full rank and is therefore an invertible linear transformation. By the Inverse Function Theorem (see Theorem 2.2), we can find a neighborhood $W_{\varphi(x)}$ of $\varphi(x)$ and a neighborhood $W_{\psi(y)}$ such that the restriction of $\psi \circ f \circ \varphi^{-1}$ onto $W_{\varphi(x)}$ is a diffeomorphism. Consequently, $\left(\psi \circ f \circ \varphi^{-1}\right)|_{W_{\varphi(x)}}$ is invertible and its inverse $\left[\left(\psi \circ f \circ \varphi^{-1}\right)|_{W_{\varphi(x)}}\right]^{-1}$ is a smooth function as well. Next, $\left(\varphi^{-1}\left(W_{\varphi(x)}\right), \varphi|_{\varphi^{-1}(W_{\varphi(x)})}\right)$ and $\left(\psi^{-1}\left(W_{\psi(y)}\right), \varphi|_{\psi^{-1}(W_{\psi(y)})}\right)$ are charts on M, N, respectively, around x, y, respectively. Additionally, $f^{-1}\left(\psi^{-1}\left(W_{\psi(y)}\right)\right)$ is contained in $\varphi^{-1}\left(W_{\varphi(x)}\right)$ and

$$\left(\varphi|_{\varphi^{-1}(W_{\varphi(x)})} \circ f^{-1} \circ \left[\psi|_{\psi^{-1}(W_{\psi(y)})}\right]^{-1}\right) = \varphi \circ f^{-1} \circ \psi^{-1}|_{\psi^{-1}(W_{\psi(y)})}$$

is differentiable. In summary, f^{-1} is differentiable and we conclude that f is a diffeomorphism.

Proposition 2.10 Let M, N be two differentiable manifolds such that $\dim M = m$ and $\dim N = n$. Suppose that $f : M \to N$ is a differentiable map of constant rank r at every point p in M. Then there exist (1) a chart (O, φ) on M such that $p \in O$, (2) a chart (Q, ψ) on N such that $f(p) \in Q$ and (3) $f(O)$ is a subset of Q and (4)

$$\psi \circ f \circ \varphi^{-1}(x_1, \dots, x_n) = (x_1, \dots, x_r, 0, \dots, 0). \tag{2.7}$$

Proof: Our proof strategy hinges on the following arguments. Pairing the fact that f is of constant rank at any arbitrary point p in the manifold M, together with the Euclidean version of the Constant Rank Theorem, allows us to construct charts around p and its image, with respect to which a local representative of f is as described in (2.7). Precisely, let p be a point in M. By assumption, there exist charts $\left(O', \varphi'\right)$ on M such that $p \in O'$, a chart $\left(Q', \psi'\right)$ on N such that $f(p) \in Q'$ and $f\left(O'\right)$ is a subset of Q'. Applying the Constant Rank Theorem (see Theorem 2.4) to the map $\psi' \circ f \circ \left[\varphi'\right]^{-1}$, there exist open neighborhoods O'' of $\varphi'(p)$ and Q'' of $\psi'(f(p))$ and a diffeomorphism $s : O'' \to s\left(O''\right)$ and $r : Q'' \to \mu\left(Q''\right)$ such that

$$\left[r \circ \left(\psi' \circ f \circ \left[\varphi'\right]^{-1}\right) \circ s^{-1}\right](x_1, \dots, x_n) = (x_1, \dots, x_r, 0, \dots, 0).$$

Letting $\varphi = s \circ \varphi'$ and $\psi = r \circ \psi'$ gives

$$\left[\psi \circ f \circ \varphi^{-1}\right]\left(x_1, \dots, x_n\right) = \left(x_1, \dots, x_r, 0, \dots, 0\right)$$

as desired.

2.2.2 Submanifolds

Definition 2.11 Let M, N be two differentiable manifolds and let $f : M \to N$ be a differentiable (or smooth or C^∞) map. Then

1. f is an immersion if the rank of f is equal to the dimension of M at every point in M.

2. f is an injective immersion if f is an immersion which is injective.
3. f is an embedding if f is an injective immersion and additionally, f defines a homeomorphism between M and its range.
4. f is a closed embedding if f is an embedding and additionally $f(M)$ is closed in N.

Definition 2.12 Let M be a differentiable manifold. A submanifold or an immersed submanifold of M is a subset $X \subseteq M$ equipped with a differentiable and injective map $f : N \to X$ such that the rank of f is equal to the dimension of X everywhere and $f(N) = X$.

Definition 2.13 Let U be an open subset of $\mathbb{R}^n$. We say that $V \subseteq U$ is an m-slice of U if $m \leq n$ and

$$V = \left\{ \left(x_1, \dots, x_n\right) \in U : x_{m+1} = c_{m+1}, \dots, x_n = c_n \right\}$$

for some constants $c_{m+1}, \dots, c_n$. Moreover, $N \subseteq M$ is an embedded or regular submanifold of dimension s if for each $x \in N$, there exists a smooth chart (U, φ) in M such that $\varphi(U \cap N)$ is an s-slice of $\varphi(U)$.

Proposition 2.11 Let $N \subseteq M$ be an embedded or regular submanifold of dimension s. When endowed with the subspace topology, N is a topological manifold of dimension s. Additionally, N admits a differentiable structure with respect to which the inclusion map of N into M is a smooth embedding.

Proof: Since N is given the subspace topology of M, which is assumed to be a manifold, it is clear that N is a second countable and Hausdorff space. It remains next, to establish that N is locally Euclidean. To this end, we will appeal to the fact that N is an embedded submanifold to construct an open set around any arbitrarily fixed point $p \in N$ and an open subset of some Euclidean space. Precisely, let $\pi : \left(x_1, \dots, x_n\right) \mapsto \left(x_1, \dots, x_s\right)$ be a linear projection of rank s. Next, let (U, φ) be a chart in M such that $\varphi(U \cap N)$ is an s-slice of $\varphi(U)$. Let $V = U \cap N$, set $W = \pi(\varphi(V))$ and let ψ be the map obtained by taking the restriction of $\pi \circ \varphi$ on V. Note that ψ is a homeomorphism given by the mapping

$$\varphi^{-1}\left(x_1, \dots, x_s, c_{s+1}, \dots, c_n\right) \mapsto \left(x_1, \dots, x_s\right),$$

where $c_{s+1}, \dots, c_n$ are some real constants. This shows that N is locally Euclidean as well. Consequently, the inclusion mapping N into M is a topological embedding.

We will now aim to describe smooth structure on N. To this end, we consider the collection $\mathcal{A}$ of pairs $\left(\varphi(U \cap N), \varphi|_{U \cap N}\right)$ such that $\varphi(U \cap N)$ is an s-slice of

$\varphi(U)$. Suppose that (U_0, φ_0) and (U_1, φ_1) are two elements of this collection such that U_0 and U_1 overlap. By computing the pointwise image of every element in the domain of the change of coordinate map $\varphi_1 \circ \varphi_0^{-1}$, it is not hard to verify that $\varphi_1 \circ \varphi_0^{-1}$ is a composition of smooth maps. As such, $\varphi_1 \circ \varphi_0^{-1}$ is smooth, and $\mathcal{A}$ is a smooth atlas on N. Moreover, with respect to this smooth structure, the inclusion map of N into M is a differentiable function.

We will next provide sufficient conditions under which a subset of a manifold is a regular submanifold.

Definition 2.14 Let $f : M \to N$ be a differentiable map.

1. A point p in M is called a regular point of f if the rank of f at p coincides with the dimension of N.
2. A point p in M is called a critical point of f if the rank of f at p is less than the dimension of N.
3. A point q in N is a critical value of f if $q = f(p)$ for some critical point p in M.
4. A point q in N is a regular value of f if q is not a critical value of f.

Note that it is vacuously true that for a differentiable map $f : M \to N$, points in N which are not located in the range of f are regular values.

Example 2.4 Let $f : \mathbb{R} \to \mathbb{R}^2$ such that $f(x, y) = x^2 + y^2$. Then the Jacobian of f at (x, y) is given by

$$Jf(x, y) = \begin{bmatrix} 2x \\ 2y \end{bmatrix}$$

and the set of all regular points of f takes the form

$$S = \{(x, y) \in \mathbb{R}^2 : x^2 + y^2 \neq 0\}.$$

Consequently, $(0, 0)$ is the only critical point of f.

Example 2.5 Let $f : \mathbb{R}^2 \to \mathbb{R}^2$ such that $f(x, y) = (x^2 + y^2, x^2 - y^2)$. Then the Jacobian of f is given by the matrix

$$Jf(x, y) = 2\begin{bmatrix} x & y \\ x & -y \end{bmatrix}.$$

Since $\det Jf(x, y) = -8xy$, the set of all regular points of f takes the form

$$S = \{(x, y) \in \mathbb{R}^2 : xy \neq 0\}$$

and any critical point belongs to the complement of S in $\mathbb{R}^2$.

Theorem 2.7 Let $f : M \to N$ be a smooth map and let q in N be a regular value of f. Then $f^{-1}(q)$ can be given a differentiable structure such that $f^{-1}(q)$ is a closed embedded submanifold of M of dimension $\dim M - \dim N$.

Proof: As a starting point, we give $P = f^{-1}(q)$ the subspace topology by viewing P as a subset of M. Clearly, P inherits is both Hausdorff and second countable as well. Moreover, by continuity of f, P is a closed subset of M.

Fix a point $p \in f^{-1}(q)$ so that $f(p) = q$. Suppose additionally that the rank of f at p coincides with the dimension of N. By the Rank Theorem, there exists a local coordinate (U, ϕ) on M and (V, ψ) on N such that $p \in U$, $\phi(p) = (0, \dots, 0) \in \mathbb{R}^n$, $\psi(q) = (0, \dots, 0) \in \mathbb{R}^m$ and additionally,

$$\left[\psi \circ f \circ \phi^{-1}\right]\left(x_1, \dots, x_m\right) = \left(x_1, \dots, x_n, 0, \dots, 0\right),$$

where $m = \dim M \geq \dim N = n$. Letting $u \in U \cap f^{-1}(q)$, we have

$$\psi(f(u)) = \psi(q) = (0, \dots, 0) \in \mathbb{R}^m$$

and as such, for $u = \phi^{-1}\left(x_1, \dots, x_m\right)$,

$$\left[\psi \circ f \circ \phi^{-1}\right]\left(x_1, \dots, x_m\right) = (0, \dots, 0, 0, \dots, 0).$$

This shows that

$$U \cap P = \left\{u \in U : [\phi(u)]_1 = \cdots = [\phi(u)]_n = 0\right\}.$$

By naturally identifying $\mathbb{R}^m$ with the Cartesian product $\mathbb{R}^n \times \mathbb{R}^{m-n}$, without loss of generally, we may assume that $\phi(U) = U' \times U''$, $U' \subseteq \mathbb{R}^n$, and $U'' \subseteq \mathbb{R}^{m-n}$ are open subsets and $\phi(P \cap U) = \{0\} \times U'' \subseteq \mathbb{R}^n \times \mathbb{R}^{m-n}$. Next, we define a chart $\left(U \cap P, \phi'\right)$ where $\phi' : U \cap P \to U''$ is uniquely determined by the equation

$$\phi(x) = \left(0, \dots, 0, \phi'(x)\right) \in \{(0, \dots, 0)\} \times U''.$$

Then $U \cap P$ is open in P and ϕ' is a homeomorphism (since it is obtained as a restriction of a homeomorphism). Consequently, $f^{-1}(q)$ is a locally Euclidean space of dimension $m - n$.

To complete the proof of this result, it remains to show that the collection

$$\mathcal{A} = \left\{\left(U \cap P, \phi'\right) : (U, \phi) \text{ is in a smooth atlas of } M\right\}$$

is a smooth atlas on $f^{-1}(q)$. To this end, consider two elements $\left(U_1 \cap P, \phi_1'\right)$ and $\left(U_2 \cap P, \phi_2'\right)$ in the given collection $\mathcal{A}$. Since $\phi_2' \circ \phi_1'^{-1}$ is smooth on $\phi_1'\left(U_1 \cap U_2\right)$, it follows that $\phi_2' \circ \phi_1'^{-1}$ is smooth on $U_1 \cap U_2 \cap \mathcal{A}$.

2.2.3 Derivations

We recall that an algebra $\mathcal{A}$ over a field of real numbers is a vector space equipped a ring structure. Moreover, an algebra is said to be commutative if its multiplicative operation is commutative.

Example 2.6 Let M be a smooth manifold of dimension n. Then clearly, $C^{\infty}(M)$, the set of smooth functions on M equipped with pointwise function addition operation is an infinite-dimensional vector space over $\mathbb{R}$. By endowing $C^{\infty}(M)$ with the group operation which takes two functions to their product (pointwise multiplication), it is easy to verify that $C^{\infty}(M)$ is an algebra over the reals.

Definition 2.15 A derivation of the algebra of smooth functions on a smooth manifold M is a linear operator $X : C^{\infty}(M) \to C^{\infty}(M)$ satisfying the following (product) rule: $X(fg) = fX(g) + gX(f)$ for all $f \in C^{\infty}(M)$.

Definition 2.16 Let $\mathfrak{X}(M)$ be the set of all derivations on $C^{\infty}(M)$.

1. For arbitrary $X, Y \in \mathfrak{X}(M)$, define $X + Y$ as a linear operator acting as follows. For any function $f \in C^{\infty}(M)$, $[X + Y](f) = X(f) + Y(f)$.
2. For $r \in C^{\infty}(M)$, define the linear operator rX such that $[rX](f) = r(X(f))$.

Moreover, the following holds true.

Proposition 2.12 Given $X, Y \in \mathfrak{X}(M)$ and $\phi \in C^{\infty}(M)$,

1. $X + Y \in \mathfrak{X}(M)$.
2. $\phi X \in \mathfrak{X}(M)$.

Proof: **Proof of Part 1**: Given smooth functions $f, g \in C^{\infty}(M)$ and $\lambda \in \mathbb{R}$, we verify that (a)

$$\begin{aligned}[X+Y](\lambda f) &= X(\lambda f) + Y(\lambda f)\\ &= \lambda X(f) + \lambda Y(f)\\ &= \lambda(X(f) + Y(f))\\ &= \lambda[X+Y](f)\end{aligned}$$

and (b)

$$\begin{aligned}[X+Y](f+g) &= X(f+g) + Y(f+g)\\ &= X(f) + X(g) + Y(f) + Y(g)\\ &= [X+Y](f) + [X+Y](g).\end{aligned}$$

Finally, to show that $X + Y$ satisfies the product rule, it suffices to verify that $[X+Y](fg) = f[X+Y](g) + g[X+Y](f)$ for all $f \in C^{\infty}(M)$. Indeed,

$$\begin{aligned}[X+Y](fg) &= X(fg) + Y(fg)\\ &= fX(g) + gX(f) + fY(g) + gY(f)\end{aligned}$$

and

$$
\begin{aligned}
[X+Y](fg) &= (fX(g)+fY(g)) + (gX(f)+gY(f)) \\
&= f(X(g)+Y(g)) + g(X(f)+Y(f)) \\
&= f[X+Y](g) + g[X+Y](f).
\end{aligned}
$$

Proof of Part 2: To show that ϕX is a linear operator, let $f, g \in C^{\infty}(M)$ and $\lambda \in \mathbb{R}$. Then by definition, $[\phi X](\lambda f) = \phi X(\lambda f) = \lambda([\phi X](f))$. Next, since X is a linear operator, it follows that

$$[\phi X](f+g) = \phi X(f) + \phi X(g) = [\phi X](f) + [\phi X](g).$$

Finally, to show that ϕX satisfies the product rule, we verify that for $f, g \in C^{\infty}(M)$,

$$[\phi X](fg) = \phi(fX(g) + gX(f)) = f[\phi X](g) + g[\phi X](f).$$

Proposition 2.13 Any constant function belongs to the kernel of derivation. In other words, if $X \in \mathfrak{X}(M)$ and if f is a constant function then $Xf = 0$.

Proof: Let $X \in \mathfrak{X}(M)$, let f be a constant function, and let 1_M be the indicator function of the manifold M. Since the indicator function is equal to its own square, appealing to the product rule, it follows that the image of the indicator function of M under X is equal to its double. Precisely,

$$X(1_M) = X(1_M \cdot 1_M) = 1_M X(1_M) + X(1_M) 1_M = 2X(1_M).$$

Solving the equation $X(1_M) = 2X(1_M)$ yields $X(1_M) = 0$. Finally, for arbitrary $f \in C^{\infty}(M)$, we obtain $X(f) = X(f 1_M) = fX(1_M) = 0$ as desired.

Definition 2.17 Given derivations $X, Y \in \mathfrak{X}(M)$, the Lie bracket of X, Y is defined as the map $[X, Y] : C^{\infty}(M) \to C^{\infty}(M)$ satisfying the following properties: $[X, Y](f) = X(Y(f)) - Y(X(f))$ for all $f \in C^{\infty}(M)$.

Proposition 2.14 $\mathfrak{X}(M)$ is closed under the Lie bracket operation. In other words, given $X, Y \in \mathfrak{X}(M)$, $[X, Y] \in \mathfrak{X}(M)$.

Proof: Since the set of all linear operators on $C^{\infty}(M)$ is a vector space, it is clear that $[X, Y]$ is a linear operator as well. To show that $[X, Y]$ satisfies the product rule, let $f, g \in C^{\infty}(M)$. Since Y and X satisfy the product rule,

$$
\begin{aligned}
[X,Y](fg) &= X(Y(fg)) - Y(X(fg)) \\
&= X(fY(g) + gY(f)) - Y(fX(g) + gX(f)).
\end{aligned}
$$

Next,

$$[X,Y](fg) = f(X(Y(g))) + X(f)Y(g) + g(X(Y(f))) + X(g)Y(f)$$
$$- fY(X(g)) - Y(f)X(g) - gY(X(f)) - Y(g)X(f)$$

and

$$\begin{aligned}[X,Y](fg) &= f(X(Y(g))) + g(X(Y(f))) - fY(X(g)) - gY(X(f)) \\ &= f(X(Y(g)) - Y(X(g))) + g(X(Y(f)) - Y(X(f))) \\ &= f[X,Y](g) + g[X,Y](f)\end{aligned}$$

as desired.

Proposition 2.15 Let $[\cdot,\cdot] : \mathfrak{X}(M) \times \mathfrak{X}(M) \to \mathfrak{X}(M)$ be the Lie bracket given by $[X,Y](f) = X(Y(f)) - Y(X(f))$ for all $f \in C^\infty(M)$. Then

1. $[\cdot,\cdot]$ is skew-symmetric: for all $X, Y \in \mathfrak{X}(M)$, $[X,Y] = -[Y,X]$.
2. $[\cdot,\cdot]$ is bilinear: for a real number λ and $X_1, X_2, Y_1, Y_2, X, Y \in \mathfrak{X}(M)$,

$$[X_1 + X_2, Y] = [X_1, Y] + [X_2, Y],$$
$$[X, Y_1 + Y_2] = [X, Y_1] + [X, Y_2],$$

   and $[\lambda X, Y] = [X, \lambda Y] = \lambda [X,Y]$.
3. $[\cdot,\cdot]$ satisfies Jacobi's identity: for all $X, Y, Z \in \mathfrak{X}(M)$

$$[[X,Y],Z] + [[Y,Z],X] + [[Z,X],Y] = 0.$$

4. If $f, g \in C^\infty(M)$ and $X, Y \in \mathfrak{X}(M)$ then

$$[fX, gY] = fg[X,Y] + fX(g)Y - gY(f)X.$$

Proof: Since the first three parts of the results are straightforward applications of the definition of the Lie bracket, we omit their proofs. To show that $[fX, gY] = fg[X,Y] + fX(g)Y - gY(f)X$, let $h \in C^\infty(M)$. By definition

$$\begin{aligned}[fX, gY](h) &= [(fX)\circ(gY)](h) - [(gY)\circ(fX)](h) \\ &= [fX]((gY)(h)) - [gY]((fX)(h)).\end{aligned}$$

Since fX and gY must satisfy the product rule, it follows that

$$[fX](gY(h)) = g \cdot [fX](Y(h)) + [fX](g) \cdot Y(h)$$
$$[gY](fX(h)) = f \cdot [gY](X(h)) + [gY](f) \cdot X(h)$$

and

$$\begin{aligned}
[fX, gY](h) &= g[fX](Y(h)) + [fX](g)Y(h) - f[gY](X(h)) - [gY](f)X(h) \\
&= (fg)(XY(h) - YX(h)) + [fX](g)Y(h) - [gY](f)X(h) \\
&= fg[X, Y](h) + [fX](g)Y(h) - [gY](f)X(h) \\
&= (fg[X, Y] + fX(g)Y - gY(f)X)(h).
\end{aligned}$$

Definition 2.18 Let V be a vector space over the reals and let $[\cdot,\cdot] : V \times V \to V$ be a map which is skew-symmetric, bilinear, and also satisfies Jacobi's identity. Then the structure obtained by equipping the vector space V with the map $[\cdot,\cdot]$ is called a Lie algebra.

Lemma 2.5 ***(Local Operator Lemma)*** Let M be a differentiable manifold of dimension n. The elements of $\mathfrak{X}(M)$ can be viewed as local operators. More precisely, given smooth functions $f, g \in C^{\infty}(M)$ and $p \in M$, let V be an open subset of M containing p as a point. If the restrictions of f and g on V coincides then $(Xf)(p) = (Xg)(p)$.

Proof: By assumption, the function $f - g$ is the zero map on V. Next, let W be an open subset of M containing a point p such that the closure of W is properly contained in V. Next, let $\rho : M \to [0, 1]$ be a smooth function such that ρ is equal to 1 on the complement of V in M and the support of ρ is contained in the complement of the closure of W in M. In other words, $\rho(p) = 0$. Moreover, $\rho \cdot (f - g) = f - g$. Indeed, given $x \in V$, $\rho(x) \cdot (f - g)(x) = \rho(x) \cdot 0 = 0$ and for x not in V, $\rho(x) \cdot (f - g)(x) = (f - g)(x)$. Next, by linearity, one obtains that

$$X(f - g) = X(f) - X(g) = X(\rho \cdot (f - g)) = \rho \cdot X(f - g) + X(\rho) \cdot (f - g).$$

Consequently,

$$\begin{aligned}
[X(f)](p) - [X(g)](p) &= \rho(p) \cdot [X(f - g)](p) + [X(\rho)](p) \cdot (f - g)(p) \\
&= 0
\end{aligned}$$

showing that $(Xf)(p) = (Xg)(p)$.

Proposition 2.16 Let U be an open and convex subset of $\mathbb{R}^n$. Then

$$\mathfrak{X}(U) = \left\{ \sum_{k=1}^{n} f_k \frac{\partial}{\partial x_k} : f_1, \ldots, f_n \in C^{\infty}(U) \right\}.$$

Proof: Given $X \in \mathfrak{X}(U)$, let $f_k = X(x_k)$. We aim to prove that given a smooth function $\phi \in C^{\infty}(U)$,

$$[X\phi](p) = \sum_{k=1}^{n} f_k(p) \left[\frac{\partial \phi}{\partial x_k} \right](p)$$

for any point $p \in M$. Appealing to the first-order Taylor's formula with remainder [1, Theorem A.58], we have

$$\phi(x) = \phi(p) + \sum_{k=1}^{n} \frac{\partial \phi}{\partial x_k}(p)\left(x_k - p_k\right) + \sum_{k=1}^{n} \varphi_k(x)\left(x_k - p_k\right) \tag{2.8}$$

for some $\varphi_1, \ldots, \varphi_n \in C^\infty(U)$ such that $\varphi_k(p) = 0$. Next, applying the derivation X on each side of Equation (2.8), we obtain

$$[X\phi](p) = \sum_{k=1}^{n} \left[X\left(\frac{\partial \phi}{\partial x_k}(p)\left(x_k - p_k\right)\right)\right](p) + \sum_{k=1}^{n} X\left[\left(\varphi_k(x)\left(x_k - p_k\right)\right)\right](p).$$

Linearity together with the fact that X satisfies the product rule gives:

$$X\left(\frac{\partial \phi}{\partial x_k}(p)\left(x_k - p_k\right)\right) = \frac{\partial \phi}{\partial x_k}(p) \cdot X\left(x_k - p_k\right) = \frac{\partial \phi}{\partial x_k}(p) \cdot X\left(x_k\right)$$

and

$$X\left[\left(\varphi_k(x)\left(x_k - p_k\right)\right)\right](p) = \varphi_k(p) \cdot \left[X\left(x_k - p_k\right)\right](p) = 0.$$

Therefore,

$$[X\phi](p) = \sum_{k=1}^{n} \left(\left[X\left(x_k\right)\right](p)\right) \cdot \frac{\partial \phi}{\partial x_k}(p)$$

as desired.

Definition 2.19 Given $X \in \mathfrak{X}(M)$ and letting U be an open subset of M, the operator $X|U$ obtained by restricting X to U is the operator $X|U$ defined as follows. Given $f \in C^\infty(U)$, and $p \in U$,

$$(X|U)(f)(p) = X\left(f_{\text{ext}}\right)(p),$$

where $f_{\text{ext}} \in C^\infty(M)$ and $f_{\text{ext}}(q) = f(q)$ for all $q \in U$.

It is not difficult to verify that given $X, Y \in \mathfrak{X}(M)$,

$$(X+Y)|\,U = X|\,U + Y|\,U$$
$$(fX)|\,U = (f|\,U)(X|\,U)$$

and additionally,

$$[X|\,U, Y|\,U] = [X, Y]|\,U.$$

Furthermore, we show that restricting a derivation on an open subset of a manifold results in a derivation of the ring of smooth functions defined on this subset.

Proposition 2.17 Let U be an open subset of a manifold M and let $X \in \mathfrak{X}(M)$. Then $X|U \in \mathfrak{X}(U)$.

Proof: First, we show that the assignment $f \mapsto (X|U)(f)$ is well defined. To this end, let f_1, f_2 be two smooth functions agreeing with f in a neighborhood V_1 and V_2 of p, respectively. Since, f_1, f_2 must also agree on the overlap of these neighborhoods, appealing to the Local Operator Lemma established earlier in Lemma 2.5, we obtain $X(f_1)(p) = X(f_2)(p)$.

Secondly, we will show that if $f \in C^\infty(U)$ then $(X|U)f \in C^\infty(U)$. Indeed, let $p \in U$ and let $f_{\text{ext}} \in C^\infty(M)$ such that $f_{\text{ext}}(q) = f(q)$ for all $q \in V$ where V is an open subset of U. Then, for arbitrary $v \in V$, since V is an open set around all of the points that belong to it, $(X|U)(f)(v) = X(f_{\text{ext}})(v)$. Consequently, $(X|U)f$ is smooth on V, and thus smooth on U.

Since $X|U$ is clearly a linear operator acting on $C^\infty(U)$, it remains to show that $X|U$ satisfies the product rule. Indeed, given $f, g \in C^\infty(U)$,

$$(X|U)(fg)(v) = X\left((fg)_{\text{ext}}\right)(v),$$

where $(fg)_{\text{ext}} \in C^\infty(M)$ and $(fg)_{\text{ext}}(q) = (fg)(q)$ for all $q \in U$. Applying the product rule $X\left((fg)_{\text{ext}}\right)$, we obtain

$$\begin{aligned}(X|U)(fg)(v) &= X\left((fg)_{\text{ext}}\right)(v) \\ &= \left(f_{\text{ext}} X g_{\text{ext}} + g_{\text{ext}} X f_{\text{ext}}\right)(v) \\ &= f_{\text{ext}}(v) X g_{\text{ext}}(v) + g_{\text{ext}}(v) X f_{\text{ext}}(v) \\ &= f(v)(X|U)(g)(v) + g(v)(X|U)(f)(v)\end{aligned}$$

as desired.

Definition 2.20 Suppose $\varphi : M \to N$ is a diffeomorphism of two manifolds. We define the pushforward $\varphi_* : \mathfrak{X}(M) \to \mathfrak{X}(N)$ of derivations such that given $X \in \mathfrak{X}(M)$ and $f \in C^\infty(N)$,

$$\varphi_*(X)f = X(f \circ \varphi).$$

Example 2.7 Let $M = \mathbb{R}^2 \backslash (\{0,0\})$ and define $\tau : M \to M$ such that

$$\tau(t, \theta) = \left(e^t \cos(t+\theta), e^t \sin(t+\theta)\right) = (x, y).$$

Next, let $X = \frac{\partial}{\partial t}$. Then $\tau_* X = f\frac{\partial}{\partial x} + g\frac{\partial}{\partial y}$ and

$$\begin{aligned}f = \tau_* X(x) = X(x \circ \tau) &= \frac{\partial}{\partial t}\left(e^t \cos(t+\theta)\right) \\ &= (\cos(t+\theta))e^t - (\sin(t+\theta))e^t \\ &= x - y,\end{aligned}$$

$$
\begin{aligned}
g = \tau_* X(y) = X(y \circ \tau) &= \frac{\partial}{\partial t}\left(e^t \sin(t+\theta)\right) \\
&= e^t(\cos(t+\theta) + \sin(t+\theta)) \\
&= x + y.
\end{aligned}
$$

Therefore,

$$
\tau_* X = (x-y)\frac{\partial}{\partial x} + (x+y)\frac{\partial}{\partial y}.
$$

Example 2.8 Let $f : \mathbb{R}^2 \to M(2, \mathbb{R})$ where

$$
f(t_1, t_2) = \begin{bmatrix} e^{t_1} & e^{t_1}t_2 \\ 0 & e^{t_1} \end{bmatrix} = \begin{bmatrix} x_1 & x_2 \\ 0 & x_1 \end{bmatrix}.
$$

Let $\varphi : f(M) \to (0, \infty) \times \mathbb{R}$ be a global chart for the manifold $f(M)$ such that

$$
\varphi\left(\begin{bmatrix} x_1 & x_2 \\ 0 & x_1 \end{bmatrix}\right) = (x_1, x_2).
$$

Then

$$
\varphi \circ f : (t_1, t_2) \mapsto (e^{t_1}, e^{t_1}t_2) = (y_1, y_2).
$$

Let $X = \frac{\partial}{\partial t_1}$ and write $[\varphi \circ f]_* (X) = f\frac{\partial}{\partial y_1} + g\frac{\partial}{\partial y_2}$. Then the pushforward of X under $f \circ \varphi$ can be computed explicitly by finding the components functions of $[\varphi \circ f]_* (X)$. Since

$$
\begin{aligned}
f &= [f \circ \varphi]_* (X)(y_1) = X(y_1 \circ f \circ \varphi) = \frac{\partial}{\partial t_1}(e^{t_1}) = y_1, \\
g &= [f \circ \varphi]_* (X)(y_2) = X(y_2 \circ f \circ \varphi) = \frac{\partial}{\partial t_1}(e^{t_1}t_2) = t_2 e^{t_1} = y_2,
\end{aligned}
$$

we obtain

$$
[f \circ \varphi^{-1}]_* (X) = y_1 \frac{\partial}{\partial y_1} + y_2 \frac{\partial}{\partial y_2}.
$$

Alternatively, computing the Jacobian of $f \circ \varphi$ gives

$$
\begin{bmatrix} \frac{\partial(e^{t_1})}{\partial t_1} & \frac{\partial(e^{t_1})}{\partial t_2} \\ \frac{\partial(e^{t_1}t_2)}{\partial t_1} & \frac{\partial(e^{t_1}t_2)}{\partial t_2} \end{bmatrix} = \begin{bmatrix} e^{t_1} & 0 \\ t_2 e^{t_1} & e^{t_1} \end{bmatrix},
$$

and the vector

$$
\begin{bmatrix} e^{t_1} & 0 \\ t_2 e^{t_1} & e^{t_1} \end{bmatrix} \begin{bmatrix} 1 \\ 0 \end{bmatrix} = \begin{bmatrix} e^{t_1} \\ t_2 e^{t_1} \end{bmatrix} = \begin{bmatrix} y_1 \\ y_2 \end{bmatrix}
$$

returns the components of $[f \circ \varphi]_* (X)$.

Choosing a different coordinate system such that $\psi : f(M) \to \mathbb{R} \times \mathbb{R}$ where

$$\psi\left(\begin{bmatrix} x_1 & x_2 \\ 0 & x_1 \end{bmatrix}\right) = \left(\log\left(x_1\right), x_2\right) = \left(y_1, y_2\right),$$

we verify that

$$\psi \circ f : \left(t_1, t_2\right) \mapsto \left(t_1, e^{t_1} t_2\right) = \left(z_1, z_2\right).$$

Thus, if $\left[f \circ \psi\right]_* (X) = f\frac{\partial}{\partial z_1} + g\frac{\partial}{\partial z_2}$ then

$$f = \left[f \circ \psi\right]_* (X)\left(z_1\right) = X\left(z_1 \circ f \circ \psi\right) = \frac{\partial}{\partial t_1}\left(t_1\right) = 1,$$

$$g = \left[f \circ \psi\right]_* (X)\left(z_2\right) = X\left(z_2 \circ f \circ \psi\right) = \frac{\partial}{\partial t_1}\left(e^{t_1} t_2\right) = t_2 e^{t_1} = z_2,$$

and $\left[f \circ \psi\right]_* (X) = \frac{\partial}{\partial z_1} + z_2 \frac{\partial}{\partial z_2}$. Again, computing the Jacobian of $f \circ \psi$, we obtain

$$\begin{bmatrix} \frac{\partial(t_1)}{\partial t_1} & \frac{\partial(t_1)}{\partial t_2} \\ \frac{\partial(e^{t_1} t_2)}{\partial t_1} & \frac{\partial(e^{t_1} t_2)}{\partial t_2} \end{bmatrix} = \begin{bmatrix} 1 & 0 \\ t_2 e^{t_1} & e^{t_1} \end{bmatrix},$$

and

$$\begin{bmatrix} 1 & 0 \\ t_2 e^{t_1} & e^{t_1} \end{bmatrix} \begin{bmatrix} 1 \\ 0 \end{bmatrix} = \begin{bmatrix} 1 \\ t_2 e^{t_1} \end{bmatrix} = \begin{bmatrix} 1 \\ z_2 \end{bmatrix}$$

returns the components of $\left[f \circ \psi\right]_* (X)$ as desired.

Proposition 2.18 If $\varphi : M \to N$ is a diffeomorphism and $\psi : N \to P$, then $(\varphi \circ \psi)_* = \varphi_* \circ \psi_*$.

Proof: For arbitrary $X \in \mathfrak{X}(M)$ and $f \in C^\infty(N)$,

$$\varphi_*\left(\psi_* X\right)(f) = \left[\psi_* X\right](f \circ \varphi) = X(f \circ \varphi \circ \psi) = (\varphi \circ \psi)_*(X)(f).$$

Letting (U, φ) be a smooth chart on M and letting $X \in \mathfrak{X}(M)$, we have $\varphi_* X \in \mathfrak{X}(\varphi(U))$. Applying Proposition 2.16, there exists a unique sequence of smooth functions $\left(f_1^{(U)}, \ldots, f_n^{(U)}\right) \in C^\infty(\varphi(U)) \times \cdots \times C^\infty(\varphi(U))$ such that

$$\varphi_*(X|U) = \sum_{k=1}^{n} f_k^{(U)} \frac{\partial}{\partial x_k}.$$

We say that $\left(f_1^{(U)}, \ldots, f_n^{(U)}\right)$ are the local components of X relative to the chart (U, φ).

Proposition 2.19 Assume the following,

1. (V, ψ) is a chart such that $U \cap V$ is not empty.
2. $\left(f_1^{(V)}, \ldots, f_n^{(V)}\right)$ are the local components of $X \in \mathfrak{X}(M)$ relative to (V, ψ).
3. $\left(f_1^{(U)}, \ldots, f_n^{(U)}\right)$ are the local components $X \in \mathfrak{X}(M)$ relative to (U, φ).
4. $D\left(\psi \circ \varphi^{-1}\right)$ is a matrix representation of the Jacobian of $\psi \circ \varphi^{-1}$.
   Then

$$D\left(\psi \circ \varphi^{-1}\right)\begin{bmatrix} f_1^{(U)} \\ \vdots \\ f_n^{(U)} \end{bmatrix} = \begin{bmatrix} f_1^{(V)} \\ \vdots \\ f_n^{(V)} \end{bmatrix}.$$

Proof: By assumption, we have

$$\left(\psi \circ \varphi^{-1}\right)_* \left(\varphi_* \left(X|U \cap V\right)\right) = \psi_* \left(X|U \cap V\right) = \sum_{k=1}^{n} f_k^{(V)} \frac{\partial}{\partial y_k}.$$

Moreover,

$$\left(\psi \circ \varphi^{-1}\right)_* \left(\varphi_* \left(X|U \cap V\right)\right) = \left(\psi \circ \varphi^{-1}\right)_* \left(\sum_{k=1}^{n} f_k^{(U)} \frac{\partial}{\partial x_k}\right)$$

and

$$\begin{aligned} f_\ell^{(V)} &= \left(\sum_{k=1}^{n} f_k^{(V)} \frac{\partial}{\partial y_k}\right)(y_\ell) \\ &= \left(\psi \circ \varphi^{-1}\right)_* \left(\sum_{k=1}^{n} f_k^{(U)} \frac{\partial}{\partial x_k}\right)(y_\ell) \\ &= \sum_{k=1}^{n} f_k^{(U)} \frac{\partial \left(y_\ell \circ \psi \circ \varphi^{-1}\right)}{\partial x_k}. \end{aligned}$$

As a result,

$$\left(\psi \circ \varphi^{-1}\right)_* \left(\varphi_* \left(X|U \cap V\right)\right) = \sum_{\ell=1}^{n} \sum_{k=1}^{n} f_k^{(U)} \frac{\partial \left(y_\ell \circ \psi \circ \varphi^{-1}\right)}{\partial x_k} \frac{\partial}{\partial y_k}.$$

Therefore

$$D\left(\psi \circ \varphi^{-1}\right)\begin{bmatrix} f_1^{(U)} \\ \vdots \\ f_n^{(U)} \end{bmatrix} = \begin{bmatrix} f_1^{(V)} \\ \vdots \\ f_n^{(V)} \end{bmatrix}$$

as desired.

2.2.4 Tangent Vectors

As a motivational tool, we will explore a number of toy examples. These will provide some illustrations for the concepts of tangent vectors and bundles.

Let's start with a helix curve in three-dimensional space, represented by the function $c : \mathbb{R} \to \mathbb{R}^3$, where $c(t) = (t, \cos(t), \sin(t))$. Here, $c(t)$ denotes the position of a particle in space at a given time t. For a fixed time $t_0 = \frac{\pi}{3}$, the position of the particle, $c(\frac{\pi}{3})$, is $\left(\frac{\pi}{3}, \frac{1}{2}, \frac{\sqrt{3}}{2}\right)$, and its velocity, $c'(\frac{\pi}{3})$, is $\left(1, -\frac{\sqrt{3}}{2}, \frac{1}{2}\right)$.

Now, let's introduce the concept of a "tangent space." The ordered pair of the particle's position and velocity, $\left(c\left(\frac{\pi}{3}\right), c'\left(\frac{\pi}{3}\right)\right)$, is an element of the tangent space $T(\mathbb{R}^3)$ of $\mathbb{R}^3$. This space can be thought of as $\mathbb{R}^3 \times \mathbb{R}^3$, essentially doubling the dimension of $\mathbb{R}^3$. We denote the velocity $c'\left(\frac{\pi}{3}\right)$ as an element of the tangent space at the position $c\left(\frac{\pi}{3}\right)$, symbolically represented as $c'\left(\frac{\pi}{3}\right) \in T_{c\left(\frac{\pi}{3}\right)}(\mathbb{R}^3)$.

The tangent space allows us to consider any ordered pair of vectors (x, v) in $T(\mathbb{R}^3)$ as the vector v, but with its tail positioned at the location indicated by the coordinates of x. In essence, we are attaching a three-dimensional plane to every fixed point in space. Each of these planes, or "fibers," is denoted as $T_x(\mathbb{R}^3)$, and the collection of all these fibers forms what we call a "tangent bundle": $T(\mathbb{R}^3) = \bigcup_{x \in \mathbb{R}^3} T_x(\mathbb{R}^3)$.

Each fiber in the tangent bundle is naturally equipped with a vector space structure. This allows us to define operations within each fiber. For example, for any (x, v) and (x, w) in $T(\mathbb{R}^3)$, and any real number a, we can define addition as $(x, v) + (x, w) = (x, v + w)$, and scalar multiplication as $a(x, v) = (x, av)$. However, it is important to note that these operations are only meaningful within a single fiber. Attempting to add (x, v) to (y, w) where $x \neq y$ would not make sense, as they belong to different vector spaces.

As a second illustration, consider a differentiable function $f : \mathbb{R}^3 \to \mathbb{R}^2$, defined by $f(x, y, z) = (x^2 + y^2, z(x^2 - y^2))$. In this case, the derivative of f at (x, y, z) is a linear map $Df(x, y, z) : \mathbb{R}^3 \to \mathbb{R}^2$, which can be explicitly represented by the matrix:

$$Df\,(x, y, z) = \begin{bmatrix} 2x & 2y & 0 \\ 2xz & -2yz & x^2 - y^2 \end{bmatrix}.$$

This derivative induces a natural map that transforms the tangent space at (x, y, z) in $\mathbb{R}^3$, denoted $T_{(x,y,z)}(\mathbb{R}^3)$, into the tangent space at $f(x, y, z)$ in $\mathbb{R}^2$, denoted $T_{f(x,y,z)}(\mathbb{R}^2)$. To illustrate this, consider a vector in the tangent space at (x, y, z). This vector is mapped to a new vector in the tangent space at $f(x, y, z)$, given by:

$$\begin{bmatrix} v_1 \\ v_2 \\ v_3 \end{bmatrix}_{(x,y,z)} \mapsto \left(\begin{bmatrix} 2x & 2y & 0 \\ 2xz & -2yz & x^2 - y^2 \end{bmatrix} \begin{bmatrix} v_1 \\ v_2 \\ v_3 \end{bmatrix} \right)_{(x^2+y^2, x(x^2-y^2))}$$

$$= \left(\begin{bmatrix} 2xv_1 + 2yv_2 \\ v_3x^2 + 2zv_1x - v_3y^2 - 2zv_2y \end{bmatrix} \right)_{(x^2+y^2, x(x^2-y^2))}.$$

We denote this map as the pushforward of f at (x, y, z). Since this pushforward has a concrete expression for each vector (x, y, z) in $\mathbb{R}^3$, we can define a map $f_* : T(\mathbb{R}^3) \to T(\mathbb{R}^2)$ between the tangent bundles of $\mathbb{R}^3$ and $\mathbb{R}^2$, such that:

$$f_* \left((x, y, z), \begin{bmatrix} v_1 \\ v_2 \\ v_3 \end{bmatrix} \right) = \left(\left(x^2 + y^2, z\left(x^2 - y^2\right)\right), \begin{bmatrix} 2xv_1 + 2yv_2 \\ v_3x^2 + 2zv_1x - v_3y^2 - 2zv_2y \end{bmatrix} \right).$$

This example underscores the usefulness of the map f_*, as it encapsulates information about both the function f and its derivative. Furthermore, if we let π denote the restriction map that retains the first argument, we find that the composition $\pi \circ f_*$ equals $f \circ \pi$, a relation that can prove quite useful.

The examples provided should serve to underscore the practicality and efficiency of the pushforward concept in the computation of derivatives for smooth maps. Following the examination of these examples, attention will now be directed toward the main objective: the comprehensive and detailed exposition of the concepts of tangent vectors and tangent bundles as they pertain to abstract manifolds. It should be noted that there exists a multitude of approaches to defining tangent vectors on a smooth manifold. Nevertheless, regardless of the specific approach employed for the realization of tangent vectors, the ensuing principles maintain their validity.

1) **Vector space structure**: Imagine you have a smooth surface (a manifold) and you pick a point on it. You can draw all possible arrows (vectors) that just touch that point without cutting into the surface. These arrows form a collection, which is called a vector space. We denote this vector space as T_pM. The dimension of this vector space is the same as the dimension of the surface itself.
2) **Functionality**: Given $p \in M$ and a smooth function $f : M \to N$ between two manifolds, f has a "derivative" at p. That is a linear mapping $f_* : T_pM \to T_{f(p)}N$ such that if $g : N \to Q$ is a smooth map then

$$\left[g \circ f\right]_* = g_* \circ f_* : T_pM \to T_{g(f(p))}Q.$$

3) **Naturality**: If U is an open subset of $\mathbb{R}^n$ and p is a point in U then there exists a natural isomorphism $\lambda_U : T_p U \to \mathbb{R}^n$ such that if $V \subseteq \mathbb{R}^m$ and $f : U \to V$ is a smooth map then the following diagram:

$$\begin{array}{ccc} T_p(U) & \xrightarrow{f_*} & T_{f(p)}(V) \\ \lambda_U \downarrow & & \downarrow \lambda_V \\ \mathbb{R}^n & \xrightarrow{Df(p)} & \mathbb{R}^n \end{array}$$

commutes.

The connection between the vector space structure and functorality lies in the way they interact with each other in the context of manifolds and functions that map between them. The vector space structure is a way of describing the set of all possible directions at a given point on a manifold. This gives us a way to talk about "movement" or "direction" at a specific point on the surface. Functorality, on the other hand, is about how these directions change when we apply a function that maps one manifold to another. Specifically, it is about how the derivative of a function at a point (which can be thought of as a linear transformation) acts on the vectors at that point. So, the connection is that the vector space structure provides the "raw material" (the vectors) that the functorality (the function's derivative) acts upon. When you apply a function to a manifold, it does not just move the points around; it also changes the vectors inside the tangent space at each point in a way that's determined by the derivative of the function at that point. This is what functorality is all about: it's a rule that tells you how to update the vectors when you apply a function.

The connection between the vector space structure and functorality lies in the way they interact with each other in the context of manifolds (smooth surfaces) and functions that map between them. The vector space structure is a way of describing the set of all possible directions (vectors) at a given point on a manifold. This gives us a way to talk about "movement" or "direction" at a specific point on the surface. Functorality, on the other hand, is about how these directions change when we apply a function that maps one manifold to another. Specifically, it is about how the derivative of a function at a point (which can be thought of as a linear transformation) acts on the vectors at that point. So, the connection is that the vector space structure provides the "raw material" (the vectors) that the functorality (the function's derivative) acts upon. When you apply a function to a manifold, it does not just move the points around; it also changes the vectors at each point in a way that is determined by the derivative of the function at that point. This is what functorality is all about: it is a rule that tells you how to update the vectors when you apply a function. The connection between naturality and the vector space structure and functorality lies in how they all work together harmoniously when dealing with transformations on manifolds.

Naturality, in this context, ensures that the relationship between vectors at different points on a manifold is consistent and does not depend on the specific choice of coordinates or charts used to represent the manifold. It guarantees that the transformation of vectors under a smooth function is done in a "natural" and coordinate-independent way.

When we consider the vector space structure and functorality, they are concerned with the behavior of vectors at individual points on a manifold and how they transform under smooth functions. The vector space structure allows us to talk about the collection of all possible tangent vectors at a single point, while functorality deals with how these vectors are mapped from one manifold to another through a smooth function.

The connection with naturality ensures that the transformation of vectors under the derivative of a smooth function is consistent with the way vectors are naturally associated with points on the manifolds. It guarantees that when we move from one point to another on the manifold and then apply a smooth function's derivative, the result is the same as directly applying the function's derivative and then moving to the corresponding point on the target manifold.

2.2.4.1 Tangent Vector As Equivalent Classes of Smooth Curves

Let M be an n-dimensional manifold. Given two smooth curves $a, b : \mathbb{R} \to M$ such that $a(0) = b(0) = p$, we say that $a \sim b$ if and only if $(\varphi \circ a)'(0) = (\varphi \circ b)'(0)$ for some chart (U, φ) around p. Letting $[c]$ be the equivalence class of a curve c, define

$$K_p(M) = \{[c] \quad \text{where } c : (-\epsilon, \epsilon) \to M \text{ is a smooth curve and } c(0) = p\}.$$

Note that since the derivative of a smooth function between two manifolds at a fixed point is independent of the choice of a chart, this definition of tangent vectors is also independent of the choice of a chart.

Example 2.9 Let $\alpha : (-1, 1) \to \mathbb{R}^2$ such that $\alpha(t) = (\sin 2t, \cos 3t)$ and let $\beta : (-1, 1) \to \mathbb{R}^2$ such that $\beta(t) = (2t, 1)$. Then $\alpha(0) = \beta(0) = (0, 1)$ and

$$\frac{d}{dt}((\sin 2t, \cos 3t))\bigg|_{t=0} = \begin{bmatrix} 2 & 0 \end{bmatrix},$$

$$\frac{d}{dt}((2t, 1))\bigg|_{t=0} = \begin{bmatrix} 2 & 0 \end{bmatrix}.$$

This shows that α, β are equivalent curves in $K_p\left(\mathbb{R}^2\right)$.

Example 2.10 Let $M(4, \mathbb{R})$ be the manifold of square matrices of order 4. $M(4, \mathbb{R})$ is naturally identified with $\mathbb{R}^4$ equipped with its usual smooth structure.

Next, let $GL(4,\mathbb{R})$ be a submanifold of $M(4,\mathbb{R})$ consisting of invertible matrices. Next, let $\alpha : \mathbb{R} \to GL(4,\mathbb{R})$ and $\beta : \mathbb{R} \to GL(4,\mathbb{R})$ such that

$$\alpha(t) = \begin{bmatrix} 1 & t & \frac{1}{2}t^2 & \frac{1}{6}t^3 \\ 0 & 1 & t & \frac{1}{2}t^2 \\ 0 & 0 & 1 & t \\ 0 & 0 & 0 & 1 \end{bmatrix} \quad \text{and} \quad \beta(t) = \begin{bmatrix} e^t & te^t & 0 & 0 \\ 0 & e^t & 0 & 0 \\ 0 & 0 & 1 & t \\ 0 & 0 & 0 & 1 \end{bmatrix}.$$

Then

$$\alpha(0) = \beta(0) = \begin{bmatrix} 1 & 0 & 0 & 0 \\ 0 & 1 & 0 & 0 \\ 0 & 0 & 1 & 0 \\ 0 & 0 & 0 & 1 \end{bmatrix}.$$

For any real number h,

$$\alpha'(0) = \lim_{h\to 0} \frac{\alpha(h) - \alpha(0)}{h} = \lim_{h\to 0} \begin{bmatrix} 0 & 1 & \frac{1}{2}h & \frac{1}{6}h^2 \\ 0 & 0 & 1 & \frac{1}{2}h \\ 0 & 0 & 0 & 1 \\ 0 & 0 & 0 & 0 \end{bmatrix} = \begin{bmatrix} 0 & 1 & 0 & 0 \\ 0 & 0 & 1 & 0 \\ 0 & 0 & 0 & 1 \\ 0 & 0 & 0 & 0 \end{bmatrix}$$

and

$$\beta'(0) = \lim_{h\to 0} \frac{\beta(h) - \beta(0)}{h} = \lim_{h\to 0} \begin{bmatrix} \frac{1}{h}\left(e^h - 1\right) & e^h & 0 & 0 \\ 0 & \frac{1}{h}\left(e^h - 1\right) & 0 & 0 \\ 0 & 0 & 0 & 1 \\ 0 & 0 & 0 & 0 \end{bmatrix} = \begin{bmatrix} 1 & 1 & 0 & 0 \\ 0 & 1 & 0 & 0 \\ 0 & 0 & 0 & 1 \\ 0 & 0 & 0 & 0 \end{bmatrix}.$$

Since $\alpha'(0) \neq \beta'(0)$ then $[\alpha] \neq [\beta]$. On the other hand, letting $\gamma : \mathbb{R} \to GL(4,\mathbb{R})$ such that

$$\gamma(t) = \begin{bmatrix} 1 & t & 0 & 0 \\ 0 & 1 & t & 0 \\ 0 & 0 & 1 & t \\ 0 & 0 & 0 & 1 \end{bmatrix},$$

since

$$\gamma(0) = \begin{bmatrix} 1 & 0 & 0 & 0 \\ 0 & 1 & 0 & 0 \\ 0 & 0 & 1 & 0 \\ 0 & 0 & 0 & 1 \end{bmatrix} \quad \text{and} \quad \gamma'(0) = \begin{bmatrix} 0 & 1 & 0 & 0 \\ 0 & 0 & 1 & 0 \\ 0 & 0 & 0 & 1 \\ 0 & 0 & 0 & 0 \end{bmatrix},$$

then $[\alpha] = [\gamma]$.

Proposition 2.20 $K_p(M)$ can be equipped with a vector space structure.

Proof: Let (U, φ) be a smooth chart around $p \in U$ such that $\varphi(p) = 0$. Next, let $[a], [b] \in K_p(M)$. Since M does not generally carry a vector space structure, we may define

$$[a] + [b] = t \mapsto \varphi^{-1}(\varphi(a(t)) + \varphi(b(t))).$$

Letting $\kappa(t) = \varphi^{-1}(\varphi(a(t)) + \varphi(b(t)))$, we verify that

$$\begin{aligned}\kappa(0) &= \varphi^{-1}(\varphi(a(0)) + \varphi(b(0)))\\ &= \varphi^{-1}(\varphi(p) + \varphi(p))\\ &= \varphi^{-1}(0)\\ &= p.\end{aligned}$$

Similarly, we define scaling as follows. Letting $\lambda \in \mathbb{R}$, $\lambda[a] = t \mapsto \varphi^{-1}(\lambda\varphi(a(t)))$. Then for $\kappa(t) = \varphi^{-1}(\lambda\varphi(a(t)))$, we have

$$\kappa(0) = \varphi^{-1}(\lambda\varphi(a(0))) = \varphi^{-1}(\lambda\varphi(p)) = \varphi^{-1}(0) = p.$$

Moreover, it is not hard to verify that the operations defined above are independent of a choice of a chart and a choice of representatives for $[a], [b] \in K_p(M)$. In addition, the structure obtained by endowing $K_p(M)$ with the operations defined above satisfy all axioms of a vector space.

Let $f : M \to N$ be a smooth map between two smooth manifolds. It is often of interest to linearize f near some point $p \in M$. This may be done by exploiting the vector space structures of $T_p(M)$ and $T_{f(p)}(N)$. This idea is carried out by the push-forward at a point p which maps $T_p(M)$ into $T_{f(p)}(N)$.

Definition 2.21 Let $f : M \to N$ be a smooth map between manifolds M, N. Define $f_* : K_p(M) \to K_{f(p)}(N)$ such that $f_*[c] = [f \circ c]$.

Example 2.11 Let $M(2, \mathbb{R})$ be the manifold of square matrices of order 3. Define $f : M(2, \mathbb{R}) \to M(2, \mathbb{R})$ such that

$$\begin{bmatrix} a & b \\ c & d \end{bmatrix} \overset{f}{\mapsto} \begin{bmatrix} a & b \\ c & d \end{bmatrix}^2 = \begin{bmatrix} a^2 + bc & b(a+d) \\ c(a+d) & d^2 + bc \end{bmatrix}.$$

Next, letting

$$c(t) = \begin{bmatrix} e^t & \sinh(t) \\ 0 & e^{-t} \end{bmatrix}$$

so that

$$c'(0) = \begin{bmatrix} 1 & 1 \\ 0 & -1 \end{bmatrix} \quad \text{and} \quad c(0) = \begin{bmatrix} 1 & 0 \\ 0 & 1 \end{bmatrix},$$

we obtain

$$f_*[c] = [f \circ c] = \left[t \mapsto \begin{bmatrix} e^{2t} & \sinh(2t) \\ 0 & e^{-2t} \end{bmatrix} \right].$$

Example 2.12 Let $M(2, \mathbb{R})$ be the manifold of square matrices of order 2. Define $f : M(2, \mathbb{R}) \to M(2, \mathbb{R})$ such that

$$\begin{bmatrix} a & b \\ c & d \end{bmatrix} \overset{f}{\mapsto} \begin{bmatrix} 1 & 2 \\ 0 & 1 \end{bmatrix} \begin{bmatrix} a & b \\ c & d \end{bmatrix} \begin{bmatrix} 1 & 2 \\ 0 & 1 \end{bmatrix}^{-1} = \begin{bmatrix} a + 2c & b - 2a - 4c + 2d \\ c & d - 2c \end{bmatrix}.$$

Next, letting

$$c(t) = \begin{bmatrix} e^t & \sinh(t) \\ 0 & e^{-t} \end{bmatrix}$$

so that

$$c'(0) = \begin{bmatrix} 1 & 1 \\ 0 & -1 \end{bmatrix} \quad \text{and} \quad c(0) = \begin{bmatrix} 1 & 0 \\ 0 & 1 \end{bmatrix}$$

since

$$\begin{bmatrix} 1 & 2 \\ 0 & 1 \end{bmatrix} \begin{bmatrix} e^t & \sinh(t) \\ 0 & e^{-t} \end{bmatrix} \begin{bmatrix} 1 & 2 \\ 0 & 1 \end{bmatrix}^{-1} = \begin{bmatrix} e^t & -3\sinh(t) \\ 0 & e^{-t} \end{bmatrix}$$

it follows that

$$f_*[c] = [f \circ c] = \left[t \mapsto \begin{bmatrix} e^t & -3\sinh(t) \\ 0 & e^{-t} \end{bmatrix} \right].$$

Proposition 2.21 Let $f : M \to N$ be a smooth map between the manifolds M, N. Then the map f_* (see Definition 2.21) is well defined.

Proof: Given two curves c_1 and c_2 such that $[c_1] = [c_2]$, one verifies that given $g \in C^\infty(N)$, since $(f \circ c_1)'(0) = (f \circ c_2)'(0)$,

$$(g \circ f \circ c_1)'(0) = (g \circ f \circ c_2)'(0).$$

This shows that $[f \circ c_1] = [f \circ c_2]$.

Proposition 2.22 Let $f : M \to N$ be a smooth map between the manifolds M, N and let $g : N \to Q$ be a smooth map. Then $(g \circ f)_*[c] = g_* f_*[c]$.

Proof: We verify that given $[c] \in K_p(M)$,

$$(g \circ f)_* [c] = [g \circ f \circ c] = g_* [f \circ c] = g_* f_* [c].$$

Proposition 2.23 Let $f : M \to N$ be a smooth map between the manifolds M, N. Then $f_* : K_p(M) \to K_{f(p)}(N)$ is a linear operator.

Proof: Let (U, φ) be a smooth chart around $p \in U$ such that $\varphi(p) = 0$ and let (V, ψ) be a smooth chart around $f(p) \in V$ such that $\psi(f(p)) = 0$. Suppose additionally that $f(U)$ is a subset of V. Next, let $[a], [b] \in K_p(M)$. Then

$$\begin{aligned} f_* ([a] + [b]) &= f_* \left[t \mapsto \varphi^{-1}(\varphi(a(t)) + \varphi(b(t)))\right] = [f \circ a + f \circ b] \\ &= t \mapsto f\left(\varphi^{-1}(\varphi(a(t)) + \varphi(b(t)))\right) \\ &= t \mapsto \left(\psi^{-1}\left(\psi \circ f \circ \varphi^{-1}(\varphi(a(t)) + \varphi(b(t)))\right)\right) \\ &= t \mapsto \psi^{-1}(\psi(f(a(t))) + \psi(f(b(t)))) \\ &= [f \circ a] + [f \circ b]. \end{aligned}$$

In a similar fashion, for any fixed real number λ, we verify that

$$\begin{aligned} f_* (\lambda [a]) &= t \mapsto f\left(\varphi^{-1}(\lambda \varphi(a(t)))\right) \\ &= t \mapsto \left(\psi^{-1}\left(\psi \circ f \circ \varphi^{-1}(\lambda \varphi(a(t)))\right)\right) \\ &= t \mapsto \psi^{-1}(\lambda \psi(f(a(t)))) \\ &= \lambda [f \circ a] \\ &= \lambda f_* ([a]). \end{aligned}$$

Definition 2.22 Given a derivation $X \in \mathfrak{X}(M)$, for each coordinate system (U, φ) on M, define the pushforward of $X|_U$ as follows: for arbitrary $f \in C^\infty(\varphi(U))$, $\varphi_* (X|_U) f = [X|_U] (f \circ \varphi)$.

Proposition 2.24 Let U be an open subset of $\mathbb{R}^n$ and let p be a point in U. Then the map $\lambda_U : K_p(U) \to \mathbb{R}^n$ given by $\lambda_U [c] = c'(0) = \left(c_1'(0), \ldots, c_n'(0)\right)$ is a vector space isomorphism.

Proof: By construction, λ_U is linear and clearly injective as well. To prove that λ_U is surjective, let $v = (v_1, \ldots, v_n) \in \mathbb{R}^n$ and let c be a curve defined as follows: $c(t) = (p_1 + tv_1, \ldots, p_n + v_n)$. Since $c'(0) = (v_1, \ldots, v_n)$, it follows that $\lambda_U([c]) = v$.

Example 2.13 Let $M(2, \mathbb{R})$ be the manifold of square matrices of order 2. Define $c : \mathbb{R} \to M(2, \mathbb{R})$ such that

$$c(t) = \begin{bmatrix} 1 & 0 \\ 0 & 1 \end{bmatrix} + t \begin{bmatrix} 0 & -1 \\ 1 & 0 \end{bmatrix} = \begin{bmatrix} 1 & -t \\ t & 1 \end{bmatrix}.$$

Then $c'(0) = \begin{bmatrix} 0 & -1 \\ 1 & 0 \end{bmatrix}$ and letting $I_2 = \begin{bmatrix} 1 & 0 \\ 0 & 1 \end{bmatrix}$, it follows that

$$\lambda_{M(2,\mathbb{R})} : T_{I_2}(M(2,\mathbb{R})) \to M(2,\mathbb{R})$$

and

$$\lambda_{M(2,\mathbb{R})}([c]) = \begin{bmatrix} 0 & -1 \\ 1 & 0 \end{bmatrix}.$$

Example 2.14 Let

$$M = \left\{ \begin{bmatrix} \cos\theta & -\sin\theta \\ \sin\theta & \cos\theta \end{bmatrix} \begin{bmatrix} a & 0 \\ 0 & a^{-1} \end{bmatrix} \begin{bmatrix} 1 & t \\ 0 & 1 \end{bmatrix} : \theta \in \mathbb{R}, a > 0, t \in \mathbb{R} \right\}.$$

Next, let $\varphi : M \to [0, 2\pi) \times (0, \infty) \times \mathbb{R}$ such that

$$\varphi\left(\begin{bmatrix} \cos\theta & -\sin\theta \\ \sin\theta & \cos\theta \end{bmatrix} \begin{bmatrix} a & 0 \\ 0 & a^{-1} \end{bmatrix} \begin{bmatrix} 1 & t \\ 0 & 1 \end{bmatrix} \right) = (\theta, a, t).$$

Put

$$a(t) = \begin{bmatrix} \cos t & -\sin t \\ \sin t & \cos t \end{bmatrix}, \quad b(t) = \begin{bmatrix} t & 0 \\ 0 & t^{-1} \end{bmatrix},$$

and

$$c(t) = \begin{bmatrix} 1 & t \\ 0 & 1 \end{bmatrix}.$$

Then

$$\varphi(a(t)) = (t, 0, 0), \varphi(b(t)) = (0, t, 0) \text{ and } \varphi(c(t)) = (0, 0, t).$$

Moreover,

$$\lambda_{\varphi(M)}([a]) = (1, 0, 0),$$
$$\lambda_{\varphi(M)}([b]) = (0, 1, 0),$$
$$\lambda_{\varphi(M)}([c]) = (0, 0, 1)$$

This shows that the tangent space $T_{I_2}(M)$ is a three-dimensional vector space spanned by $\{[a], [b], [c]\}$.

Remark 2.4 Let $f : U \to V \subseteq \mathbb{R}^n$. Appealing to Chain rule, $(f \circ \alpha)'(0) = Df(p)c'(0)$ and consequently, $\lambda_U f_*([c]) = Df(p)\lambda_U([c])$.

2.2.4.2 Tangent Vectors As Derivations at a Point

Although defining tangent vectors at a point on a manifold as equivalent classes of smooth curves is acceptable and useful, this collection of equivalent curves does not have an intrinsic vector space structure. To remedy this situation, we will provide an alternative definition of tangent vectors.

Definition 2.23 Let p be a point contained in a smooth manifold M. A linear functional $\ell : C^{\infty}(M) \to \mathbb{R}$ is called a derivation at p if ℓ satisfies the product rule at $p \in M$. In other words, given smooth functions $f, g \in C^{\infty}(M)$, one has

$$\ell(fg) = f(p) \cdot \ell(g) + \ell(f) \cdot g(p).$$

Example 2.15 Let $M = \mathbb{R}^n$. Recall that given a multivariable function $f : \mathbb{R}^n \to \mathbb{R}$ and a vector $a \in \mathbb{R}^n$, the directional derivative of f is formally defined as:

$$\left[D_v(f)\right](a) = \sum_{k=1}^{n} v_k \frac{\partial f(a)}{\partial x_k}.$$

Moreover, for $f, g \in C^{\infty}(\mathbb{R}^n), a \in \mathbb{R}^n$, and for a real scalar λ, it is not difficult to verify that

$$D_v(\lambda f + g)(a) = \lambda D_v(f)(a) + D_v(g)(a)$$

and

$$\left[D_v(fg)\right](a) = D_v(f)(a) g(a) + f(a) D_v(g)(a).$$

Therefore, for any fixed point $a \in \mathbb{R}^n$, the mapping $f \mapsto \left[D_v(f)\right](a)$ defines a derivation at a.

Proposition 2.25 The following holds true

1. Let $\ell : C^{\infty}(M) \to \mathbb{R}$ be a derivation at point p. Then $\ell\left(1_M\right) = 0$.
2. If two smooth functions $f, g \in C^{\infty}(M)$ agree on an open subset of M containing $p \in M$ then $\ell(f) = \ell(g)$.

Proof: First, since $1_M = 1_M \cdot 1_M$, appealing to the fact that ℓ satisfies the product rule, the following is immediate:

$$\ell\left(1_M\right) = \ell\left(1_M \cdot 1_M\right) = \ell\left(1_M\right) + \ell\left(1_M\right) = 2\ell\left(1_M\right).$$

Therefore, $\ell\left(1_M\right) = 0$.

Secondly, suppose $f = g$ on U. Next, let ρ be a smooth function defined on M which is equal to 1 on the complement of U in M and assume that the support of ρ is contained in $M \setminus \{p\}$. Then

$$\begin{aligned} \ell(f) - \ell(g) &= \ell(f - g) = \ell(\rho(f - g)) \\ &= \rho(p) \cdot \ell(f - g) + \ell(\rho) \cdot (f - g)(p) \\ &= 0. \end{aligned}$$

This shows that $\ell(f) = \ell(g)$.

Define

$$T_p(M) = \{\ell : \ell \text{ is a derivation at a point } p\}.$$

Proposition 2.26 ***(Vector Space Structure)*** $T_p(M)$ is a vector space.

Proof: Given two derivations ℓ_1, ℓ_2 at p, and $f \in C^\infty(M)$, since

$$\begin{aligned}(\ell_1 + \ell_2)(fg) &= \ell_1(fg) + \ell_2(fg)\\ &= f(p)\cdot \ell_1(g) + \ell_1(f)\cdot g(p) + f(p)\cdot \ell_2(g) + \ell_2(f)\cdot g(p)\\ &= f(p)\cdot \left(\ell_1(g) + \ell_2(g)\right) + g(p)\cdot\left(\ell_1(f) + \ell_2(f)\right)\\ &= f(p)\cdot\left(\ell_1 + \ell_2\right)(g) + g(p)\left(\ell_1 + \ell_2\right)(f).\end{aligned}$$

Similarly, for any real number λ,

$$\begin{aligned}\lambda\ell(fg) &= \lambda(f(p)\cdot\ell(g) + \ell(f)\cdot g(p))\\ &= f(p)\cdot\lambda\ell(g) + \lambda\ell(f)\cdot g(p).\end{aligned}$$

Proposition 2.27 Let (U, φ) be a smooth chart on a manifold M containing a point p. Then the set

$$\left\{\delta_k : f \mapsto \delta_k(f) = \left.\frac{\partial\left(f\circ\varphi^{-1}\right)}{\partial x_k}\right|_{\varphi(p)} : k = 1, 2, \ldots, n\right\}$$

is a basis for the vector space $T_p(M)$. Moreover, $\dim\left(T_p(M)\right) = \dim(M) = n$.

Proof: First, let us prove that $\left\{\delta_k : k = 1, 2, \ldots, n\right\}$ is linearly independent. To this end, suppose that $\sum_{k=1}^{n} \alpha_k\delta_k = 0$. Let $\widetilde{x}_k$ be a smooth function in M associated to the kth coordinate function $\left(x_1, \ldots, x_n\right) \mapsto x_k$ in a fixed neighborhood of a point p in M. In other words, $\widetilde{x}_k = x_k \circ \varphi$. Evaluating the linear functional δ_j at $\widetilde{x}_k$, we obtain

$$\delta_j\left(\widetilde{x}_k\right) = \left.\frac{\partial\left(\left[x_k\circ\varphi\right]\circ\varphi^{-1}\right)}{\partial x_k}\right|_{\varphi(p)} = \left.\frac{\partial\left(x_k\right)}{\partial x_k}\right|_{\varphi(p)} = \begin{cases} 1, & \text{if } k = j,\\ 0, & \text{otherwise.}\end{cases}$$

Thus, for any $j = 1, 2, \ldots, n$, $0 = \sum_{k=1}^{n} \alpha_k\delta_k\left(x_j\right) = a_j$.

Secondly, to show that $\left\{\delta_k : k = 1, 2, \ldots, n\right\}$ spans an n-dimensional vector space, let ℓ be a derivation at some point p. Given a function $f \in C^\infty(M)$, let $f_U : \varphi(U) \to \mathbb{R}$ such that $f_U = \left[f\cdot\rho\right]\circ\varphi^{-1}$ where $\rho \in C^\infty(M)$ and is constantly equal to 1 on U. Using Taylor's Theorem, we may write

$$f_U = f_U(\varphi(p)) + \sum_{k=1}^{\infty} g_k(x)\left(x_k - \varphi(p)_k\right)$$

for some smooth functions g_k such that

$$g_k(p) = \frac{\partial f_U}{\partial x_k}(\varphi(p)) = \left.\frac{\partial\left(f \circ \varphi^{-1}\right)}{\partial x_k}\right|_{\varphi(p)} = \delta_k(f).$$

Then,

$$\begin{aligned}\ell\left(f_U\right) &= \sum_{k=1}^{\infty} \ell\left(g_k(x)\left(x_k - \varphi(p)_k\right)\right)\\ &= \sum_{k=1}^{\infty} g_k(p) \cdot \ell\left(x_k\right)\\ &= \sum_{k=1}^{\infty} \delta_k(f) \cdot \ell\left(x_k\right).\end{aligned}$$

Letting $\alpha_k = \ell\left(x_k\right)$, we obtain $\ell\left(f_U\right) = \sum_{k=1}^{\infty} \alpha_k \cdot \delta_k(f)$.

Definition 2.24 ***(Functoriality)*** Given a smooth function $f : M \to N$, and $\ell \in T_p(M)$, define

$$\left[f_*(\ell)\right](g) = \ell(g \circ f)$$

for $g \in C^{\infty}(N)$.

Proposition 2.28 The following statements holds true:

1. $f_*(\ell) \in T_{f(p)}(N)$ for every $\ell \in T_p(M)$.
2. The map $f_* : T_p(M) \to T_{f(p)}(N)$, which assigns to $\ell \in T_p(M)$ the element $f_*(\ell) \in T_{f(p)}(N)$ is a linear assignment.
3. $\left[g \circ f\right]_* = g_* \circ f_*$.

Proof: First, let $u, v \in C^{\infty}(N)$. Then

$$\begin{aligned}\left[f_*\ell\right](u+v) &= \ell((u+v)\circ f)\\ &= \ell(u \circ f + v \circ f)\\ &= \ell(u \circ f) + \ell(v \circ f)\\ &= \left[f_*\ell\right](u) + \left[f_*\ell\right](v),\end{aligned}$$

and similarly, for any real number r, we have

$$\begin{aligned}\left[f_*\ell\right](ru) &= \ell((ru)\circ f)\\ &= r\ell(u \circ f)\\ &= r\left[f_*\ell\right](u).\end{aligned}$$

Moreover,

$$\begin{aligned} f_*\ell\,(uv) &= \ell\,(uv\circ f) = \ell\,((u\circ f)\,(v\circ f)) \\ &= (u\circ f)\,(p)\cdot \ell\,(v\circ f) + \ell\,(u\circ f)\cdot (v\circ f)\,(p) \\ &= (u\circ f)\,(p)\,\left[f_*\ell\right](v) + (v\circ f)\,(p)\,\left[f_*\ell\right](v) \\ &= u\,(f\,(p))\,\left[f_*\ell\right](v) + v\,(f\,(p))\,\left[f_*\ell\right](v)\,. \end{aligned}$$

Thus, $f_*\ell \in T_{f(p)}\,(N)$ for every $\ell \in T_p\,(M)$.

Secondly, for arbitrary $a, b \in T_p\,(M)$ and $u \in C^\infty\,(N)$

$$\begin{aligned} f_*\,(a+b)\,(u) &= (a+b)\,(u\circ f) \\ &= (a)\,(u\circ f) + (b)\,(u\circ f) \\ &= \left[f_* a\right](u) + \left[f_* a\right](b)\,. \end{aligned}$$

Similarly, given a real number r,

$$f_*\,(ra)\,(u) = (ra)\,(u\circ f) = r\left[f_*\,(a)\right](u)\,.$$

This shows that $\ell \mapsto f_*\ell$ is indeed a linear map.

For a suitable linear functional a and a smooth function u, we have

$$\begin{aligned} \left[g\circ f\right]_*(a)\,u &= a\,(u\circ g\circ f) \\ &= f_*\,(a)\,(u\circ g) \\ &= g_*\left(f_*\,(a)\right)u \\ &= \left[g_*\circ f_*\right](a)\,u. \end{aligned}$$

Remark 2.5 ***(Naturality)*** Let U be an open subset of an n-dimensional Euclidean space. According to Proposition 2.27, given any linear functional $\ell \in T_p\,(U)$, there exist unique scalars $\alpha_1\,(\ell), \ldots, \alpha_n\,(\ell)$ such that

$$\ell = \sum_{k=1}^{n} \alpha_k\,(\ell)\left.\frac{\partial}{\partial x_k}\right|_p.$$

Next, we define $\lambda_U : T_p\,(U) \to \mathbb{R}^n$ such that

$$\lambda_U\,(\ell) = \left(\alpha_1\,(\ell), \ldots, \alpha_n\,(\ell)\right).$$

Then the mapping $\ell \mapsto \lambda_U\,(\ell)$ is clearly a linear isomorphism.

Example 2.16 Let

$$M = \left\{ p\left(t_1, t_2\right) = \begin{bmatrix} e^{t_1} & 0 & 0 \\ 0 & e^{t_1} & 0 \\ 0 & 0 & e^{t_1} \end{bmatrix} \begin{bmatrix} 1 & t_2 & \frac{1}{2}t_2^2 \\ 0 & 1 & t_2 \\ 0 & 0 & 1 \end{bmatrix} : t_1, t_2 \in \mathbb{R} \right\}$$

and define $f : M \to M$ such that $f(p(t_1, t_2)) = p(t_1, t_2)^{-1}$. Since $p(t_1, 0)$ commutes with $p(0, t_2)$ for every real number t_1, t_2, we verify that $(p(t_1, t_2))^{-1} = p(-t_1, -t_2)$ as follows:

$$\begin{aligned}(p(t_1, t_2))^{-1} &= (p(t_1, 0)\, p(0, t_2))^{-1} \\ &= p(0, t_2)^{-1} p(t_1, 0)^{-1} \\ &= p(0, -t_1)\, p(0, -t_2) \\ &= p(-t_1, -t_2)\,.\end{aligned}$$

Letting $\varphi : M \to \mathbb{R}^2$ such that $\varphi((p(t_1, t_2))) = (t_1, t_2)$, be a global chart for the manifold M, it follows that

$$\widetilde{f}(t_1, t_2) = [\varphi \circ f \circ \varphi^{-1}](t_1, t_2) = (-t_1, -t_2)\,.$$

Consequently,

$$(\widetilde{f})_* \left(\left. \frac{\partial}{\partial t_1} \right|_{(t_1, t_2)} \right) = - \left. \frac{\partial}{\partial t_1} \right|_{(-t_1, -t_2)}$$

and

$$(\widetilde{f})_* \left(\left. \frac{\partial}{\partial t_2} \right|_{(t_1, t_2)} \right) = - \left. \frac{\partial}{\partial t_2} \right|_{(-t_1, -t_2)}.$$

Proposition 2.29 Let $f : U \to V$ be a smooth function where U and V are some open subsets of $\mathbb{R}^n$ and $\mathbb{R}^m$, respectively. If ℓ is a linear functional in $T_p(U)$ then $\lambda_V f_* (\ell) = Df(p)\, \lambda_U(\ell)$.

Proof: First, we write $f(x_1, \ldots, x_n) = (t_1, \ldots, t_m)$. Secondly, let

$$\ell = \sum_{k=1}^{n} \alpha_k \left. \frac{\partial}{\partial x_k} \right|_p \quad \text{and} \quad f_*(\ell) = \sum_{k=1}^{m} \beta_k \left. \frac{\partial}{\partial t_k} \right|_{f(p)}$$

for some real scalars α_k and β_j. Then

$$\beta_j = f_* \ell(t_j) = \ell(t_j \circ f) = \sum_{k=1}^{n} \alpha_k \left. \frac{\partial (t_j \circ f)}{\partial x_k} \right|_p.$$

As a result,

$$\begin{bmatrix} \beta_1 \\ \vdots \\ \beta_m \end{bmatrix} = Df(p) \begin{bmatrix} \alpha_1 \\ \vdots \\ \alpha_n \end{bmatrix}.$$

Remark 2.6 There is a natural one-to-one correspondence η between $K_p(M)$ and $T_p(M)$ such that given $[c] \in K_p(M)$, $\eta([c])$ is a derivation acting in $C^\infty(M)$ as follows:

$$\eta([c])f = (f \circ c)'(0) \quad \text{for } f \in C^\infty(M).$$

Finally, a vector space structure on $K_p(M)$ can be obtained by defining η to be linear bijection $T_p(M)$.

2.2.5 Tangent Bundles

Let M be an n-dimensional smooth manifold. The tangent bundle

$$TM = \bigcup_{p \in M} T_p M$$

is the collection of all vectors belonging to the tangent space of every point in M. In this section, we will describe how to equip the tangent bundle with a topology and a smooth atlas that will turn TM into a smooth manifold.

Let $\pi : TM \to M$ be the projection mapping defined as follows. Given a tangent vector X_p which belongs to the vector space T_pM, $\pi\left(X_p\right) = p$. The projection map returns the point associated to the tangent vector. Moreover, letting (U, ω) be a chart on M, define

$$\overline{\omega} : TM \supseteq \pi^{-1}(U) \to \omega(U) \times \mathbb{R}^n \subseteq \mathbb{R}^n \times \mathbb{R}^n$$

such that

$$\overline{\omega}\left(X_p\right) = \left(\omega(p), \lambda_U\left(\omega_*\left(X_p\right)\right)\right).$$

Example 2.17 Let $M = \left\{p(x, y, z) : (x, y, z) \in \mathbb{R}^3\right\}$ such that

$$p(x, y, z) = \begin{bmatrix} 1 & 0 & z \\ 0 & 1 & 0 \\ 0 & 0 & 1 \end{bmatrix} \begin{bmatrix} 1 & 0 & 0 \\ 0 & 1 & y \\ 0 & 0 & 1 \end{bmatrix} \begin{bmatrix} 1 & x & 0 \\ 0 & 1 & 0 \\ 0 & 0 & 1 \end{bmatrix}.$$

Let (M, φ) be a global chart on M such that $\varphi(p(x, y, z)) = (x, y, z)$. Put

$$c(t) = (x + ut, y + vt, z + wt)$$

such that $c(0) = (x, y, z)$ and $c'(0) = (u, v, w)$. Then

$$t \mapsto \begin{bmatrix} 1 & x + tu & z + tw \\ 0 & 1 & y + tv \\ 0 & 0 & 1 \end{bmatrix}$$

is a curve passing through $p(x, y, z)$ and

$$TM = \bigcup_{(x,y,z) \in \mathbb{R}^3} T_{p(x,y,z)} M,$$

where

$$T_{p(x,y,z)}M = \left\{ t \mapsto \begin{bmatrix} 1 & x+tu & z+tw \\ 0 & 1 & y+tv \\ 0 & 0 & 1 \end{bmatrix} : (u,v,w) \in \mathbb{R}^3 \right\}.$$

Example 2.18 Let $M = \{p(x) : x \in \mathbb{R}\}$ such that

$$p(x) = \begin{bmatrix} \cos x & -\sin x & 0 & 0 \\ \sin x & \cos x & 0 & 0 \\ 0 & 0 & e^x & xe^x \\ 0 & 0 & 0 & e^x \end{bmatrix}.$$

Let (M, φ) be a global chart on M such that $\varphi(p(x)) = x$. Put $c_{(x,y)}(t) = x + yt$ so that $c'_{(x,y)}(0) = y$ and $c_{(x,y)}(0) = x$. Then

$$p\left(c_{(x,y)}(t)\right) = \begin{bmatrix} \cos(x+ty) & -\sin(x+ty) & 0 & 0 \\ \sin(x+ty) & \cos(x+ty) & 0 & 0 \\ 0 & 0 & e^{x+ty} & e^{x+ty}(x+ty) \\ 0 & 0 & 0 & e^{x+ty} \end{bmatrix}$$

and $TM = \bigcup_{p(x)\in M} T_{p(x)}M$, where

$$T_{p(x)}M = \left\{ t \mapsto p\left(c_{(x,y)}(t)\right) : y \in \mathbb{R} \right\}.$$

Example 2.19 Let $M = \mathbb{T}$ be the one-dimensional torus. Then

$$TM = \left\{ (\cos t, \sin t, -x\sin t, x\cos t) \in \mathbb{R}^4 : t, x \in \mathbb{R} \right\}$$

is a realization of the tangent bundle of the torus.

Proposition 2.30 Let (U, ω) and (V, ψ) be charts on M. Then $\overline{\psi} \circ \overline{\omega}^{-1} : \omega(U \cap V) \times \mathbb{R}^n \to \psi(U \cap V) \times \mathbb{R}^n$ is a smooth function.

Proof: Let us suppose that

$$X = \left(\omega^{-1}\right)_* \left(\sum_{k=1}^{n} f_k \frac{\partial}{\partial x_k} \right) \in \mathfrak{X}(U)$$

for some smooth functions $f_k \in C^\infty(\omega(U))$ and $\sum_{k=1}^n f_k \frac{\partial}{\partial x_k} \in \mathfrak{X}(\omega(U))$. Then

$$\overline{\omega}\left(p, X_p\right) = \left(\omega(p)_1, \dots, \omega(p)_n, f_1(\omega(p)), \dots, f_n(\omega(p))\right)$$

and

$$\overline{\omega}^{-1}\left(t_1,\ldots,t_n,a_1,\ldots,a_n\right)=\left(\omega^{-1}(t),\sum_{k=1}^{n}a_k\left.\frac{\partial}{\partial x_k}\right|_{(t_1,\ldots,t_n)}\right).$$

In light of the above, it is not difficult to verify that

$$\overline{\psi}\circ\overline{\omega}^{-1}:\omega(U\cap V)\times\mathbb{R}^n\to\psi(U\cap V)\times\mathbb{R}^n$$

is defined as follows:

$$\left(\overline{\psi}\circ\overline{\omega}^{-1}\right)(t,a)=\left(\psi\circ\omega^{-1}(t),D\left(\psi\circ\omega^{-1}\right)(t)a\right).$$

This shows that the mapping $\overline{\psi}\circ\overline{\omega}^{-1}$ is smooth.

Definition 2.25 Let (U,ω) be a smooth chart on M. Given a set $O\subseteq TU$, we say that O is open in M if and only if $\overline{\omega}(O)$ is open in $\omega(U)\times\mathbb{R}^n$.

Proposition 2.31 Given a smooth chart (U,ω) on M, the following holds true:

1. $\pi^{-1}(U)=\overline{\omega}^{-1}\left(\omega^{-1}(U)\times\mathbb{R}^n\right)$ is open in TM.
2. $\overline{\omega}:TM\supseteq\pi^{-1}(U)\to\omega(U)\times\mathbb{R}^n\subseteq\mathbb{R}^n\times\mathbb{R}^n$ is a homeomorphism.
3. $\pi:TM\to M$ is a continuous map.
4. TM is Hausdorff.
5. TM is second countable.

Proof: Note that if $O\subseteq\omega(U)\times\mathbb{R}^n$ is open then $\overline{\omega}^{-1}(O)$ is an open subset of TM. Indeed, letting (V,ψ) be a smooth chart defined on M, the set

$$\overline{\psi}\left(\overline{\omega}^{-1}(O)\cap\pi^{-1}(V)\right)=\left[\psi\circ\omega^{-1}\right]\left(O\cap\omega(U\cap V)\times\mathbb{R}^n\right)$$

is open since $\psi\circ\omega^{-1}$ is a homeomorphism. As a direct consequence of this observation, the statements in Part (1), (2), and (3) are immediate.

Proof of (4): Let X,Y be two distinct elements in TM. Assuming first that $\pi(X)$ is not equal to $\pi(Y)$ as points in M, let (U,ω) and (V,ψ) such that U and V do not overlap. Additionally, let us assume that $\pi(X),\pi(Y)$ belongs to U,V respectively. Clearly, $\pi^{-1}(U)\cap\pi^{-1}(V)=\emptyset$. Next, assuming that $\pi(X)$ and $\pi(Y)$ are equal to each other, let (U,ω) be a smooth chart around the point $\pi(X)$. Since the Cartesian product space $\omega(U)\times\mathbb{R}^n$ is Hausdorff, we can find disjoint open sets $\Omega_1,\Omega_2\subseteq\omega(U)\times\mathbb{R}^n$ such that $\overline{\omega}(X)\in\omega\left(\Omega_1\right)$ and $\overline{\omega}(Y)\in\omega\left(\Omega_2\right)$. This shows that $\overline{\omega}^{-1}\left(\Omega_1\right)$ and $\overline{\omega}^{-1}\left(\Omega_2\right)$ are disjoint open subset around X and Y, respectively.

Proof of (5): Let $\left\{W_k:k\in\mathbb{N}\right\}$ be a countable basis for the standard topology in the n-dimensional Euclidean space. Additionally, let $\left\{\left(U_k,\omega_k\right):k\in\mathbb{N}\right\}$ be countable basis for $\mathbb{R}^n$ such that each pair $\left(U_k,\omega_k\right)$ is a smooth chart defined on $\mathbb{R}^n$.

Then,

$$\left\{\overline{\omega}_j^{-1}\left(\omega_j\left(U_k\right)\times W_k\right)\ :\ (j,k)\in\mathbb{N}\times\mathbb{N}\right\}$$

forms a countable basis for the topology on TM.

In light of Proposition 2.31, the following is immediate

Corollary 2.1 Let M be an n-dimensional smooth manifold with atlas

$$\left\{\left(U_\alpha,\omega_\alpha\right)\ :\ \alpha\in I\right\}.$$

Then TM is a $2n$-dimensional manifold and atlas $\left\{\left(\pi^{-1}\left(U_\alpha\right),\overline{\omega}_\alpha\right)\ :\ \alpha\in I\right\}$.

Proposition 2.32 $\pi\ :\ TM\to M$ is a smooth map.

Proof: Since a local representation of the map π in smooth charts (U,ω) on M and $\left(\pi^{-1}(U),\overline{\omega}\right)$, is given by $(x,y)\mapsto x$, it is clear that π is smooth.

Definition 2.26 A section of TM is a function $X\ :\ M\to TM$ such that $\pi\circ X$ is the identity map on M.

Let X be a section of TM and let (U,φ) and $\left(\pi^{-1}(U),\overline{\varphi}\right)$ be charts on M and TM, respectively. Then the image of U under X is a subset of $\pi^{-1}(U)$. Next,

$$\overline{\varphi}\circ X\circ\varphi^{-1}\ :\ \varphi(U)\to\varphi(U)\times\mathbb{R}^n$$

such that

$$\overline{\varphi}\circ X\circ\varphi^{-1}\left(x_1,\dots,x_n\right)=\left(x_1,\dots,x_n,f_1\left(x_1,\dots,x_n\right)\dots,f_n\left(x_1,\dots,x_n\right)\right).$$

The objects $f_1,\dots,f_n$ are called the components of X in the chart (U,φ). Moreover, a section is called smooth if and only if every component in every chart on M is smooth as well.

For each fixed point $p\in M$, since T_pM has the structure of a vector space, we can define the sum of two sections and the pointwise product of a real-valued-function with a section of TM. Precisely, for arbitrary $X,Y\in TM$, and for a suitable smooth function ρ, we define

$$(X+Y)_p=X_p+Y_p\quad\text{and}\quad(\rho X)_p=\rho(p)X_p.$$

Next, if $f_1,\dots,f_n$ are the components of X and if $g_1,\dots,g_n$ then one verifies that $f_1+g_1,\dots,f_n+g_n$ and $\rho f_1,\dots,\rho f_n$ are the components of $X+Y$ and ρX, respectively. Under these operations, it is then clear that the collection of smooth sections on TM forms a $C^\infty(M)$-module.

To conclude our discussion on tangent bundles, we will now describe a natural identification ($C^{\infty}(M)$-module isomorphism) made between derivations of the ring $C^{\infty}(M)$ and smooth sections of the tangent bundle TM.

First, given a derivation $X \in \mathfrak{X}(M)$ and a point p in the manifold M, let us define the linear functional $X_p : C^{\infty}(M) \to \mathbb{R}$ was a derivation at p: $X_p(f) = (Xf)(p)$ for all real-valued smooth functions f defined on M. By identifying T_pM with the vector space of derivations at p, we can then view X_p as an element of T_pM. Therefore, the map which sends p to X_p is a section of TM, and in fact, it is not difficult to verify that $p \mapsto X_p$ is indeed a smooth section.

Conversely, given a smooth section $s : M \to TM$, for each $p \in M$, we may identify $s(p)$ with a derivation. As such, for a suitable real-valued smooth function f, the equation $s(f)(p) = s(p)(f)$ makes sense. On the other hand, regarding $s(p)$ as a class of equivalent curves, so that $s(p) = [c]$, then $s(f)(p)$ is equal to the differential of $f \circ c$ evaluated at zero. In fact, it can be shown that $p \mapsto s(p)$ is a smooth function and consequently, the mapping $f \mapsto s(f)$ is a derivation.

From here on, we will make no distinction between derivations of the ring $C^{\infty}(M)$ and smooth sections of TM. Regardless of the choice made (on the basis of whichever option is convenient) we shall call such an object a vector field defined on M, and we will denote the collection of all vector fields on M by $\mathfrak{X}(M)$.

2.2.6 1-Forms

Let V be a finite-dimensional real vector space let V^* be the vector space of linear functionals defined V. Given $\ell \in V^*$ and $v \in V$, we denote $\ell(v)$ obtained by letting ℓ acts on the vector v by the pairing $\langle \ell, v \rangle_{V^* \times V}$. Moreover, any linear transformation $T : V \to V$, induces a dual map $T^* : V^* \to V^*$ such that

$$\langle T^*\ell, v \rangle_{V^* \times V} = T^*\ell(v) = \ell(Tv) = \langle \ell, Tv \rangle_{V^* \times V}$$

for all $v \in V$.

When T is injective then T^* is surjective and when T is surjective then T^* is injective. Moreover $(T \circ J)^* = J^* \circ T^*$.

Let M be an n-dimensional manifold, and let TM be the tangent bundle of M. Viewing $T_p(M)$ as an n-dimensional real vector space identified with the real line $\mathbb{R}$, let $\left[T_p(M)\right]^*$ be the dual vector space of $T_p(M)$. Then the collection

$$\bigcup_{p \in M} \left[T_p(M)\right]^*$$

is a bundle of linear functionals. We will be particularly interested in covector fields $\omega \in \bigcup_{p \in \mathbb{R}} \left[T_p(M)\right]^*$ where (1) $\omega_p \in \left[T_p(M)\right]^*$, $X \in T(M)$, $X_p \in T_p(M)$, and (2) $p \mapsto \omega_p\left(X_p\right) \in \mathbb{R}$ is a smooth function defined on M.

Definition 2.27 A 1-form on a smooth manifold M is a $C^\infty(M)$-module homomorphism $\omega : \mathfrak{X}(M) \to C^\infty(M)$ mapping derivations to smooth functions. In other words, for all $X, Y \in \mathfrak{X}(M)$ and for any real-valued smooth function $\rho \in C^\infty(M)$, the following holds true: $\omega(X+Y) = \omega(X) + \omega(Y)$ and $\omega(\rho X) = \rho\omega(X)$. We will denote the set of all 1-forms by $T'(M)$.

Given $f \in C^\infty(M)$ and a point p in M, for a fixed derivation X_p, we define df_p as the linear functional in $[T_p(M)]^*$ satisfying

$$df_p(X_p) = \langle df_p, X_p\rangle_{[T_p(M)]^* \times [T_p(M)]} = X_p(f).$$

If $X : M \to T(M)$ is a vector field on M, then the formula above allows us to define a function df on $\mathfrak{X}(M)$ such that

$$df(X) = X(f), \tag{2.9}$$

where $X(f)(p) = X_p(f)$.

Definition 2.28 df as defined in (2.9) is called the differential of f.

In the case where $M = \mathbb{R}^n$ or M is an open subset of M and $p \in \mathbb{R}^n$, suppose $X_p = \left.\frac{\partial}{\partial x_k}\right|_p$. Then letting $f : \mathbb{R}^n \to \mathbb{R}$ such that $f(x_1, \dots, x_n) = x_j$,

$$X_p(f) = \left.\frac{\partial}{\partial x_k}\right|_p \left((x_1, \dots, x_n) \mapsto x_j\right) = \begin{cases} 1, & \text{if } k = j \\ 0, & \text{otherwise} \end{cases} = (dx_j)_p.$$

The calculations above show that the collection of linear functionals $\left\{(dx_1)_p, \dots, (dx_n)_p\right\}$ is a basis of the dual vector space $[T_p(M)]^*$. More generally if

$$X_p = \sum_{k=1}^{n} \alpha_k \left.\frac{\partial}{\partial x_k}\right|_p$$

then

$$df(X_p) = \sum_{k=1}^{n} \alpha_k \left.\frac{\partial(f)}{\partial x_k}\right|_p = \sum_{k=1}^{n} \left.\frac{\partial(f)}{\partial x_k}\right|_p (dx_k)_p(X_p).$$

This proves that

$$(df)_p = \sum_{k=1}^{n} \left.\frac{\partial(f)}{\partial x_k}\right|_p (dx_k)_p.$$

Proposition 2.33 The following holds true:

1. Let $f \in C^{\infty}(M)$. Then the differential $df : \mathfrak{X}(M) \to C^{\infty}(M)$ of the function f is a 1-form defined as follows:
$$df(X) = X(f). \tag{2.10}$$
2. If ω is a 1-form and $Y \in \mathfrak{X}(M)$, then the operator $X \mapsto Y(\omega(X)) - \omega([Y,X])$ is a 1-form on M.
3. Let U be an open subset of $\mathbb{R}^n$ and let $x_1, \dots, x_n$ be the coordinate functions on U. Then
$$T'(U) = \left\{ \sum_{k=1}^{n} f_k dx_k : f_k \in C^{\infty}(M) \quad \text{for } k = 1, \dots, n \right\}.$$
4. Let $\omega \in T'(M)$. Then ω is a local operator. In other words, if X and Y agree in some neighborhood of a point p in the manifold M then $\omega(X)(p) = \omega(Y)(p)$.

Proof: For the first part, given $X, Y \in \mathfrak{X}(M), \rho \in C^{\infty}(M)$
$$df(X+Y) = (X+Y)(f) = X(f) + Y(f) = df(X) + df(Y)$$
and $df(\rho X) = \rho \cdot [X(f)] = \rho\, df(X)$.

Secondly, the additive property is easy to verify and we shall omit the detail and $Y(\omega(\rho X)) - \omega([Y, \rho X]) = \rho(Y(\omega(X)) - \omega([Y,X]))$ as shown below
$$\begin{aligned} Y(\omega(\rho X)) - \omega([Y, \rho X]) &= Y(\rho)\omega(X) + \rho Y(\omega(X)) - \rho\omega([Y,X]) - Y(\rho)\omega(X) \\ &= \rho Y(\omega(X)) - \rho\omega([Y,X]) \\ &= \rho(Y(\omega(X)) - \omega([Y,X])). \end{aligned}$$

Thirdly, let $\omega \in T'(U)$. Then for $a_k = \omega\left(\frac{\partial}{\partial x_k}\right)$. We shall prove that $\omega = \sum_{k=1}^{n} a_k\, dx_k$. To this end, let $X \in \mathfrak{X}(M)$ and write $X = \sum_{k=1}^{n} f_k \frac{\partial}{\partial x_k}$. Thus,
$$\omega(X) = \omega\left(\sum_{k=1}^{n} \alpha_k \frac{\partial}{\partial x_k}\right) = \sum_{k=1}^{n} \alpha_k \omega\left(\frac{\partial}{\partial x_k}\right) = \sum_{k=1}^{n} a_k \alpha_k.$$

Appealing to (2.10) we obtain
$$\begin{aligned} \sum_{j=1}^{n} a_j\, dx_j(X) &= \sum_{j=1}^{n} a_j(X)(x_j) \\ &= \sum_{j=1}^{n} a_j \sum_{k=1}^{n} \alpha_k \frac{\partial(x_j)}{\partial x_k} \\ &= \sum_{j=1}^{n} a_j \alpha_k. \end{aligned}$$

Finally, to establish this result, it is enough to show that if X is the zero operator in some open set U around p then $\omega(X)(p)$ must be equal to zero. To this end, let $f \in C^\infty(M)$, where f is constantly equal to 1 on the complement of U such that the support of f is contained in the complement of the singleton containing p. Then $f(p) = 0$ and $fX = X$ since $f \equiv 1$ on $M \backslash U$ and X is the zero operator on $C^\infty(U)$. Therefore, $\omega(X)(p) = \omega(fX)(p) = 0$.

Example 2.20 Let $dy \in T'(\mathbb{R}^3)$ and let $X = \frac{\partial}{\partial x} + 2\frac{\partial}{\partial y} + \frac{\partial}{\partial z}$. Then

$$\begin{aligned} dy(X) &= dy\left(\frac{\partial}{\partial x} + 2\frac{\partial}{\partial y} + \frac{\partial}{\partial z}\right) \\ &= dy\left(\frac{\partial}{\partial x}\right) + 2dy\left(\frac{\partial}{\partial y}\right) + dy\left(\frac{\partial}{\partial z}\right) = 2. \end{aligned}$$

Example 2.21 Let $f : \mathbb{R}^3 \to \mathbb{R}$ such that $f(x, y, z) = z^2 e^{y-x} + z e^{x+y}$. Next, write $df = u(x, y, z)\, dx + v(x, y, z)\, dy + w(x, y, z)\, dz$. Then

$$u(x, y, z) = df\left(\frac{\partial}{\partial x}\right) = \frac{\partial}{\partial x}\left(z^2 e^{y-x} + z e^{x+y}\right) = z e^{x+y} - z^2 e^{y-x},$$

$$v(x, y, z) = df\left(\frac{\partial}{\partial y}\right) = \frac{\partial}{\partial y}\left(z^2 e^{y-x} + z e^{x+y}\right) = e^{y-x} z^2 + e^{x+y} z,$$

$$w(x, y, z) = df\left(\frac{\partial}{\partial z}\right) = \frac{\partial}{\partial z}\left(z^2 e^{y-x} + z e^{x+y}\right) = e^{x+y} + 2z e^{y-x}.$$

Therefore,

$$df = \left(z e^{x+y} - z^2 e^{y-x}\right) dx + \left(e^{y-x} z^2 + e^{x+y} z\right) dy + \left(e^{x+y} + 2z e^{y-x}\right) dz.$$

Proposition 2.34 Let $\omega \in T'(M)$. If $X_p = Y_p$ then $\omega(X)(p) = \omega(Y)(p)$.

Proof: To prove this result, it is enough to show that if X_p is trivial then $\omega(X)(p) = 0$. To this end, let (U, φ) be a smooth chart on M such that p belongs to the set U. Let

$$\varphi_*\left(X|_U\right) = \sum_{k=1}^{n} f_k \frac{\partial}{\partial x_k}.$$

Since φ_* is a linear isomorphism, it must be the case that f_k is equal to zero for all k. Next, let $\rho : M \to [0, 1]$ be a smooth function such that ρ is constantly equal to 1 in some neighborhood and assume additionally that ρ is supported inside U. Let

$$\overline{f_k}(x) = \begin{cases} \rho \cdot f_k \circ \varphi(x), & \text{if } x \text{ belongs to } U, \\ 0, & \text{otherwise,} \end{cases}$$

and set

$$\overline{X_k} = \begin{cases} \rho \cdot \varphi_*^{-1}\left(\dfrac{\partial}{\partial x_k}\right), & \text{if } x \text{ belongs to } U, \\ 0, & \text{otherwise.} \end{cases}$$

By construction, $\overline{f_k}$ is a real-valued smooth function on M and $\overline{X_k}$ is a derivation. Next, $\sum_{k=1}^{n} \overline{f_k}\,\overline{X_k} = \rho^2 X$ is equal to X in some neighborhood of p. Since ω is a local operator, it follows that since all the $\overline{f_k}(p)$ are equal to zero,

$$\omega(X)(p) = \omega\left(\sum_{k=1}^{n} \overline{f_k}\,\overline{X_k}\right)(p) = \sum_{k=1}^{n} \overline{f_k}(p)\,\omega\left(\overline{X_k}\right)(p) = 0.$$

Remark 2.7 For a 1-form $\omega \in T'(M)$, the quantity $\omega(X)(p)$ only depends on the value of X at p. As such, we may define a function $\omega_p : T_pM \to \mathbb{R}$ such that the following holds true. For every vector $v \in T_pM$, and a vector field X, we define $X_p = v$ and $\omega_p(v) = \omega(X)(p)$ is well defined since it is independent of our choice of X. Since ω is an $C^\infty(M)$ module homomorphism, we obtain that ω_p is a linear functional on T_pM.

Given $\omega_p : T_pM \to \mathbb{R}$ a linear functional defined for each point p in M, we define $\overline{\omega} : TM \to \mathbb{R}$ as follows: $\overline{\omega}(v) = \omega_p(v)$ if v is assumed to belong to the tangent space at p.

Proposition 2.35 The map $\overline{\omega}$ is smooth if and only if $\omega(X) = \overline{\omega} \circ X$ is a real valued and smooth function defined on M for every $X \in \mathfrak{X}(M)$.

Proof: Let us assume that $\overline{\omega} : TM \to \mathbb{R}$ is smooth for all $X \in \mathfrak{X}(M)$. Since $X : M \to TM$ is a smooth function, it follows that $\overline{\omega}(X)$ is smooth as well. For the converse of the result, let us assume that $\omega(X) = \overline{\omega}(X)$ is a real-valued and smooth function defined on M for every $X \in \mathfrak{X}(M)$. Next, let (U, φ) be a smooth chart on M such that p belongs to the set U. Next, we select a neighborhood of p such that the closure of V is contained in U and ρ is a real-valued function defined on M which is equal to 1 on the closure of V and is supported in U. Next, we define vector fields $Y_1, \dots, Y_n$ such that

$$Y_k = \begin{cases} \rho\left(\varphi_*^{-1}\frac{\partial}{\partial x_k}\right) & \text{on } U, \\ 0 & \text{on } M \backslash U. \end{cases}$$

In other words, the pushforward of each Y_k via φ_* is equal to $\frac{\partial}{\partial x_k}$ on W. Consequently, $\overline{(\varphi|V)^{-1}} : \varphi(V) \times \mathbb{R} \to \pi^{-1}(V) \subseteq TM$ takes the form

$$\overline{(\varphi|V)^{-1}}(t, a) = \sum_{k-1}^{n} a_k X_k \circ \varphi^{-1}(t)$$

and

$$\overline{\omega}\left(\overline{(\varphi|V)^{-1}}\right)(t,a) = \sum_{k=1}^{n} a_k \overline{\omega}\left(X_k\right) \circ \varphi^{-1}(t).$$

The smoothness of $\overline{\omega} \circ \overline{(\varphi|V)^{-1}}$ in local coordinates follows immediately from the assumption that $\overline{\omega}\left(X_k\right) \circ \varphi^{-1}$ is smooth.

2.2.7 Pull-Backs

Let $f : M \to N$ be a smooth function and let ω be a 1-form defined on N. Then f induces a linear operator $f_* : T_p(M) \to T_{f(p)}(N)$. This in turns defines a dual map

$$\left(f_*\right)^* : \left(T_{f(p)}(N)\right)^* \to \left(T_p(M)\right)^*$$

such that given $X_p \in T_p(M)$,

$$\left(f_*\right)^*\left(\omega_{f(p)}\right) X_p = \left\langle \left(f_*\right)^*\left(\omega_{f(p)}\right), X_p \right\rangle = \left\langle \omega_{f(p)}, f_* X_p \right\rangle = \omega_{f(p)}\left(f_* X_p\right).$$

To avoid cluster of notation, simply write

$$\left(f_*\right)^*\left(\omega_{f(p)}\right) = f^*\left(\omega_{f(p)}\right) = \omega_{f(p)}\left(f_*\right).$$

The observations above lead to the following definition.

Definition 2.29 The pull-back of a 1-form ω is the map $f^* : T'(N) \to T'(M)$ defined as follows: $f^*\omega = \omega \circ f_*$.

2.2.8 Tensor Fields

A covariant tensor field of order r or tensor field of the type $\binom{r}{0}$ is a mapping

$$T : \underbrace{\mathfrak{X}(M) \times \cdots \times \mathfrak{X}(M)}_{r \text{ times}} \to C^\infty(M),$$

which is r-linear over the vector space of real-valued smooth functions defined on M. In other words, for all

$$\left(X_1, \ldots, X_{k-1}, X_k, X_{k+1}, \ldots, X_r\right) \in \mathfrak{X}(M) \times \cdots \times \mathfrak{X}(M)$$

1. For $X \in \mathfrak{X}(M)$,

$$\begin{aligned} &T\left(X_1, \ldots, X_{k-1}, X_k + X, X_{k+1}, \ldots, X_r\right) \\ &\quad = T\left(X_1, \ldots, X_{k-1}, X_k, X_{k+1}, \ldots, X_r\right) + T\left(X_1, \ldots, X_{k-1}, X, X_{k+1}, \ldots, X_r\right). \end{aligned}$$

2. For $\rho \in C^\infty(M)$,

$$T\left(X_1, \ldots, X_{k-1}, \rho X_k, X_{k+1}, \ldots, X_r\right) = \rho T\left(X_1, \ldots, X_{k-1}, X_k, X_{k+1}, \ldots, X_r\right).$$

A tensor field of type $\binom{r}{1}$ is a mapping

$$T : \underbrace{\mathfrak{X}(M) \times \cdots \times \mathfrak{X}(M)}_{r \text{ times}} \to \mathfrak{X}(M),$$

which is r-linear over the ring of smooth functions $C^\infty(M)$.

- The collection of all covariant tensor fields of order r is a $C^\infty(M)$-module denoted by $T^r(M) = T^r_0(M)$. Moreover, the tensor field of type $\binom{r}{1}$ is also a $C^\infty(M)$-module, which we denote by $T^r_1(M)$.
- Since every 1-form is a $C^\infty(M)$-module homomorphism of the type

$$\omega : \mathfrak{X}(M) \to C^\infty(M),$$

  we may denote the set of all 1-forms by $T^1_0(M)$.
- Using a similar analogy, we also adopt the following conventions:

Scalar fields	$C^\infty(M) = T^0_0(M)$
Vector fields	$\mathfrak{X}(M) = T^0_1(M)$
1-forms	$T'(M) = T^1_0(M)$

Next, we define a tensor product $\otimes : T^r_1(M) \times T^s_1(M) \to T^{r+s}_1(M)$ as $\otimes(T, S) = T \otimes S$ such that

$$[T \otimes S]\left(X_1, \dots, X_{r+s}\right) = T\left(X_1, \dots, X_r\right) S\left(X_{r+1}, \dots, X_{r+s}\right).$$

It is easy to verify that the tensor product is noncommutative, associative, and distributive over addition. In other words,

$$(T \otimes S) \otimes R = T \otimes (S \otimes R)$$

and

$$\left(T_1 + T_2\right) \otimes S = T_1 \otimes S + T_1 \otimes S,$$
$$T \otimes \left(S_1 + S_2\right) = T \otimes S_1 + T \otimes S_2.$$

Example 2.22 Let $U \subseteq \mathbb{R}^n$ be an open set. Then

$$T^r(U) = \left\{ \sum_{(k_1,\dots,k_r)} T_{(k_1,\dots,k_r)} dx_{k_1} \otimes \cdots \otimes dx_{k_r} : T_{(k_1,\dots,k_r)} \in C^\infty(M) \right\}.$$

Given $T \in T^r(U)$, then the component functions of T are uniquely determined as follows:

$$T\left(\frac{\partial}{\partial x_1}, \dots, \frac{\partial}{\partial x_n}\right) = T_{(k_1,\dots,k_r)}.$$

Thus,

$$T = \sum_{(k_1,\dots,k_r)} T\left(\frac{\partial}{\partial x_1}, \dots, \frac{\partial}{\partial x_n}\right) dx_{k_1} \otimes \dots \otimes dx_{k_r}.$$

Similarly,

$$T_1^r(U) = \left\{ \sum_{(k_1,\dots,k_r,l)} T_{(k_1,\dots,k_r,l)} \left(dx_{k_1} \otimes \cdots \otimes dx_{k_r}\right) \otimes \frac{\partial}{\partial x_l} : \right.$$

$$\left. T_{(k_1,\dots,k_r,l)} \in C^\infty(M) \right\}$$

and letting $T = \sum_{(k_1,\dots,k_r,l)} T_{(k_1,\dots,k_r,l)} \left(dx_{k_1} \otimes \cdots \otimes dx_{k_r}\right) \otimes \frac{\partial}{\partial x_l}$,

$$T_{(k_1,\dots,k_r,l)} = dx_j\left(T\left(\frac{\partial}{\partial x_1}, \dots, \frac{\partial}{\partial x_n}\right)\right).$$

Using similar arguments used to show that 1-forms are local operators, we can also prove that elements of $T^r(M)$ and $T_1^r(M)$ are also local operators. In other words, the values of $T(X_1, \dots, X_n)$ at p in M are only dependent on the values of the derivations $X_1, \dots, X_n$ at p. This implies that if $T \in T^r(M)$ or $T_1^r(M)$, then there exists a multilinear function

$$T_p : T_pM \times \cdots \times T_pM \to \mathbb{R}$$

or

$$T_p : T_pM \times \cdots \times T_pM \to T_pM$$

such that

$$T(X_1, \dots, X_r)(p) = T_p\left(\left(X_1\right)_p, \dots, \left(X_r\right)_p\right).$$

Moreover, if for each point p in the manifold, there exit a multilinear map T_p such that the assignment that takes p into $T_p\left(\left(X_1\right)_p, \dots, \left(X_r\right)_p\right)$ is smooth for all $X_1, \dots, X_r \in \mathfrak{X}(M)$ then

$$(X_1, \dots, X_r) \mapsto T_p\left(\left(X_1\right)_p, \dots, \left(X_r\right)_p\right)$$

defines a tensor field.

Example 2.23 We will show here that the dot product on $\mathbb{R}^3$ is a covariant tensor field of order 2. Let

$$\omega : \mathfrak{X}(M) \times \mathfrak{X}(M) \to C^\infty\left(\mathbb{R}^3\right)$$

such that

$$\omega\left(\sum_{k=1}^{3} f_k \frac{\partial}{x_k}, \sum_{k=1}^{3} g_k \frac{\partial}{x_k}\right) = \sum_{k=1}^{3} f_k g_k.$$

Letting

$$\begin{aligned}\omega = {} & \omega_{11}\left(dx_1 \otimes dx_1\right) + \omega_{12}\left(dx_1 \otimes dx_2\right) + \omega_{13}\left(dx_1 \otimes dx_3\right) \\ & + \omega_{21}\left(dx_2 \otimes dx_1\right) + \omega_{22}\left(dx_2 \otimes dx_2\right) + \omega_{23}\left(dx_1 \otimes dx_3\right) \\ & + \omega_{31}\left(dx_3 \otimes dx_1\right) + \omega_{32}\left(dx_3 \otimes dx_2\right) + \omega_{33}\left(dx_3 \otimes dx_3\right),\end{aligned}$$

then

$$\omega = dx_1 \otimes dx_1 + dx_2 \otimes dx_2 + dx_3 \otimes dx_3$$

since

$$\omega\left(\frac{\partial}{x_k}, \frac{\partial}{x_j}\right) = \begin{cases} 1, & \text{if } k = j, \\ 0, & \text{if } k \neq j. \end{cases}$$

Definition 2.30 Given a smooth function $f : M \to N$ and $T \in T^r(N)$. The pull-back of T by f is defined as $f^*(T) \in T^r(M)$ as follows:

$$f^*(T)_p\left(v_1, \ldots, v_r\right) = T_{f(p)}\left(f_* v_1, \ldots, f_* v_r\right)$$

for $p \in M$ and $v_1, \ldots, v_r \in T_pM$.

Let V be a vector space over the reals and let T be a multilinear map from $V \times \cdots \times V$ into $\mathbb{R}$. We say that T is alternating if

$$T\left(X_1, \ldots, X_i, \ldots, X_j, \ldots, X_k\right) = -T\left(X_1, \ldots, X_j, \ldots, X_i, \ldots, X_k\right).$$

Furthermore, we define $\Lambda^k(V)$ to be the set of all multilinear maps $T : V \times \cdots \times V \to \mathbb{R}$ which are alternating. Appealing to well-known fact of multilinear algebra, one verifies that the assignment which takes a tensor $T \in T^k(V)$ to $\operatorname{alt}(T)$ such that

$$\operatorname{alt}(T)\left(X_1, \ldots, X_k\right) = \frac{1}{k!} \sum_{\sigma \in S_k} \left(\operatorname{sgn}(\sigma)\right) T\left(X_{\sigma(1)}, \ldots, X_{\sigma(k)}\right)$$

is an alternating map. That is $\operatorname{alt} : T^k(V) \to \Lambda^k(V)$.

Example 2.24 Let $\operatorname{alt} : T^3(V) \to \Lambda^3(V)$ where V is a real vector space with basis $\{e_1, e_2, e_3\}$. Let $S_3 = \{\sigma : \sigma \text{ is a bijection of } \{1, 2, 3\}\}$. Then S_3 contains the following functions:

$$\sigma_1 = \begin{cases} 1 \mapsto 1 \\ 2 \mapsto 2, \\ 3 \mapsto 3 \end{cases} \quad \sigma_2 = \begin{cases} 1 \mapsto 1 \\ 2 \mapsto 3, \\ 3 \mapsto 2 \end{cases} \quad \sigma_3 = \begin{cases} 1 \mapsto 2 \\ 2 \mapsto 1, \\ 3 \mapsto 3 \end{cases}$$

$$\sigma_4 = \begin{cases} 1 \mapsto 2 \\ 2 \mapsto 3\,, \\ 3 \mapsto 1 \end{cases} \quad \sigma_5 = \begin{cases} 1 \mapsto 3 \\ 2 \mapsto 1\,, \\ 3 \mapsto 2 \end{cases} \quad \sigma_6 = \begin{cases} 1 \mapsto 3 \\ 2 \mapsto 2\,. \\ 3 \mapsto 1 \end{cases}$$

In cycle notation, we have

$$\sigma_1 = (1)\,,\ \sigma_2 = (23)\,,\ \sigma_3 = (12)\,,\ \sigma_4 = (123)\,,\ \sigma_5 = (132)\,,\ \sigma_6 = (13)$$

and as a result,

$$\begin{aligned} &\text{alt}\left(e_1 \otimes e_2 \otimes e_3\right) \\ &\quad = \text{sgn}\left(\sigma_1\right) e_1 \otimes e_2 \otimes e_3 + \text{sgn}\left(\sigma_2\right) e_1 \otimes e_3 \otimes e_2 \\ &\qquad + \text{sgn}\left(\sigma_3\right) e_2 \otimes e_1 \otimes e_3 + \text{sgn}\left(\sigma_4\right) e_2 \otimes e_3 \otimes e_1 \\ &\qquad + \text{sgn}\left(\sigma_5\right) e_3 \otimes e_1 \otimes e_2 + \text{sgn}\left(\sigma_6\right) e_3 \otimes e_2 \otimes e_1 \\ &\quad = e_1 \otimes e_2 \otimes e_3 - e_1 \otimes e_3 \otimes e_2 - e_2 \otimes e_1 \otimes e_3 \\ &\qquad + e_2 \otimes e_3 \otimes e_1 + e_3 \otimes e_1 \otimes e_2 - e_3 \otimes e_2 \otimes e_1. \end{aligned}$$

Let M be a smooth manifold. Suppose that for each point p in the manifold, we choose $\omega_p \in \Lambda^k\left(T_pM^*\right)$ such that $p \mapsto \omega_p$ is a smooth assignment. Each tensor ω_p is alternating and we define a differential k-form ω as a smooth global section of $\Lambda^k\left(TM'\right)$.

Given a collection of one-forms $\omega_1, \ldots, \omega_k$ on a smooth manifold M of dimension n, we may form a differential k form by taking the wedge product $\omega_1 \wedge \cdots \wedge \omega_k$ such that

$$\left(\omega_1 \wedge \cdots \wedge \omega_k\right)\left(\left(\frac{\partial}{\partial x_1}, \ldots, \frac{\partial}{\partial x_k}\right)\right) = \det \begin{bmatrix} \omega_1\left(\frac{\partial}{\partial x_1}\right) & \cdots & \omega_1\left(\frac{\partial}{\partial x_k}\right) \\ \vdots & \ddots & \vdots \\ \omega_k\left(\frac{\partial}{\partial x_1}\right) & \cdots & \omega_k\left(\frac{\partial}{\partial x_k}\right) \end{bmatrix}. \tag{2.11}$$

Example 2.25 If $M = \mathbb{R}^2$ then

$$dx \wedge dy = u\,dx \otimes dx + v\,dx \otimes dy + w\,dy \otimes dx + z\,dy \otimes dy$$

for some smooth functions u, v, w, z and

$$u = \det \begin{bmatrix} dx\left(\frac{\partial}{\partial x}\right) & dx\left(\frac{\partial}{\partial y}\right) \\ dx\left(\frac{\partial}{\partial x}\right) & dx\left(\frac{\partial}{\partial y}\right) \end{bmatrix} = \det \begin{bmatrix} 1 & 0 \\ 1 & 0 \end{bmatrix} = 0,$$

$$u = \det \begin{bmatrix} dx\left(\frac{\partial}{\partial x}\right) & dx\left(\frac{\partial}{\partial y}\right) \\ dy\left(\frac{\partial}{\partial x}\right) & dy\left(\frac{\partial}{\partial y}\right) \end{bmatrix} = \det \begin{bmatrix} 1 & 0 \\ 0 & 1 \end{bmatrix} = 1,$$

$$w = \det \begin{bmatrix} dy\left(\frac{\partial}{\partial x}\right) & dy\left(\frac{\partial}{\partial y}\right) \\ dx\left(\frac{\partial}{\partial x}\right) & dx\left(\frac{\partial}{\partial y}\right) \end{bmatrix} = \det \begin{bmatrix} 0 & 1 \\ 1 & 0 \end{bmatrix} = -1,$$

and

$$z = \det \begin{bmatrix} dy\left(\frac{\partial}{\partial x}\right) & dy\left(\frac{\partial}{\partial y}\right) \\ dy\left(\frac{\partial}{\partial x}\right) & dy\left(\frac{\partial}{\partial y}\right) \end{bmatrix} = \det \begin{bmatrix} 0 & 1 \\ 0 & 1 \end{bmatrix} = 0.$$

Therefore, $dx \wedge dy = dx \otimes dy - dy \otimes dx$.

Let (U, φ) be a smooth chart around a fixed point p in M and $\Omega^k(U)$ be the set of all differential k-forms on U. Then $\Omega^k(U)$ is a vector space and additionally, the collection

$$\left\{dx_{i_1} \wedge \cdots \wedge dx_{i_k} : 1 \leq i_1 < i_2 < \cdots < i_k \leq m\right\}$$

is a basis for $\Omega^k(U)$. More generally, in local coordinates, an r-form T is given by the summation

$$\sum_{(k_1,\cdots,k_r)} f_{(k_1,\cdots,k_r)} dx_{k_1} \wedge \cdots \wedge dx_{k_r},$$

where each $f_{(k_1,\ldots,k_r)}$ is a smooth function and

$$dx_{k_1} \wedge \cdots \wedge dx_{k_r}\left(\left(\frac{\partial}{\partial x_{i_1}}, \ldots, \frac{\partial}{\partial x_{i_r}}\right)\right) = \begin{cases} 1, & \text{if } (k_1, \ldots, k_r) = (i_1, \ldots, i_r), \\ 0, & \text{otherwise.} \end{cases}$$

Proposition 2.36 Let U, V be open subsets of $\mathbb{R}^m, \mathbb{R}^n$, respectively. Next, let $f : U \to V$ be a smooth function. Then, the following holds on $U \cap f^{-1}(V)$:

$$f^*\left(h\, dy_1 \wedge \cdots \wedge dy_n\right) = (h \circ f)\left(\det f_*\right)\left(dx_1 \wedge \cdots \wedge dx_m\right),$$

where h is a smooth real-valued function defined on V.

Proof: Let p be a point in the manifold M. Then

$$\left[f^*\left(h\, dy_1 \wedge \cdots \wedge dy_n\right)\right]_p \left(\frac{\partial}{\partial x_1}, \ldots, \frac{\partial}{\partial x_m}\right)$$
$$= \left(h\, dy_1 \wedge \cdots \wedge dy_n\right)_{f(p)} \left(f_*\left(\frac{\partial}{\partial x_1}\right), \ldots, f_*\left(\frac{\partial}{\partial x_m}\right)\right)$$
$$= h(f(p))\left(\det f_*\right).$$

References

1 John M. Lee. *Introduction to smooth manifolds*, volume 218 of *Graduate Texts in Mathematics*. Springer, New York, second edition, 2013.

2 Michael Spivak. *Calculus on manifolds. A modern approach to classical theorems of advanced calculus*. W. A. Benjamin, Inc., New York, Amsterdam, 1965.

3 Frank W. Warner. *Foundations of differentiable manifolds and Lie groups*, volume 94. Springer Science & Business Media, 1983.

3

Lie Theory

Now that we have laid down some foundational work on the theory of manifolds, we are in a position to develop the concepts of Lie theory satisfactorily. To this end, we will introduce the notions of vector fields and explore connections between Lie derivatives and Lie brackets of vector fields. We will next define a Lie group as a set equipped with differentiable and group structures that are compatible with each other in a manner that will be explicitly described in this chapter. Moreover, we will introduce the concept of left and right invariant vector fields by exploiting the fact that left and right translations on a Lie group induce a diffeomorphism between the group and itself. As such, any Lie group is naturally associated with a Lie algebra, which we will first define as the set of all left-invariant vector fields on the underlying manifold structure of the given Lie group. Alternatively, the Lie algebra of the Lie group can also be identified with the tangent space at the group's neutral element (that is, the identity). This chapter also introduces the exponential map as a differentiable function between the Lie algebra and its Lie group obtained by evaluating the maximal integral curve associated with a left-invariant vector field on the Lie group at the unit element in the domain of this curve. The exponential map will prove to be of central import in Lie theory. For instance, the exponential map can be used to prove that every continuous homomorphism from the real line into a Lie group is necessarily smooth. Later in this chapter, we will establish that every Lie group admits up to multiplication by a constant, an invariant measure under the group's left action. Such a measure is often called a left-invariant or Haar measure. This will allow us to prove that any Lie group admits a unique Hilbert function space for which we will aim to construct basis-like objects such as frames in subsequent chapters. Next, given a Lie group G with a closed subgroup H, the quotient space G/H gives rise to a homogeneous space, regarded as a G-space equipped with a natural transitive and smooth action. Finally, we will introduce the concept of a matrix Lie group, formally defined as a closed subgroup of the groups of invertible complex matrices of a fixed order. Such Lie groups contain the classical Lie groups (the orthogonal, symplectic, and special linear

A Bridge Between Lie Theory and Frame Theory: Applications of Lie Theory to Harmonic Analysis, First Edition. Vignon Oussa.

Lie groups) and admit more straightforward definitions for the exponential functions (a power series of matrices in this case), the adjoint map, and its derivative. Finally, we will conclude this chapter by describing a relevant (in the context of this book) class of Lie groups called solvable. The contents of this chapter find inspiration in lecture notes taken as a graduate student at Saint Louis University (a year-long course in Dr. James Hebda's differential geometry course and a reading course in Lie theory taught by Dr. Bradley Currey). Additionally, I have drawn from Lee and Hall's textbooks [6, 10] important ideas for proving various significant results.

3.1 Lie Derivatives

Let $U \subseteq \mathbb{R} \times \mathbb{R}^n$ be an open set, and let $f : U \to \mathbb{R}^n$ be a continuous function. We consider the following ordinary differential equation:

$$\frac{dx}{dt} = f(t, x)$$

with unknown function $t \mapsto x(t)$.

A function $x : (a, b) \to \mathbb{R}^n$ of class C^1 is said to be a solution to the differential equation $\frac{dx}{dt} = f(t, x)$ if the following conditions are satisfied. First, for all $t \in (a, b)$, $(t, x(t)) \in U$ and secondly, $\frac{dx}{dt} = f(t, x(t))$.

Given a point (t_0, x_0) in U, the initial value problem aims to solve the following system of equations

$$\begin{cases} \dfrac{d(x(t))}{dt} = f(t, x(t)) \\ x(t_0) = x_0 \end{cases} \tag{3.1}$$

for unknown function x.

Example 3.1 Let us consider the following ordinary differential equation

$$\begin{cases} \dfrac{d(x_1(t))}{dt} = x_1(t) + x_2(t), \\ \dfrac{d(x_2(t))}{dt} = x_2(t). \end{cases} \tag{3.2}$$

Define $x : \mathbb{R} \to \mathbb{R}^2$ such that $x(t) = \left(e^t\left(c_1 + tc_2\right), c_2e^t\right)$ for some constants c_1, c_2. Then

$$x_1(t) + x_2(t) = e^t\left(c_1 + tc_2\right) + c_2e^t = e^t\left(c_1 + c_2 + tc_2\right) = \frac{d\left(e^t\left(c_1 + tc_2\right)\right)}{dt}$$

and

$$x_2(t) = c_2 e^t = \frac{d\left(c_2 e^t\right)}{dt} = \frac{d\left(x_2(t)\right)}{dt}.$$

Therefore, x is a solution to the differential equation (3.2).

To prove the fundamental theorem of ordinary differential equations (see also [9, Theorem 4.2, 4.3] for a more elementary approach), we will need the following lemma.

Lemma 3.1 ***(Contraction mapping principle)*** If $(\mathcal{M}, d)$ is a nonempty complete metric space and if $T : \mathcal{M} \to \mathcal{M}$ is a contraction (a Lipschitz continuous function with Lipschitz constant less than 1) then T has a unique fixed point in $\mathcal{M}$.

Proof: Suppose that for arbitrary $u, v \in \mathcal{M}$, $d(T(u), T(v)) \leq \lambda d(u, v)$ for some constant $\lambda < 1$. Next, define a sequence $\left(v_k\right)_{k=0}^{\infty}$ in $\mathcal{M}$ such that v_0 is a fixed point in $\mathcal{M}$ and $v_{k+1} = Tv_k$. We aim to prove that $\left(v_k\right)_{k=0}^{\infty}$ is a Cauchy sequence. First, we observe that

$$d\left(T^k v_0, T^{k-1} v_0\right) \leq \lambda d\left(T^{k-1} v_0, T^{k-2} v_0\right) \leq \cdots \leq \lambda^{k-1} d\left(Tv_0, v_0\right).$$

Secondly, letting $k \geq m$, we verify that

$$\begin{aligned} d\left(v_k, v_m\right) &= d\left(T^k v_0, T^m v_0\right) \\ &\leq \lambda d\left(T^{k-1} v_0, T^{m-1} v_0\right) \\ &\leq \lambda^m d\left(T^{k-m} v_0, v_0\right). \end{aligned}$$

Using Triangle inequality, together with the inequality $d\left(T^k v_0, T^{k-1} v_0\right) \leq \lambda^{k-1} d\left(Tv_0, v_0\right)$, it follows that

$$\begin{aligned} d\left(v_k, v_m\right) &\leq \lambda^m \sum_{j=1}^{k-m} d\left(T^{n-m-j+1} v_0, T^{n-m-j} v_0\right) \\ &\leq \lambda^m \sum_{j=1}^{k-m} \lambda^{n-m-j} d\left(Tv_0, v_0\right) \\ &= \lambda^m d\left(Tv_0, v_0\right) \sum_{j=1}^{k-m} \lambda^{k-m-j}. \end{aligned}$$

Since

$$\sum_{j=1}^{k-m} \lambda^{k-m-j} = \frac{\lambda^m - \lambda^k}{\lambda^m (1 - \lambda)} = \frac{\lambda^m \left(1 - \lambda^{k-m}\right)}{\lambda^m (1 - \lambda)} = \frac{1 - \lambda^{k-m}}{1 - \lambda},$$

it follows that

$$d\left(v_k, v_m\right) < \frac{\lambda^m}{1-\lambda} d\left(Tv_0, v_0\right).$$

Given $\epsilon > 0$, let N be large enough so that

$$\frac{\lambda^N}{1-\lambda} d\left(Tv_0, v_0\right) < \frac{\epsilon}{2}.$$

Assuming that $k, m \geq N$

$$\begin{aligned} d\left(v_k, v_m\right) &\leq d\left(v_k, v_N\right) + d\left(v_N, v_m\right) \\ &\leq \frac{\lambda^N}{1-\lambda} d\left(Tv_0, v_0\right) + \frac{\lambda^N}{1-\lambda} d\left(Tv_0, v_0\right) \\ &= \frac{2\lambda^N}{1-\lambda} d\left(Tv_0, v_0\right) < \epsilon. \end{aligned}$$

Therefore $\left(v_k\right)_{k=0}^{\infty}$ is a Cauchy sequence and the completeness of $\mathcal{M}$ implies that $\left(v_k\right)_{k=0}^{\infty}$ converges to some vector v. However, since $\left(Tv_k\right)_{k=0}^{\infty}$ is a subsequence of $\left(v_k\right)_{k=0}^{\infty}$, and since T is a continuous operator, it follows that $Tv = v$. To prove the uniqueness, suppose there exists another fixed point w such that $\left(v_k\right)_{k=0}^{\infty}$ converges to w. Then

$$d(v, w) = d(Tv, Tw) \leq \lambda d(v, w).$$

Assuming that $d(v, w)$ is nonzero, we reach the absurd fact that $1 \leq \lambda < 1$. Therefore, $d(v, w) = 0$ and $v = w$.

Theorem 3.1 ***(Fundamental theory of O.D.E.)*** Let $a, b > 0$ and let

$$C\left(x_0, b\right) = \left\{y \in \mathbb{R}^n : \left\|y - x_0\right\| \leq b\right\}.$$

Suppose that

$$f : \left[t_0 - a, t_0 + a\right] \times C\left(x_0, b\right) \to \mathbb{R}^n$$

is continuous and bounded by some positive constant M. If additionally, f is Lipschitz continuous in its second argument with Lipschitz constant L, then the initial value problem

$$\frac{d}{dt}(x(t)) = f(t, x) \quad \text{and} \quad x\left(t_0\right) = x_0$$

has a unique solution defined on the interval $I = [t_0 - \min\{a, bM^{-1}\}, t_0 + \min\{a, bM^{-1}\}]$.

Proof: Let $\mathcal{M}$ be the set of functions from I into $C\left(x_0, b\right)$, which are continuous. We equip $\mathcal{M}$ with the norm

$$\|f\|_e = \sup\left\{e^{-2L\left|t-t_0\right|}\|f(t)\| : t \in I\right\}.$$

It can be shown that the structure $(\mathcal{M}, \|\|_e)$ obtained by endowing $\mathcal{M}$ with the norm $\|\|_e$ is a complete normed space. Moreover, given $c \in \mathcal{M}$, let Tc be the function defined on I such that

$$Tc(t) = x_0 + \int_{t_0}^{t} f(\tau, c(\tau))\, d\tau.$$

We claim that $\mathcal{M}$ is invariant under the action of T. Indeed, given $c \in \mathcal{M}$, the image of c under T is differentiable and since $\|f(t,x)\| \leq M$ for all $(t,x) \in [t_0 - a, t_0 + a] \times C(x_0, b)$, it follows that

$$\|Tc(t) - x_0\| = \left\| \int_{t_0}^{t} f(\tau, c(\tau))\, d\tau \right\| \leq \int_{t_0}^{t} \|f(\tau, c(\tau))\|\, d\tau \leq M\left|t - t_0\right|.$$

Since $t \in I$, it follows that

$$M\left|t - t_0\right| \leq M\left|t - t_0\right| \leq M \min\left\{a, bM^{-1}\right\} \leq MbM^{-1} = b.$$

Therefore,

$$\|Tc(t) - x_0\| \leq M \min\left\{a, bM^{-1}\right\} \leq b.$$

This shows that $Tc \in C(x_0, b)$ and as such $Tc \in \mathcal{M}$. Next, we will prove that T is a contraction map. To this end, let $c, s \in \mathcal{M}$. By definition

$$\begin{aligned}
\|Tc - Ts\|_e &= \sup\left\{e^{-2L|t-t_0|}\, \|Tc(t) - Ts(t)\| \;:\; t \in I\right\} \\
&= \sup\left\{ e^{-2L|t-t_0|} \left\| \int_{t_0}^{t} (f(\tau, c(\tau)) - g(\tau, s(\tau)))\, d\tau \right\| \;:\; t \in I \right\}.
\end{aligned}$$

Next, for arbitrary $t \in I$,

$$\begin{aligned}
&e^{-2L|t-t_0|} \left\| \int_{t_0}^{t} (f(\tau, c(\tau)) - g(\tau, s(\tau)))\, d\tau \right\| \\
&\quad \leq e^{-2L|t-t_0|} \int_{t_0}^{t} \|f(\tau, c(\tau)) - g(\tau, s(\tau))\|\, d\tau.
\end{aligned}$$

Since L is a Lipschitz constant, it follows that

$$\begin{aligned}
&e^{-2L|t-t_0|} \left\| \int_{t_0}^{t} (f(\tau, c(\tau)) - g(\tau, s(\tau)))\, d\tau \right\| \\
&\quad \leq e^{-2L|t-t_0|} L \int_{t_0}^{t} \|c(\tau) - s(\tau)\|\, d\tau \\
&\quad \leq e^{-2L|t-t_0|} L \|c - s\|_e \int_{t_0}^{t} e^{2L|\tau - t_0|}\, d\tau \\
&\quad \leq e^{-2L|t-t_0|} L \|c - s\|_e \left(\frac{e^{2L|t-t_0|} - 1}{2L} \right) \\
&\quad = \left(\frac{1 - e^{-2L|t-t_0|}}{2} \right) \|c - s\|_e \leq \frac{\|c - s\|_e}{2}.
\end{aligned}$$

Consequently, $\|Tc - Ts\|_e \leq \frac{\|c-s\|_e}{2}$. Appealing to the contraction mapping principle (Lemma 3.1), it follows that the initial value problem has a unique solution in $\mathcal{M}$.

Definition 3.1 Let $X \in \mathfrak{X}(M)$. An integral curve of X is a smooth curve $c : (a, b) \to M$ satisfying

$$\frac{d\left(c\left(t\right)\right)}{dt} = c_*\left(\left.\frac{d}{dt}\right|_t\right) = X_{c(t)}$$

for all $t \in (a, b)$. In other words, an integral curve of a vector field is a smooth curve whose tangent vector at any point is equal to the vector obtained by evaluating the vector field at the given point.

Example 3.2 Let $X = y\frac{\partial}{\partial x} - x\frac{\partial}{\partial y}$ and let $c : \mathbb{R} \to \mathbb{R}^2$ be a curve described as $c(t) = (x(t), y(t))$. Then the derivative of c is given by $c'(t) = \left(x'(t), y'(t)\right)$. Moreover, c is an integral curve for the vector field X if and only if

$$\begin{aligned} & y(t) \left.\frac{\partial}{\partial x}\right|_{(x(t),y(t))} - x(t) \left.\frac{\partial}{\partial y}\right|_{(x(t),y(t))} \\ & \quad = x'(t) \left.\frac{\partial}{\partial x}\right|_{(x(t),y(t))} + y'(t) \left.\frac{\partial}{\partial y}\right|_{(x(t),y(t))}. \end{aligned}$$

In other words, c is an integral curve for the given vector field if and only if $c(t) = (x(t), y(t))$ is a solution to the following system of differential equations

$$\begin{cases} \dfrac{d\left(x\left(t\right)\right)}{dt} = y(t), \\ \dfrac{d\left(y\left(t\right)\right)}{dt} = -x(t). \end{cases}$$

For example, $c(t) = (\cos(t), \sin(t))$ is an integral curve for $X = y\frac{\partial}{\partial x} - x\frac{\partial}{\partial y}$.

Example 3.3 Let $y\frac{\partial}{\partial x} + x\frac{\partial}{\partial y}$ be a vector field defined on a two-dimensional plane. Solving the system of equations

$$\begin{cases} x'(t) = y(t) \\ y'(t) = x(t) \end{cases} \Leftrightarrow \begin{bmatrix} 0 & 1 \\ 1 & 0 \end{bmatrix} \begin{bmatrix} x(t) \\ y(t) \end{bmatrix} = \begin{bmatrix} x'(t) \\ y'(t) \end{bmatrix},$$

we obtain

$$\begin{bmatrix} x(t) \\ y(t) \end{bmatrix} = \begin{bmatrix} x_0 \cosh(t) + y_0 \sinh(t) \\ y_0 \cosh(t) + x_0 \sinh(t) \end{bmatrix}.$$

This shows that

$$c(t) = \begin{bmatrix} x_0 \cosh(t) + y_0 \sinh(t) \\ y_0 \cosh(t) + x_0 \sinh(t) \end{bmatrix}$$

is an integral curve at $p = \begin{bmatrix} x_0 \\ y_0 \end{bmatrix}$.

The following result provides a connection between integral curves and solutions of ordinary differential equations.

Lemma 3.2 Let $X \in \mathfrak{X}(M)$ and let $c : (a, b) \to M$ be an integral curve of X. Let $c_* : T_t((a, b)) \to T_{\alpha(t)}(M)$ and let (U, φ) be a chart on the manifold M around a point p such that $\varphi(p) = (x_1, \dots, x_n)$ and

$$\varphi_*(X|U) = \sum_{k=1}^{n} f_k(x) \frac{\partial}{\partial x_k}.$$

1) Let $c(a', b')$ be a subset of U and define $f : \varphi(U) \to \mathbb{R}^n$, such that $f = (f_1, \dots, f_n)$. Then $\varphi \circ c : (a', b') \to \mathbb{R}^n$ is a solution of the differential equation $\frac{d(x(t))}{dt} = f(x(t))$.
2) If $\omega : (a, b) \to \varphi(U)$ is a solution of $\frac{d(x(t))}{dt} = f(x(t))$ then $\varphi^{-1} \circ \omega : (a, b) \to M$ is an integral curve.

Proof: For the first part, let us suppose that t is a real number in (a', b'). Then

$$\frac{d([\varphi \circ c])}{dt}(t) = [\varphi \circ c]_* \left(\left. \frac{d}{dt} \right|_t \right) = \varphi_* \left(c_* \left(\left. \frac{d}{dt} \right|_t \right) \right).$$

Next, since $c : (a, b) \to M$ is an integral curve of X,

$$\varphi_* \left(c_* \left(\left. \frac{d}{dt} \right|_t \right) \right) = \varphi_* \left(X_{c(t)} \right) = \sum_{k=1}^{n} f_k(\varphi(c(t))) \frac{\partial}{\partial x_k}.$$

Finally, the latter expression is naturally identified with the vector

$$\left(f_1(\varphi(c(t))), \dots, f_n(\varphi(c(t))) \right) = f(\varphi(c(t))).$$

For the second part of the result, let t be a point in (a, b). Then

$$\begin{aligned} \frac{d\left(\left[\varphi^{-1} \circ \omega\right]\right)}{dt}(t) &= \left[\varphi^{-1} \circ \omega\right]_* \left(\left. \frac{d}{dt} \right|_t \right) \\ &= \varphi_*^{-1} \left(\omega_* \left(\left. \frac{d}{dt} \right|_t \right) \right) \\ &= \varphi_*^{-1} \left(\omega'(t) \right). \end{aligned}$$

Since $d : (a, b) \to \varphi(U)$ is a solution of $\frac{d(x(t))}{dt} = f(x(t))$, $\omega'(t) = f(\omega(t))$ and as a result,

$$\frac{d\left(\left[\varphi^{-1} \circ \omega\right]\right)}{dt}(t) = \varphi_*^{-1}(f(\omega(t))) = X_{\varphi^{-1}(\omega(t))}.$$

Proposition 3.1 Let $X \in \mathfrak{X}(M)$. Given any point p in some manifold M, there exists a unique maximal integral curve through p. In other words, there exist real numbers a_p, b_p satisfying $-\infty \leq a_p < 0 < b_p \leq \infty$ and an integral curve $c_p : (a_p, b_p) \to M$ of X such that

1) $c_p(0) = p$.
2) If $\omega : (a, b) \to M$ is any integral curve of X which satisfies the condition $\omega(0) = p$ then (a, b) is a subset of (a_p, b_p) and additionally, ω and c must agree on (a, b). In other words, ω is a restriction of c_p onto (a, b).

Proof: Let

$$L = \{(a, b) : \text{There exists an integral curve } \beta : (a, b) \to M \text{ of } X \text{ and } \beta(0) = p\}.$$

First, the set L is nonempty since by due to the local existence theorem for ordinary differential equations, which guarantees the existence of a solution in a neighborhood of the initial condition. That is, there exists a positive real number ϵ and an integral curve $\beta : (-\epsilon, \epsilon) \to M$ such that $\beta(0) = p$. Suppose next that $\beta : (a, b) \to M$ and $\gamma : (c, d) \to M$ are two integral curves of a vector field X such that $\beta(0) = \gamma(0) = p$. Let $(a', b') = (a, b) \cap (c, d)$. We shall prove that the functions β and γ must agree on (a', b'). To this end, let

$$I = \{t \in (a', b') : \beta(t) = \gamma(t)\}.$$

To show that $I = (a', b')$, it suffices to prove that I is a closed, open and connected subset of (a', b'). We observe that I is a nonempty set since it contains at least zero. Moreover, appealing to the fact that β and γ are continuous, I must be a closed subset of (a', b'). Let t_0 be a point in I. By assumption, there exists a point q in M such that $\beta(t_0) = \gamma(t_0) = q$. Next, let (U, φ) be a chart around q such that U contains q as a point. Let ϵ be positive real number, small enough to ensure that $\beta(t_0 - \epsilon, t_0 + \epsilon)$ is contained in U. Then $\varphi \circ \beta$ and $\varphi \circ \gamma$ are solutions of the same differential equation with the same initial condition in $\varphi(U)$. By the uniqueness of solutions of differential equation, it follows that the functions $\varphi \circ \beta$ and $\varphi \circ \gamma$ agree on $(t_0 - \epsilon, t_0 + \epsilon)$. Since φ is a local diffeomorphism, it follows that $\beta = \gamma$ on the open interval $(t_0 - \epsilon, t_0 + \epsilon)$. Thus, I is an open and connected subset of (a', b'). This shows that $\beta(t) = \gamma(t)$ for all points t in the set $(a', b') = (a, b) \cap (c, d)$. Finally, let

$$(a_p, b_p) = \bigcup_{(a,b) \in I} (a, b).$$

Next, we define $c_p(t) = \beta(t)$ if t belongs to (a, b) and $\beta : (a, b) \to M$ is an integral curve with $\beta(0) = p$. Then c_p is well defined, $c_p(0) = p$ and by construction c_p is an integral curve of the vector field X which is maximal.

Definition 3.2 Let $X \in \mathfrak{X}(M)$. We say that X is a complete vector field if for every point $p \in M$, every maximal integral curve through p is defined on the set of all real numbers.

Let $X \in \mathfrak{X}(M)$ where M is a smooth manifold. For each point $p \in M$, we define $c_p : (a_p, b_p) \to M$ to be the maximal integral curve of X through p (that is, $c_p(0) = p$).

Theorem 3.2 Let $\Omega = \{(t,p) \in \mathbb{R} \times M : a_p < t < b_p\}$ be an open set in $\mathbb{R} \times M$ and let $\varphi : \Omega \to M$ defined by $\varphi(t,p) = c_p(t)$. Then φ is smooth.

Proof: Let

$$S_p = \{s \geq 0 : \text{there exists a neighborhood } V \text{ of } p \text{ such that } \varphi \text{ is smooth on } [0,s] \times V \subseteq \Omega\}.$$

We will first prove that $\sup(S_p) = b_p$. By definition, since b_p is an upperbound of the domain of the maximal integral curve of X through p, b_p must be an upper bound for S_p. To prove that $\sup(S_p) = b_p$, it remains to show that $\sup(S_p) \geq b_p$. Suppose by contradiction that $\sup(S_p) < b_p$. Then applying the fundamental theorem of solutions of ordinary differential equations, there exists a neighborhood U of the point $c_p(\sup(S_p))$ and a positive real number $\epsilon > 0$ such that for all $z \in U$, and for all $t \in [-\epsilon, \epsilon]$, $c_q(t)$ is defined and depends smoothly on t and q. Next, since c_p is a continuous function, there exists $r > 0$ such that $\sup(S_p) - \epsilon < r < \sup(S_p)$ and $c_q(r)$ belong to U. Since $r < \sup(S_p)$, it follows that r belongs to S_p. As such, there exists a neighborhood V' of p such that $[0,r] \times V'$ is a subset of Ω and φ is smooth on $[0,r] \times V'$. Next, since $\varphi(r,p) = c_p(r) \in U$, continuity of φ implies that there exists a neighborhood V of p such that V is a subset of V', and $\varphi(r,q)$ belongs to U for all q in V. Next, the uniqueness of the solutions of ordinary differential equations gives that if q belongs to V then

$$c_q(t) = \begin{cases} \varphi(q,t) \text{ for } t \in (0,r), \\ c_{\varphi(q,r)}(t-r) \text{ if } t \in (r, r+\epsilon). \end{cases}$$

Therefore, $[0, r+\epsilon] \times V$ is a subset of Ω and the restriction of φ to $[0, r+\epsilon] \times V$ is smooth. As a result, $\sup(S_p) < r + \epsilon \in S_p$ and this is absurd. Next, let (t,p) be a point in Ω. Suppose first that $t = 0$. By the fundamental theorem of solutions of ordinary differential equations, there exists a neighborhood U of p and $\epsilon > 0$ such that $c_p(s)$ depends smoothly on all point (s,q) in $(-\epsilon, \epsilon) \times U$. Thus, $(-\epsilon, \epsilon) \times U$ is a subset of Ω and φ is smooth on $(-\epsilon, \epsilon) \times U$. Suppose secondly, that t is positive. Then $t \in (0, a_p)$. If (t, a_p) then there exists a neighborhood V of p such that $[0,s] \times V$ is a subset of Ω and φ is smooth on $[0,s] \times V$. Therefore, there exists an open subset of (t,p) which is contained in Ω on which φ is smooth. Finally, if t is assumed to be negative, we arrive to the same conclusion by applying the preceding arguments to the vector field $-X$. This completes the proof.

Let X be a complete vector field. In other words, $\Omega = \{(t,p) \in \mathbb{R} \times M : a_p < t < b_p\} = \mathbb{R} \times M$. Next, for any fixed real parameter t, we define the smooth map $\rho_t : M \to M$ such that $\rho_t(p) = c_p(t)$.

Proposition 3.2 Let X be a complete vector field on M.

1) Then for all $t, s \in \mathbb{R}$, $\rho_{t+s} = \rho_t \circ \rho_s$.
2) For every $t \in \mathbb{R}$, $\rho_t : M \to M$ is a diffeomorphism with inverse ρ_{-t}.

Proof: First, fix a real parameter $s \in \mathbb{R}$ and $p \in M$. Next, define $\omega : \mathbb{R} \to M$ where $\omega(t) = \rho_{t+s}(p) = c_p(t+s)$. Then $\omega(0) = \rho_s(p) = c_p(s)$ and ω is an integral curve of X. Consequently,

$$\omega(t) = c_{\omega(0)}(t) = c_{c_p(s)}(t) = \rho_t\left(c_p(s)\right) = \rho_t\left(\rho_s(p)\right).$$

Secondly, to prove the second statement, note that

$$\rho_{-t} \circ \rho_t = \rho_t \circ \rho_{-t} = \rho_{t-t} = \rho_0 = c_p(0).$$

Definition 3.3 Note that the collection of maps $\{\rho_t : t \in \mathbb{R}\}$ forms a group when equipped with the composition operation. This group is called the one-parameter group of diffeomorphisms or the flow generated by the vector field X.

Example 3.4 Let $X = (x-y)\frac{\partial}{\partial x} + (x+y)\frac{\partial}{\partial y} \in \mathfrak{X}\left(\mathbb{R}^2\right)$. Next, for any point $(x,y) \in \mathbb{R}^2$, we define the curve

$$c_{(x,y)}(t) = \left(e^t(x\cos t - y\sin t), e^t(y\cos t + x\sin t)\right).$$

Then, $c_{(x,y)}(0) = (x,y)$ and additionally,

$$\begin{aligned} c'_{(x,y)}(t) &= \left(\frac{d}{dt}\left(e^t(x\cos t - y\sin t)\right), \frac{d}{dt}\left(e^t(y\cos t + x\sin t)\right)\right) \\ &= \left(-e^t(y\cos t - x\cos t + x\sin t + y\sin t), x(\cos t)e^t + y(\cos t)e^t + xe^t\sin t - ye^t\sin t\right) \end{aligned}$$

and $c'_{(x,y)}(0) = (x-y, x+y)$ can be naturally identified with the vector $X_{(x,y)}$. This shows that

$$t \mapsto c_{(x,y)}(t) = \left(e^t(x\cos t - y\sin t), e^t(y\cos t + x\sin t)\right)$$

is a maximal integral curve for X.

Given $X, Y \in \mathfrak{X}(M)$, we define the Lie bracket of X and Y as the element $[X,Y] \in \mathfrak{X}(M)$ acting on $C^\infty(M)$ as follows:

$$[X,Y]f = XY(f) - YX(f).$$

Next, let p be a point in M and let ρ_t be a local one-parameter group of diffeomorphisms generated by X in a neighborhood U of p.

Definition 3.4 Let $Y \in \mathfrak{X}(M)$ and define formally the Lie derivative of Y with respect to X at a point p as

$$\left(L_X Y\right)_p = \lim_{h\to 0}\frac{1}{h}\left(\left[\rho_{-h}\right]_* Y_{\rho_h(p)} - Y_p\right).$$

Theorem 3.3 Given $X, Y \in \mathfrak{X}(M)$, $\left(L_X Y\right)_p = [X, Y]_p$.

Proof: To prove this result, it suffices to show that for arbitrary $f \in C^\infty(M)$,

$$\lim_{h\to 0}\frac{1}{h}\left(\left[\rho_{-h}\right]_* Y_{\rho_h(p)} - Y_p\right) f = [X, Y]_p f.$$

In other words, we aim to prove that

$$\lim_{h\to 0}\frac{\left[\rho_{-h}\right]_* Y_{\rho_h(p)} f - Y_p f}{h} = \lim_{h\to 0}\frac{Y_{\rho_h(p)}\left(f\circ\rho_{-h}\right) - Y_p f}{h} = X_p(Y(f)) - Y_p(X(f)).$$

To this end, let $\theta(s,h) = Y_{\rho_h(p)}\left(f\circ\rho_s\right)$ and $\phi(h) = \theta(-h,h)$. Next, since $\phi(0) = \theta(0,0) = Y_p f$, using the chain rule, since $\phi(h) = \theta(-h,h)$, we obtain that

$$\phi'(0) = -\frac{\partial\theta}{\partial s}(0,0) + \frac{\partial\theta}{\partial h}(0,0).$$

On the one hand,

$$\frac{\partial\theta}{\partial h}(0,0) = \lim_{t\to 0}\frac{\theta(0,t) - \theta(0,0)}{t} = \lim_{t\to 0}\frac{Y_{\rho_t(p)}(f) - Y_p(f)}{t} = \frac{d}{dt}\left(Yf\left(\rho_t(p)\right)\right)\Big|_{t=0}.$$

On the other, using the fact that $h \mapsto \rho_h$ is an integral curve of X through p at zero,

$$X_p(Y(f)) = \frac{d}{dt}\left(Yf\left(\rho_t(p)\right)\right)\Big|_{t=0} = \frac{\partial\theta}{\partial h}(0,0)$$

together with chain rule, we obtain $\frac{\partial\theta}{\partial h}(0,0) = X_p(Y(f))$. Next, we will prove that $\frac{\partial\phi}{\partial s}(0,0) = Y_p(X(f))$. To this end, observe that

$$\phi(s,0) - \phi(0,0) = Y_p\left(f\circ\rho_s\right) - Y_p f.$$

Appealing next to Taylor's theorem, we may write $f\circ\rho_s = f(q) + g(s,q)s$ where g is smooth on $(-\epsilon,\epsilon)\times U$ and $g(0,(q)) = Xf(q)$. Consequently,

$$\begin{aligned}\theta(s,0) - \theta(0,0) &= Y_p(f + g(s,\cdot)s) - Y_p f\\ &= Y_p f - Y_p f + Y_p g((s,\cdot)s)\\ &= Y_p g((s,\cdot)s).\end{aligned}$$

Therefore,

$$\lim_{s\to 0}\frac{\theta(s,0)-\theta(0,0)}{s}=\lim_{s\to 0}\frac{Y_p g((s,\cdot)s)}{s}=Y_p(g(0,\cdot))=Y_p(Xf).$$

In summary, we have shown that

$$\begin{aligned}\lim_{h\to 0}\frac{1}{h}\left(\left[\rho_{-h}\right]_* Y_{\rho_h(p)}-Y_p\right)f &= \lim_{h\to 0}\frac{\phi(h)-\phi(0)}{h}\\ &= -\frac{\partial\theta}{\partial s}(0,0)+\frac{\partial\theta}{\partial h}(0,0)\\ &= X_p(Y(f))-Y_p(Xf).\end{aligned}$$

3.2 Lie Groups and Lie Algebras

3.2.1 Lie Groups and Examples

A Lie group G is a set equipped with a differentiable and a group structure which are compatible with each other in the sense that multiplication and inverse maps on G are smooth. In other words, $\mu : G\times G\to G$, defined by $\mu(g,h)=gh$ and $\iota : G\to G$, defined by $\iota(g)=g^{-1}$ are smooth maps.

Proposition 3.3 Let $\alpha : G\times G\to G$ such that $\alpha(g,h)=gh^{-1}$. Then α is smooth if and only if μ and ι are smooth as well.

Proof: Assume that the multiplication and inverse maps are smooth. Define $\psi : G\times G\to G$ such that $\psi(g,h)=\left(g,h^{-1}\right)$. Then $\alpha=\mu\circ\psi$ as shown below

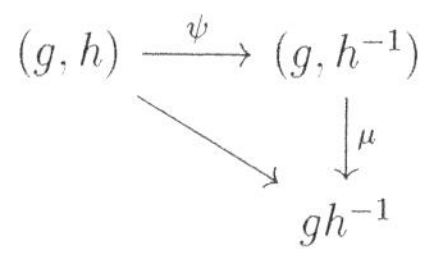

Therefore, as a composition of smooth maps, α is smooth as well.

Conversely, let us assume that α is smooth and let e be the identity element in the group G. Next, let $\eta : G\to G\times G$ be a closed embedding map such that $\eta(x)=(e,x)$. Clearly η is smooth. Next observe that

$$[\alpha\circ\eta](x)=\alpha(e,x)=x^{-1}=\iota(x).$$

Since $\iota=\alpha\circ\eta$

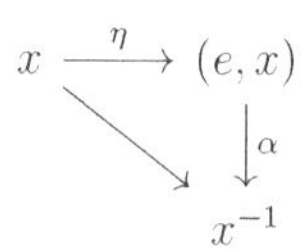

it follows that the inverse map is smooth. Moreover, given $x, y \in G$, we verify that

$$[\alpha \circ \psi](x,y) = \alpha(\psi(x,y)) = \alpha\left(x, y^{-1}\right) = x\left(y^{-1}\right)^{-1} = xy = \mu(x,y).$$

Thus $\mu = \alpha \circ \psi$ is smooth, since it is a composition of smooth maps.

Example 3.5 Let $G = \mathbb{R}$ be the smooth manifold of real numbers equipped with the usual addition. Since the maps $\mu : \mathbb{R} \times \mathbb{R} \to \mathbb{R}$ and $\iota : \mathbb{R} \to \mathbb{R}$ defined as $\mu(x,y) = x + y$ and $\iota(x) = -x$, respectively are smooth, it follows that $(\mathbb{R}, +)$ is a Lie group.

Example 3.6 Let $\mathbb{S} = \{z \in \mathbb{C} : |z| = 1\}$ be the set of all complex numbers of unit modulus together with multiplication of complex numbers. Since the maps $\mu : \mathbb{S} \times \mathbb{S} \to \mathbb{S}$ and $\iota : \mathbb{S} \to \mathbb{S}$ defined as $\mu(z,w) = z \cdot w$ and $\iota(z) = z^{-1}$, respectively are smooth, it follows that $(\mathbb{S}, \cdot)$ is a Lie group.

Example 3.7 Let $G = \mathbb{R} \times \mathbb{R}$ be a group equipped with multiplication $\mu : (\mathbb{R} \times \mathbb{R}) \times (\mathbb{R} \times \mathbb{R}) \to \mathbb{R} \times \mathbb{R}$ where

$$\mu((x,t),(y,s)) = \left(x + e^t y, t + s\right).$$

Then the inverse map $\iota : \mathbb{R} \times \mathbb{R} \to \mathbb{R} \times \mathbb{R}$ is given by $\iota(x,t) = \left(-xe^{-t}, e^{-t}\right)$. Since the map

$$\alpha : ((x,t),(y,s)) \mapsto \left(x - ye^{t-s}, t - s\right)$$

as described in Proposition 3.3 is smooth, it follows that G is a Lie group.

Example 3.8 We equip the set $\mathbb{R}^3 \times \mathbb{R}^3$ with the noncommutative operation $(v,t) \bullet (w,s) = (v + \beta(t)w, t + s)$ where

$$\beta(t)w = \begin{bmatrix} e^{t_3} & t_2 e^{t_3} & \frac{1}{2}e^{t_3}\left(t_2^2 + 2t_1\right) \\ 0 & e^{t_3} & t_2 e^{t_3} \\ 0 & 0 & e^{t_3} \end{bmatrix} \begin{bmatrix} w_1 \\ w_2 \\ w_3 \end{bmatrix} = \begin{bmatrix} \frac{1}{2}w_3 e^{t_3} t_2^2 + w_2 e^{t_3} t_2 + w_1 e^{t_3} + t_1 w_3 e^{t_3} \\ e^{t_3}\left(w_2 + t_2 w_3\right) \\ w_3 e^{t_3} \end{bmatrix}.$$

Then the inverse of (v,t) can be computed as follows:

$$\begin{aligned} (v,t)^{-1} &= ((v,0) \bullet (0,t))^{-1} = (0,-t) \bullet (-v,0) = (-\beta(-t)v, -t) \\ &= \left(\begin{bmatrix} -\frac{1}{2}v_3 e^{-t_3} t_2^2 + v_2 e^{-t_3} t_2 - v_1 e^{-t_3} + t_1 v_3 e^{-t_3} \\ -e^{-t_3}\left(v_2 - t_2 v_3\right) \\ -v_3 e^{-t_3} \end{bmatrix}, \begin{bmatrix} t_1 \\ t_2 \\ t_3 \end{bmatrix}\right). \end{aligned}$$

Since the function

$$\begin{aligned} &\left(\left(v_1, v_2, v_3, t_1, t_2, t_3\right), \left(w_1, w_2, w_3, s_1, s_2, s_3\right)\right) \\ &\mapsto \left(\frac{1}{2}w_3 e^{t_3} t_2^2 + w_2 e^{t_3} t_2 + w_1 e^{t_3} + t_1 w_3 e^{t_3}, e^{t_3}\left(w_2 + t_2 w_3\right), w_3 e^{t_3}, t_1 + s_1, t_2 + s_2, t_3 + s_3\right) \end{aligned}$$

and

$$(v_1, v_2, v_3, t_1, t_2, t_3) \mapsto \left(-\frac{1}{2}v_3e^{-t_3}t_2^2 + v_2e^{-t_3}t_2 - v_1e^{-t_3} + t_1v_3e^{-t_3},\right.$$
$$\left. -e^{-t_3}\left(v_2 - t_2v_3\right), -v_3e^{-t_3}, t_1, t_2, t_3\right)$$

is a differentiable map from $\mathbb{R}^6 \times \mathbb{R}^6$ into $\mathbb{R}^6$, and $\mathbb{R}^6$ into $\mathbb{R}^6$ respectively, it follows that $\mathbb{R}^3 \times \mathbb{R}^3$ together with the noncommutative operation $\bullet$ is a Lie group.

Remark 3.1 Referring to Remark 2.3, it is not hard to verify that if G_1, G_2, then the group $G_1 \times G_1$ equipped with the Cartesian product of manifold is also a Lie group.

Proposition 3.4 Let G be a Lie group and let H be a subgroup of G, which is a closed embedded submanifold of G. When equipped with the smooth structure of G, H is a Lie group as well.

Proof: Let H be a closed embedded submanifold of G, which is also a subgroup of G. Then H inherits the smooth structure of G (see Proposition 2.11). Being a subgroup of G, H is a group in its own right and by definition. Moreover, it is clear that $H \times H$ is a closed embedded submanifold of $G \times G$ and both multiplication and inversion operations are smooth. As a result, H is a Lie group.

Example 3.9 We endow the collection

$$G = \left\{ \begin{bmatrix} x & y \\ z & w \end{bmatrix} : x, y, z, w \in \mathbb{R} \text{ and } wx - yz \neq 0 \right\}$$

of invertible matrices of order 2 with matrix multiplication:

$$\begin{bmatrix} x_1 & y_1 \\ z_1 & w_1 \end{bmatrix} \begin{bmatrix} x_2 & y_2 \\ z_2 & w_2 \end{bmatrix} = \begin{bmatrix} x_1x_2 + y_1z_2 & w_2y_1 + x_1y_2 \\ w_1z_2 + x_2z_1 & w_1w_2 + y_2z_1 \end{bmatrix}.$$

The inverse of an arbitrary element of G is computed as follows:

$$\begin{bmatrix} x & y \\ z & w \end{bmatrix}^{-1} = \begin{bmatrix} \frac{w}{wx - yz} & -\frac{y}{wx - yz} \\ -\frac{z}{wx - yz} & \frac{x}{wx - yz} \end{bmatrix}.$$

G is given the standard smooth structure of an open subset of $\mathbb{R}^4$. Since the function α as described in Proposition 3.3

$$\alpha\left(\begin{bmatrix} x_1 & y_1 \\ z_1 & w_1 \end{bmatrix}, \begin{bmatrix} x_2 & y_2 \\ z_2 & w_2 \end{bmatrix}\right) = \frac{1}{w_2x_2 - y_2z_2}\begin{bmatrix} w_2x_1 - y_1z_2 & x_2y_1 - x_1y_2 \\ w_2z_1 - w_1z_2 & w_1x_2 - y_2z_1 \end{bmatrix}$$

is smooth, it follows that G is a Lie group. Next, let

$$H = \left\{ p(x, y, z, w) := \begin{bmatrix} x & y \\ z & w \end{bmatrix} : x, y, z, w \in \mathbb{R} \text{ and } wx - yz = 1 \right\}.$$

We will show that H is a closed embedded submanifold of G. To this end, we consider the map $\phi : G \to \mathbb{R}\setminus\{0\}$ such that

$$\phi(p(x,y,z,w)) = \det(p(x,y,z,w)) = wx - yz.$$

A matrix representation of the derivative of ϕ at $p_{(x,y,z,w)}$ takes the form

$$\begin{aligned} J\phi(p(x,y,z,w)) &= \begin{bmatrix} \dfrac{\partial(wx-yz)}{\partial x} & \dfrac{\partial(wx-yz)}{\partial y} & \dfrac{\partial(wx-yz)}{\partial z} & \dfrac{\partial(wx-yz)}{\partial w} \end{bmatrix} \\ &= \begin{bmatrix} w & -z & -y & x \end{bmatrix}. \end{aligned}$$

Since $\det(p(x,y,z,w)) = 1$, it follows that $J\phi(p(x,y,z,w))$ has full rank at every point in H. Thus, 1 is a regular value of the map ϕ. Since H is also a subgroup, it follows that H is a Lie group as well.

Example 3.10 The general linear group: $GL(n,\mathbb{R})$, which is the group of invertible matrices of order n equipped with the standard usual smooth structure of a Zariski open subset of $\mathbb{R}^{n^2}$ is a Lie group. Indeed, a straightforward application of Proposition 3.3 shows that the function which takes a pair (A,B) of invertible matrices to AB^{-1} is smooth (it is a polynomial map).

Example 3.11 Let us consider the map $\phi : GL(n,\mathbb{R}) \to \mathbb{R}\setminus\{0\}$ where ϕ takes an arbitrary invertible matrix to its determinant. Note that ϕ defines a group homomorphism between the general linear group and the group obtained by endowing $\mathbb{R}\setminus\{0\}$ with the usual multiplication. Next, let $SL(n,\mathbb{R}) = \ker\phi$ be the special linear group. That is, $SL(n,\mathbb{R})$ is the subgroup of $GL(n,\mathbb{R})$ consisting of invertible matrices with one as a determinant. We aim to show that $SL(n,\mathbb{R})$ is a closed embedded submanifold of $GL(n,\mathbb{R})$ (see Theorem 2.18). To this end, it suffices to show that 1 is a regular value of the determinant map ϕ. Let $P \in SL(n,\mathbb{R})$ and let E_{jk} be the matrix which has 1 in the (j,k)-entry and zero everywhere else. Put

$$P = \begin{bmatrix} a_{11} & & & a_{1n} \\ \vdots & & \vdots & \vdots \\ & \cdots & a_{jk} & \cdots \\ \vdots & & \vdots & \vdots \\ a_{n1} & \cdots & \cdots & a_{nn} \end{bmatrix} \quad \text{and} \quad E_{jk} = \begin{bmatrix} 0 & \cdots & \cdots & 0 \\ \vdots & & \vdots & \vdots \\ & \cdots & 1 & \cdots \\ \vdots & & \vdots & \vdots \\ 0 & \cdots & \cdots & 0 \end{bmatrix}.$$

Given a matrix M, let $M_{(j,k)}$ be the submatrix of M of order $n-1$ obtained by removing the jth row and the kth column of M. Applying Laplace expansion's theorem for determinants, we compute

$$\begin{aligned} &\det\left(P + hE_{jk}\right) - \det(P) \\ &= \left((-1)^{k+1}a_{1k}\cdot\det\left[\left(P+hE_{jk}\right)\right]_{(1,k)} + \cdots + (-1)^{k+n}a_{nk}\cdot\det\left[\left(P+hE_{jk}\right)\right]_{(n,k)}\right) \\ &\quad - \left((-1)^{k+1}a_{1k}\cdot\det\left(P_{(1,k)}\right) + \cdots + (-1)^{k+n}a_{nk}\cdot\det\left(P_{(n,k)}\right)\right) \end{aligned}$$

$$= (-1)^{k+j}\left(a_{jk}+h\right)\cdot \det\left(P_{(j,k)}\right) - (-1)^{k+j}\left(a_{jk}\right)\cdot \det\left(P_{(j,k)}\right)$$
$$= (-1)^{k+j}\left(a_{jk}+h-a_{jk}\right)\cdot \det\left(P_{(j,k)}\right)$$
$$= (-1)^{k+j}\cdot h\cdot \det\left(P_{(j,k)}\right)$$

and

$$\lim_{h\to 0}\frac{\det\left(P+hE_{jk}\right)-\det(P)}{h} = \lim_{h\to 0}\frac{(-1)^{k+j}h\cdot \det\left(P_{(j,k)}\right)}{h} = \lim_{h\to 0}(-1)^{k+j}\cdot \det\left(P_{(j,k)}\right).$$

This shows that a matrix representation of the derivative of the determinant function is given by the following vector

$$\left[(-1)^{1+1}\mathrm{det}P_{(1,1)} \quad \cdots \quad (-1)^{n+n}\mathrm{det}P_{(n,n)}\right]$$

with n^2 entries. Since $P \in SL(n,\mathbb{R})$, the rank of the matrix above cannot be zero. This shows that 1 is a regular value of ϕ and as a result, the special linear group is a closed embedded submanifold of the general linear group. Finally, since $SL(n,\mathbb{R})$ is a subgroup of $GL(n,\mathbb{R})$, it follows that the special linear group is a Lie group.

The following example shows that it is not generally true that every subgroup of a Lie group is a closed subset.

Example 3.12 Let

$$G = \left\{\begin{bmatrix} e^{2\pi i t} & 0 \\ 0 & e^{2\pi i\sqrt{2}t}\end{bmatrix} : t \in \mathbb{R}\right\}.$$

Since, for any real numbers s and t,

$$\begin{bmatrix} e^{2\pi i t} & 0 \\ 0 & e^{2\pi i\sqrt{2}t}\end{bmatrix}\begin{bmatrix} e^{2\pi i s} & 0 \\ 0 & e^{2\pi i\sqrt{2}s}\end{bmatrix} = \begin{bmatrix} e^{2i\pi(t+s)} & 0 \\ 0 & e^{2i\sqrt{2}\pi(s+t)}\end{bmatrix}$$

and

$$\begin{bmatrix} e^{2\pi i t} & 0 \\ 0 & e^{2\pi i\sqrt{2}t}\end{bmatrix}^{-1} = \begin{bmatrix} e^{-2i\pi t} & 0 \\ 0 & e^{-2i\sqrt{2}\pi t}\end{bmatrix}.$$

G is closed under matrix multiplication and taking inverses. Consequently, G equipped with matrix operation has the structure of a group. Note also that if G is given the smooth structure of the real line, then G can be regarded as a Lie group. Next, given real numbers x, y, we let

$$\alpha(x,y) = \begin{bmatrix} e^{2\pi i x} & 0 \\ 0 & e^{2\pi i y}\end{bmatrix}.$$

One may view α as a smooth function between $\mathbb{R}^2$ and $GL(2,\mathbb{C})$. Fix an element $\alpha\left(\frac{1}{3},\frac{1}{3}\right)$ in the range of α but not in G. Letting f be the square of the Euclidean distance of any arbitrary element of G and $\alpha\left(\frac{1}{3},\frac{1}{3}\right)$, for any real number t, we obtain

$$f(t) = \left\| \begin{bmatrix} e^{2\pi i t} & 0 \\ 0 & e^{2\pi i\sqrt{2}t} \end{bmatrix} - \begin{bmatrix} e^{\frac{2\pi i}{3}} & 0 \\ 0 & e^{\frac{2\pi i}{3}} \end{bmatrix} \right\|_2^2 = \left| e^{2i\pi t} - e^{\frac{2\pi i}{3}} \right|^2 + \left| e^{2i\sqrt{2}\pi t} - e^{\frac{2\pi i}{3}} \right|^2$$
$$= 4 + \cos(2\pi t) - \sqrt{3}\sin(2\pi t) + \cos\left(2\pi\sqrt{2}t\right) - \sqrt{3}\sin\left(2\sqrt{2}\pi t\right).$$

Since the sequence $(f(n))_{n\in\mathbb{R}}$ is dense in the range of f, there exists a sequence $(x_n)_{n\in\mathbb{N}}$ of real numbers such that $(f(x_n))_{n\in\mathbb{N}}$ converges to zero. This shows that there exists a sequence

$$\left(\begin{bmatrix} e^{2\pi i x_n} & 0 \\ 0 & e^{2\pi i\sqrt{2}x_n} \end{bmatrix} \right)_{n\in\mathbb{N}}$$

of matrices contained in G, which is convergent to the element $\alpha\left(\frac{1}{3},\frac{1}{3}\right) \notin G$. In other words, when G is equipped with the subspace topology of $GL(2,\mathbb{C})$, G is not a closed subgroup.

Proposition 3.5 Let G be a Lie group and let G_0 be the connected component of G containing the identity in G. Then G_0 is a Lie group.

Proof: Let x, y be two elements in G_0. Next, let $\alpha_x, \alpha_y : [0,1] \to G_0$ be two curves in G_0 such that $\alpha_x(0) = e$, $\alpha_x(1) = x$ and $\alpha_y(0) = e$, $\alpha_y(1) = y$. Next, $t \mapsto \alpha_x(t)\alpha_t(t)$ defines a curve joining the identity in G to the element $xy = \alpha_x(1)\alpha_t(1) \in G_0$. Similarly, let $\beta_x : t \mapsto \alpha_x(t)^{-1}$. Then β_x is a curve that joins the identity in G to the inverse of x. Thus, $x^{-1} = \beta_x(1)$ belongs to G_0. Thus, G_0 is a subgroup of G_0 which is also a closed embedded submanifold of G. As a result, G_0 is a Lie group.

Proposition 3.6 Let G be a Lie group and let G_0 be the connected component of G containing the identity in G. Then any connected neighborhood of the identity generates G.

Proof: Let V be an open subset of G which is connected and contains the identity element. Next, let

$$U = \bigcup_{k=1}^{\infty} V^k,$$

where V^k is the subset of G consisting of elements of the form $x_1 \cdots x_k$ where $x_1, \ldots, x_k \in V$. Since U is obtained as the union of an increasing sequence of open sets, it is open in G. Moreover, U is connected as well. Letting $W = V \cap V^{-1}$, it follows that

$$U_1 = \bigcup_{k=1}^{\infty} W^k$$

is an open subgroup of G and consequently, it must be closed. Next, since U_1 is a subset of the connected component of the identity element and since $U_1 \subset U$, it follows that U_1 coincides with the connected component of the identity element in G. Since G is connected, U_1 must generate G.

3.2.2 Left and Right Translations

Let G be a Lie group and let x be an element in G. A left translation map is a function $L_x : G \to G$, which takes every element g in G to xg. Similarly, the right translation map $R_x : G \to G$ is the function that takes every element g in G to gx.

Proposition 3.7 For any fixed element x in G, the left translation and the right translation maps are both diffeomorphisms.

Proof: The fact that L_x and R_x are diffeomorphisms follows from the smoothness of the multiplication map. Moreover, it is straightforward to check that the inverse of L_x, and R_x is given by the smooth function $L_{x^{-1}}$, and $R_{x^{-1}}$, respectively.

Definition 3.5 Let X be a vector field in $\mathfrak{X}(G)$.

1) We say that X is a left-invariant vector field if for $x, y \in G$, $(L_x)_*(X_y) = X_{xy}$.
2) We say that X is a right-invariant vector field if for $x, y \in G$, $(R_x)_*(X_y) = X_{yx}$.

Proposition 3.8 Let X be a vector field in $\mathfrak{X}(G)$ and let e be the trivial element in G. Then, the following holds.

1) Let $Y \in \mathfrak{X}(G)$ such that $Y_x = (L_x)_*(X_e)$ for every $x \in G$. Then, Y_x is a left-invariant vector field.
2) If X is a left-invariant vector field, then $X_x = (L_x)_*(X_e)$ for every $x \in G$.

Proof: For (1), let $x, y \in G$. Then

$$\begin{aligned}(L_x)_*(Y_y) &= (L_x)_*(L_y)_*(X_e)\\ &= (L_x \circ L_y)_*(X_e)\\ &= (L_{xy})_* X_e = Y_{xy}.\end{aligned}$$

For (2), let us assume that X is a left-invariant vector field. By definition, given $x \in G$, it is immediately seen that $(L_x)_*(X_e) = X_{xe} = X_x$.

Example 3.13 ***(The ax + b Lie group)*** Let $G = \mathbb{R}^2$ be a group equipped with the following noncommutative operation

$$(y, s) \bullet (x, t) = \left(y + e^s x, s + t\right).$$

A matrix representation of the derivative of $L_{(y,s)}$ at the identity $(0, 0)$ in G is given by

$$\left[\begin{array}{cc} \dfrac{\partial\,(y+e^s x)}{\partial x} & \dfrac{\partial\,(y+e^s x)}{\partial t} \\ \dfrac{\partial\,(t+s)}{\partial x} & \dfrac{\partial\,(t+s)}{\partial t} \end{array}\right]\Bigg|_{(x,t)=(0,0)} = \begin{bmatrix} e^s & 0 \\ 0 & 1 \end{bmatrix}.$$

Appealing to Proposition 3.8, the following vector fields

$$\left(L_{(x,t)}\right)_*\left(\frac{\partial}{\partial x}\bigg|_{(0,0)}\right) = e^t\,\frac{\partial}{\partial x}\bigg|_{(x,t)} = e^t\frac{\partial}{\partial x} \quad \text{and} \quad \left(L_{(x,t)}\right)_*\left(\frac{\partial}{\partial t}\bigg|_{(0,0)}\right) = \frac{\partial}{\partial t}\bigg|_{(x,t)} = \frac{\partial}{\partial t}$$

are left-invariant on G.

Example 3.14 ***(The Heisenberg Lie group)*** We equip $\mathbb{R}^3$ with the following noncommutative group operation:

$$\left(x_1, y_1, z_1\right) \bullet \left(x_2, y_2, z_2\right) = \left(x_1 + x_2, y_1 + y_2, z_1 + z_2 + x_1 y_2\right).$$

Observing that $(x, y, z)^{-1} = (-x, -y, xy - z)$, $\left(\mathbb{R}^3, \bullet\right)$ becomes a Lie group when $\mathbb{R}^3$ is given its usual smooth structure. Next, we verify that the left translation map acts as follows on $\mathbb{R}^3$:

$$L_{(u,v,w)}\,(x, y, z) = (u + x, v + y, w + z + uy).$$

Moreover, a matrix representation of the derivative of $L_{(u,v,w)}$ at the identity $(0, 0, 0)$ is given by

$$\left[\begin{array}{ccc} \dfrac{\partial\,(u+x)}{\partial x} & \dfrac{\partial\,(u+x)}{\partial y} & \dfrac{\partial\,(u+x)}{\partial z} \\ \dfrac{\partial\,(v+y)}{\partial x} & \dfrac{\partial\,(v+y)}{\partial y} & \dfrac{\partial\,(v+y)}{\partial z} \\ \dfrac{\partial\,(w+z+uy)}{\partial x} & \dfrac{\partial\,(w+z+uy)}{\partial y} & \dfrac{\partial\,(w+z+uy)}{\partial z} \end{array}\right] = \begin{bmatrix} 1 & 0 & 0 \\ 0 & 1 & 0 \\ 0 & u & 1 \end{bmatrix}$$

and the following vector fields

$$\left(L_{(x,y,z)}\right)_*\left(\frac{\partial}{\partial x}\bigg|_{(0,0,0)}\right) = \frac{\partial}{\partial x}$$

$$\left(L_{(x,y,z)}\right)_*\left(\frac{\partial}{\partial y}\bigg|_{(0,0,0)}\right) = \frac{\partial}{\partial y} + x\frac{\partial}{\partial z}$$

$$\left(L_{(x,y,z)}\right)_*\left(\frac{\partial}{\partial z}\bigg|_{(0,0,0)}\right) = \frac{\partial}{\partial z}$$

are all left-invariant on G.

Example 3.15 We equip $\mathbb{R}^2 \times \mathbb{R}$ with the following noncommutative group operation:

$$(v, t) \bullet (w, s) = \left(v + M_t(w), t + s\right),$$

where

$$M_t(w) = \begin{bmatrix} \cos t & -\sin t \\ \sin t & \cos t \end{bmatrix} \begin{bmatrix} w_1 \\ w_2 \end{bmatrix} = \begin{bmatrix} w_1 \cos(t) - w_2 \sin(t) \\ w_2 \cos(t) + w_1 \sin(t) \end{bmatrix}.$$

Next, we verify that the left translation map on $\mathbb{R}^2 \times \mathbb{R}$ acts as follows:

$$L_{(v,t)}(w, s) = \left(\begin{bmatrix} v_1 + w_1 \cos t - w_2 \sin t \\ v_2 + w_2 \cos t + w_1 \sin t \end{bmatrix}, t + s \right).$$

A matrix representation of the derivative of $L_{(v,t)}$ at the identity in the group is given by

$$\begin{bmatrix} \frac{\partial(v_1 + w_1\cos(t) - w_2\sin(t))}{\partial w_1} & \frac{\partial(v_1 + w_1\cos(t) - w_2\sin(t))}{\partial w_2} & \frac{\partial(v_1 + w_1\cos(t) - w_2\sin(t))}{\partial s} \\ \frac{\partial(v_2 + w_2\cos(t) + w_1\sin(t))}{\partial w_1} & \frac{\partial(v_2 + w_2\cos(t) + w_1\sin(t))}{\partial w_2} & \frac{\partial(v_2 + w_2\cos(t) + w_1\sin(t))}{\partial s} \\ \frac{\partial(t+s)}{\partial w_1} & \frac{\partial(t+s)}{\partial w_2} & \frac{\partial(t+s)}{\partial s} \end{bmatrix}$$

$$= \begin{bmatrix} \cos(t) & -\sin(t) & 0 \\ \sin(t) & \cos(t) & 0 \\ 0 & 0 & 1 \end{bmatrix}.$$

As a result,

$$\left(L_{(v,t)}\right)_* \left(\left.\frac{\partial}{\partial v_1}\right|_{(0,0,0)} \right) = \cos(t)\frac{\partial}{\partial v_1} + \sin(t)\frac{\partial}{\partial v_2},$$

$$\left(L_{(v,t)}\right)_* \left(\left.\frac{\partial}{\partial v_2}\right|_{(0,0,0)} \right) = -\sin(t)\frac{\partial}{\partial v_1} + \cos(t)\frac{\partial}{\partial v_2},$$

$$\left(L_{(v,t)}\right)_* \left(\left.\frac{\partial}{\partial t}\right|_{(0,0,0)} \right) = \frac{\partial}{\partial t}$$

are left-invariant vector fields on G.

3.2.3 Lie Algebras

Let G be a Lie group. Define

$$\mathfrak{g} = \{X \in \mathfrak{X}(G) : X \text{is a left-invariant vector field on } G\}$$

and let $T_e(G)$ be the tangent space at the identity element of G. We will prove in this section that any fixed vector $X_e \in T_e(G)$ determines in a unique fashion a smooth vector field on the Lie group, which is invariant under left translations.

Theorem 3.4 As vector spaces $\mathfrak{g}$ and $T_e(G)$ are isomorphic.

The proof of Theorem 3.4 follows from Lemmas 3.3–3.6 established below.

Lemma 3.3 $\mathfrak{g}$ is a vector subspace of $\mathfrak{X}(G)$.

Proof: Given $X, Y \in \mathfrak{g}$, by definition, for $x, y \in G$, $(L_x)_*(X_y) = X_{xy}$ and $(L_x)_*(Y_y) = Y_{xy}$. Next, the linearity of L_x gives:

$$\begin{aligned}(L_x)_*((X+Y)_y) &= (L_x)_*(X_y + Y_y) = (L_x)_*(X_y) + (L_x)_*(Y_y)\\ &= X_{xy} + Y_{xy}\\ &= (X+Y)_{xy}.\end{aligned}$$

Similarly, given any real number r,

$$(L_x)_*((rX)_y) = (L_x)_*(rX_y) = r(L_x)_*(X_y) = rX_{xy} = (rX)_{xy}.$$

We define $J : \mathfrak{g} \to T_e(G)$ such that $J(X) = X_e$. In other words, J returns the tangent vector at the identity, obtained by evaluating the vector field X at the identity of the Lie group.

Lemma 3.4 J is a linear map.

Proof: Given $X, Y \in \mathfrak{g}$,

$$J(X+Y) = (X+Y)_e = X_e + Y_e = J(X) + J(Y)$$

and for a fixed real number r, $J(rX) = (rX)_e = rX_e = rJ(X)$.

Lemma 3.5 J is an injective map.

Proof: To prove this result, it suffices to show that the kernel of J is trivial. To this end, fix $X \in \mathfrak{g}$ such that $J(X) = X_e$ is the trivial vector. Since X is left invariant, we know that for $x, y \in G$, $(L_x)_*(X_y) = X_{xy}$. Next, let g be an arbitrary element in the Lie group. Then $X_g = X_{ge} = (L_g)_*(X_e)$. Since X_e is trivial, X_g must be the zero vector. Consequently, X_g is the zero vector at any point g in G. Therefore, J is injective.

Lemma 3.6 J is a surjective map.

Proof: Let $v \in T_e(G)$. Define $X_g = (L_g)_*(v) \in T_g(G)$ for arbitrary g in G. For arbitrary, $x, y \in G$,

$$\begin{aligned}\left(L_x\right)_*\left(X_y\right) &= \left(L_x\right)_*\left(\left(L_y\right)_*(v)\right)\\ &= \left[\left(L_x\right)_*\circ\left(L_y\right)_*\right](v)\\ &= \left(L_{xy}\right)_*(v)\\ &= X_{xy}.\end{aligned}$$

This shows that $X : g \mapsto X_g$ is a left-invariant vector field.

Secondly, we need to show that X as constructed above is smooth. Regarding X as a map between G and $T(G)$, we define $\tilde{v} : G \to T(G) \oplus T(G) \simeq T(G\times G)$ such that $\tilde{v}(g) = \left(0_g, v\right)$, respectively where 0_g is the zero vector in $T(G)$ and $v \in T_e(G)$. We aim next to show that $\tilde{v}$ is a smooth function. To achieve this, fix $g_0 \in G$ and let $\left(U_{g_0}, \varphi_{g_0}\right), \left(U_e, \varphi_e\right)$ be smooth charts around g and e, respectively. Assuming also that $\varphi_e(e) = 0 \in \mathbb{R}^n$, we may write

$$\left(\varphi_e\right)_* v = \sum_{k=1}^{n} a_k \left.\frac{\partial}{\partial y_k}\right|_0$$

for some real constants $a_1, \dots, a_n$. Next, letting $\left(U_{g_0} \times U_e, \varphi_{g_0} \times \varphi_e\right)$ be a smooth chart around the point $\left(g_0, e\right) \in G \times G$, we compute

$$\left[\left(\varphi_{g_0} \times \varphi_e\right)\circ\tilde{v}\circ\varphi_{g_0}^{-1}\right]\left(x_1, \dots, x_n\right) = \left(x_1, \dots, x_n, \overbrace{0, \dots, 0}^{n \text{ times}}, \overbrace{0, \dots, 0}^{n \text{ times}}, a_1, \dots, a_n\right).$$

Therefore, $\tilde{v}$ is a smooth map. Next, we will show that $X = \mu_* \circ \tilde{v}$. To this end, we fix an arbitrary element g in G. Let $t \mapsto (g, \alpha(t))$ where $t \mapsto \alpha(t)$ is a curve through e in G. Then the curve $t \mapsto \mu(g, \alpha(t)) = g\alpha(t) = L_g\alpha(t)$ represents the vector $\mu_*(\tilde{v}(g))$. This shows that $\mu_*(\tilde{v}(g)) = L_{g_*} v = X_g$.

Definition 3.6 Let $f : M \to N$ be a smooth map and let $X \in \mathfrak{X}(M)$ and let $Y \in \mathfrak{X}(N)$. We say that Y is f-related to X if for every point p in the manifold M, $f_*\left(X_p\right) = Y_{f(p)}$.

Example 3.16 Let $f : \mathbb{R}^2 \to \mathbb{R}$ be the smooth function sending the point (x, y) to x. Next, let $X = \frac{\partial}{\partial x} + x\frac{\partial}{\partial y}$ and $Y = \frac{d}{dx}$. A matrix representation of the derivative of f at the point (x, y) is given by $\begin{bmatrix}1 & 0\end{bmatrix}$. Next, since

$$\begin{bmatrix}1 & 0\end{bmatrix}\begin{bmatrix}1\\ x\end{bmatrix} = 1,$$

it follows that

$$f_*\left(\frac{\partial}{\partial x} + x\frac{\partial}{\partial y}\right) = \frac{d}{dx}.$$

This shows that Y is f-related to X. Next, we compute the integral curves of X by considering the equation

$$c'(t) = X_{c(t)} \Leftrightarrow \begin{cases} \dfrac{dx}{dt} = 1 \\ \dfrac{dy}{dt} = x \end{cases} \Leftrightarrow \begin{cases} x = C + t \\ \dfrac{dy}{dt} = t^2 + Ct + K \end{cases}$$

for some real constants C, K, and the integral curves of Y are $\{C + t : C \in \mathbb{R}\}$. Note that all integral curves of Y are of the form $f(c(t))$ where c is an integral curve of X.

Proposition 3.9 Let $f : M \to N$ be a smooth map and let $X \in \mathfrak{X}(M)$ and let $Y \in \mathfrak{X}(N)$. Then the following statements are equivalent:

1) Y is f-related to X.
2) f sends every integral curve of X to an integral curve of Y.
3) $\rho_t^{(Y)} \circ f = f \circ \rho_t^{(X)}$ if $\rho_t^{(X)}$ is the flow generated by X (see Definition 3.3) on an open subset U of M and $\rho_t^{(Y)}$ is the flow generated by Y on an open set V of N which contains $f(U)$.

Proof: ((1) ⇒ (2)) Suppose that Y is f-related to X and let us assume that α is an integral curve of X. Put $\beta = f \circ \alpha$. We may regard the derivative β' of β as $f_* X_{\alpha(t)} = Y_{f(\alpha(t))} = Y_{\beta(t)}$. Therefore, β' is an integral curve of Y. ((2) ⇒ (3)) Suppose next that f sends every integral curve of X to an integral curve of Y. Let $\rho_t^{(X)}$ be the flow generated by X on an open subset U of M and let $\rho_t^{(Y)}$ be the flow generated by Y on an open set V of N which contains $f(U)$. Write $\rho_t^{(X)}(p) = c_p(t)$ where c_p is a maximal integral curve of X through p. Then,

$$\left[f \circ \rho_t^{(X)}\right](p) = f\left(c_p(t)\right) = \rho_t^{(Y)}(f(p)).$$

((3) ⇒ (1)) Finally, let p be a point in M assume that $\rho_t^{(Y)} \circ f = f \circ \rho_t^{(X)}$. Then

$$f_*\left(X_p\right) = \left.\frac{d}{dt}\right|_{t=0} f \circ \rho_t^{(X)} = \left.\frac{d}{dt}\right|_{t=0} \rho_t^{(Y)}(f(p)) = Y_{f(p)}.$$

Therefore, Y is related to X.

Proposition 3.10 Let $f : M \to N$ be a smooth map and let $X \in \mathfrak{X}(M)$ and let $Y \in \mathfrak{X}(N)$.

1) Then Y is f-related to X if and only if for any real-valued smooth function u defined on N, $Y(u) \circ f = X(u \circ f)$.
2) If X_1 is f-related to X and Y_1 is f-related to Y then $\left[X_1, Y_1\right]$ is f-related to $[X, Y]$.

Proof: For (1), we need to verify that $Y(u)\circ f = X(u\circ f)$ if and only if for every point $p \in M$, $Y_{f(p)}(u) = X_p(u\circ f)$. By identifying $X_p(u\circ f)$ with $f_*(X_p)\,u$, it follows that $Y(u)\circ f = X(u\circ f)$ if and only if for every point p, $Y_{f(p)}(u)$ is given by the vector $f_*(X_p)\,u$. For (2), let us assume that X_1 is f-related to X and Y_1 is f-related to Y and let $u \in C^\infty(N)$. Then

$$\begin{aligned}\left(\left[X_1, Y_1\right](u)\right)\circ f &= \left(\left(X_1Y_1 - Y_1X_1\right)(u)\right)\circ f\\ &= X_1\left(Y_1(u)\right)\circ f - Y_1\left(X_1(u)\right)\circ f\\ &= X_1\left(Y_1(u\circ f)\right) - Y_1\left(X_1(u\circ f)\right)\\ &= \left[X_1, Y_1\right](u\circ f).\end{aligned}$$

Appealing next to Part (1), it follows that $\left[X_1, Y_1\right]$ is f-related to $[X, Y]$.

Proposition 3.11 Let $X, Y \in \mathfrak{g}$. Then $[X, Y] \in \mathfrak{g}$.

Proof: Let $X, Y \in \mathfrak{g}$. For every point $x, y \in G$, since

$$\left(L_x\right)_* X_y = \left(L_x\right)_*\left(L_y\right)_* X_e = L_{xy}X_e = X_{xy} = X_{L_x(y)}$$

then X is L_x-related to itself and similarly, Y is L_x-related to itself. By Proposition 3.10, $[X, Y]$ is L_x-related to itself and consequently, $[X, Y]$ is indeed, left invariant.

Definition 3.7

1) A Lie algebra V is a vector space equipped with nonassociative operation $(v, w) \mapsto b(v, w)$ which is bilinear, alternating, and satisfies Jacobi identity. That is, for $u, v, w \in V$,

$$b(u, b(v, w)) + b(v, b(w, u)) + b(w, b(u, v)) = 0.$$

2) Let V be a finite-dimensional Lie algebra with basis $\mathcal{B} = \left\{e_1, \dots, e_n\right\}$. Then for $e_i, e_j \in \mathcal{B}$, there exists a finite sequence of constant scalars $\left(c_{(i,j,k)}\right)_{k=1}^{n}$ called the structure constants relative to B and determined by the matrix

$$\left(b\left(e_i, e_j\right)\right)_{1\le i,j\le n} = \left(\sum_{k=1}^{n} c_{(i,j,k)}e_k\right)_{1\le i,j\le n}.$$

Referring to Proposition 2.15, it is easy to verify all the following facts.

Remark 3.2 The following is worth noting.

1) The vector space $\mathfrak{g}$ of left-invariant vector fields equipped with the operation $(X, Y) \mapsto b(X, Y) = [X, Y] = XY - YX$ is a finite-dimensional Lie algebra. In other words, the following statements hold true:
   a) $\left[X_1 + X_2, Y\right] = \left[X_1, Y\right] + \left[X_2, Y\right]$ for all $X_1, X_2, Y \in \mathfrak{g}$.
   b) $[aX, Y] = a[X, Y]$ and $[X, aY] = a[X, Y]$ for any real number a and for $X, Y \in \mathfrak{g}$.

c) $[X, Y_1 + Y_2] = [X, Y_1] + [X, Y_2]$ for all $X, Y_1, Y_2 \in \mathfrak{g}$.
d) $[X, Y] = -[Y, X]$ for $X, Y \in \mathfrak{g}$.
e) $[[X, Y], Z] + [[Y, Z], X] + [[Z, X], Y] = 0$ for all $X, Y, Z \in \mathfrak{g}$.

2) Let $T_e(G)$ be the tangent space at the identity element of the Lie group G. Next, given $X, Y \in T_e(G)$ let $\tilde{X}$ and $\tilde{Y}$ be left-invariant vector fields on G such that $X = \tilde{X}_e$ and $Y = \tilde{Y}_e$. Then $[\tilde{X}, \tilde{Y}]$ is a left-invariant vector field on G and we define the bracket of X, Y as follows:

$$[X, Y] = [\tilde{X}, \tilde{Y}]_e. \tag{3.3}$$

Then, it is easy to verify that the vector space $T_e(G)$ endowed with this bracket is a Lie algebra as well.

3) Let $\mathcal{B} = \{X_1, \ldots, X_n\}$ be a basis for $\mathfrak{g}$. Write $X = \sum_{k=1}^{n} x_k X_k$ and $Y = \sum_{k=1}^{n} y_k X_k$. By the multilinearity of the Lie bracket,

$$[X, Y] = \left[\sum_{k=1}^{n} x_k X_k, \sum_{j=1}^{n} y_j X_j\right] = \sum_{k=1}^{n}\sum_{j=1}^{n} x_k y_j [X_k, X_j].$$

Next, letting $[X_k, X_j] = \sum_{m=1}^{n} c_{(k,j,m)} X_m$, it follows that

$$[X, Y] = \sum_{m=1}^{n}\left(\sum_{k=1}^{n}\sum_{j=1}^{n} x_k y_j c_{(k,j,m)}\right) X_m.$$

Definition 3.8

1) A Lie group homomorphism $\phi : G \to H$ is a smooth map which is also a group homomorphism. In other words, for $x, y \in G$, $\phi(xy) = \phi(x)\phi(y)$.
2) A Lie algebra homomorphism $J : \mathfrak{g} \to \mathfrak{h}$ is a linear map that is Lie bracket preserving. In other words, $J([X, Y]) = [J(X), J(Y)]$ for all $X, Y \in \mathfrak{g}$.

Proposition 3.12 $\mathfrak{g}$ and $T_e(G)$ are Lie isomorphic. In other words,

$$J([\tilde{X}, \tilde{Y}]) = [J(\tilde{X}), J(\tilde{Y})]$$

for all $\tilde{X}, \tilde{Y} \in \mathfrak{g}$.

Proof: Appealing to Theorem 3.4, we know that the mapping $J : \mathfrak{g} \to T_e(G)$ such that $J(\tilde{X}) = \tilde{X}_e$ defines a vector space isomorphism. Next, given $\tilde{X}, \tilde{Y} \in \mathfrak{g}$, we have

$$\begin{aligned} J([\tilde{X}, \tilde{Y}]) &= ([\tilde{X}, \tilde{Y}])_e \\ &\overset{(3.3)}{=} [X, Y] \\ &= [\tilde{X}_e, \tilde{Y}_e] \\ &= [J(\tilde{X}), J(\tilde{Y})]. \end{aligned}$$

This shows that J defines a Lie algebra isomorphism.

In light of Proposition 3.12, it will often be convenient to identify the Lie algebra $\mathfrak{g}$ of a Lie group with the tangent space at the identity element of the group.

Example 3.17 Let G be the group $\mathbb{R}^2$ endowed with the usual vector addition. We shall identify $\mathfrak{g}$ naturally with $T_0\left(\mathbb{R}^2\right) \simeq \mathbb{R}^2$. Next, let $\mathcal{B} = \left\{X_1, X_2\right\}$ be a basis for $\mathfrak{g}$. Given $\left(x_1, x_2\right) \in \mathbb{R}^2$, $L_{(x_1,x_2)} : \left(y_1, y_2\right) \mapsto \left(x_1 + y_1, x_2 + y_2\right)$. As a result, the derivative of $L_{(x_1,x_2)}$ at the identity is:

$$\begin{bmatrix} \dfrac{\partial\left(x_1+y_1\right)}{\partial x_1} & \dfrac{\partial\left(x_1+y_1\right)}{\partial x_2} \\ \dfrac{\partial\left(x_2+y_2\right)}{\partial x_1} & \dfrac{\partial\left(x_2+y_2\right)}{\partial x_2} \end{bmatrix} = \begin{bmatrix} 1 & 0 \\ 0 & 1 \end{bmatrix}.$$

This shows

$$\left(L_{(x_1,x_2)}\right)_* \left.\frac{\partial}{\partial x_1}\right|_{(0,0)} = \left.\frac{\partial}{\partial x_1}\right|_{(x_1,x_2)}, \quad \left(L_{(x_1,x_2)}\right)_* \left.\frac{\partial}{\partial x_2}\right|_{(0,0)} = \left.\frac{\partial}{\partial x_2}\right|_{(x_1,x_2)}$$

and with the identification $\mathfrak{g} \simeq T_0\left(\mathbb{R}^2\right)$,

$$\mathcal{B} = \left\{ \left.\frac{\partial}{\partial x_1}\right|_{(0,0)}, \left.\frac{\partial}{\partial x_2}\right|_{(0,0)} \right\}$$

can be viewed as a basis for $\mathfrak{g}$. Moreover, since

$$\left[\frac{\partial}{\partial x_1}, \frac{\partial}{\partial x_2}\right] = \frac{\partial}{\partial x_1}\frac{\partial}{\partial x_2} - \frac{\partial}{\partial x_2}\frac{\partial}{\partial x_1} = 0,$$

it follows that $\mathfrak{g}$ is a commutative Lie algebra.

Example 3.18 Coming back to Example 3.9, let $G = \mathbb{R}^2$ be a group equipped with the following noncommutative operation

$$(y, s) \bullet (x, t) = \left(y + e^s x, s + t\right).$$

Let

$$\mathcal{B} = \left\{ \left(L_{(x,t)}\right)_* \left(\left.\frac{\partial}{\partial x}\right|_{(0,0)} \right), \left(L_{(x,t)}\right)_* \left(\left.\frac{\partial}{\partial t}\right|_{(0,0)} \right) \right\} = \left\{ e^t \frac{\partial}{\partial x}, \frac{\partial}{\partial t} \right\}$$

be a basis for the Lie algebra $\mathfrak{g}$ of left-invariant vector fields in $\mathfrak{X}\left(\mathbb{R}^2\right)$. For arbitrary $f \in C^\infty\left(\mathbb{R}^2\right)$,

$$\left[e^t \frac{\partial}{\partial x}, \frac{\partial}{\partial t}\right](f) = e^t \frac{\partial}{\partial x}\left(\frac{\partial f}{\partial t}\right) - \frac{\partial}{\partial t}\left(e^t \frac{\partial f}{\partial x}\right).$$

Applying the product rule to the expression $\frac{\partial}{\partial t}\left(e^t \frac{\partial f}{\partial x}\right)$, we derive that

$$\left[e^t \frac{\partial}{\partial x}, \frac{\partial}{\partial t}\right](f) = e^t \frac{\partial^2}{\partial x \partial t}(f) - e^t \frac{\partial f}{\partial x} - e^t \frac{\partial^2}{\partial x \partial t}(f) = -e^t \frac{\partial f}{\partial x}.$$

Letting $A = \frac{\partial}{\partial t}$ and $X = e^t \frac{\partial}{\partial x}$, the following equations determine all Lie brackets of $\mathfrak{g}$:

$$[A, X] = X, [A, A] = 0, [X, A] = -X, [X, X] = 0.$$

Example 3.19 We endow $\mathbb{R}^3$ with the following noncommutative group operation:

$$(x_1, y_1, z_1) \bullet (x_2, y_2, z_2) = (x_1 + x_2, y_1 + y_2, z_1 + z_2 + x_1 y_2)$$

and the following vector fields

$$X = \left(L_{(x,y,z)}\right)_* \left(\left. \frac{\partial}{\partial x} \right|_{(0,0,0)} \right) = \frac{\partial}{\partial x}$$

$$Y = \left(L_{(x,y,z)}\right)_* \left(\left. \frac{\partial}{\partial y} \right|_{(0,0,0)} \right) = \frac{\partial}{\partial y} + x \frac{\partial}{\partial z}$$

$$Z = \left(L_{(x,y,z)}\right)_* \left(\left. \frac{\partial}{\partial z} \right|_{(0,0,0)} \right) = \frac{\partial}{\partial z}$$

are left-invariant on G. Given $f \in C^\infty\left(\mathbb{R}^2\right)$, we verify that

$$\begin{aligned}
[X, Y](f) &= \frac{\partial}{\partial x}\left(\frac{\partial f}{\partial y} + x \frac{\partial f}{\partial z} \right) - \frac{\partial^2 f}{\partial x \partial y} - x \frac{\partial}{\partial z}\left(\frac{\partial f}{\partial x} \right) \\
&= \frac{\partial^2 f}{\partial x \partial y} + \frac{\partial f}{\partial z} + x \frac{\partial^2 f}{\partial x \partial z} - \frac{\partial^2 f}{\partial x \partial y} - x \frac{\partial^2 f}{\partial x \partial z} \\
&= \frac{\partial^2 f}{\partial x \partial y} - \frac{\partial^2 f}{\partial x \partial y} + x \frac{\partial^2 f}{\partial x \partial z} - x \frac{\partial^2 f}{\partial x \partial z} + \frac{\partial f}{\partial z} = \frac{\partial f}{\partial z}.
\end{aligned}$$

Therefore, the matrix of the Lie brackets of $\mathfrak{g}$ with respect to the basis elements X, Y, Z is given by

	X	Y	Z
X	0	Z	0
Y	Z	0	0
Z	0	0	0

The table above shows that the center of the algebra $\mathfrak{g}$ is the one-dimensional vector space spanned by the left-invariant vector field $\frac{\partial}{\partial z}$.

Example 3.20 The general linear group $GL(2, \mathbb{R})$ can be viewed as a subset of the set

$$\left\{ p(x) = \begin{bmatrix} x_{11} & x_{12} \\ x_{21} & x_{22} \end{bmatrix} : x = (x_{11}, x_{12}, x_{21}, x_{22}) \in \mathbb{R}^4 \right\},$$

which is a four-dimensional vector space over the reals. Put

$$E_{11} = \begin{bmatrix} 1 & 0 \\ 0 & 0 \end{bmatrix}, \quad E_{12} = \begin{bmatrix} 0 & 1 \\ 0 & 0 \end{bmatrix},$$
$$E_{21} = \begin{bmatrix} 0 & 0 \\ 1 & 0 \end{bmatrix}, \quad E_{22} = \begin{bmatrix} 0 & 0 \\ 0 & 1 \end{bmatrix}.$$

For each E_{jk}, let $c_{jk} : (-\epsilon, \epsilon) \to GL(2, \mathbb{R})$ be a curve in $GL(2, \mathbb{R})$ through the identity element I defined as follows: $c_{jk}(t) = I + t \cdot E_{jk}$ for some positive real number ϵ. Then each matrix $c'_{jk}(0) = E_{jk}$ can be viewed as an element of the tangent space to the Lie group $GL(2, \mathbb{R})$ at the identity. Via this identification, $\mathcal{B} = \{E_{11}, E_{12}, E_{21}, E_{22}\}$ is an ordered basis for the vector space $T_I(GL(n, \mathbb{R}))$ and a matrix representation of the derivative of the left translation map $L_{p(x)} : p(y) \mapsto p(x)p(y)$ at the identity, with respect to $\mathcal{B}$ is given by

$$\begin{bmatrix} x_{11} & 0 & x_{12} & 0 \\ 0 & x_{11} & 0 & x_{12} \\ x_{21} & 0 & x_{22} & 0 \\ 0 & x_{21} & 0 & x_{22} \end{bmatrix}.$$

By slightly abusing notation, we obtain a basis for the vector space of left-invariant vector fields on $GL(2, \mathbb{R})$ as follows:

$$\left(L_{p(x)}\right)_* (E_{11}) = \begin{bmatrix} x_{11} & x_{12} \\ x_{21} & x_{22} \end{bmatrix} \begin{bmatrix} 1 & 0 \\ 0 & 0 \end{bmatrix} = \begin{bmatrix} x_{11} & 0 \\ x_{21} & 0 \end{bmatrix},$$
$$\left(L_{p(x)}\right)_* (E_{12}) = \begin{bmatrix} x_{11} & x_{12} \\ x_{21} & x_{22} \end{bmatrix} \begin{bmatrix} 0 & 1 \\ 0 & 0 \end{bmatrix} = \begin{bmatrix} 0 & x_{11} \\ 0 & x_{21} \end{bmatrix},$$

and

$$\left(L_{p(x)}\right)_* (E_{21}) = \begin{bmatrix} x_{11} & x_{12} \\ x_{21} & x_{22} \end{bmatrix} \begin{bmatrix} 0 & 0 \\ 1 & 0 \end{bmatrix} = \begin{bmatrix} x_{12} & 0 \\ x_{22} & 0 \end{bmatrix},$$
$$\left(L_{p(x)}\right)_* (E_{22}) = \begin{bmatrix} x_{11} & x_{12} \\ x_{21} & x_{22} \end{bmatrix} \begin{bmatrix} 0 & 0 \\ 0 & 1 \end{bmatrix} = \begin{bmatrix} 0 & x_{12} \\ 0 & x_{22} \end{bmatrix}.$$

Note that $\left(L_{p(x)}\right)_* (E_{11}), \left(L_{p(x)}\right)_* (E_{12}), \left(L_{p(x)}\right)_* (E_{21})$, and $\left(L_{p(x)}\right)_* (E_{22})$ corresponds to the left-invariant vector fields:

$$x_{11}\frac{\partial}{\partial x_{11}} + x_{21}\frac{\partial}{\partial x_{21}}, x_{11}\frac{\partial}{\partial x_{12}} + x_{21}\frac{\partial}{\partial x_{22}}, x_{12}\frac{\partial}{\partial x_{11}} + x_{22}\frac{\partial}{\partial x_{21}},$$

and $x_{12}\frac{\partial}{\partial x_{12}} + x_{22}\frac{\partial}{\partial x_{22}}$, respectively.

Proposition 3.13 generalizes the construction given in Example 3.20 by describing explicitly, the Lie algebra of the general linear group and its Lie brackets.

Proposition 3.13 For every natural number n, the Lie algebra of the general linear group $GL(n,\mathbb{R})$ can be identified with the vector space of all square matrices of order n equipped with the Lie bracket $(A,B)\mapsto [A,B]=AB-BA$.

Proof: Let $T_I(GL(n,\mathbb{R}))$ be the tangent space to the identity of $GL(n,\mathbb{R})$, where I is the identity square matrix of order n. Given $A\in T_I(GL(n,\mathbb{R}))$, let $\tilde{A}$ be a left-invariant vector field on $GL(n,\mathbb{R})$ such that $\tilde{A}_I=A$. Next, given an invertible matrix X, $\tilde{A}_X=(L_X)_*A$. Put

$$X=\begin{bmatrix} x_{11} & \cdots\cdots & x_{1n} \\ \vdots & \vdots & \vdots \\ & \cdots\ x_{jk}\ \cdots & \\ \vdots & \vdots & \vdots \\ x_{n1} & \cdots\cdots & x_{nn}\end{bmatrix} \quad \text{and} \quad E_{jk}=\begin{bmatrix} 0 & \cdots\cdots & 0 \\ \vdots & \vdots & \vdots \\ & \cdots\ 1\ \cdots & \\ \vdots & \vdots & \vdots \\ 0 & \cdots\cdots & 0\end{bmatrix},$$

where E_{jk} is the matrix satisfying

$$\left(E_{jk}\right)_{(j',k')}=\begin{cases}0, \text{if } (j',k')\neq(j,k),\\ 1, \text{if } (j',k')=(j,k).\end{cases}$$

Clearly, the collection $\{E_{jk} : 1\leq j,k\leq n\}$ is a basis for the vector space of square matrices of order n and

$$\begin{aligned}\lim_{h\to 0}\frac{1}{h}\left(L_X\left(I+hE_{jk}\right)-L_X(I)\right) &= \lim_{h\to 0}\frac{1}{h}\left(L_X\left(I+hE_{jk}-I\right)\right)\\ &= \lim_{h\to 0}\frac{1}{h}\left(L_X\left(hE_{jk}\right)\right)\\ &= XE_{jk}.\end{aligned}$$

This shows that $\tilde{A}_X=(L_X)_*A=XA$ and

$$XA=\sum_{k=1}^{n}\sum_{j=1}^{n}\left(\sum_{l=1}^{n}x_{jl}a_{jk}\right)E_{jk}.$$

Via the identification $E_{jk}\mapsto\frac{\partial}{\partial x_{jk}}$, we may view $\tilde{A}_X$ as

$$\sum_{k=1}^{n}\sum_{j=1}^{n}\left(\sum_{l=1}^{n}x_{jl}a_{jk}\right)\frac{\partial}{\partial x_{jk}}$$

and

$$\tilde{A}\left(x_{jk}\right)=\sum_{k=1}^{n}\sum_{j=1}^{n}\left(\sum_{l=1}^{n}x_{jl}a_{jk}\right)\frac{\partial\left(x_{jk}\right)}{\partial x_{jk}}=\sum_{l=1}^{n}x_{jl}a_{lk}.$$

Given $A, B \in T_I(GL(n,\mathbb{R}))$, we define left-invariant vector fields $\tilde{A}$ and $\tilde{B}$ on $GL(n,\mathbb{R})$ such that $\tilde{A}_X = (L_X)_* A$ and $\tilde{B}_X = (L_X)_* B$. Letting $A = (a_{ij})_{1\le i,j\le n}$ and $B = (b_{ij})_{1\le i,j\le n}$, we obtain:

$$\begin{aligned}
\left[\tilde{A}_X, \tilde{B}_X\right](x_{ij}) &= \tilde{A}\tilde{B}(x_{ij}) - \tilde{B}\tilde{A}(x_{ij}) \\
&= \tilde{A}\left(\sum_{l=1}^{n} x_{il} b_{lj}\right) - \tilde{B}\left(\sum_{l=1}^{n} x_{il} a_{lj}\right) \\
&= \sum_{l=1}^{n} \tilde{A}(x_{il})\, b_{lj} - \sum_{l=1}^{n} \tilde{B}(x_{il})\, a_{lj} \\
&= \sum_{l=1}^{n}\sum_{k=1}^{n} x_{ik} a_{kl} b_{lj} - \sum_{l=1}^{n}\sum_{k=1}^{n} x_{ik} b_{kl} a_{lj} \\
&= \widetilde{[A,B]}(x_{ij}).
\end{aligned}$$

Moreover,

$$\begin{aligned}
\left[\tilde{A}, \tilde{B}\right](x_{ij}) &= \sum_{k=1}^{n} x_{ik}\left(\sum_{l=1}^{n} a_{kl} b_{lj}\right) - \sum_{k=1}^{n} x_{ik}\left(\sum_{l=1}^{n} b_{kl} a_{lj}\right) \\
&= \widetilde{AB}(x_{ij}) - \widetilde{BA}(x_{ij}) \\
&= \left(\widetilde{AB} - \widetilde{BA}\right)(x_{ij}) \\
&= \widetilde{[A,B]}(x_{ij}).
\end{aligned}$$

This shows that $\widetilde{[A,B]} = \left[\tilde{A}, \tilde{B}\right]$, and as a result,

$$\widetilde{[A,B]}_I = [A,B] = AB - BA.$$

Finally, denoting the Lie algebra $\mathfrak{g}$ of the general linear group by $\mathfrak{gl}(n,\mathbb{R})$, it follows that $\mathfrak{gl}(n,\mathbb{R})$ can be regarded as the vector space of all square matrices of order n equipped with the Lie bracket $(A,B) \mapsto [A,B] = AB - BA$ as desired.

Proposition 3.14 Let $\phi : G \to H$ be a Lie group homomorphism. Then ϕ induces a Lie algebra homomorphism $\phi_* : \mathfrak{g} \to \mathfrak{h}$.

Proof: Let $X \in \mathfrak{g}$ be a left-invariant vector field on G. Then $\phi_*(X_e)$ is an element of $T_e(H)$. Next, let Y be a left-invariant vector field on H such that Y_e is equal to $\phi_*(X_e)$. We will show that Y is ϕ-related to X. To this end, let g be an element of G. Then

$$\phi_*(X_g) = \phi_*(L_g)_*(X_e) = (\phi \circ L_g)_* X_e.$$

Since ϕ is a homomorphism,

$$(\phi \circ L_g)(y) = \phi(gy) = \phi(g)\phi(y) = L_{\phi(g)}\phi(y).$$

That is $\phi \circ L_g = L_{\phi(g)}\phi$,

$$\left(\phi \circ L_g\right)_* X_e = \left(L_{\phi(g)} \circ \phi\right)_* X_e = \left(L_{\phi(g)}\right)_* \left(\phi_* X_e\right) = \left(L_{\phi(g)}\right)_{*-} Y_e = Y_{\phi(g)}$$

and $\phi_* \left(X_g\right) = Y_{\phi(g)}$. Appealing to Proposition 3.10, Y is ϕ-related to X. Consequently, the vector field $\left[\phi_* X, \phi_* X'\right]$ is ϕ-related to $\left[X, X'\right]$ for all $X, X' \in \mathfrak{g}$.

Proposition 3.15 Let $\det : GL(n, \mathbb{R}) \to \mathbb{R}^*$ be the determinant function. Then det is a Lie group homomorphism and $\det_* : \mathfrak{gl}(n, \mathbb{R}) \to \mathbb{R}$ is the trace function given by $\det_*(X) = \sum_{k=1}^{n} X_{kk} = \mathrm{tr}(X)$.

Proof: Since det and tr are both linear maps, to prove the result, we only need to make sure that both functions agree on the basis of the Lie algebra of $GL(n, \mathbb{R})$. Let $\left\{E_{jk} : j, k \in \mathbb{R}\right\}$ be a basis element of $\mathfrak{gl}(n, \mathbb{R})$ such that

$$\left(E_{jk}\right)_{j'k'} = \begin{cases} 1, \text{if } (j,k) = \left(j', k'\right) \\ 0, \text{if } (j,k) \neq \left(j', k'\right) \end{cases}$$

and $\mathrm{tr}\left(E_{jk}\right) = \begin{cases} 1, \text{if} j = k, \\ 0, \text{if} j \neq k. \end{cases}$ Computing the derivative of det at the identity element in G, we obtain

$$\frac{\det\left(I + hE_{jk}\right) - \det(I)}{h} = \begin{cases} \dfrac{1+h}{h}, \text{if} j = k, \\ 0, \text{if} j \neq k, \end{cases}$$

and

$$\lim_{h\to 0} \frac{\det\left(I + hE_{jk}\right) - \det(I)}{h} = \begin{cases} \lim\limits_{h\to 0} \dfrac{1+h}{h}, & \text{if} j = k \\ 0, & \text{if} j \neq k \end{cases} = \mathrm{tr}\left(E_{jk}\right).$$

This shows that $\det_* = \mathrm{tr}$.

Proposition 3.16 Let $SL(n, \mathbb{R}) = \{A \in GL(n, \mathbb{R}) : \det A = 1\}$. Then the Lie algebra of $SL(n, \mathbb{R})$ is the vector space $\mathfrak{sl}(n, \mathbb{R}) = \{X \in \mathfrak{gl}(n, \mathbb{R}) : \mathrm{tr}(X) = 0\}$.

Proof: Let $\iota : SL(n, \mathbb{R}) \to GL(n, \mathbb{R})$ be the inclusion map. Since $SL(n, \mathbb{R})$ is the inverse image of 1 through the determinant function, and 1 is a regular value of the determinant function, it follows that ι is a closed embedding and Lie group homomorphism. Moreover,

$$SL(n, \mathbb{R}) \xrightarrow{\iota} GL(n, \mathbb{R}) \xrightarrow{\det} \mathbb{R}^*$$

and $\det \circ \iota$ defines a trivial homomorphism between $SL(n, \mathbb{R})$ and $\mathbb{R}^*$. Next, computing the differential of $\det \circ \iota$, we obtain

$$(\det \circ \iota)_* = (\det)_* \circ \iota_* = \mathrm{tr} \circ \iota_* = 0.$$

The calculations above show that $\iota_* (\mathfrak{sl}(n, \mathbb{R}))$ belongs to the kernel of $(\det)_*$. Since $\ker((\det)_*) = \ker(\mathrm{tr})$, $\dim(SL(n, \mathbb{R})) = n^2 - 1$ and ι is a smooth embedding, it follows that $\iota_* (\mathfrak{sl}(n, \mathbb{R}))$ must be an $(n^2 - 1)$-dimensional subspace of $\mathfrak{gl}(n, \mathbb{R})$. However, $\dim \ker(\mathrm{tr}) = n^2 - 1$. This shows that $\iota_* (\mathfrak{sl}(n, \mathbb{R})) = \ker(\mathrm{tr})$.

Proposition 3.17 Let $O(n) = \{A \in GL(n, \mathbb{R}) : A^T = A^{-1}\}$. Then the Lie algebra of $O(n)$ is the vector space $\mathfrak{o}(n) = \{X \in \mathfrak{gl}(n, \mathbb{R}) : X^T = -X\}$ of all skew-symmetric square matrices of order n.

Proof: Let $\iota : O(n) \to GL(n, \mathbb{R})$ be the inclusion map. Then ι is a closed embedding and Lie group homomorphism. Next, let $t \mapsto A(t)$ be a C^∞ curve in $O(n)$ such that $A(0) = I$. By construction, for every real number t, $A(t)^T A(t) = I$. Computing the derivative of $t \mapsto A(t)^T A(t)$ at zero gives

$$\begin{aligned} 0 &= A'(0)^T A(0) + A(0)^T A'(0) \\ &= A'(0)^T I + IA'(0) \\ &= A'(0)^T + A'(0). \end{aligned}$$

This shows that $A'(0)$ is skew-symmetric. Moving forward, we shall identify the Lie algebra of $O(n)$ with the tangent space at the identity. Since (1) the dimension of the tangent space at the identity of the orthogonal group is equal to the dimension of $O(n)$ and (2), the vector space of skew-symmetric matrices forms a subspace of dimension $\frac{n(n-1)}{2}$; it follows that

$$\iota_* (O(n)) = \{X \in \mathfrak{gl}(n, \mathbb{R}) : X^T = -X\}.$$

Let M be a smooth manifold and let $X \in \mathfrak{X}(M)$. We recall that X is a complete vector field if for every point $p \in M$, every maximal integral curve through p is defined on the set of all real numbers.

Example 3.21 Returning to Example 3.19, we consider $\mathbb{R}^3$ with the noncommutative group operation:

$$(x_1, y_1, z_1) \bullet (x_2, y_2, z_2) = (x_1 + x_2, y_1 + y_2, z_1 + z_2 + x_1 y_2).$$

Let $Y = \frac{\partial}{\partial y} + x\frac{\partial}{\partial z}$ be a left-invariant vector field. Next, solving the system of differential equation

$$\begin{cases} \dfrac{dx(t)}{dt} = 0, \\ \dfrac{dy(t)}{dt} = 1, \\ \dfrac{dz(t)}{dt} = x(t), \end{cases}$$

we obtain

$$x(t) = x_0, y(t) = t + y_0, z(t) = x_0 t + z_0.$$

This shows that $c(t) = \left(x_0, t + y_0, x_0 t + z_0\right)$ is an integral curve for Y through the point $\left(x_0, y_0, z_0\right)$ and Y is a complete vector field. We will show in Proposition 3.18 that this is not a mere fluke and that in fact, all left-invariant vector fields on Lie groups are actually complete.

Proposition 3.18 Let G be a Lie group and let $X \in \mathfrak{g}$. Then X defines a complete vector field on the Lie group G.

Proof: Let $c : (-\epsilon, \epsilon) \to G$ be an integral curve of X through the identity element where ϵ is some positive real number. By assumption, $c(0) = e$ and since X is L_x-related to itself, it is clear that the image of an integral curve under L_x is an integral curve as well. Thus, for any element $x \in G$, $t \mapsto xc(t)$ is an integral curve of X through x. Next, given real numbers $s, t \in (-\epsilon, \epsilon)$ such that $s + t \in (-\epsilon, \epsilon)$, observe that $c(s+t) = c(s)c(t)$. Indeed, fixing $s \in (-\epsilon, \epsilon)$, let $d(t) = c(t+s)$. Clearly, $d(0) = c(s)$ and $d'(t) = X_{d(t)}$. Since, $t \mapsto u(t) = c(s)c(t)$ is an integral curve of X through $c(s)$, it is clear that $c(s+t) = c(s)c(t)$. Next, let $\mu(t) = c\left(\frac{\epsilon}{2}\right)c\left(t - \frac{\epsilon}{2}\right)$ for $t \in \left(-\frac{\epsilon}{2}, \frac{3\epsilon}{2}\right)$. Then

$$\mu(0) = c\left(\frac{\epsilon}{2} - \frac{\epsilon}{2}\right) = c(0) = e.$$

Using the fact that X is left-invariant, we regard μ as a composition of $L_{c\left(\frac{\epsilon}{2}\right)}$ and the function $t \mapsto c\left(t - \frac{\epsilon}{2}\right)$. A straightforward application of chain rule shows that the derivative of μ is given by $X_{c\left(\frac{\epsilon}{2}\right)c\left(t-\frac{\epsilon}{2}\right)} = X_{\mu(t)}$. Therefore, μ and c agree on the overlap of their respective domain. Moving forward, we let

$$c_{\text{ext}}(t) = \begin{cases} c(t), \text{if}\, t \in (-\epsilon, \epsilon) \\ \mu(t), \text{if}\, t \in \left[\epsilon, \frac{3\epsilon}{2}\right) \end{cases}$$

then c_{ext} extends the integral curve c. By iterating this process indefinitely, we can extend c to $(0, \infty)$. Similar arguments applied to $\kappa(t) = c\left(\frac{\epsilon}{2}\right) c\left(t + \frac{\epsilon}{2}\right)$ show that c can be extended to $(-\infty, \infty)$ as well.

3.3 Exponential Map

Let G be a Lie group with Lie algebra $\mathfrak{g}$ and let $X_e = \mathfrak{g}$, where $\mathfrak{g}$ is identified with the tangent space of at the identity of G. Next, let X be a left invariant vector field on G, which is generated by X_e. Next, let c_X be the maximal integral curve of X through the identity element in G.

Definition 3.9 The exponential map is defined as $\exp : \mathfrak{g} \to G$ such that for $X_e \in \mathfrak{g}$, $\exp\left(X_e\right) = c_X(1)$.

Example 3.22 Let

$$G = \left\{ \begin{bmatrix} 1 & x & \frac{1}{2}x^2 \\ 0 & 1 & x \\ 0 & 0 & 1 \end{bmatrix} : x \in \mathbb{R} \right\}$$

be a commutative Lie group. Next, let $c : \mathbb{R} \to G$ be a smooth curve in G passing through the identity element such that

$$c(t) = \begin{bmatrix} 1 & t & \frac{1}{2}t^2 \\ 0 & 1 & t \\ 0 & 0 & 1 \end{bmatrix}.$$

Computing the derivative of c at zero, verify that

$$X_e = \lim_{h \to 0} \frac{c(h) - c(0)}{h} = \lim_{h \to 0} \begin{bmatrix} 0 & 1 & \frac{1}{2}h \\ 0 & 0 & 1 \\ 0 & 0 & 0 \end{bmatrix} = \begin{bmatrix} 0 & 1 & 0 \\ 0 & 0 & 1 \\ 0 & 0 & 0 \end{bmatrix}.$$

Therefore, $\{X_e\}$ is a basis for the Lie algebra of G, when viewed as an element of $T_e(G)$. Moreover, $\left(L_{c(t)}\right)_* X_e$ is a left-invariant vector field and c is a maximal integral curve of c through the identity element e. Applying the definition of the exponential function given in Definition 3.9, we obtain

$$\exp\left(X_e\right) = c(1) = \begin{bmatrix} 1 & 1 & \frac{1}{2} \\ 0 & 1 & 1 \\ 0 & 0 & 1 \end{bmatrix}.$$

Example 3.23 Let

$$G = \left\{ p\left(x_1, x_2\right) = \begin{bmatrix} e^{x_1} & e^{x_1} x_2 \\ 0 & e^{x_1} \end{bmatrix} : \left(x_1, x_2\right) \in \mathbb{R}^2 \right\}$$

be a two-dimensional noncommutative Lie group and let $c : \mathbb{R} \to G$ be a smooth curve where

$$c(t) = \begin{bmatrix} e^{ta_1} & e^{ta_1}\left(b_2 t\right) \\ 0 & e^{ta_1} \end{bmatrix}$$

and $c(0) = e$. Since $\lim_{h \to 0} \frac{1}{h}\left(e^{ha_1} - 1\right) = a_1$ and $\lim_{h \to 0} b_2 e^{ha_1} = b_2$, we verify that

$$\lim_{h \to 0} \frac{c(h) - c(0)}{h} = \lim_{h \to 0} \begin{bmatrix} \frac{1}{h}\left(e^{ha_1} - 1\right) & b_2 e^{ha_1} \\ 0 & \frac{1}{h}\left(e^{ha_1} - 1\right) \end{bmatrix} = \begin{bmatrix} a_1 & b_2 \\ 0 & a_1 \end{bmatrix}.$$

This shows that the tangent space at the identity element of G is the two-dimensional vector space spanned by matrices

$$A_e = \begin{bmatrix} 1 & 0 \\ 0 & 1 \end{bmatrix}, \quad X_e = \begin{bmatrix} 0 & 1 \\ 0 & 0 \end{bmatrix}.$$

Any left-invariant vector field on G takes the form $X = \left(L_{p(x_1, x_2)}\right)_* \left(a_1 A_e + b_1 X_e\right)$ and the maximal integral curve of X passing through the identity element is given by

$$c_{a_1, b_1}(t) = \begin{bmatrix} e^{ta_1} & e^{ta_1}\left(b_2 t\right) \\ 0 & e^{ta_1} \end{bmatrix}.$$

Appealing to the definition of the exponential function given in Definition 3.9, we obtain

$$\exp\left(a_1 A_e + b_1 X_e\right) = c_{a_1, b_1}(1) = \begin{bmatrix} e^{a_1} & e^{a_1} b_2 \\ 0 & e^{a_1} \end{bmatrix}.$$

Proposition 3.19 Let $X_e \in \mathfrak{g}$. Then for any real number t, $\exp\left(tX_e\right) = c_X(t)$.

Proof: First, given a real number r,

$$\frac{d}{dt}\left(c_X(rt)\right) = r\frac{d}{dt}\left(c_X(rt)\right) = rX_{c_X(rt)}$$

and as such, $t \mapsto c_X(rt)$ is an integral curve of rX passing through the identity of G. In other words, $c_X(rt) = c_{rX}(t)$ and $\exp\left(tX_e\right) = c_{tX}(1) = c_X(t)$ as desired.

Proposition 3.20 Given real numbers s, t and $X_e \in \mathfrak{g}$, the following holds true $\exp\left((s+t)X_e\right) = \left(\exp sX_e\right)\left(\exp tX_e\right)$.

Proof: By definition, we have $\exp\left((s+t)X_e\right) = c_X(s+t)$. Next,

$$\frac{d}{dt}\left(\left(\exp sX_e\right)\left(\exp tX_e\right)\right) = \frac{d}{dt}\left(L_{\exp sX_e}\left(\exp tX_e\right)\right)$$
$$= \left(L_{\exp sX_e}\right)_* X_{\exp tX_e}.$$

Since X is left-invariant, $\frac{d}{dt}\left(\left(\exp sX_e\right)\left(\exp tX_e\right)\right) = X_{(\exp sX_e)(\exp tX_e)}$. This shows that the map $t \mapsto \left(\exp sX_e\right)\left(\exp tX_e\right)$ is an integral curve of X passing through $\exp sX_e$. Next,

$$\frac{d}{dt}\left(\exp\left((s+t)X_e\right)\right) = \frac{d}{dt}\left(c_X(s+t)\right) = X_{c_X(s+t)}$$

and $t \mapsto \exp\left((s+t)X_e\right)$ is also an integral curve of X passing through $\exp sX_e$. By uniqueness, it follows that $\exp\left((s+t)X_e\right) = \left(\exp sX_e\right)\left(\exp tX_e\right)$.

Proposition 3.21 The exponential map is smooth.

Proof: Define a vector field W on $G \times \mathfrak{g}$ by $W_{(x,X_e)} = \left(X_x, 0\right) = \left(\left(L_x\right)_* X_e, 0\right)$. Next, it is easy to verify that the integral curve at x of the left-invariant vector field X is given by $t \mapsto x\exp\left(tX_e\right)$. Note also that an integral curve of W through the point $\left(x, X_e\right) \in G \times \mathfrak{g}$ is given by $t \mapsto c(t) = \left(x\exp\left(tX_e\right), X_e\right)$. Define $f : \mathbb{R} \times G \times \mathfrak{g} \to G \times \mathfrak{g}$ such that $f\left(t, x, X_e\right) = \left(x\exp\left(tX_e\right), X_e\right)$. Then f is the flow of the smooth vector field W. By the theory of ordinary differential equations, the flow of a smooth vector field is smooth and as such, f is a smooth function. Next, let $\pi : G \times \mathfrak{g} \to G$ such that $\pi\left(x, X_e\right) = x$. Since $\exp X_e = \left[\pi \circ f\right]\left(1, e, X_e\right)$, it follows that the exponential map is smooth since it can be described as a composition of smooth functions.

Proposition 3.22 The differential at zero of the exponential map is the identity map.

Proof: Given $X_e \in \mathfrak{g}$, and $\left.\frac{d}{dt}\right|_{t=0} \exp\left(tX_e\right) = \left.\frac{d}{dt}\right|_{t=0}\left(c_X(t)\right) = X_e$.

Proposition 3.23 For the general linear group, the exponential map $\exp : \mathfrak{gl}(n,\mathbb{R}) \to GL(n,\mathbb{R})$ is given by $\exp(Y) = \sum_{k=0}^{\infty} \frac{Y^k}{k!}$.

Proof: Let $\|\|: \mathfrak{gl}(n,\mathbb{R}) \to \mathbb{R}$ such that $\|Y\| = \left(\sum_{k=1}^{n}\sum_{j=1}^{n}\left|Y_{jk}\right|^2\right)^{1/2}$ is the Euclidean norm on $\mathfrak{gl}(n,\mathbb{R})$. Let $\|Y\|_{Op}$ be the operator norm of Y. Since all norms on a finite-dimensional vector space are equivalent, there exists a universal constant C such that $\|Y\| \leq C\|Y\|_{Op}$. As such,

$$\left\|\sum_{k=0}^{\infty}\frac{Y^k}{k!}\right\| \leq \sum_{k=0}^{\infty}\frac{1}{k!}\left\|Y^k\right\| \leq C\sum_{k=0}^{\infty}\frac{1}{k!}\|Y\|_{Op}^k = Ce^{\|Y\|}.$$

This shows that the infinite series $\sum_{k=0}^{\infty} \frac{Y^k}{k!}$ is absolutely convergent for any $Y \in \mathfrak{gl}(n, \mathbb{R})$. Since every absolutely convergent series in $\mathfrak{gl}(n, \mathbb{R})$ is also convergent, it follows that $\sum_{k=0}^{\infty} \frac{Y^k}{k!}$ converges in $\mathfrak{gl}(n, \mathbb{R})$. Next, let $c : \mathbb{R} \to \mathfrak{gl}(n, \mathbb{R})$ such that $c(t) = \sum_{k=0}^{\infty} \frac{t^k Y^k}{k!}$. Observe that $c(0)$ is the identity matrix and

$$c'(t) = \frac{d}{dt}\left(\sum_{k=0}^{\infty} \frac{t^k Y^k}{k!}\right) = \sum_{k=1}^{\infty} \frac{kt^{k-1}Y^k}{k!} = \sum_{k=1}^{\infty} \frac{kt^{k-1}Y^{k-1}Y}{k!}.$$

Since $k! = (k-1)!k$, it follows that

$$c'(t) = \sum_{k=1}^{\infty} \frac{t^{k-1}Y^{k-1}}{(k-1)!} Y = \left(\sum_{j=0}^{\infty} \frac{t^j Y^j}{j!}\right) Y = c(t)\, Y.$$

Next, let W be a left-invariant vector field on $GL(n, \mathbb{R})$, which is generated by some vector $Y \in \mathfrak{gl}(n, \mathbb{R})$ as follows: $W_Q = QY$. Then $c'(t) = W_{c(t)}$. This shows that $c(t)$ is an integral curve of the vector field W passing through the identity element of the general linear group. Appealing to Proposition 3.19, we obtain $\exp Y = c(1) = \sum_{k=0}^{\infty} \frac{Y^k}{k!}$.

In the result below, we will prove that a continuous homomorphism between two Lie groups is necessarily a Lie group homomorphism.

Theorem 3.5 Let H and G be Lie groups. Suppose $\psi : H \to G$ is a continuous group homomorphism. Then ψ is smooth. That is, ψ is a Lie group homomorphism.

The proof of Theorem 3.5 hinges on a Lemmas 3.7 and 3.8, which we present below.

Lemma 3.7 Let H and G be a Lie groups. Let $\psi : H \to G$ be a group homomorphism. If there exists an open set U of the identity element e_H in H and an open set V of the identity element e_G in G such that $\psi(U)$ is a subset of V and the restriction of ψ onto U is a smooth map, then ψ is smooth as well.

Proof: Let $h \in H$. Then hU is a neighborhood of h and $\psi(h)V$ is a neighborhood of $\psi(h)$. Since

$$hU \xrightarrow{L_{h^{-1}}} U \xrightarrow{\psi} V \xrightarrow{L_{\psi(h)}} \psi(h)\, V.$$

Given $x \in hU$,

$$\begin{aligned} L_{\psi(h)} \circ \psi \circ L_{h^{-1}}(x) &= \psi(h)\, \psi\left(h^{-1}x\right) \\ &= \psi(h)\, \psi\left(h^{-1}\right) \psi(x) \\ &= e_G \psi(x) \\ &= \psi(x). \end{aligned}$$

This shows that ψ is smooth in an open set around h. As a result, $\psi : H \to G$ is smooth.

Lemma 3.8 Every continuous homomorphism from the real line into G is smooth.

Proof: Let $\mathfrak{g}$ be the Lie algebra of G and let U be an open subset of $\mathfrak{g}$ around the zero vector in $\mathfrak{g}$ such that the restriction of the exponential map to U is a diffeomorphism (see Proposition 3.22). In other words, there exists an open subset V of G around the identity element in G such that $\exp|U : U \to V$ is a diffeomorphism. Without loss of generality, we may assume that U is star-like. That is, if X is an element of U, then tX belongs to U for all $t \in [0,1]$. Next, let $U_{1/2} = \frac{1}{2}U$ and let $V_{1/2} = \exp U_{1/2}$. We shall prove that every element in $V_{1/2}$ has a unique square root. In other words, for any $v \in V_{1/2}$, there exists a unique element $w \in V_{1/2}$ such that $v = w^2$. Let $X \in U_{1/2}$ such that $v = \exp X$. Next, let $w = \exp\left(\frac{1}{2}X\right)$. Clearly, w belongs to $V_{1/2}$ and $w^2 = \exp\left(\frac{1}{2}X\right)\exp\left(\frac{1}{2}X\right) = \exp(X) = v$. To prove the uniqueness of square root, let $u \in V_{1/2}$ such that $u^2 = v$. In other words, there exists X_u in $U_{1/2}$ such that $u = \exp X_u$. Moreover, $u^2 = \exp\left(2X_u\right) = v = \exp(X)$. Since $\exp|U : U \to V$ is an injective map, $2X_u = X$. Consequently, $X_u = \frac{1}{2}X$ and this shows that $u = w$. Let $\psi : \mathbb{R} \to G$ be a continuous homomorphism. Then there exists a positive real number ϵ such that $\psi(t)$ belongs to $V_{1/2}$ for all $t \in [-\epsilon, \epsilon]$. Put $\psi(\epsilon) = \exp X$. By taking successive square roots, for any nonnegative integer, we have $\psi\left(\frac{\epsilon}{2^k}\right) = \exp\left(\frac{1}{2^k}X\right)$. Next, given $0 \le m \le 2^k$,

$$\exp\left(\frac{m}{2^k}X\right) = \left(\exp\left(\frac{1}{2^k}X\right)\right)^m = \left(\psi\left(\frac{\epsilon}{2^k}\right)\right)^m = \psi\left(\frac{m\epsilon}{2^k}\right).$$

Since $t \mapsto \exp tX$ and $t \mapsto \psi(t\epsilon)$ are continuous and agree on a dense subset of $[0,1]$, it follows that $\exp tX = \psi(t\epsilon)$ for all $t \in [0,1]$. As a result, $\psi(t) = \exp\left(\frac{t}{\epsilon}X\right)$ for all $t \in [-\epsilon, \epsilon]$. This shows that ψ is smooth on an open set of zero, and by Lemma 3.7, ψ is smooth.

Proof of Theorem 3.5 First, let $X_1, X_2, \dots, X_n$ be a basis of the Lie algebra $\mathfrak{h}$. Define $\alpha : \mathbb{R}^n \to H$ by

$$\alpha\left(t_1, \dots, t_n\right) = \exp\left(t_1X_1\right) \cdots \exp\left(t_nX_n\right).$$

Then clearly, α is a smooth function. Moreover, the differential map of α is a nonsingular at zero and by the inverse function theorem, there exists a neighborhood W of zero in $\mathbb{R}^n$ and U of the identity element in H such that the restriction of α to W defines a diffeomorphism between W and U.

Suppose $\psi : H \to G$ is a continuous group homomorphism. Since each $t \mapsto \psi\left(\exp tX_k\right)$ is a continuous homomorphism, according to Lemma 3.8, $t \mapsto \psi\left(\exp tX_k\right)$ is smooth as well. Consider the map

$$F : \left(t_1, \dots, t_n\right) \mapsto \psi\left(\exp\left(t_1 X_1\right)\right) \cdots \psi\left(\exp\left(t_n X_n\right)\right).$$

Since

$$\begin{aligned}\psi\left(\exp\left(t_1 X_1\right)\right) \cdots \psi\left(\exp\left(t_n X_n\right)\right) &= \psi\left(\exp\left(t_1 X_1\right) \cdots \exp\left(t_n X_n\right)\right) \\ &= \psi\left(\alpha\left(t_1, \dots, t_n\right)\right) \\ &= [\psi \circ \alpha]\left(t_1, \dots, t_n\right),\end{aligned}$$

it follows that F is smooth. Consequently $\psi = F \circ (\alpha|W)^{-1}$ is smooth on U and by Lemma 3.7, ψ is smooth.

Proposition 3.24 Two Lie groups that are isomorphic as topological groups are isomorphic as Lie groups.

Proof: Given a bi-continuous group isomorphism between two topological groups, a straightforward application of Theorem 3.5 shows that such a map must necessarily be a Lie group isomorphism as well.

3.4 Invariant Measure on Lie Groups

As a motivational example, we consider the noncommutative Lie group $G = \mathbb{R} \times \mathbb{R}$ endowed with the operation

$$(x, t) \bullet (y, s) = \left(x + e^t y, t + s\right).$$

Next, let f be a compactly supported and smooth function defined on $\mathbb{R} \times \mathbb{R}$. Then for a fixed point $(x, t) \in \mathbb{R} \times \mathbb{R}$ and a smooth function $\tau : \mathbb{R} \times \mathbb{R} \to \mathbb{R}$

$$I_W(f) = \int_{\mathbb{R}} \int_{\mathbb{R}} f((x, t) \bullet (y, s))\, \iota\, (y, s)\, dy\, ds = \int_{\mathbb{R}} \int_{\mathbb{R}} f\left(\left(x + e^t y, t + s\right)\right) \tau\, (y, s)\, dy\, ds.$$

Put $F_{(x,t)}(y, s) = \left(x + e^t y, t + s\right)$. Letting $z = (y, s)$,

$$I_\tau(f) = \int_{\mathbb{R}^2} f\left(F_{(x,t)}(z)\right) \tau(z)\, dz.$$

The change of variables $\xi = F_{(x,t)}(z)$ gives

$$\begin{aligned}I_\tau(f) &= \int_{\mathbb{R}^2} f(\xi)\, \tau\left(F^{-1}_{(x,t)}(\xi)\right) d\left(F^{-1}_{(x,t)}(\xi)\right) \\ &= \int_{\mathbb{R}^2} f(\xi)\, \tau\left(F^{-1}_{(x,t)}(\xi)\right) e^{-t}\, d\xi \\ &= \int_{\mathbb{R}^2} f(\xi)\, \tau\left(-e^{-t}\left(x - \xi_1\right), -t + \xi_2\right) e^{-t}\, d\xi.\end{aligned}$$

We wish to find a smooth function τ such that $I_\tau(f) = I_\tau(f)$. That is,

$$\tau\left(-e^{-t}\left(x-\xi_1\right), -t+\xi_2\right)e^{-t} = \tau\left(\xi_1, \xi_2\right)$$

for all $\left(\xi_1, \xi_2\right) \in \mathbb{R}^2$. If we were to define $\tau(y, s) = e^{-s}$ then

$$\tau\left(-e^{-t}\left(x-\xi_1\right), -t+\xi_2\right)e^{-t} = e^{-\xi_2+t}e^{-t} = e^{-\xi_2} = \tau\left(\xi_1, \xi_2\right).$$

By fixing τ as defined above, $I_\tau(f) = I_\tau(f)$ for all smooth nonnegative functions f of compact support. Therefore, the weighted Lebesgue measure $e^{-s}dy\,ds$ defines a left-invariant measure on the Lie group G.

Our main objective in this section is to generalize these calculations for arbitrary Lie groups. In fact, we will prove that up to multiplication by a constant, every Lie group admits a unique left-invariant measure.

Definition 3.10 Let M be an m-dimensional manifold. We say that M is an oriented manifold if it is equipped with an atlas for its smooth structure such that the matrix representation of the derivative of any of its change of coordinates map has a determinant that is positive everywhere.

For a smooth map $f : M \to N$, and an r-form ω, the pullback of ω by f is the alternating r form $f^*(\omega) \in \Lambda^r(M)$ defined as follows:

$$f^*(\omega)_p\left(X_1, \dots, X_r\right) = \omega_{f(p)}\left(f_*X_1, \dots, f_*X_r\right)$$

for $p \in M$ and $X_1, \dots, X_r \in T_pM$.

Proposition 3.25 The following holds true

1) Let $f : M \to N$ where M, N are open subsets of $\mathbb{R}^m$ and

$$f\left(y_1, \dots, y_m\right) = \left(f_1\left(y_1, \dots, y_m\right), \dots, f_m\left(y_1, \dots, y_m\right)\right).$$

   If $\varphi = dx_k$ is a differential 1-form on N then

$$f^*(\varphi) = \sum_{j=1}^{m} \frac{\partial f_k}{\partial y_j}\left(y_1, \dots, y_m\right) dy_j.$$

2) Let $f : M \to N$ where M, N are open subsets of $\mathbb{R}^m$ and

$$f\left(y_1, \dots, y_m\right) = \left(f_1\left(y_1, \dots, y_m\right), \dots, f_m\left(y_1, \dots, y_m\right)\right).$$

   If $\varphi = \sum_{k=1}^{m} a_k\left(x_1, \dots, x_m\right) dx_k$ is a differential 1-form on N then

$$f^*(\varphi) = \sum_{j=1}^{m}\left(\sum_{k=1}^{m}\left(a_k f\left(x_1, \dots, x_m\right)\right) \frac{\partial f_k}{\partial y_j}\left(y_1, \dots, y_m\right)\right) dy_j.$$

Proof: To prove (1), let $p \in M$. Since $f^*(\varphi) = \sum_{j=1}^{m} b_j(y_1, \dots, y_m)\, dy_j$ and

$$b_j(y_1, \dots, y_m) = f^*(\varphi) \left.\frac{\partial}{\partial y_j}\right|_p = dx_k \left(f_* \left.\frac{\partial}{\partial y_j}\right|_p \right),$$

the element $f_* \left.\frac{\partial}{\partial y_j}\right|_p$ can be identified with the tangent vector

$$\begin{bmatrix} \dfrac{\partial f_1(p)}{\partial y_j} \\ \vdots \\ \dfrac{\partial f_m(p)}{\partial y_j} \end{bmatrix}_{f(p)} \in T_{f(p)}(N)$$

and $dx_k \left(f_* \left.\frac{\partial}{\partial y_j}\right|_p \right) = \frac{\partial f_k(p)}{\partial y_j}$. Therefore,

$$f^*(\varphi) = \sum_{j=1}^{m} b_j(y_1, \dots, y_m)\, dy_j = \sum_{j=1}^{m} \frac{\partial f_1(y_1, \dots, y_m)}{\partial y_j} dy_j.$$

For the second part of the proof

$$\begin{aligned} f^*(\varphi) &= f^*\left(\sum_{k=1}^{m} a_k(x_1, \dots, x_m)\, dx_k \right) \\ &= \sum_{k=1}^{m} (a_k f(x_1, \dots, x_m))\, f^*(dx_k) \\ &= \sum_{k=1}^{m} (a_k f(x_1, \dots, x_m)) \sum_{j=1}^{m} \frac{\partial f_k}{\partial y_j}(y_1, \dots, y_m)\, dy_j \\ &= \sum_{j=1}^{m} \left(\sum_{k=1}^{m} (a_k f(x_1, \dots, x_m)) \frac{\partial f_k}{\partial y_j}(y_1, \dots, y_m) \right) dy_j. \end{aligned}$$

In light of Propositions 3.25 and (2.11), the following is immediate.

Proposition 3.26 Let $f : M \to N$ be a smooth map where M, N are open subsets of $\mathbb{R}^m$ and

$$f(y_1, \dots, y_m) = (f_1(y_1, \dots, y_m), \dots, f_m(y_1, \dots, y_m)).$$

If $\varphi = F(x_1, \dots, x_m)\, dx_1 \wedge \cdots \wedge dx_m$ is a differential m-form on N then

$$f^*(\varphi) = \left((F \circ f) \cdot \det \begin{bmatrix} \dfrac{\partial f_1}{\partial y_1} & \cdots & \dfrac{\partial f_1}{\partial y_m} \\ \vdots & \ddots & \vdots \\ \dfrac{\partial f_m}{\partial y_1} & \cdots & \dfrac{\partial f_m}{\partial y_m} \end{bmatrix} \right) dy_1 \wedge \cdots \wedge dy_m.$$

Let us assume that M is an m-dimensional manifold and let T be an m differential form on M. Letting (U, φ) be a chart on M, we may write

$$\left(\varphi^{-1}\right)^{*}(\omega) = F\left(x_1, \dots, x_m\right) dx_1 \wedge \cdots \wedge dx_m,$$

where F is a smooth function defined on $\varphi(U)$.

We say that $F\left(x_1, \dots, x_m\right) dx_1 \wedge \cdots \wedge dx_m$ is the representation of ω in local coordinates.

Definition 3.11 Let $u : M \to \mathbb{R}$ be a smooth function which is compactly supported on U. Then

$$\int_M u\omega = \int_{\varphi(U)} \left[u \circ \varphi^{-1}\right]\left(x_1, \dots, x_m\right) F\left(x_1, \dots, x_m\right) dx_1 \wedge \cdots \wedge dx_m.$$

Proposition 3.27 $\int_M u\omega$ is well defined for every smooth and compactly supported function u.

Proof: Let (V, ψ) be a smooth chart defined on M and assume that $U \cap V$ is nonempty and that f is compactly supported on $U \cap V$. Letting

$$\left(\psi^{-1}\right)^{*}(\omega) = F_V\left(y_1, \dots, y_m\right) dy_1 \wedge \cdots \wedge dy_m,$$

the change of variable $\left(y_1, \dots, y_m\right) = \left[\psi \circ \varphi^{-1}\right]\left(x_1, \dots, x_m\right)$ gives

$$\begin{aligned}
&\int_{\psi(U\cap V)} \left[u \circ \psi^{-1}\right]\left(y_1, \dots, y_m\right) F_V\left(y_1, \dots, y_m\right) dy_1 \wedge \cdots \wedge dy_m \\
&= \int_{\psi(U\cap V)} \left[u \circ \psi^{-1} \circ \psi \circ \varphi^{-1}\right]\left(x_1, \dots, x_m\right) \left[F_V \circ \psi \circ \varphi^{-1}\right]\left(x_1, \dots, x_m\right) \\
&\quad \left|\det\left(\left(\psi \circ \varphi^{-1}\right)_*\right)\right| dx_1 \wedge \cdots \wedge dx_m.
\end{aligned}$$

Noting that

$$\begin{aligned}
F\left(dx_1 \wedge \cdots \wedge dx_m\right) &= \left(\varphi^{-1}\right)^{*}(\omega) \\
&= \left(\psi^{-1} \circ \psi \circ \varphi^{-1}\right)^{*}(\omega) \\
&= \left(\psi \circ \varphi^{-1}\right)^{*}\left(\psi^{-1}\right)^{*}(\omega) \\
&= \left(\psi \circ \varphi^{-1}\right)^{*}\left(F_V dy_1 \wedge \cdots \wedge dy_m\right) \\
&= \left(F_V \circ \psi \circ \varphi^{-1} \left|\det\left(\left(\psi \circ \varphi^{-1}\right)_*\right)\right|\right) dx_1 \wedge \cdots \wedge dx_m,
\end{aligned}$$

it follows that

$$F = F_V \circ \psi \circ \varphi^{-1} \left|\det\left(\left(\psi \circ \varphi^{-1}\right)_*\right)\right|.$$

As a result,

$$\begin{aligned}&\int_{\psi(U\cap V)} \left[u \circ \psi^{-1}\right](y_1, \dots, y_m)\, F_V(y_1, \dots, y_m)\, dy_1 \wedge \cdots \wedge dy_m \\ &= \int_{\psi(U\cap V)} \left[u \circ \varphi^{-1}\right](x_1, \dots, x_m)\, F(x_1, \dots, x_m)\, dx_1 \wedge \cdots \wedge dx_m \\ &= \int_M u\omega.\end{aligned}$$

This shows that $\int_M u\omega$ is well defined.

Let $(\psi_\alpha)_{\alpha\in I}$ be a partition of unity subordinated to $((U_\alpha, \varphi_\alpha))_{\alpha\in I}$ such that each ψ_α is compactly supported in U_α and only finitely many ψ_α are nonzero on each set. Then given any real-valued function u, we have

$$\int_M u\omega = \sum_{\alpha\in I} \int_M (u \cdot \psi_\alpha)\, \omega.$$

Proposition 3.28 The definition of $\int_M uT$ does not depend on the choice of oriented charts or partition of unity.

Proof: Let $(\psi'_\alpha)_{\alpha\in I}, (\psi_\alpha)_{\alpha\in I}$ be partitions of unity subordinated to $((U_\alpha, \varphi_\alpha))_{\alpha\in I}$ such that each $\psi_\alpha, \psi'_\alpha$ is compactly supported in U_α and only finite many $\psi_\alpha, \psi'_\alpha$ are nonzero on each set. Since $\sum_{\beta\in I} \psi'_\beta \equiv 1$,

$$\begin{aligned}\int_M (u \cdot \psi_\alpha)\, \omega &= \int_M \left(\sum_{\beta\in I} \psi'_\beta\right) (u \cdot \psi_\alpha)\, \omega \\ &= \sum_{\beta\in I} \int_M \psi'_\beta \cdot (u \cdot \psi_\alpha)\, \omega.\end{aligned}$$

Next, summing over $\alpha \in I$,

$$\begin{aligned}\sum_{\alpha\in I} \int_M (u \cdot \psi_\alpha)\, \omega &= \sum_{\alpha\in I} \left(\sum_{\beta\in I} \int_M \psi'_\beta (u \cdot \psi_\alpha)\, \omega\right) \\ &= \sum_{\alpha\in I} \sum_{\beta\in I} \int_M \psi'_\beta \cdot u \cdot \psi_\alpha \omega.\end{aligned}$$

Similarly,

$$\begin{aligned}\sum_{\beta\in I} \int_M \left(u \cdot \psi'_\beta\right) \omega &= \sum_{\beta\in I} \int_M \left(\sum_{\alpha\in I} \psi_\alpha\right) \left(u \cdot \psi'_\beta\right) \omega \\ &= \sum_{\beta\in I} \sum_{\alpha\in I} \int_M \psi_\alpha \left(u \cdot \psi'_\beta\right) \omega \\ &= \int_M (u \cdot \psi_\alpha)\, \omega\end{aligned}$$

Let ω be a differential n-form. Then the mapping $u \mapsto \int_M u\omega$ is a linear functional on the space of a compactly supported smooth function.

Definition 3.12 We say that the form ω is positive with respect to an atlas $\mathcal{A}$ if for any chart $(U, \varphi) \in \mathcal{A}$ on M, such that

$$\left(\varphi^{-1}\right)^* (\omega) = F_{(U,\varphi)} \left(x_1, \dots, x_m\right) dx_1 \wedge \dots \wedge dx_m$$

then $F_{(U,\varphi)}$ is a smooth function that is also positive on $\varphi(U)$.

Proposition 3.29 Let M be an m-dimensional smooth manifold. If M admits a nowhere nonvanishing m-form ω then there exists an atlas on M with respect to which ω is positive.

Proof: Let $\left(\left(U_\alpha, \varphi_\alpha\right)\right)_{\alpha \in I}$ be an atlas for the manifold M. Without loss of generality, we may assume that every open set U_α is connected. Next, let

$$\left(\varphi_\alpha^{-1}\right)^* (\omega) = \left(F_{\left(U_\alpha, \varphi_\alpha\right)} \left(x_1, \dots, x_m\right)\right) dx_1 \wedge \dots \wedge dx_m.$$

Since it is assumed that ω is nonvanishing, and because every set U_α is connected, then $F_{\left(U_\alpha, \varphi_\alpha\right)}$ has a constant sign. That is, either $F_{\left(U_\alpha, \varphi_\alpha\right)} \left(x_1, \dots, x_m\right) > 0$ for all $\left(x_1, \dots, x_m\right) \in \varphi_\alpha \left(U_\alpha\right)$ or $F_\alpha \left(x_1, \dots, x_m\right) < 0$ for all $\left(x_1, \dots, x_m\right) \in \varphi_\alpha \left(U_\alpha\right)$. Let

$$J = \left\{\alpha \in I : F_{\left(U_\alpha, \varphi_\alpha\right)} \left(x_1, \dots, x_m\right) < 0 \text{ for all } \left(x_1, \dots, x_m\right) \in \varphi_\alpha \left(U_\alpha\right)\right\}.$$

Then for each $\alpha \in J$, we may replace φ_α with the function

$$\left(\left(x_1, \dots, x_m\right) \mapsto \left(-x_1, \dots, x_m\right)\right) \circ \varphi_\alpha$$

to ensure that $F_{\left(U_\alpha, \varphi_\alpha\right)} \left(x_1, \dots, x_m\right) > 0$ for all $\left(x_1, \dots, x_m\right) \in \varphi_\alpha \left(U_\alpha\right)$.

Proposition 3.30 Let M be an m-dimensional smooth and connected manifold which is positively oriented. If ω is a differential m-form which is nonvanishing on M, then either ω is positive or $-\omega$ is positive.

Proof: Let $\left(\left(U_\alpha, \varphi_\alpha\right)\right)_{\alpha \in I}$ be an atlas for the manifold M and let

$$A = \left\{x \in M : F_{\left(U_\alpha, \varphi_\alpha\right)} \left(\varphi_\alpha (x)\right) > 0, \text{if} x \in U_\alpha\right\},$$

where $\left(\varphi_\alpha^{-1}\right)^* (\omega) = F_{\left(U_\alpha, \varphi_\alpha\right)} \left(x_1, \dots, x_m\right) dx_1 \wedge \dots \wedge dx_m$. Without loss of generality, we may assume that the set A is nonempty. Otherwise, we may simply replace ω by its opposite. Since by assumption M is connected, to establish the statement of the proposition, it is sufficient to prove that A is both open and closed in M. **Step 1**: Showing that A is open. Letting a be a point in A contained in U_α such that $\left(U_\alpha, \varphi_\alpha\right)$ is an element of the atlas on M, by assumption, we know that $F_{\left(U_\alpha, \varphi_\alpha\right)} \left(\varphi_\alpha (a)\right) > 0$. Since $F_\alpha \circ \varphi_\alpha$ is continuous, it follows that $F_\alpha \circ \varphi_\alpha$ is positive on a neighborhood of a. This proves that A is indeed open. **Step 2**: Showing that A is closed. Let us assume that $\left(a_n\right)_{n \in \mathbb{N}}$ is a sequence of points which is converging to a in M.

Suppose that a is an element of U_α for some $\alpha \in I$. For each natural number $n \in \mathbb{N}$, we have $F_{(U_\alpha,\varphi_\alpha)}\left(\varphi_\alpha\left(a_n\right)\right) > 0$ and $F_{(U_\alpha,\varphi_\alpha)}\left(\varphi_\alpha(a)\right)$ is nonzero. By continuity, $\lim_{n\to\infty} F_{(U_\alpha,\varphi_\alpha)}\left(\varphi_\alpha\left(a_n\right)\right) = F_{(U_\alpha,\varphi_\alpha)}\left(\varphi_\alpha(a)\right) > 0$. This shows that $a \in A$ and consequently, A is closed. Finally, since by assumption M is connected, the only nonempty subset of M which is both open and closed in M is M itself. Therefore, $A = M$.

Let M be an m-dimensional smooth, which is positively oriented, and let us assume that ω is a positive m-form and let $\left\{\psi_\alpha : \alpha \in I\right\}$ be a partition of unity such that each ψ_α is supported in U_α. Then

$$f = \sum_{\alpha\in I} \psi_\alpha f.$$

Next, define a measure μ_ω associated with ω such that if f is a smooth and compactly supported function defined on M, then

$$\int_M (f\cdot\omega)\, d\mu_\omega = \sum_{\alpha\in I}\int_{U_\alpha}\left(\psi_\alpha\cdot f\right)\omega.$$

Proposition 3.31 Let M, N be two m-dimensional manifolds that are positively oriented. Let $f : M \to N$ be a diffeomorphism such that for every smooth chart $\left(U_\lambda, \varphi_\lambda\right)$ in an atlas $\left(\left(U_\alpha, \varphi_\alpha\right)\right)_{\alpha\in I}$ for M, the chart $\left(f\left(U_\lambda\right), \varphi_\lambda\circ f^{-1}\right)$ is positive relative to every element in $\left(\left(U_\alpha, \varphi_\alpha\right)\right)_{\alpha\in I}$ in the sense that each smooth map $\varphi_\lambda\circ\varphi_\alpha^{-1}$ has a positive determinant. If ω is an n-form, then for every function u, which is smooth and compactly supported on N,

$$\int_N u\cdot\omega = \int_M (u\circ f)\cdot f^*\omega.$$

Proof: We shall equip M and N with charts $\left(\left(U_\alpha, \varphi_\alpha\right)\right)_{\alpha\in I}$ and $\left(\left(f\left(U_\alpha\right), \varphi_\alpha\circ f^{-1}\right)\right)_{\alpha\in I}$ respectively. Next, let u be a function, which is smooth and compactly supported on U_α. On the one hand, we have

$$\int_N u\cdot\omega = \int_{[\varphi_\alpha\circ f^{-1}](f(U_\alpha))}\left[u\circ f\circ\varphi_\alpha^{-1}\right]\left(x_1,\ldots,x_m\right)F_\alpha\left(x_1,\ldots,x_m\right)dx_1\wedge\cdots\wedge dx_m,$$

where F_α is a smooth function defined as follows:

$$\left(\left(\varphi_\alpha\circ f^{-1}\right)^{-1}\right)^*(\omega) = F_\alpha\left(x_1,\ldots,x_m\right)dx_1\wedge\cdots\wedge dx_m.$$

One the other hand, since $\left(\left(\varphi_\alpha\circ f^{-1}\right)^{-1}\right)^* = \left(f\circ\varphi_\alpha^{-1}\right)^* = \left(\varphi_\alpha^{-1}\right)^* f^*$, it follows that

$$\left(\varphi_\alpha^{-1}\right)^*\left(f^*\omega\right) = \left(\left(\varphi_\alpha\circ f^{-1}\right)^{-1}\right)^*\omega = F_\alpha\left(x_1,\ldots,x_m\right)dx_1\wedge\cdots\wedge dx_m$$

and consequently,

$$\int_M (u \circ f) \cdot f^* \omega = \int_{\varphi_\alpha(U_\alpha)} \left[u \circ f \circ \varphi_\alpha\right]^{-1} (x_1, \dots, x_m)\, F_\alpha (x_1, \dots, x_m)\, dx_1 \wedge \cdots \wedge dx_m$$
$$= \int_N u \cdot \omega.$$

By choosing a suitable partition of unity on M, the results above can be extended to arbitrary compactly supported and smooth functions defined on N.

Definition 3.13 Let G be a Lie group with Lie algebra $\mathfrak{g}$. A nonzero Borel measure μ on G is called a left Haar measure (or left-invariant measure) if μ is finite on compact sets of G, and $\mu(gE) = \mu(E)$ for all $g \in G$ and all Borel subsets g of G.

Definition 3.14 We say that ω is a left-invariant form if $\omega = L_g^*(\omega)$ for all $g \in G$ where $L_g : g_1 \mapsto gg_1$ is a diffeomorphism on G. Similarly, we say that ω is a right-invariant form if $\omega = R_g^*(\omega)$ for all $g \in G$ where $R_g : g_1 \mapsto g_1 g$ is a diffeomorphism on G.

Letting $X_1, \dots, X_n$ be left-invariant vector fields on G, we define $\ell_1, \dots, \ell_n$ to be 1-forms satisfying

$$\left(\ell_j\right)_g \left(X_k\right)_g = \begin{cases} 1, & \text{if } j = k, \\ 0, & \text{Otherwise.} \end{cases} \tag{3.4}$$

In other words, for each fixed $g \in G$, $\left\{ \left(\ell_1\right)_g, \dots, \left(\ell_n\right)_g \right\}$ is a canonical dual basis for $\left\{ \left(X_1\right)_g, \dots, \left(X_n\right)_g \right\}$.

Proposition 3.32 If $\ell = \ell_1 \wedge \cdots \wedge \ell_n \in T^n(G)$ then T is left-invariant form.

Proof: Appealing to Proposition 3.30, we may choose an orientation of the Lie group G such that ℓ is positive. Next, for $g, h \in G$, the pullback of ℓ by L_g is given by

$$\left[L_g^*\right](\ell)_h \left(\left(X_1\right)_h, \dots, \left(X_m\right)_h\right) = \ell_{gh} \left(\left(L_g\right)_* \left(X_1\right)_h, \dots, \left(L_g\right)_* \left(X_m\right)_h\right)$$
$$= \ell_{gh} \left(\left(X_1\right)_{gh}, \dots, \left(X_m\right)_{gh}\right).$$

This shows that $\ell = L_a^*(\ell)$ is a nonvanishing and left-invariant n-form.

Proposition 3.33 Let G be a Lie group and let μ_ℓ be a measure on G associated with ℓ as defined in (3.4). Then μ_ℓ is a left Haar measure on G.

Proof: Let f be a smooth function which is compactly supported on G. Given $a \in G$,

$$\int_G f(ax)\, d\mu_\ell(x) = \int_G (f \circ L_a)\, \ell$$
$$= \int_G (f \circ L_a)\, (L_{a^{-1}} \circ L_a)^* \ell$$
$$= \int_G (f \circ L_a)\, L_a^* \left(L_{a^{-1}}^* \ell\right).$$

Since ℓ is left-invariant, $L_{a^{-1}}^* \ell = \ell$ and as a result,

$$\int_G f(ax)\, d\mu_\ell(x) = \int_G (f \circ L_a)\, L_a^* \ell.$$

Appealing to Proposition 3.31, we obtain

$$\int_G (f \circ L_a)\, L_a^* \ell = \int_G f \cdot \ell = \int_G f(x)\, d\mu_\ell(x)$$

as desired.

Proposition 3.34 Let G be a Lie group and let μ, μ' be two left Haar measures defined on G. Then there exists a positive real constant c such that $\mu = c\mu'$.

Proof: Let $\mu'' = \mu + \mu'$. Appealing to Radon–Nikodym's theorem, there exists a nonnegative function z such that $\mu = z\mu''$. Then for every positive Borel function b,

$$\int_G b(x)\, z\left(a^{-1}x\right) d\mu''(x) = \int_G b(ax)\, z(x)\, d\mu''(ax).$$

Since the sum of two left-invariant measures is also left-invariant, it follows that

$$\int_G b(x)\, z\left(a^{-1}x\right) d\mu''(x) = \int_G b(ax)\, z(x)\, d\mu''(x)$$
$$= \int_G b(ax)\, d\mu(x)$$
$$= \int_G b(x)\, d\mu\left(a^{-1}x\right).$$

Using the left invariance of μ, it follows that

$$\int_G b(x)\, z\left(a^{-1}x\right) d\mu''(x) = \int_G b(x)\, d\mu(x) = \int_G b(x)\, z(x)\, d\mu''(x).$$

In summary,

$$\int_G b(x)\, z\left(a^{-1}x\right) d\mu''(x) = \int_G b(x)\, z(x)\, d\mu''(x).$$

This shows that $z\left(a^{-1}x\right) d\mu''(x) = z(x)\, d\mu''(x)$ for almost every x in G. Since A is arbitrarily selected, it follows that z must be constant almost everywhere.

This shows that μ and μ'' are proportional and consequently, μ and μ' are proportional as well.

Fix a left Haar measure μ on a Lie group G. For a fixed element a in G, let us define μ_a such that $\mu_a(E) = \mu(Ea)$. Then

$$\begin{aligned}\mu_a(bE) = \mu(bEa) &= \int_G 1_{bEa}(x)\, d\mu(x)\\ &= \int_G 1_{Ea}(b^{-1}x)\, d\mu(x)\\ &= \int_G 1_{Ea}(x)\, d\mu(bx)\\ &= \int_G 1_{Ea}(x)\, d\mu(x)\\ &= \mu_a(E).\end{aligned}$$

This shows that μ_a is left-invariant.

Next, let us define $D : G \to (0, \infty)$ such that $\mu_a = D(a)^{-1}\mu$.

Proposition 3.35 For every $g \in G$, $D(g) = |\det Ad(g)|$.

Proof: Let X be an element in the Lie algebra of G and let $\tilde{X}$ be a left-invariant vector field on G such that $\tilde{X}_e = X$. Letting q, p be the elements in G,

$$\begin{aligned}\left(R(Q^{-1})_*\right)_p \tilde{X}_p\phi = \tilde{X}_p\left(\phi \circ R\left(q^{-1}\right)\right) &= \frac{d}{dt}\left(\phi(p\exp(tX))q^{-1}\right)\Big|_{t=0}\\ &= \frac{d}{dt}\left(\phi\left(pq^{-1}q\exp(tX)\right)q^{-1}\right)\Big|_{t=0}\\ &= \frac{d}{dt}\left(\phi\left(pq^{-1}\exp\left(tqXq^{-1}\right)\right)\right)\Big|_{t=0}\\ &= \frac{d}{dt}\left(\phi\left(pq^{-1}\exp(tAd(q)X)\right)\right)\Big|_{t=0}.\end{aligned}$$

Letting $\widetilde{Ad(q)X}$ be a left-invariant vector field on G such that $\left(\widetilde{Ad(q)X}\right)_e = Ad(q)X$, we obtain

$$\left(R(Q^{-1})_*\right)_P \tilde{X}_P\phi = \left(\widetilde{Ad(Q)X}\right)_{PQ^{-1}}$$

and

$$\begin{aligned}&\left(R(q^{-1})_*T\right)_p\left(\left(\tilde{X}_1\right)_p, \dots, \left(\tilde{X}_n\right)_p\right)\\ &= T_{pq^{-1}}\left(\left(R(q^{-1})_*\right)_p\left(\tilde{X}_1\right)_p, \dots, \left(R(q^{-1})_*\right)_p\left(\tilde{X}_n\right)_p\right)\end{aligned}$$

$$= T_{pq^{-1}}\left(\left(\widetilde{Ad(q)X_1}\right)_{pq^{-1}}, \dots, \left(\widetilde{Ad(q)X_n}\right)_{pq^{-1}}\right)$$
$$= \det(Ad(q))\, T_{pq^{-1}}\left(\left(\tilde{X}_1\right)_{pq^{-1}}, \dots, \left(\tilde{X}_n\right)_{pq^{-1}}\right).$$

This proves that $R\left(q^{-1}\right)_* T = \det(Ad(q))\, T$. Next, if $\det(Ad(q))$ is positive then

$$\det(Ad(q)) \int_G uT = \int_G uR\left(q^{-1}\right)_* T$$
$$= \int_G \left(u \circ R(q) \circ R\left(q^{-1}\right)\right) R\left(q^{-1}\right)_* T.$$

By Proposition 3.31,

$$\int_G \left(u \circ R(q) \circ R\left(q^{-1}\right)\right) R\left(q^{-1}\right)_* T = \int_G (u \circ R(q))\, T$$
$$= D(A) \int_G uT.$$

Otherwise if $\det(Ad(q))$ is not positive,

$$-\det(Ad(q)) \int_G uT = \int_G u\left(-R\left(q^{-1}\right)_* T\right) = D(q) \int_G uT.$$

This shows that $D(q)$ is equal to the absolute value of $\det(Ad(q))$.

Definition 3.15 Given a Lie group G, define the modular function of G as follows: $\Delta : G \to (0, \infty)$ such that $\Delta(g) = D(g)^{-1} = |\det Ad(g)|^{-1}$.

Proposition 3.36 Let G be a Lie group with Lie algebra $\mathfrak{g}$. Let μ_G be a fixed Haar measure defined on G and let dX be the Lebesgue measure given by identifying $\mathfrak{g}$ with $\mathbb{R}^n$. Suppose that the exponential map defines a diffeomorphism between a neighborhood $\mathfrak{o}$ of the zero vector in $\mathfrak{g}$ and some open subset O around the identity e in G. Then there exists a constant $C > 0$ such that

$$\int_G f(x)\, d\mu_G(x) = C \int_{\mathfrak{g}} f(\exp X)\, \Theta(X)\, dX$$

for all $f \in C^\infty(G)$ supported in O where Θ is a smooth and positive function satisfying $\ell_{\exp X}\left((D\exp)_X X_1, \dots, (D\exp)_X X_n\right) = \Theta(X)\, \ell_e\left(X_1, \dots, X_n\right)$, where $(D\exp)_X$ stands for the differential of the exponential function at X.

Proof: Let ℓ be a left-invariant differential n-form on G. Then its pullback through exp is $(\exp^* \ell)_X$ where

$$(\exp^* \ell)_X\left(X_1, \dots, X_n\right) = \ell_{\exp X}\left((D\exp)_X X_1, \dots, (D\exp)_X X_n\right)$$
$$= \Theta(X)\, \ell_e\left(X_1, \dots, X_n\right).$$

3.5 Homogeneous Spaces

Let G be a Lie group, let H be a closed subgroup of G and let π be the canonical quotient map which sends every element g in G to gH. Note that

$$\begin{aligned} xH \cap yH \neq \emptyset &\Leftrightarrow xh = ys \quad \text{for some } h, s \in H \\ &\Leftrightarrow x = ysh^{-1} \quad \text{for some } h, s \in H. \end{aligned}$$

Consequently, $xH \cap yH \neq \emptyset$ if and only if $xH = ysh^{-1}H = yH$. Thus, the set of all cosets of H defines a partition of G.

Let $\pi : G \to G/H$ be the function which sends every element $g \in G$ to the coset gH. We shall endow $G/H = \{gH : g \in G\}$ with the quotient topology. In other words, a subset U of G/H is said to be open in such a topology if and only if the inverse image of U under π is open in G.

Lemma 3.9 When G/H is equipped with the quotient topology, then π is an open map. Moreover, G/H is second countable.

Proof: Given an open set O in G, since

$$\pi^{-1}(\pi(O)) = \pi^{-1}(OH) = \bigcup_{h \in H} Oh$$

is open in G then $\pi(O)$ is open in G/H. To show that G/H is second countable, let $\{O_j : j \in \mathbb{N}\}$ be a countable base for the topology which we endow G with. Then clearly, $\{\pi(O_j) : j \in \mathbb{N}\}$ is also a countable base for the topology of G/H.

Lemma 3.10 G/H is a Hausdorff space.

Proof: Given aH and bH, we need to prove that there exist open sets U_{aH} and U_{bH} in G/H around aH and bH, respectively, so that $U_{aH} \cap U_{bH}$ is empty. To this end, let

$$E = \{(a, b) \in G \times G : a^{-1}b \in H\}.$$

Letting $\alpha : (a, b) \mapsto a^{-1}b$, it follows that $E = \alpha^{-1}(H)$ and consequently, E is a closed subset of $G \times G$. Suppose next that (a, b) is not an element of E. Then, there exists an open set $V_a \times V_b$ of $G \times G$ such that $(V_a \times V_b) \cap E$ is empty and additionally, V_a and V_b are open subsets of a and b, respectively. Consequently, $\pi(V_a)$ and $\pi(V_b)$ are open subsets of aH and bH respectively such that $\pi(V_a) \cap \pi(V_b)$ is empty.

Example 3.24 Let

$$G = \left\{ p(x,y,z) = \begin{bmatrix} 1 & x & z \\ 0 & 1 & y \\ 0 & 0 & 1 \end{bmatrix} : z, y, x \in \mathbb{R} \right\}$$

be a three-dimensional Lie group and let

$$H = \left\{ p(0,y,z) = \begin{bmatrix} 1 & 0 & z \\ 0 & 1 & y \\ 0 & 0 & 1 \end{bmatrix} : z, y, x \in \mathbb{R} \right\}$$

be a closed subgroup of G. Since for every real number x, y, z,

$$\begin{aligned} p(x,y,z) &= p(0,y,z)\,p(x,0,0) \\ &= p(x,0,0)\,p(x,0,0)^{-1}p(0,y,z)\,p(x,0,0) \\ &= p(x,0,0)\,p(0,y,z-xy) \end{aligned}$$

then $p(x,y,z)H = p(x,0,0)\,p(0,y,z-xy)H = p(x,0,0)H$ and

$$G/H = \{p(x,0,0)H : x \in \mathbb{R}\}.$$

Therefore, the canonical map $\pi : G \to G/H$ is explicitly given by $\pi(p(x,y,z)) = p(x,0,0)H$. Next, G/H can be equipped with the standard smooth structure of the real line via the identification map $p(x,0,0)H \mapsto x$. Moreover, there exist global coordinates systems on G and G/H such that, with respect to such coordinates, π is given by the smooth and surjective function $(x,y,z) \mapsto x$.

Example 3.25 Let $G = \mathrm{SL}(2,\mathbb{R})$ be the set of invertible matrices of order 2 with unit determinant and let

$$K = \mathrm{SO}(2,\mathbb{R}) = \left\{ \begin{bmatrix} \cos\theta & -\sin\theta \\ \sin\theta & \cos\theta \end{bmatrix} : \theta \in \mathbb{R} \right\}$$

be the rotation subgroup of K. We aim to show that

$$G/K = \left\{ \begin{bmatrix} r & t \\ 0 & \frac{1}{r} \end{bmatrix} K : r, t \in \mathbb{R} \quad \text{and} \quad r > 0 \right\}.$$

To this end, we fix

$$\begin{bmatrix} x & y \\ z & w \end{bmatrix} \in \mathrm{SL}(2,\mathbb{R}) \quad \text{such that } xw - yz = 1.$$

If $z = 0$ then $x \neq 0$ and

$$\begin{bmatrix} x & y \\ z & w \end{bmatrix} = \begin{bmatrix} x & y \\ 0 & w \end{bmatrix} = \begin{bmatrix} x & y \\ 0 & x^{-1} \end{bmatrix}.$$

Suppose on the other hand that z is nonzero. Observing that

$$\begin{bmatrix} \cos\theta & -\sin\theta \\ \sin\theta & \cos\theta \end{bmatrix} \begin{bmatrix} x & y \\ z & w \end{bmatrix} = \begin{bmatrix} x\cos\theta - z\sin\theta & y\cos\theta - w\sin\theta \\ z\cos\theta + x\sin\theta & w\cos\theta + y\sin\theta \end{bmatrix},$$

there exists a rotation matrix

$$\begin{bmatrix} \cos\theta & -\sin\theta \\ \sin\theta & \cos\theta \end{bmatrix} \in K$$

such that $\cot\theta = -\frac{x}{z}$ and

$$\begin{bmatrix} \cos\theta & -\sin\theta \\ \sin\theta & \cos\theta \end{bmatrix} \begin{bmatrix} x & y \\ z & w \end{bmatrix} = \begin{bmatrix} u & v \\ 0 & u^{-1} \end{bmatrix}.$$

As a result,

$$G/K = \left\{ \begin{bmatrix} r & t \\ 0 & \frac{1}{r} \end{bmatrix} K \, : \, r, t \in \mathbb{R} \quad \text{and} \quad r > 0 \right\}$$

as desired. Therefore, the quotient map $\pi : \mathrm{SL}(2,\mathbb{R}) \to \mathrm{SL}(2,\mathbb{R})/\mathrm{SO}(2,\mathbb{R})$ is explicitly given by

$$\begin{bmatrix} r & t \\ 0 & \frac{1}{r} \end{bmatrix} \begin{bmatrix} \cos\theta & -\sin\theta \\ \sin\theta & \cos\theta \end{bmatrix} \mapsto \begin{bmatrix} r & t \\ 0 & \frac{1}{r} \end{bmatrix} K.$$

Via the bijective function

$$\begin{bmatrix} r & t \\ 0 & \frac{1}{r} \end{bmatrix} K \mapsto \begin{bmatrix} r & t \\ 0 & \frac{1}{r} \end{bmatrix},$$

we may endow G/K with the smooth structure of $(0,\infty) \times \mathbb{R}$. The above illustrates that there exist smooth structures on $\mathrm{SL}(2,\mathbb{R})$ and $\mathrm{SL}(2,\mathbb{R})/\mathrm{SO}(2,\mathbb{R})$ such that π is a smooth and surjective function.

Our main objective in this section is to generalize the observations made in the examples above. Precisely, we aim to prove that if G is a Lie group and if H be a closed Lie subgroup of G, then the quotient space G/H can be equipped with a smooth structure such that the canonical quotient map is smooth. To this end, we will first need to establish a result coined as the slice theorem.

Theorem 3.6 (Slice theorem) Let G be a Lie group with Lie algebra $\mathfrak{g}$ and let H be a closed Lie subgroup of G with Lie algebra $\mathfrak{h}$. Next, let $\mathfrak{k} \oplus \mathfrak{h}$ be a direct sum decomposition of $\mathfrak{g}$ and define $\lambda : \mathfrak{k} \times H \to G$ such that $\lambda(X,h) = \exp(X)h$. Then there exists a neighborhood U_0 of the zero vector in $\mathfrak{k}$ such that λ defines a diffeomorphism between $U_0 \times H$ and $\lambda(U_0 \times H)$.

Proof: As a starting point, we will identify the tangent space of $\mathfrak{k} \times H$ at $(0,e)$ with the direct sum $\mathfrak{k} \oplus \mathfrak{h}$, which is naturally identified with the Lie algebra of the

Lie group G. There exist smooth charts around $(0, e)$ and $\lambda(0, e) = e$ such that the Jacobian of the derivative of $\lambda : (X, h) \mapsto \exp(X) h$ in local coordinates is equal to the identity matrix. Indeed, appealing to Proposition 3.22, we have established that generally, for arbitrary $X_e \in \mathfrak{g}$, and $\frac{d}{dt}\Big|_{t=0} \exp(tX_e) = \frac{d}{dt}\Big|_{t=0} (c_X(t)) = X_e$. Thus, the differential of $(X, h) \mapsto \exp(X) h$ at $(0, e)$ is the identity, and by the inverse function theorem (see Theorem 2.7), there exists a neighborhood U of $(0, e)$ such that λ defines a diffeomorphism between U and $\lambda(U) \subset G$. Next, let U_1 be an open set around 0 in $\mathfrak{k}$ and let U_2 be an open set around e in H such that the Cartesian product $U_1 \times U_2$ is contained in U and the differential of λ at (X, e) is invertible for all elements X in the set U_1. Note that given h in H and X in $\mathfrak{k}$,

$$R_h \lambda(X, e) = \lambda(X, e) h = \exp(X) eh = \exp(X) h = \lambda(X, h).$$

By chain rule, it follows that

$$\left(\left(R_h\right)_*\right)_{\lambda(X,e)} \circ \left(\lambda_*\right)_{(X,e)} = \left(\lambda_*\right)_{(X,h)}.$$

Since R_h is a diffeomorphism, it follows that $\left(\left(R_h\right)_*\right)_{\lambda(X,e)}$ and $\left(\lambda_*\right)_{(X,e)}$ are both invertible linear operators and consequently, $\left(\lambda_*\right)_{(X,h)}$ is an invertible operator as well for all $(X, h) \in \mathfrak{k} \times H$. Appealing again to the inverse function theorem, it follows that for each point in the open set $U_1 \times H$, there is an open set containing such a point such that λ maps the given neighborhood diffeomorphically into its image. This implies that $\lambda\left(U_1 \times H\right)$ is an open subset of G. Next, let U_1, U_2 be as constructed such that $\lambda : U_1 \times U_2 \to \lambda\left(U_1 \times U_2\right)$ is a diffeomorphism. Let V' be an open subset of e in G such that $V' \cap H$ coincides with U_2. The fact that V' exists is a direct consequence of the assumption that H is a closed embedded submanifold of G. Let H be a subgroup of G that is both a closed subset in the topological sense and an embedded submanifold of G. Then H inherits a smooth structure from G, making it a Lie group. Since the exponential map is continuous, there exists an open subset U of U_1, which is small enough to ensure that for any X, Y in U, $\exp(-Y) \exp X$ belongs to V'. Note also that $\exp(-Y) \exp X \in V' \cap H = U_2$. Clearly λ is a one-to-one function. To see this, suppose that $(X, h), (Y, s)$ belong to $U \times H$ and $\lambda(X, h) = \lambda(Y, s)$. That is, $\exp(X) h = \exp(Y) s$ and consequently, $sh^{-1} = \exp(-Y) \exp X$ belongs to U_2. Therefore,

$$\lambda(X, e) = \exp X = \exp Y \left(sh^{-1}\right) = \lambda\left(Y, sh^{-1}\right).$$

Since (X, e) and $\left(Y, sh^{-1}\right)$ belong to $U_1 \times U_2 \subset V$ where λ is injective, it follows that $X = Y$ and $s = h$. In summary, we have shown that λ is a local diffeomorphism on $U_1 \times H$ and we have also proved that λ is injective on $U \times H$. This implies that λ is a diffeomorphism of $U \times H$ onto $\lambda(U \times H)$.

Lemma 3.11 Let M and N be smooth manifolds and let $f : M \to N$ be a smooth surjective map that has full rank everywhere. Then $l : N \to P$ is smooth if and only if $l \circ f$ is smooth.

Proof: Suppose first that $l : N \to P$ is smooth. Then $l \circ f$ is smooth as the composition of smooth functions. For the converse, let us assume that $l \circ f$ defines a smooth map between M and P. Given q in N, since f is assumed to be surjective, there exists a point p in the manifold M such that f maps p to q. Then there exist smooth charts (U, φ) and (V, ψ) centered at p and q, respectively, and

$$\left[\psi \circ f \circ \varphi^{-1}\right]\left(x_1, \dots, x_n, x_{n+1}, \dots, x_m\right) \mapsto \left(y_1, \dots, y_n\right).$$

Next, let ϵ be a sufficiently small positive real number such that given

$$B_{(\epsilon,m)} = \prod_{k=1}^{m} (-\epsilon, \epsilon) \quad \text{and} \quad B_{(\epsilon,n)} = \prod_{k=1}^{n} (-\epsilon, \epsilon),$$

we have

$$\left[\psi \circ f \circ \varphi^{-1}\right]\left(B_{(\epsilon,m)}\right) = B_{(\epsilon,n)}.$$

Define $\tilde{\sigma} : B_{(\epsilon,n)} \to B_{(\epsilon,m)}$ such that $\sigma\left(x_1, \dots, x_n\right) = \left(x_1, \dots, x_n, 0, \dots, 0\right)$ is a smooth function and let $\sigma = \varphi^{-1} \circ \tilde{\sigma} \circ \psi|_{\psi^{-1}(B_{(\epsilon,n)})}$. Then σ maps p to q and additionally, $f \circ \sigma$ is the identity map (we say that σ is a local section) on $\psi^{-1}\left(B_{(\epsilon,n)}\right)$. Next,

$$l|_{\psi^{-1}(B_{(\epsilon,n)})} = l|_{\psi^{-1}(B_{(\epsilon,n)})} \circ \mathrm{id}_{\psi^{-1}(B_{(\epsilon,n)})} = l|_{\psi^{-1}(B_{(\epsilon,n)})} \circ (f \circ \sigma) = (l \circ \pi) \circ \sigma.$$

Therefore, f is smooth on $\psi^{-1}\left(B_{(\epsilon,n)}\right)$. Since q was selected arbitrarily, we may conclude that f is indeed a smooth map.

Proposition 3.37 Let G be a Lie group, let H be a closed matrix subgroup of G. Then the quotient topology on G/H equips it with a unique smooth structure on G/H such that $\dim(G/H) = \dim G - \dim H$.

Proof: **Step 1**: We will prove the uniqueness of a smooth structure on G/H. Suppose first that G/H has two smooth structures, which we shall denote by S_1, S_2. Suppose that $\pi : G \to S_j$ for $j \in \{1, 2\}$. Next, let $\mathrm{id} : S_1 \to S_2$ be the map which sends every point to itself. By Lemma 3.11, if $\mathrm{id} \circ \pi : G \to S_2$ is smooth then id must be smooth. In a similar fashion the map between S_2 and S_1, which sends every point to itself must be smooth as well. This shows that id is a diffeomorphism. **Step 2**: We will now prove that there exists a smooth structure on the quotient space G/H. Recall λ as defined in Theorem 3.6. Given g in G, we let $\lambda_g = L_g \circ \lambda : U \times H \to G$ such that $\lambda_g(X, h) = g\exp(X)h$ where $\mathfrak{g} = \mathfrak{k} \oplus \mathfrak{h}$ and U is an open subset of $\mathfrak{k}$ containing the zero element in $\mathfrak{k}$. Clearly, each λ_g is a diffeomorphism since it is defined as a composition of diffeomorphisms. Next, let $W_g = \lambda_g(U \times H)$ and let V_g be the image of W_g under the canonical quotient map π. In other words,

$$V_g = \{(g\exp X)H \in G/H : X \in U\}.$$

Since W_g is open in G and G/H is given the quotient topology, it follows that V_g is an open subset of G/H. Moreover, note that the inverse image of V_g under π coincides with W_g and the map $X \mapsto (g \exp X) H$ is injective as well. Suppose indeed that $(g \exp X) H = (g \exp Y) H$. Then $(\exp X)^{-1} g^{-1} g \exp Y = \exp(-X) \exp Y$ must be in H. But this is only possible if $X = Y$. Next, the map $X \mapsto (g \exp X) H$ is an injective a continuous map from U to V_g. Moreover, let φ_g be its inverse and consider the function $\varphi_g \circ \pi : W_g \to U$. Then for any element $\lambda_g(X, h)$ in W_g, we verify that

$$\begin{aligned}(\varphi_g \circ \pi)(\lambda_g(X, h)) &= (\varphi_g \circ \pi)(g \exp X h) \\ &= \varphi_g(g \exp(X) hH) \\ &= \varphi_g(g \exp(X) H) = X.\end{aligned}$$

In other words, $\varphi_g \circ \pi = \pi_1 \circ \lambda_g^{-1}$ where π_1 is the map defined on $U \times H$ which returns the first coordinate of its argument. Thus, $\varphi_g \circ \pi$ is a continuous map and each φ_g is continuous as well. Indeed, if E is an open subset of U, we obtain that $\pi^{-1}(\varphi_g^{-1}(E)) = (\varphi_g \circ \pi)^{-1}(E)$ is open in W_g. Thus, for each g in G, there exists a neighborhood V_g of the coset gH and a homeomorphism $\varphi_g : V_g \to U \subseteq \mathfrak{k}$. Since G/H is Hausdorff, it follows that G/H is locally Euclidean and the dimension of G/H as a topological manifold is $\dim G - \dim H$. Moving forward, let us consider the change of coordinate map: $\varphi_s \circ \varphi_g^{-1} : \varphi_g(V_g \cap V_s) \to \varphi_s(V_g \cap V_s)$ which may be computed quite explicitly as follows:

$$\varphi_s \circ \varphi_g^{-1} = \pi_1 \circ \lambda_s^{-1} \circ L_g \circ \exp .$$

Thus, $\varphi_s \circ \varphi_g^{-1}$ is smooth and consequently, $\{(V_g, \varphi_g) : g \in G\}$ is a smooth atlas for G/H.

Proposition 3.38 Let G be a Lie group and let H be a closed subgroup of G. Then the canonical map $\pi : g \mapsto gH$ is a smooth map with full rank everywhere.

Proof: Given $g \in G$ and letting W_g be an open subset of G containing g, we have $\varphi_g \circ \pi = \pi_1 \circ \lambda_g^{-1}$. Therefore, $\varphi_g \circ \pi$ is a smooth map and consequently, π is smooth as well. Next, by chain rule applied to $\varphi_g \circ \pi = \pi_1 \circ \lambda_g^{-1}$, we obtain

$$\begin{aligned}\left((\varphi_g)_*\right)_{\pi(s)} \circ (\pi_*)_s &= \left((\varphi_g \circ \pi)_*\right)_s \\ &= \left((\pi_1)_*\right)_{\lambda_g^{-1}(s)} \circ \left((\lambda_g^{-1})_*\right)_s .\end{aligned}$$

Next, since φ_g and λ_g^{-1} are diffeomorphisms, we obtain that each linear map $(\pi_*)_s$ must have the same rank as $\left((\pi_1)_*\right)_{\lambda_g^{-1}(s)}$, which is clearly constant and equal to the dimension of the manifold G/H. This gives the desired result.

Example 3.26 Let

$$G = \left\{ p(x,y,z) = \begin{bmatrix} 1 & x & z \\ 0 & 1 & y \\ 0 & 0 & 1 \end{bmatrix} : x,y,z \in \mathbb{R} \right\}$$

be the Heisenberg group and let

$$H = \left\{ p(0,y,z) = \begin{bmatrix} 1 & 0 & z \\ 0 & 1 & y \\ 0 & 0 & 1 \end{bmatrix} : x,y,z \in \mathbb{R} \right\}$$

be a normal subgroup of G. Then the canonical quotient map associated with G and H is $\pi : p(x,y,z) \mapsto p(x,0,0)H$. In local coordinates, this map is of the form: $(x,y,z) \mapsto x$ and its Jacobian is

$$\begin{bmatrix} 1 \\ 0 \\ 0 \end{bmatrix},$$

showing that π has full rank 1 everywhere.

Proof:

Proposition 3.39 The map $(g, sH) \mapsto gsH$ is a smooth map between $G \times G/H$ and G/H.

Proof: Let $\tau : G \times G/H \to G/H$ such that $\tau(g, sH) = gsH = \pi(gs)$. For a fixed point g_1 in the Lie group G, let $\left(V_{g_1}, \varphi_{g_1}\right)$ be a smooth chart around g_1. Next, let us define $\tau_1 : G \times U \to G/H$ such that

$$\tau_1(g, X) = \tau\left(g, \varphi_{g_1}^{-1}(X)\right).$$

Since $\varphi_{g_1}^{-1}(X) = g_1 \exp(X) H$, we obtain

$$\tau_1(g, X) = gg_1 \exp(X) H = \pi\left(gg_1 \exp(X)\right).$$

In other words,

$$\tau_1 : (g, X) \mapsto gg_1 \exp X \mapsto \pi\left(gg_1 \exp(X)\right).$$

Thus τ_1 is smooth since it is obtained as the composition of smooth maps and consequently, τ is smooth as well.

Definition 3.16 Let M be a smooth manifold on which a Lie group acts by a smooth action, which maps every element (g, x) in $G \times M$ to an element $g \bullet x \in M$

such that for all $g, s \in G$, and x in M, (a) $s \bullet (g \bullet x) = (sg) \bullet x$ and (b) $e \bullet x = x$. If the action on G on M is transitive in the sense that for any $x \in M$, the set

$$\{g \bullet x : g \in G\} = G \bullet x = M$$

then we say that M is a homogeneous space.

Example 3.27 Let

$$G = \left\{ \begin{bmatrix} a & x \\ 0 & 1 \end{bmatrix} : (a, x) \in \mathbb{R}^2 \text{ and } a \neq 0 \right\}$$

be the ax+b Lie group. Then the multiplicative action of G on itself defines the following action:

$$\left(\begin{bmatrix} a & x \\ 0 & 1 \end{bmatrix}, \begin{bmatrix} b & y \\ 0 & 1 \end{bmatrix} \right) \mapsto \begin{bmatrix} ab & x + ay \\ 0 & 1 \end{bmatrix}.$$

This induces an action of G on $\mathbb{R}^2$ of the form

$$\begin{bmatrix} a & x \\ 0 & 1 \end{bmatrix} \bullet (b, y) = (ab, x + ay).$$

Example 3.28 Let

$$G = \left\{ \begin{bmatrix} 1 & x & z \\ 0 & 1 & y \\ 0 & 0 & 1 \end{bmatrix} : x, y, z \in \mathbb{R} \right\}$$

be the Heisenberg Lie group. Then the multiplicative action of G on itself defines the following action:

$$\left(\begin{bmatrix} 1 & x & z \\ 0 & 1 & y \\ 0 & 0 & 1 \end{bmatrix}, \begin{bmatrix} 1 & a & c \\ 0 & 1 & b \\ 0 & 0 & 1 \end{bmatrix} \right) \mapsto \begin{bmatrix} 1 & a + x & c + z + bx \\ 0 & 1 & b + y \\ 0 & 0 & 1 \end{bmatrix}.$$

This induces an action of the Heisenberg group $\mathbb{R}^3$ of the form

$$\begin{bmatrix} 1 & x & z \\ 0 & 1 & y \\ 0 & 0 & 1 \end{bmatrix} \bullet (a, b, c) = (a + x, b + y, c + z + bx).$$

In light of the definition above, since G acts transitively on the set G/H equipped with its unique smooth structure, it follows that G/H is a homogeneous space.

Proposition 3.40 Given $g \in G$, let $\tau_g : G/H \to G/H$ such that $\tau_g(sH) = gsH$. Then each τ_g is a diffeomorphism.

Proof: For each $g \in G$, the map τ_g is a bijection of constant rank everywhere. Therefore, each τ_g is a diffeomorphism.

Let G be a Lie group acting smoothly on some manifold M. Next, given $x \in M$, let G_x be the stabilizer of x in G. That is,

$$G_x = \{g \in G : g \bullet x = x\}.$$

Then G_x is a closed Lie subgroup of G and there exists a unique smooth structure on G/G_x such that G/G_x is a homogeneous space.

3.6 Matrix Lie Theory

We begin by fixing notation as follows:

- The general linear group $GL(n, \mathbb{C})$ is the set of invertible matrices with complex entries.
- $GL(n, \mathbb{R})$ is the set of invertible matrices with real entries.
- $M(n, \mathbb{C})$ is the set of all square matrices of order n.

By identifying every element in $M(n, \mathbb{C})$ with the vector space $\mathbb{C}^{n^2}$, the topology on $M(n, \mathbb{C})$ is induced by the metric

$$d : M(n, \mathbb{C}) \times M(n, \mathbb{C}) \to [0, \infty),$$

where

$$d(X, Y) = \left(\sum_{k=1}^{n} \sum_{i=1}^{n} \left| X_{i,k} - Y_{i,k} \right|^2 \right)^{1/2}$$

and put $\|X\|_2 = d(X, X)$. Then $X \mapsto \|X\|_2$ defines a norm on $M(n, \mathbb{C})$. Additionally, the operator norm of X is given by

$$\|X\|_{op} = \sup \left\{ \|Xv\| : v \in \mathbb{C}^n \text{ and } \|v\| = 1 \right\}.$$

Proposition 3.41 (Submultiplicative property of $\|\|_2$) Given $X, Y \in M(n, \mathbb{C})$, $\|XY\|_2 \leq \|X\|_2 \|Y\|_2$.

Proof: First, we observe that

$$\left| (XY)_{jk} \right| = \left| \sum_{l=1}^{n} X_{jl} Y_{lk} \right| \leq \sum_{l=1}^{n} \left| X_{jl} \right| \cdot \left| Y_{lk} \right|.$$

Secondly, by Cauchy–Schwarz Inequality,

$$\sum_{l=1}^{n} \left| X_{jl} \right| \cdot \left| Y_{lk} \right| \leq \left(\sum_{l=1}^{n} \left| X_{jl} \right|^2 \right)^{1/2} \left(\sum_{l=1}^{n} \left| Y_{jk} \right|^2 \right)^{1/2}$$

and

$$\begin{aligned}
\|XY\|_2^2 &= \sum_{k=1}^{n}\sum_{j=1}^{n}\left|(XY)_{jk}\right|^2 \\
&\leq \sum_{k=1}^{n}\sum_{j=1}^{n}\left(\sum_{l=1}^{n}\left|X_{jl}\right|^2\right)\left(\sum_{l=1}^{n}\left|Y_{jk}\right|^2\right) \\
&= \sum_{j=1}^{n}\sum_{l=1}^{n}\left|X_{jl}\right|^2\sum_{k=1}^{n}\sum_{l=1}^{n}\left|Y_{jk}\right|^2 \\
&= \|X\|_2^2 \cdot \|Y\|_2^2 .
\end{aligned}$$

Finally, taking the square-root of each side of the inequality $\|XY\|_2^2 \leq \|X\|_2^2 \cdot \|Y\|_2^2$, we obtain $\|XY\|_2 \leq \|X\|_2 \cdot \|Y\|_2$ as desired.

Proposition 3.42 *$GL(n,\mathbb{C})$ is an open subset of $M(n,\mathbb{C})$.*

Proof: Given $X \in M(n,\mathbb{C})$, since

$$\det(X) = \sum_{\sigma\in S_n}(-1)^{\mathrm{sgn}(\sigma)}\prod_{k=1}^{n}X_{k,\sigma(k)},$$

it follows that $X \mapsto \det(X)$ defines a polynomial map on $M(n,\mathbb{C})$. Moreover, $GL(n,\mathbb{C}) = \det^{-1}(\mathbb{C}\setminus\{0\})$. Since $\mathbb{C}\setminus\{0\}$ is open in $\mathbb{C}$, and since the determinant map defines a continuous function between $M(n,\mathbb{C})$ and $\mathbb{C}$, it follows that $GL(n,\mathbb{C})$ is an open subset of $M(n,\mathbb{C})$.

Definition 3.17 A matrix Lie group is a closed subgroup of $GL(n,\mathbb{C})$.

Example 3.29 Given vectors $v, w \in \mathbb{C}^n$, define $\langle v, w\rangle = \sum_{k=1}^{n} v_k\overline{w_k}$ and

$$U(n) = \left\{X \in GL(n,\mathbb{C}) : \langle Xv, Xw\rangle = \langle v, w\rangle \ \ \forall v, w \in \mathbb{C}^n\right\}.$$

First, $U(n)$ is a nonempty set since it contains at least the identity. Given $X, Y \in U(n)$, and $v, w \in \mathbb{C}^n$, it is easy to verify that

$$\langle XYv, XYw\rangle = \langle Yv, Yw\rangle = \langle v, w\rangle .$$

This shows that $XY \in U(n)$. Next, let $X \in U(n)$. Since X is unitary, the inverse of X and its adjoint coincide. As a result,

$$\left\langle X^{-1}v, X^{-1}w\right\rangle = \left\langle X^{-1}v, X^{*}w\right\rangle = \left\langle XX^{-1}v, w\right\rangle = \langle v, w\rangle .$$

Therefore $U(n)$ is closed under taking inverses. This shows that $U(n)$ is a subgroup of $GL(n,\mathbb{C})$. To prove that $U(n)$ is closed in $GL(n,\mathbb{C})$, let $\left(X_k\right)_{k\in\mathbb{N}}$ be a convergent sequence in $U(n)$. Next, let $X = \lim_{k\to\infty} X_k$. Fix $v, w \in \mathbb{C}^n$. For each natural number k,

$$\langle Xv, Xw\rangle = \lim_{k\to\infty} \left\langle X_k v, X_k w\right\rangle = \lim_{k\to\infty} \langle v, w\rangle = \langle v, w\rangle .$$

This shows that $X \in U(n)$. Therefore, $U(n)$ is a closed subgroup of $GL(n, \mathbb{C})$ and as such, it is a matrix subgroup.

Remark 3.3 It is worth noting that not every subgroup of $GL(n, \mathbb{C})$ is a matrix group. For instance, let

$$G = \left\{ \begin{bmatrix} e^{2\pi it} & 0 \\ 0 & e^{2\pi i\sqrt{2}t} \end{bmatrix} : t \in \mathbb{R} \right\}.$$

Then given positive real numbers r, t,

$$\begin{bmatrix} e^{2\pi it} & 0 \\ 0 & e^{2\pi i\sqrt{2}t} \end{bmatrix} \begin{bmatrix} e^{2\pi ir} & 0 \\ 0 & e^{2\pi i\sqrt{2}r} \end{bmatrix} = \begin{bmatrix} e^{2\pi i(t+r)} & 0 \\ 0 & e^{2\pi i\sqrt{2}(t+r)} \end{bmatrix}$$

and

$$\begin{bmatrix} e^{2\pi it} & 0 \\ 0 & e^{2\pi i\sqrt{2}t} \end{bmatrix}^{-1} = \begin{bmatrix} e^{-2i\pi t} & 0 \\ 0 & e^{-2i\sqrt{2}\pi t} \end{bmatrix}.$$

This shows that G is a subgroup of $GL(2, \mathbb{C})$. On the other hand, as a subset of $GL(2, \mathbb{C})$, G is not closed. Indeed, let us consider the following sequence of matrices

$$\left(\begin{bmatrix} e^{2\pi ik} & 0 \\ 0 & e^{2\pi i\sqrt{2}k} \end{bmatrix} \right)_{k\in\mathbb{N}}.$$

Then for any natural number k, we have

$$\begin{bmatrix} e^{2\pi ik} & 0 \\ 0 & e^{2\pi i\sqrt{2}k} \end{bmatrix} = \begin{bmatrix} 1 & 0 \\ 0 & e^{2\pi i\sqrt{2}k} \end{bmatrix}.$$

Next, let $\left(e^{2\pi i\sqrt{2}k_j}\right)_{j\in\mathbb{N}}$ be a subsequence in $\left(e^{2\pi i\sqrt{2}k}\right)_{k\in\mathbb{N}}$ such that $\lim_{j\to\infty} e^{2\pi i\sqrt{2}k_j} = -1$. Then

$$\lim_{j\to\infty} \begin{bmatrix} e^{2\pi ik_j} & 0 \\ 0 & e^{2\pi i\sqrt{2}k_j} \end{bmatrix} = \begin{bmatrix} 1 & 0 \\ 0 & -1 \end{bmatrix}.$$

But $\begin{bmatrix} 1 & 0 \\ 0 & -1 \end{bmatrix}$ is not an element of G. Otherwise, there must exist an integer j such that $e^{2\pi i\sqrt{2}j} = -1$.

Given $Y \in M(n,\mathbb{C})$, we define the power series

$$\exp(Y) = \sum_{k=0}^{\infty} \frac{Y^k}{k!}$$

(see also Proposition 3.23).

Proposition 3.43 For every $Y \in M(n,\mathbb{C})$, $\sum_{k=0}^{\infty} \frac{Y^k}{k!}$ converges in $M(n,\mathbb{C})$.

Proof: Since $M(n,\mathbb{C})$ equipped with the operator norm is a Banach space,

$$\left\| \sum_{k=0}^{\infty} \frac{Y^k}{k!} \right\|_2 \leq \sum_{k=0}^{\infty} \frac{1}{k!} \left\| Y^k \right\|_2 \leq \sum_{k=0}^{\infty} \frac{1}{k!} \|Y\|_2^k = e^{\|Y\|_2}.$$

This shows that the series $\sum_{k=0}^{\infty} \frac{Y^k}{k!}$ converges absolutely and therefore, it must converge in $M(n,\mathbb{C})$.

Proposition 3.44 Let $Y \in M(n,\mathbb{C})$. Then the following holds true

1) Given any complex numbers z, w, $\exp((z+w)Y) = \exp(zY)\exp(wY)$.
2) $\exp(Y)$ is an invertible matrix. That is, $\exp(Y) \in GL(n,\mathbb{C})$ and $(\exp Y)^{-1} = \exp(-Y)$.

Proof: For the first part, since the power series used in the definition of the exponential function is absolutely convergent, we obtain the following:

$$\begin{aligned}
\exp(zY)\exp(wY) &= \left(\sum_{k=0}^{\infty} \frac{z^k Y^k}{k!}\right)\left(\sum_{j=0}^{\infty} \frac{w^j Y^j}{j!}\right) \\
&= \sum_{k=0}^{\infty}\sum_{j=0}^{\infty} \frac{z^k w^j}{k!j!} Y^{k+j} \\
&\overset{m=k+j}{=} \sum_{m=0}^{\infty}\sum_{k=0}^{m} \frac{z^k w^{m-k}}{k!(m-k)!} Y^m \\
&= \sum_{m=0}^{\infty} \frac{1}{m!}\left(\sum_{k=0}^{m} \binom{m}{k} z^k w^{m-k}\right) Y^m.
\end{aligned}$$

Thanks to the binomial theorem,

$$\sum_{k=0}^{m} \binom{m}{k} z^k w^{m-k} = (z+w)^m$$

and

$$\exp(zY)\exp(wY) = \sum_{m=0}^{\infty} \frac{1}{m!}\left((z+w)^m\right) Y^m = \exp((z+w)Y).$$

Regarding Part (ii), we know that the identity in $GL(n,\mathbb{C})$ is equal to

$$I = \exp(0) = \exp((1+(-1))Y) = \exp(Y)\exp(-Y).$$

This shows that $(\exp(Y))^{-1} = \exp(-Y)$.

Proposition 3.45 Let X and Y be commuting matrices in $M(n,\mathbb{C})$. Then $\exp(X)\exp(Y) = \exp(X+Y)$.

Proof: Let X and Y be commuting matrices in $M(n,\mathbb{C})$. Using the fact that the power series in the definition of the exponential map is absolutely convergent, we are allowed to move the summation symbols around as shown below:

$$\begin{aligned}\exp(X)\exp(Y) &= \left(\sum_{k=0}^{\infty}\frac{X^k}{k!}\right)\left(\sum_{j=0}^{\infty}\frac{Y^j}{j!}\right)\\ &= \sum_{k=0}^{\infty}\sum_{j=0}^{\infty}\frac{1}{k!j!}X^kY^j.\end{aligned}$$

Next, letting $j = m-k$,

$$\exp(X)\exp(Y) = \sum_{m=0}^{\infty}\sum_{k=0}^{m}\frac{1}{k!\,(m-k)!}X^kY^{m-k}.$$

Since $\binom{m}{k} = \frac{m!}{k!(m-k)!}$, $\frac{1}{k!(m-k)!} = \frac{1}{m!}\binom{m}{k}$ and

$$\exp(X)\exp(Y) = \sum_{m=0}^{\infty}\frac{1}{m!}\left(\sum_{k=0}^{m}\binom{m}{k}X^kY^{m-k}\right).$$

By the binomial theorem, since X, Y are commuting, we obtain

$$\exp(X)\exp(Y) = \sum_{m=0}^{\infty}\frac{1}{m!}(X+Y)^m = \exp(X+Y).$$

Proposition 3.46 Let $X \in M(n,\mathbb{C})$ and $A \in GL(n,\mathbb{C})$. Then

$$A(\exp X)A^{-1} = \exp\left(AXA^{-1}\right).$$

Proof: Since $\exp X = \sum_{k=0}^{\infty}\frac{X^k}{k!}$,

$$\begin{aligned}A(\exp X)A^{-1} &= A\left(\sum_{k=0}^{\infty}\frac{1}{k!}X^k\right)A^{-1} = \sum_{k=0}^{\infty}\frac{1}{k!}AX^kA^{-1} = \sum_{k=0}^{\infty}\frac{1}{k!}\left(AXA^{-1}\right)^k\\ &= \exp\left(AXA^{-1}\right).\end{aligned}$$

Example 3.30 Let

$$X = \begin{bmatrix} 0 & 1 & 0 & 0 \\ 0 & 0 & 1 & 0 \\ 0 & 0 & 0 & 1 \\ 0 & 0 & 0 & 0 \end{bmatrix}.$$

Then

$$X^2 = \begin{bmatrix} 0 & 0 & 1 & 0 \\ 0 & 0 & 0 & 1 \\ 0 & 0 & 0 & 0 \\ 0 & 0 & 0 & 0 \end{bmatrix}, \quad X^3 = \begin{bmatrix} 0 & 0 & 0 & 1 \\ 0 & 0 & 0 & 0 \\ 0 & 0 & 0 & 0 \\ 0 & 0 & 0 & 0 \end{bmatrix}$$

and for any natural number $n \geq 4$, X^n is the zero matrix of order 4. This shows that for any real number t,

$$\exp tX = \begin{bmatrix} 1 & t & \frac{t^2}{2} & \frac{t^3}{6} \\ 0 & 1 & t & \frac{t^2}{2} \\ 0 & 0 & 1 & t \\ 0 & 0 & 0 & 1 \end{bmatrix}.$$

Example 3.31 Let $X = \begin{bmatrix} 1 & 2 \\ 1 & 3 \end{bmatrix}$. To compute the exponential of X, we may first diagonalize X as shown below:

$$X = \begin{bmatrix} 1 & 1 \\ \frac{1}{2}\sqrt{3} + \frac{1}{2} & \frac{1}{2} - \frac{1}{2}\sqrt{3} \end{bmatrix} \begin{bmatrix} \sqrt{3}+2 & 0 \\ 0 & 2-\sqrt{3} \end{bmatrix} \begin{bmatrix} \frac{1}{2} - \frac{1}{6}\sqrt{3} & \frac{1}{3}\sqrt{3} \\ \frac{1}{6}\sqrt{3} + \frac{1}{2} & -\frac{1}{3}\sqrt{3} \end{bmatrix}.$$

Next,

$$\exp X = \begin{bmatrix} 1 & 1 \\ \frac{\sqrt{3}+1}{2} + \frac{1}{2} & \frac{1-\sqrt{3}}{2} \end{bmatrix} \begin{bmatrix} e^{\sqrt{3}+2} & 0 \\ 0 & e^{2-\sqrt{3}} \end{bmatrix} \begin{bmatrix} 1 & 1 \\ \frac{\sqrt{3}+1}{2} + \frac{1}{2} & \frac{1-\sqrt{3}}{2} \end{bmatrix}^{-1}$$

and this shows that $\exp X$ is equal to

$$\begin{bmatrix} \frac{e^{\sqrt{3}+2} + e^{2-\sqrt{3}}}{2} + \frac{\sqrt{3}e^{2-\sqrt{3}} - \sqrt{3}e^{\sqrt{3}+2}}{6} & \frac{\sqrt{3}e^{\sqrt{3}+2} - \sqrt{3}e^{2-\sqrt{3}}}{3} \\ \frac{\sqrt{3}\left(e^{\sqrt{3}+2} - e^{2-\sqrt{3}}\right)}{6} & \frac{e^{\sqrt{3}+2} + e^{2-\sqrt{3}}}{2} + \frac{\sqrt{3}e^{\sqrt{3}+2} - \sqrt{3}e^{2-\sqrt{3}}}{6} \end{bmatrix}.$$

Remark 3.4 Since the exponential function is analytic over the real number ($\mathbb{R}$), it is necessarily infinitely smooth as well.

Proposition 3.47 The function $t \mapsto \exp(tX)$ defines a smooth curve into the general linear group and $\frac{d}{dt}(\exp(tX)) = X\exp(tX)$.

Proof: For any real parameter t,

$$\frac{d}{dt}(\exp(tX)) = \frac{d}{dt}\left(\sum_{k=0}^{\infty}\frac{t^k X^k}{k!}\right) = \sum_{k=1}^{\infty}\left(\frac{kt^{k-1}}{k!}\right)X^k.$$

Factoring X out of each summand above, we obtain

$$\begin{aligned}\frac{d}{dt}(\exp(tX)) &= X\sum_{k=1}^{\infty}\left(\frac{t^{k-1}}{(k-1)!}\right)X^{k-1}\\ &= X\left(\sum_{k=0}^{\infty}\frac{t^k}{k!}X^k\right)\\ &= X\exp(tX).\end{aligned}$$

Recall that for fixed positive real number x,

$$\begin{aligned}\log x &= \int_0^x \frac{1}{1+t}dt = \int_0^x\left(\sum_{k=0}^{\infty}(-1)^k t^k\right)dt\\ &= \sum_{k=0}^{\infty}(-1)^k\int_0^x t^k\,dt\\ &= \sum_{k=0}^{\infty}\frac{(-1)^k}{k+1}x^{k+1}\\ &= \sum_{k=1}^{\infty}\frac{(-1)^{k-1}}{k}x^k.\end{aligned}$$

Motivated by such a definition, in a formal way, we define the Logarithmic of a matrix $X \in M(n,\mathbb{C})$ whenever it makes sense as:

$$\log(X) = \sum_{k=1}^{\infty}\frac{(-1)^{k-1}}{k}(X-I)^k.$$

Note that $\mathbb{C} \ni z \mapsto \sum_{k=1}^{\infty}\frac{(-1)^{k-1}}{k}(z-1)^k$ has radius of convergence 1, since

$$\left|\sum_{k=1}^{\infty}\frac{(-1)^{k-1}}{k}(z-1)^k\right| \le \sum_{k=1}^{\infty}\frac{|z-1|^k}{k} < \infty$$

whenever, $|z-1| < 1$, it follows that the power series $\sum_{k=1}^{\infty} \frac{(-1)^{k-1}}{k}(X-I)^k$ converges whenever $\|X-I\|_2 < 1$.

Proposition 3.48 log is locally continuous around the identity in $M(n,\mathbb{C})$.

Proof: Let

$$B_1(I) = \{Y \in M(n,\mathbb{C}) : \|Y-I\|_2 < 1\}$$

and let $(X_m)_{m\in\mathbb{N}}$ be a sequence of matrices in $B_1(I)$ converging to X. Then with straightforward calculations, verify that

$$\lim_{m\to\infty} \log(X_m) = \sum_{k=1}^{\infty} \frac{(-1)^{k-1}}{k} \lim_{m\to\infty} (X_m - I)^k = \log(X).$$

Proposition 3.49 The log function is a local inverse of the exponential function

1) If $\|X-I\|_2 < 1$ then $\exp(\log X) = X$.
2) If $\|X\|_2 < \log 2$ and $\|\exp X - I\|_2 < 1$ then $\log(\exp(X)) = X$.

Proof: For Part 1, let us assume that X is diagonalizable. In other words, there exists an invertible matrix Q such that

$$QXQ^{-1} = \begin{bmatrix} \lambda_1 & & & \\ & \lambda_2 & & \\ & & \ddots & \\ & & & \lambda_n \end{bmatrix} = D$$

for some complex numbers $\lambda_1, \ldots, \lambda_n$. Consequently, $X = Q^{-1}DQ$ and

$$X - I = Q^{-1}DQ - I = Q^{-1}(D-I)Q.$$

As a result, for any nonnegative integer k,

$$(X-I)^k = \underbrace{\left(Q^{-1}(D-I)Q\right)\left(Q^{-1}(D-I)Q\right)\cdots\left(Q^{-1}(D-I)Q\right)}_{k\text{-times}}$$

and

$$(X-I)^k = Q^{-1}\begin{bmatrix} (\lambda_1-1)^k & & & \\ & (\lambda_2-1)^k & & \\ & & \ddots & \\ & & & (\lambda_n-1)^k \end{bmatrix} Q.$$

Since $\lambda_1, \lambda_2, \dots, \lambda_n$ are eigenvalues of X and since $\|X - I\|_2 < 1$, for every integer $k \in \{1, 2, \dots, n\}$, we have $|\lambda_k - 1| < 1$. As a result,

$$\log(X) = \sum_{k=1}^{\infty} \frac{(-1)^{k-1}(X-I)^k}{k} = Q^{-1}\begin{bmatrix} \log\lambda_1 & & & \\ & \log\lambda_2 & & \\ & & \ddots & \\ & & & \log\lambda_n \end{bmatrix} Q.$$

Next,

$$\exp(\log(X)) = Q^{-1}\begin{bmatrix} e^{\log\lambda_1} & & & \\ & e^{\log\lambda_2} & & \\ & & \ddots & \\ & & & e^{\log\lambda_n} \end{bmatrix} Q = Q^{-1}DQ = X.$$

Next, using the fact that diagonalizable matrices are dense in $M(n, \mathbb{C})$, if X is not diagonalizable, then there exists a sequence $(X_k)_{k\in\mathbb{N}}$ of diagonalizable matrices converging to X. Using the continuity of the Logarithmic function, it follows that $\lim_{k\to\infty} \exp(\log(X_k)) = \lim_{k\to\infty} X_k = X$. Let us assume that $\|X\|_2 < \log 2$ and $\|\exp X - I\|_2 < 1$. Assuming additionally that X is diagonalizable, there must exist an invertible matrix Q and complex numbers $\lambda_1, \dots, \lambda_n$ such that

$$\exp(X) = Q^{-1}\begin{bmatrix} e^{\lambda_1} & & & \\ & e^{\lambda_2} & & \\ & & \ddots & \\ & & & e^{\lambda_n} \end{bmatrix} Q$$

and

$$\log(\exp(X)) = \begin{bmatrix} \sum_{k=1}^{\infty} \frac{(-1)^{k-1}}{k}\left(e^{\lambda_1}-1\right)^k & & & \\ & \sum_{k=1}^{\infty} \frac{(-1)^{k-1}}{k}\left(e^{\lambda_2}-1\right)^k & & \\ & & \ddots & \\ & & & \sum_{k=1}^{\infty} \frac{(-1)^{k-1}}{k}\left(e^{\lambda_n}-1\right)^k \end{bmatrix}$$

$$= \begin{bmatrix} \log\left(e^{\lambda_1}\right) & & & \\ & \log\left(e^{\lambda_2}\right) & & \\ & & \ddots & \\ & & & \log\left(e^{\lambda_n}\right) \end{bmatrix} = X.$$

As argued earlier, we may again appeal to the fact that diagonalizable matrices are dense in $M(n, \mathbb{C})$ to conclude that if $\|X\|_2 < \log 2$ and $\|\exp X - I\|_2 < 1$, then $\log(\exp(X)) = X$.

Remark 3.5 Note that

$$\exp X = I + X + R(X) \text{ where } R(X) = \sum_{k=2}^{\infty} \frac{t^k X^k}{k!}$$

and

$$\|R(X)\| = \|X\| \left\| \sum_{k=2}^{\infty} \frac{t^k X^k}{\|X\| \, k!} \right\|$$

such that $\lim_{X \to 0} \left\| \sum_{k=2}^{\infty} \frac{t^k X^k}{\|X\| k!} \right\| = 0$. This shows that the differential of the exponential map at zero is the identity map. As a result, the exponential map defines a local diffeomorphism between an open subset of the zero matrix and the identity matrix in the general linear group. Moreover, in light of Proposition 3.9, the local inverse of the exponential map is the logarithmic function.

Given $X, Y \in M(n, \mathbb{C})$, we are interested in the following question. Can we compute the matrix $\exp(X+Y)$ in terms of $\exp X$ and $\exp Y$? To this end, let us consider two specific familiar examples.

First, let

$$X = \begin{bmatrix} 0 & 1 & 0 \\ 0 & 0 & 0 \\ 0 & 0 & 0 \end{bmatrix} \quad \text{and} \quad Y = \begin{bmatrix} 0 & 0 & 0 \\ 0 & 0 & 1 \\ 0 & 0 & 0 \end{bmatrix}$$

be two elements in the Heisenberg group. For any natural number k, note that

$$\exp\left(\frac{1}{k}X\right) \exp\left(\frac{1}{k}Y\right) = \begin{bmatrix} 1 & \frac{1}{k} & \frac{1}{k^2} \\ 0 & 1 & \frac{1}{k} \\ 0 & 0 & 1 \end{bmatrix}.$$

Moreover, observe that

$$\left(\exp\left(\frac{1}{k}X\right) \exp\left(\frac{1}{k}Y\right)\right)^k = \begin{bmatrix} 1 & \frac{1}{k} & \frac{1}{k^2} \\ 0 & 1 & \frac{1}{k} \\ 0 & 0 & 1 \end{bmatrix}^k = \begin{bmatrix} 1 & 1 & \frac{1+k}{2k} \\ 0 & 1 & 1 \\ 0 & 0 & 1 \end{bmatrix}.$$

On the other hand, we verify that

$$\exp(X+Y) = \begin{bmatrix} 1 & 1 & \frac{1}{2} \\ 0 & 1 & 1 \\ 0 & 0 & 1 \end{bmatrix} = \lim_{k \to \infty} \left(\exp\left(\frac{1}{k}X\right) \exp\left(\frac{1}{k}Y\right)\right)^k.$$

Secondly, let

$$X = \begin{bmatrix} 0 & 1 \\ 0 & 0 \end{bmatrix} \quad \text{and} \quad A = \begin{bmatrix} 1 & 0 \\ 0 & 0 \end{bmatrix}$$

be two elements in the two-dimensional affine group. Then, it is not too difficult to verify that

$$\left(\exp\left(\frac{1}{k}X\right)\exp\left(\frac{1}{k}A\right)\right)^k = \begin{bmatrix} e^{\frac{1}{k}} & \frac{1}{k} \\ 0 & 1 \end{bmatrix}^k = \begin{bmatrix} \left(e^{\frac{1}{k}}\right)^k & \frac{1}{k}\frac{\left(e^{\frac{1}{k}}\right)^k - 1}{e^{1/k}-1} \\ 0 & 1 \end{bmatrix}.$$

Next,

$$\exp(X+A) = \exp\left(\begin{bmatrix} 1 & 1 \\ 0 & 0 \end{bmatrix}\right) = \begin{bmatrix} e & e-1 \\ 0 & 1 \end{bmatrix}$$

and since $\lim_{k\to\infty}\frac{1}{k}\frac{\left(e^{\frac{1}{k}}\right)^k - 1}{e^{1/k}-1} = e-1$, it follows that

$$\exp(X+A) = \lim_{k\to\infty}\left(\exp\left(\frac{1}{k}X\right)\exp\left(\frac{1}{k}A\right)\right)^k.$$

We shall next, prove that the observations above hold for arbitrary matrices X, $Y \in M(n,\mathbb{C})$. First, we need some intermediate results.

Lemma 3.12 Suppose that $\|Y\|_2 < \frac{1}{2}$ and $\log(I+Y) = Y + Y_1$ where $Y_1 \in M(n,\mathbb{C})$ and $\|Y_1\|_2 < (4\ln 2 - 2)\cdot\|Y\|_2^2$.

Proof: Since

$$\begin{aligned}
\log(I+Y) - Y &= \sum_{k=1}^{\infty}\frac{(-1)^{k-1}}{k}(I+Y-I)^k - Y \\
&= \sum_{k=1}^{\infty}\frac{(-1)^{k-1}}{k}(Y)^k - Y \\
&= \sum_{k=2}^{\infty}\frac{(-1)^{k-1}}{k}(Y)^k \\
&= Y^2\sum_{k=2}^{\infty}\frac{(-1)^{k-1}}{k}Y^{k-2},
\end{aligned}$$

letting $Y_1 = \sum_{k=2}^{\infty}\frac{(-1)^{k-1}}{k}(Y)^k$, it follows that $\log(I+Y) = Y_1 + Y$. Moreover, appealing to the fact that $\|Y\|_2 < \frac{1}{2}$, the series $\sum_{k=2}^{\infty}\frac{(-1)^{k-1}}{k}Y^{k-2}$ is convergent to some element X in $M(n,\mathbb{C})$ and

$$\|X\|_2 \le \sum_{k=2}^{\infty}\frac{1}{k}\left(\frac{1}{2}\right)^{k-2} = 4\ln 2 - 2.$$

Letting $Y_1 = Y^2X$, it follows that

$$\|Y_1\|_2 = \left\|Y^2X\right\|_2 \le \|Y\|_2^2\cdot\|X\|_2 \le (4\ln 2 - 2)\cdot\|Y\|_2^2.$$

Proposition 3.50 (Lie product formula) For any $X, Y \in M(n, \mathbb{C})$,

$$\exp(X+Y) = \lim_{k\to\infty}\left(\exp\left(\frac{1}{k}X\right)\exp\left(\frac{1}{k}Y\right)\right)^k.$$

Proof: Given any natural number j, there exists a polynomial function p_j of X and Y such that (a)

$$\left(\sum_{l_1=0}^{j}\frac{1}{l_1!}\left(\frac{1}{k}X\right)^{l_1}\right)\left(\sum_{l_2=0}^{j}\frac{1}{l_2!}\left(\frac{1}{k}Y\right)^{l_2}\right)$$

$$=\left(I+\frac{1}{k}X+\sum_{l_1=2}^{j}\frac{1}{l_1!}\left(\frac{1}{k}X\right)^{l_1}\right)\left(I+\frac{1}{k}Y+\sum_{l_2=2}^{j}\frac{1}{l_2!}\left(\frac{1}{k}Y\right)^{l_2}\right)$$

$$=I+\frac{1}{k}Y+\sum_{l_2=2}^{j}\frac{1}{l_2!}\left(\frac{1}{k}Y\right)^{l_2}+\frac{1}{k}X+\frac{1}{k}X\left(\frac{1}{k}Y+\sum_{l_2=2}^{j}\frac{1}{l_2!}\left(\frac{1}{k}Y\right)^{l_2}\right)$$

$$+\left(\sum_{l_1=2}^{j}\frac{1}{l_1!}\left(\frac{1}{k}X\right)^{l_1}\right)\left(I+\frac{1}{k}Y+\sum_{l_2=2}^{j}\frac{1}{l_2!}\left(\frac{1}{k}Y\right)^{l_2}\right)$$

$$=I+\frac{1}{k}X+\frac{1}{k}Y+\frac{1}{k^2}p_j(X,Y,k),$$

where

$$p_j(X,Y,k)=2\sum_{l_2=2}^{j}\frac{1}{k^{l_2-2}l_2!}Y^{l_2}+XY$$

$$+\left(\sum_{l_1=2}^{j}\frac{1}{l_1!k^{l_1-2}}X^{l_1}\right)\left(I+\frac{1}{k}Y+\sum_{l_2=2}^{j}\frac{1}{l_2!}\left(\frac{1}{k}Y\right)^{l_2}\right),$$

(b) $p(X,Y,k)=\lim_{j\to\infty}p_j(X,Y,k)$ is a square-matrix of order n and (c) $\exp\left(\frac{X}{k}\right)\exp\left(\frac{Y}{k}\right)=I+\frac{1}{k}X+\frac{1}{k}Y+\frac{1}{k^2}p(X,Y,k)$. Note that

$$\|p(X,Y,k)\|_2=\left\|\lim_{j\to\infty}p_j(X,Y,k)\right\|_2.$$

Next, we choose open sets U_O of $0\in M(n,\mathbb{C})$ and V_1 of $I\in M(n,\mathbb{C})$ such that the restriction of the exponential map to U_O defines a diffeomorphism between U_O and V_1 with the logarithmic function restricted to V_1 as its inverse. By choosing k sufficiently large, we ensure that $\exp\left(\frac{X}{k}\right)\exp\left(\frac{Y}{k}\right)\in V_1$ and that the scaled element

$\frac{1}{k}X+\frac{1}{k}Y+\frac{1}{k^2}p(X,Y,k)$ satisfies $\|Z\|_2<\frac{1}{2}$ within U_0.

Appealing to Lemma 3.2,

$$\log\left(I + \frac{1}{k}X + \frac{1}{k}Y + \frac{1}{k^2}p(X,Y,k)\right) = \frac{1}{k}X + \frac{1}{k}Y + \frac{1}{k^2}p(X,Y,k) + Y_1,$$

and

$$\|Y_1\|_2 < (4\ln 2 - 2)\cdot\left\|\frac{1}{k}X + \frac{1}{k}Y + \frac{1}{k^2}p(X,Y,k)\right\|_2^2 = \frac{4\ln 2 - 2}{k^2}\left\|X + Y + \frac{1}{k}p(X,Y,k)\right\|_2^2.$$

Hence,

$$\exp\left(\frac{1}{k}X\right)\exp\left(\frac{1}{k}Y\right) = \exp\left(\frac{1}{k}X + \frac{1}{k}Y + \frac{1}{k^2}p(X,Y,k) + Y_1\right)$$

and

$$\left(\exp\left(\frac{1}{k}X\right)\exp\left(\frac{1}{k}Y\right)\right)^k = \exp\left(X + Y + \frac{1}{k}p(X,Y,k) + kY_1\right).$$

Now $\left(\frac{1}{k}p(X,Y,k) + kY_1\right)_{k\in\mathbb{N}}$ converges to the zero matrix and since the exponential map is continuous, we obtain that

$$\lim_{k\to\infty}\left(\exp\left(\frac{1}{k}X\right)\exp\left(\frac{1}{k}Y\right)\right)^k = \exp(X+Y)$$

as desired.

Proposition 3.51 Let $X \in M(n,\mathbb{C})$. Then $\det(\exp X) = e^{\mathrm{tr}(X)}$.

Proof: Given $X \in M(n,\mathbb{C})$, there exists an invertible matrix Q such that

$$QXQ^{-1} = \begin{bmatrix} J_1 & & \\ & \ddots & \\ & & J_r \end{bmatrix} = J$$

is a Jordan Canonical form of X. In other words, each block in QXQ^{-1} is of the form

$$J_k = \lambda_k \quad \text{or} \quad J_k = \begin{bmatrix} \lambda_k & 1 & & & \\ & \lambda_k & 1 & & \\ & & \lambda_k & \ddots & \\ & & & \ddots & 1 \\ & & & & \lambda_k \end{bmatrix} \text{for } \lambda_k \in \mathbb{C}.$$

Next, since $J_k = \lambda_k I + (J_k - \lambda_k I)$ and $\lambda_k I$ and $J_k - \lambda_k I$ commute with each other,

$$\exp(J_k) = \exp(\lambda_k I + (J_k - \lambda_k I)) = \exp(\lambda_k I)\exp(J_k - \lambda_k I)$$

and

$$\begin{aligned}\det(\exp X) &= \det(\exp J)\\ &= \prod_{k=1}^{r}\det(J_k)\\ &= \prod_{k=1}^{r} e^{\operatorname{tr}(J_k)}\\ &= e^{\sum_{k=1}^{r}\operatorname{tr}(J_k)} = e^{\operatorname{tr}(X)}.\end{aligned}$$

Let G be a matrix Lie group. Then G is a submanifold of the general linear group $GL(n,\mathbb{C})$. Next, let $\mathfrak{g}$ be the vector space of left-invariant vector fields on G and let $T_I(G)$ be the tangent space at the identity element of G. Let $T_I(G)$ be the tangent space at the identity element of the Lie group G. Given $X, Y \in T_I(G)$, let $\tilde{X}$ and $\tilde{Y}$ be left-invariant vector fields on G such that $X = \tilde{X}_I$ and $Y = \tilde{Y}_I$. Then $[\tilde{X}, \tilde{Y}]$ is a left-invariant vector field on G and we define the bracket of X, Y as follows: $[X, Y] = [\tilde{X}, \tilde{Y}]_I$ (see Remark 3.2).

Given an invertible matrix $A \in GL(n,\mathbb{C})$, let $L_A : M(n,\mathbb{C}) \to M(n,\mathbb{C})$ be the linear map defined by $L_A X = AX$. For a fixed $X \in T_I(G)$, the mapping $A \mapsto (L_A)_* X = AX$ defines a left-invariant vector field on the Lie group G. Next, $[\tilde{X}, \tilde{Y}] = \widetilde{[X, Y]}$ and

$$[X, Y] = [\tilde{X}, \tilde{Y}]_I = \widetilde{[X, Y]}_I = I(XY - YX) = XY - YX.$$

Moreover, the Lie algebra obtained by endowing $T_I(G)$ with the Lie bracket $(X, Y) \mapsto [X, Y] = XY - YX$ is a Lie algebra which is isomorphic to $\mathfrak{g}$ equipped with the Lie bracket $(\tilde{X}, \tilde{Y}) \mapsto [\tilde{X}, \tilde{Y}]$.

Proposition 3.52 $T_I(G) = \{X \in M(n,\mathbb{C}) : \exp(tX) \in G \text{ for all } t \in \mathbb{R}\}$.

Proof: Let $c_X : \mathbb{R} \to G$ be an integral curve of this left-invariant vector field passing through X. In other words, $c_X(0) = X$, and $c'_X(t) = c_X(t)X$. Solving this system of differential equations, we obtain that $c_X(t)$ is given by the matrix exponential of tX. In other words,

$$c_X(t) = \sum_{k=0}^{\infty}\frac{(tX)^k}{k!}.$$

This shows that the exponential map defines a smooth function between $T_I(G)$ and G and we conclude that

$$T_I(G) \subseteq \{X \in M(n,\mathbb{C}) : \exp(tX) \in G \quad \text{for all } t \in \mathbb{R}\}.$$

Next, let $Y \in M(n, \mathbb{C})$ such that $\exp(tY) \in G$ for all $t \in \mathbb{R}$. Define $c_Y : \mathbb{R} \to G$ such that $c_Y(t) = \exp tY$. Then $c_Y(0) = I$ and $c_Y'(0) = Y$. This shows that $Y \in T_I(G)$.

By identifying the $\mathfrak{g}$ with the tangent space of the identity element in G, from here on, we abuse notation by letting

$$\mathfrak{g} = \{X \in M(n, \mathbb{C}) : \exp(tX) \in G \quad \text{for all } t \in \mathbb{R}\}.$$

Example 3.32 Let G be the three-dimensional Heisenberg Lie group given by matrices

$$p(x, y, z) = \begin{bmatrix} 1 & x & z \\ 0 & 1 & y \\ 0 & 0 & 1 \end{bmatrix}, \quad x, y, z \in \mathbb{R}. \tag{3.5}$$

Since

$$\log(p(x, y, z)) = \begin{bmatrix} 0 & x & z - \frac{xy}{2} \\ 0 & 0 & y \\ 0 & 0 & 0 \end{bmatrix}$$

and

$$\exp \begin{bmatrix} 0 & x & z - \frac{xy}{2} \\ 0 & 0 & y \\ 0 & 0 & 0 \end{bmatrix} = \begin{bmatrix} 1 & x & z \\ 0 & 1 & y \\ 0 & 0 & 1 \end{bmatrix},$$

it is clear that the exponential is a diffeomorphism between the Lie algebra

$$\mathfrak{g} = \left\{ X(a, b, c) = \begin{bmatrix} 0 & a & c \\ 0 & 0 & b \\ 0 & 0 & 0 \end{bmatrix} : a, b, c \in \mathbb{R} \right\}$$

and G.

Example 3.33 Let G be the $ax + b$ Lie group consisting of matrices of the form

$$p(a, x) = \begin{bmatrix} a & x \\ 0 & 1 \end{bmatrix}, \quad a > 0, x \in \mathbb{R}.$$

Then for $a \neq 1$,

$$\log(p(a, x)) = \log\left(\begin{bmatrix} a & x \\ 0 & 1 \end{bmatrix}\right) = \begin{bmatrix} \log a & \dfrac{x \log a}{a - 1} \\ 0 & 0 \end{bmatrix}$$

and when $a = 1$

$$\log(p(1, x)) = \log\left(\begin{bmatrix} 1 & x \\ 0 & 1 \end{bmatrix}\right) = \begin{bmatrix} 0 & x \\ 0 & 0 \end{bmatrix}.$$

3.6.1 The Adjoint Maps

Proposition 3.53 Let G, H be matrix Lie groups, with Lie algebra $\mathfrak{g}$ and $\mathfrak{h}$, respectively. Next, let $f : G \to H$ be a smooth homomorphism and let df_I be the differential of f at the identity element of G. Then df_I defines a Lie algebra homomorphism between $\mathfrak{g}$ and $\mathfrak{h}$. Moreover, $f(\exp X) = \exp\left(df_I(X)\right)$ for all $X \in \mathfrak{g}$.

Proof: Observe that given $X \in \mathfrak{g}$,

$$df_I(X) = \frac{d}{dt}\left(f(\exp tX)\right)|_{t=0}.$$

Next,

$$\begin{aligned}\frac{d}{dt}\left(f(\exp(s+t)X)\right)|_{t=0} &= \lim_{h\to 0}\frac{f(\exp(s+h)X) - f(\exp(s)X)}{h} \\ &= \frac{d}{ds}\left(f(\exp sX)\right)\end{aligned}$$

and as such,

$$\begin{aligned}\frac{d}{ds}\left(f(\exp sX)\right) &= \frac{d}{dt}\left(f(\exp(s+t)X)\right)|_{t=0} \\ &= \frac{d}{dt}\left(f(\exp sX)f(\exp tX)\right)|_{t=0} \\ &= f(\exp sX)\frac{d}{dt}\left(f(\exp tX)\right)|_{t=0} \\ &= f(\exp sX)\, df_I(X).\end{aligned}$$

Letting c be the curve defined by $c(s) = f(\exp sX)$, it follows that $c'(s) = c(s)\, df_I(X)$ and additionally c evaluated at the zero element is the identity map. Therefore $c(t) = \exp\left(t\, df_I(X)\right)$ and $f(\exp tX) = \exp\left(t\, df_I(X)\right)$. To complete the proof, it remains next to show that $df_I : \mathfrak{g} \to \mathfrak{h}$ is a Lie algebra homomorphism. To this end, it suffices to show that $df_I([X,Y]) = \left[df_I X, df_I Y\right]$ for all vectors X, Y in the Lie algebra of G. Noting that

$$\begin{aligned}f(\exp tX \exp sY \exp(-tX)) &= f(\exp tX)f(\exp sY)f(\exp(-tX)) \\ &= \exp\left(t\, df_I(X)\right)\exp\left(sf_I(Y)\right)\exp\left(-t\, df_I(X)\right)\end{aligned}$$

and differentiating the map

$$s \mapsto f(\exp tX \exp sY \exp(-tX)) = \exp\left(t\, df_I(X)\right)\exp\left(sf_I(Y)\right)\exp\left(-t\, df_I(X)\right)$$

at $s = 0$, we obtain

$$f(\exp tX \exp Y \exp(-tX)) = \exp\left(t\, df_I(X)\right)\exp\left(f_I(Y)\right)\exp\left(-t\, df_I(X)\right).$$

Taking the derivative of

$$t \mapsto f(\exp tX \exp Y \exp(-tX)) = \exp\left(t\, df_I(X)\right)\exp\left(f_I(Y)\right)\exp\left(-t\, df_I(X)\right)$$

at $t = 0$ gives $df_I([X,Y]) = \left[df_I X, df_I Y\right]$ as desired.

Let G be a matrix Lie group with Lie algebra $\mathfrak{g}$ identified with $T_I(G)$. Given a smooth curve $c : \mathbb{R} \to G$ such that $c(0) = I$, the map $c_A : t \mapsto Ac(t)A^{-1}$ defines a smooth curve in G for any fixed invertible matrix A such that $c_A(0) = Ac(0)A^{-1} = I$. This shows that for each fixed $A \in G$, c_A can be viewed as an element of the Lie algebra.

Next, given $A \in G$ and $X \in \mathfrak{g}$, define $Ad(A)X = AXA^{-1}$. Then the vector space $\mathfrak{g}$ is invariant under the action of each linear operator $Ad(A)$. In other words, we may view each $Ad(A)$ as an endomorphism of the Lie algebra of G. Moreover, since $Ad(AB) = Ad(A)Ad(B)$ for all $A, B \in G$ and since $(Ad(A))^{-1} = Ad\left(A^{-1}\right)$, it follows that the mapping $A \mapsto Ad(A)$ defines a smooth homomorphism between G and the group of invertible linear operators acting on $\mathfrak{g}$.

We next consider the smooth curve $t \mapsto Ad(c(t))X$. Then

$$\frac{d}{dt}(Ad(c(t))X)|_{t=0} = \lim_{h\to 0}\frac{1}{h}(Ad(c(h))X - X) \in \mathfrak{g}.$$

Since $c(t)c(t)^{-1} = I$, by the product rule of derivative, we verify that

$$0 = \frac{d}{dt}\left(c(t)c(t)^{-1}\right) = c'(t)c(t)^{-1} + c(t)\left(c(t)^{-1}\right)'.$$

This shows that

$$\left(c(t)^{-1}\right) = -c(t)c'(t)c(t)^{-1}.$$

Next,

$$\begin{aligned}
\frac{d}{dt}(Ad(c(t))X) &= \frac{d}{dt}\left(c(t)Xc(t)^{-1}\right)\\
&= c(t)'Xc(t)^{-1} + c(t)\frac{d}{dt}\left(Xc(t)^{-1}\right)\\
&= c(t)'Xc(t)^{-1} + c(t)X\frac{d}{dt}\left(c(t)^{-1}\right)\\
&= c(t)'Xc(t)^{-1} + c(t)X\left(-c(t)c'(t)c(t)^{-1}\right)\\
&= c(t)'Xc(t)^{-1} - c(t)Xc(t)c'(t)c(t)^{-1}
\end{aligned}$$

and since $c(0)$ is the identity matrix,

$$\begin{aligned}
\frac{d}{dt}(Ad(c(t))X)|_{t=0} &= c(0)'Xc(0)^{-1} - c(0)Xc(0)c'(0)c(0)^{-1}\\
&= c(0)'X - Xc'(0)\\
&= \left[c(0)', X\right].
\end{aligned}$$

Definition 3.18 The linear map $ad : \mathfrak{g} \to \mathrm{End}(\mathfrak{g})$ such that $ad(X)$ is an endomorphism of $\mathfrak{g}$ defined as follows: $ad(X)Y = [X, Y] = XY - YX$.

Recalling that a linear operator $T \in \mathrm{End}(\mathfrak{g})$ is a derivation if

$$T([X, Y]) = [X, TY] + [TX, Y],$$

then it is clear that each endomorphism $ad\,(X)$ is a derivation of the Lie algebra $\mathfrak{g}$, and the calculations above show that

$$\frac{d}{dt}\,(Ad\,(c\,(t))\,X)|_{t=0} = ad\,\left(c(0)'\right) X.$$

Proposition 3.54 Let G be a matrix Lie group with Lie algebra $\mathfrak{g}$. Then the differential of the homomorphism $A \mapsto Ad\,(A)$ between G and the group of invertible linear operators acting on $\mathfrak{g}$ is the map $ad\,:\,\mathfrak{g} \to \mathrm{End}\,(\mathfrak{g})$.

Proof: First, observe that

$$\begin{aligned}\frac{d}{dt}\,(Ad\,(\exp tX)\,Y)\bigg|_{t=0} &= \frac{d}{dt}\,((\exp tX)\,Y)\,A\,(\exp\,(-tX))\bigg|_{t=0}\\ &= XY - YX\\ &= [X, Y]\\ &= ad\,(X)\,Y.\end{aligned}$$

The calculations in the first equality are left to be verified by the reader. Consequently, ad defines a Lie algebra homomorphism between $\mathfrak{g}$ and $\mathrm{End}\,(\mathfrak{g})$ and additionally,

$$Ad\,(\exp X) = \exp\,(adX)\,.$$

Proposition 3.55 Let G be a matrix Lie group with Lie algebra $\mathfrak{g}$. For sufficiently small $t \in \mathbb{R}$,

$$\exp tX \exp tY = \exp\left(\sum_{k=1}^{\infty}\frac{(-1)^{k-1}}{k}\left(\sum_{i,j\geq 0, i+j\geq 1}\frac{t^i t^j X^i Y^j}{i!j!}\right)^k\right).$$

Proof: First, since $t \mapsto \exp X$ and $t \mapsto \exp Y$ are smooth curves passing through the identity element of G, it is clear that $t \mapsto \exp X \exp Y$ is smooth as well. Moreover, $t \mapsto \exp X \exp Y$ evaluated at zero is the identity element in G. Since the exponential map is locally invertible at zero, it follows that for a sufficiently small neighborhood of 0, one has $\exp tX \exp tY = \exp Z\,(t)$ and

$$\begin{aligned}W\,(t) &= \log\,(\exp tX \exp tY)\\ &= \sum_{k=1}^{\infty}\frac{(-1)^{k-1}}{k}(\exp tX \exp tY - I)^k\\ &= \sum_{k=1}^{\infty}\frac{(-1)^{k-1}}{k}\left(\left(\sum_{i=0}^{\infty}\frac{t^i X^i}{i!}\right)\left(\sum_{j=0}^{\infty}\frac{t^j Y^j}{j!}\right) - I\right)^k\\ &= \sum_{k=1}^{\infty}\frac{(-1)^{k-1}}{k}\left(\sum_{i,j\geq 0, i+j\geq 1}\frac{t^i t^j X^i Y^j}{i!j!}\right)^k.\end{aligned}$$

This shows that

$$\exp tX \exp tY = \exp\left(\sum_{k=1}^{\infty}\frac{(-1)^{k-1}}{k}\left(\sum_{i,j\geq 0, i+j\geq 1}\frac{t^i t^j X^i Y^j}{i!j!}\right)^k\right).$$

Remark 3.6 Let G be a matrix Lie group with Lie algebra $\mathfrak{g}$ and let $A \in G$. Define $L_A : G \to G, R_A : G \to G$ such that $L_A(B) = AB$ and $R_A(B) = BA$, respectively. Then $Ad(A)$ is equal to the differential of the map $L_A \circ R_A$ at the identity element of G.

Example 3.34 Let G be the three-dimensional Heisenberg Lie group with Lie algebra $\mathfrak{g}$ spanned by matrices

$$Z = \begin{bmatrix} 0 & 0 & 1 \\ 0 & 0 & 0 \\ 0 & 0 & 0 \end{bmatrix}, \quad Y = \begin{bmatrix} 0 & 0 & 0 \\ 0 & 0 & 1 \\ 0 & 0 & 0 \end{bmatrix}, \quad X = \begin{bmatrix} 0 & 1 & 0 \\ 0 & 0 & 0 \\ 0 & 0 & 0 \end{bmatrix}.$$

Then regarding $ad(\mathfrak{g})$ as a vector subspace of End$(\mathfrak{g})$, with respect to the ordered basis $\{Z, Y, X\}$, the matrix representation of $ad(Z), ad(Y), ad(X)$ is

$$\begin{bmatrix} 0 & 0 & 0 \\ 0 & 0 & 0 \\ 0 & 0 & 0 \end{bmatrix}, \quad \begin{bmatrix} 0 & 0 & -1 \\ 0 & 0 & 0 \\ 0 & 0 & 0 \end{bmatrix}, \quad \begin{bmatrix} 0 & 1 & 0 \\ 0 & 0 & 0 \\ 0 & 0 & 0 \end{bmatrix}$$

respectively. This shows that

$$Ad(\exp tZ) = \begin{bmatrix} 1 & 0 & 0 \\ 0 & 1 & 0 \\ 0 & 0 & 1 \end{bmatrix}, \quad Ad(\exp tY) = \begin{bmatrix} 1 & 0 & -t \\ 0 & 1 & 0 \\ 0 & 0 & 1 \end{bmatrix}, \quad \text{and} \quad Ad(\exp tX) = \begin{bmatrix} 1 & t & 0 \\ 0 & 1 & 0 \\ 0 & 0 & 1 \end{bmatrix}.$$

Example 3.35 Let G be the two-dimensional affine Lie group with Lie algebra $\mathfrak{g}$ spanned by matrices

$$X = \begin{bmatrix} 0 & 1 \\ 0 & 0 \end{bmatrix}, \quad A = \begin{bmatrix} 1 & 0 \\ 0 & 0 \end{bmatrix}.$$

Regarding $ad(\mathfrak{g})$ as a vector subspace of End$(\mathfrak{g})$, with respect to the ordered basis $\{X, A\}$, the matrix representation of $ad(X), ad(A)$ is

$$\begin{bmatrix} 0 & -1 \\ 0 & 0 \end{bmatrix}, \quad \begin{bmatrix} 1 & 0 \\ 0 & 0 \end{bmatrix}$$

respectively. This shows that

$$Ad(\exp tX) = \begin{bmatrix} 1 & -t \\ 0 & 1 \end{bmatrix} \quad \text{and} \quad Ad(\exp tA) = \begin{bmatrix} e^t & 0 \\ 0 & 1 \end{bmatrix}.$$

Example 3.36 Let $\mathfrak{g}$ be a six-dimensional Lie algebra spanned by matrices

$$X_1 = \begin{bmatrix} 0&0&0&0\\0&0&0&0\\0&0&0&0\\0&0&0&0 \end{bmatrix}, \quad X_2 = \begin{bmatrix} 0&0&0&0\\0&0&0&0\\0&0&0&0\\0&0&0&0 \end{bmatrix}, \quad X_3 = \begin{bmatrix} 0&0&0&0\\0&0&0&0\\0&0&0&0\\0&0&0&0 \end{bmatrix}$$

$$X_4 = \begin{bmatrix} 0&0&0&0\\0&0&0&0\\0&0&0&0\\0&0&0&0 \end{bmatrix}, \quad X_5 = \begin{bmatrix} 0&0&0&0\\0&0&0&0\\0&0&0&0\\0&0&0&0 \end{bmatrix}, \quad X_6 = \begin{bmatrix} 0&0&0&0\\0&0&0&0\\0&0&0&0\\0&0&0&0 \end{bmatrix}.$$

Then with to the ordered basis $\{X_1, X_2, \ldots, X_6\}$, the matrix representation of $ad(X_1), \ldots, ad(X_6)$ is

$$\begin{bmatrix} 0&0&0&0&0&0\\0&0&0&0&0&0\\0&0&0&0&0&0\\0&0&0&0&0&0\\0&0&0&0&0&0\\0&0&0&0&0&0 \end{bmatrix}, \quad \begin{bmatrix} 0&0&0&0&0&-1\\0&0&0&0&0&0\\0&0&0&0&0&0\\0&0&0&0&0&0\\0&0&0&0&0&0\\0&0&0&0&0&0 \end{bmatrix}, \quad \begin{bmatrix} 0&0&0&-1&0&0\\0&0&0&0&-1&0\\0&0&0&0&0&0\\0&0&0&0&0&0\\0&0&0&0&0&0\\0&0&0&0&0&0 \end{bmatrix}$$

$$\begin{bmatrix} 0&0&1&0&0&0\\0&0&0&0&0&0\\0&0&0&0&0&0\\0&0&0&0&0&0\\0&0&0&0&0&0\\0&0&0&0&0&0 \end{bmatrix}, \quad \begin{bmatrix} 0&0&0&0&0&0\\0&0&1&0&0&0\\0&0&0&0&0&0\\0&0&0&0&0&-1\\0&0&0&0&0&0\\0&0&0&0&0&0 \end{bmatrix}, \quad \begin{bmatrix} 0&1&0&0&0&0\\0&0&0&0&0&0\\0&0&0&0&0&0\\0&0&0&0&1&0\\0&0&0&0&0&0\\0&0&0&0&0&0 \end{bmatrix}$$

respectively.

3.6.1.1 Lie's Theorem

Let G be a Lie group with Lie algebra $\mathfrak{g}$. Define $D(\mathfrak{g}) = [\mathfrak{g}, \mathfrak{g}]$ to be the real span of all elements of the form $[X, Y]$ where X, Y are arbitrary vectors in the Lie algebra. Then $D(\mathfrak{g})$ is an ideal of $\mathfrak{g}$. In other words, for any element $X \in D(\mathfrak{g})$ and for arbitrary $Y \in \mathfrak{g}$, $[Y, X] \in \mathfrak{g}$. Indeed, letting $X = \sum_{k=1}^{m} c_k [X_k, Y_k]$ for some real numbers c_k, it is easy to verify that

$$[Y, X] = \left[Y, \sum_{k=1}^{m} c_k [X_k, Y_k] \right] = \sum_{k=1}^{m} c_k [Y, [X_k, Y_k]] \in \mathfrak{g}.$$

Generally, define $D^{k+1}(\mathfrak{g}) = [D^k(\mathfrak{g}), D^k(\mathfrak{g})]$ and call $D^k(\mathfrak{g})$ the kth-derived algebra of $\mathfrak{g}$. We will be concerned here with the structure of Lie algebras for which the following sequence $(\dim D^k(\mathfrak{g}))_{k=0}^{\infty}$ only has finitely many nonzero terms. Such

Lie algebras are called solvable Lie algebras, and we formalize this concept with the following definition.

Definition 3.19 A Lie algebra $\mathfrak{g}$ is said to be solvable if there exists a natural number m such that $\dim(D^m(\mathfrak{g})) = 0$.

Definition 3.20 A Lie algebra is said to be simple if it has no nontrivial ideal and is not commutative.

The examples below prove that the class of solvable Lie algebras is quite large.

Example 3.37 For any commutative Lie algebra $\mathfrak{g}$, $\dim(D(\mathfrak{g})) = 0$, and as a result, every commutative Lie algebra is automatically solvable.

Example 3.38 Let $\mathfrak{g}$ be the Lie algebra spanned by

$$A = \begin{bmatrix} 1 & 0 \\ 0 & 0 \end{bmatrix}, \quad X = \begin{bmatrix} 0 & 1 \\ 0 & 0 \end{bmatrix}.$$

Then $\left[aA + xX, bA + yX\right] = (ay - bx)X$. Therefore, $D(\mathfrak{g})$ is the one-dimensional ideal spanned by X. Since any one-dimensional Lie algebra is commutative, it follows that $\dim\left(D^2(\mathfrak{g})\right) = 0$ and this shows that $\mathfrak{g}$ is a solvable Lie algebra of dimension two.

Generally, the Lie algebra $\mathfrak{ut}(n, \mathbb{R})$ of upper-triangular matrices of the form

$$\begin{bmatrix} \lambda_1 & x_{(1,1)} & x_{(1,2)} & x_{(1,3)} & \cdots & x_{(1,n-1)} \\ 0 & \lambda_2 & x_{(2,1)} & x_{(2,3)} & \cdots & x_{(2,n-2)} \\ 0 & 0 & \lambda_3 & x_{(3,1)} & \cdots & x_{(3,n-3)} \\ 0 & \ddots & \ddots & \ddots & \cdots & \vdots \\ \vdots & \cdots & 0 & 0 & \lambda_{n-1} & x_{(n-1,1)} \\ 0 & 0 & \cdots & 0 & 0 & \lambda_n \end{bmatrix}$$

is solvable since $\dim(D^n(\mathfrak{g})) = 0$.

The examples below illustrate that generally, the class of solvable Lie algebras is a proper subset of the collection of all Lie algebras.

Example 3.39 Let $\mathfrak{g}$ be a Lie algebra consisting of matrices on the form

$$X\left(x_1, x_2, x_3, x_4\right) = \sum_{k=1}^{4} x_k X_k = \begin{bmatrix} x_3 & x_1 \\ x_4 & x_2 \end{bmatrix}.$$

Then

$$[X(x), X(y)] = \begin{bmatrix} x_1 y_4 - y_1 x_4 & x_1 y_2 - x_2 y_1 - x_1 y_3 + x_3 y_1 \\ x_2 y_4 - x_4 y_2 - x_3 y_4 + x_4 y_3 & y_1 x_4 - x_1 y_4 \end{bmatrix}$$

and one can easily verify that for any natural number m, $\dim(D^m(\mathfrak{g})) > 0$. This shows that the Lie algebra of the general Lie group $GL(2,\mathbb{R})$ is not solvable. Since the subalgebra $\mathfrak{z}$ of all constant multiples of the identity commutes with every element of $\mathfrak{g}$, it follows that $\mathfrak{z}$ is a nontrivial ideal of $\mathfrak{g}$. In fact, since

$$\begin{bmatrix} 1 & 0 \\ 0 & 1 \end{bmatrix}, \quad \begin{bmatrix} 0 & 0 \\ 1 & 0 \end{bmatrix}, \quad \begin{bmatrix} 0 & 1 \\ 0 & 0 \end{bmatrix}, \quad \begin{bmatrix} 1 & 0 \\ 0 & -1 \end{bmatrix}$$

is a basis of the vector space $\mathfrak{g}$, it follows that $\mathfrak{g} \simeq \mathbb{R} \oplus \mathfrak{sl}(2,\mathbb{R})$. As a result, there exist nonsolvable Lie algebras which are not simple either.

Example 3.40 Although $\mathfrak{so}(2,\mathbb{R})$ is commutative and thus solvable, we can easily check that its counterpart $\mathfrak{so}(3,\mathbb{R})$ is not solvable. To verify that this, let

$$X_1 = \begin{bmatrix} 0 & -1 & 0 \\ 1 & 0 & 0 \\ 0 & 0 & 0 \end{bmatrix}, \quad X_2 = \begin{bmatrix} 0 & 0 & -1 \\ 0 & 0 & 0 \\ 1 & 0 & 0 \end{bmatrix}, \quad X_3 = \begin{bmatrix} 0 & 0 & 0 \\ 0 & 0 & -1 \\ 0 & 1 & 0 \end{bmatrix}$$

be a basis for $\mathfrak{so}(3,\mathbb{R})$. With respect to this ordered basis, the matrix representation of X_1, X_2, X_3 is given by

$$ad(X_1) = \begin{bmatrix} 0 & 0 & 0 \\ 0 & 0 & -1 \\ 0 & 1 & 0 \end{bmatrix}, \quad ad(X_2) = \begin{bmatrix} 0 & 0 & -1 \\ 0 & 0 & 0 \\ 1 & 0 & 0 \end{bmatrix}, \quad ad(X_3) = \begin{bmatrix} 0 & -1 & 0 \\ 1 & 0 & 0 \\ 0 & 0 & 0 \end{bmatrix}.$$

The above shows that the kernel of the Lie homomorphism ad is trivial. In other words, $\mathfrak{so}(3,\mathbb{R})$ has no proper and nontrivial ad-invariant subspace. In fact, one easily verifies that for any natural number k, $\dim\left(ad^k(\mathfrak{so}(3,\mathbb{R}))\right) = 3$. This shows that $\mathfrak{so}(3,\mathbb{R})$ is not a solvable Lie algebra. In fact $\mathfrak{so}(3,\mathbb{R})$ is a simple Lie algebra.

Proposition 3.56 The following hold true:

1) Let $\mathfrak{h}$ be an ideal of $\mathfrak{g}$. If $\mathfrak{g}$ is solvable then $\mathfrak{g}/\mathfrak{h}$ is solvable as well.
2) If $\mathfrak{h}$ is a solvable ideal of $\mathfrak{g}$ and if $\mathfrak{g}/\mathfrak{h}$ is a solvable quotient Lie algebra, then $\mathfrak{g}$ is solvable as well.
3) If $\mathfrak{h}_1$ and $\mathfrak{h}_2$ are solvable ideals then $\mathfrak{h}_1 + \mathfrak{h}_2$ is a solvable ideal.

Proof: Part 1 follows from the observation that $D^k(\mathfrak{g}/\mathfrak{h})$ is Lie isomorphic to $D^k(\mathfrak{g})/\left(\mathfrak{h} \cap D^k(\mathfrak{g})\right)$. For Part 2, suppose that $\mathfrak{h}$ and $\mathfrak{g}/\mathfrak{h}$ are solvable. Then there exists an integer m such that $\dim D^m(\mathfrak{g}/\mathfrak{h}) = 0$. Consequently, $D^m(\mathfrak{g})$ must be contained in the ideal $\mathfrak{h}$. By assumption, there exists an integer n such that $D^n(\mathfrak{h})$ is trivial as well and as a result, $\dim D^{m+n}(\mathfrak{g}) = 0$. Finally, to prove Part 3, observe that $\left(\mathfrak{h}_1 + \mathfrak{h}_2\right)/\mathfrak{h}_2$ is Lie isomorphic to $\mathfrak{h}_1/\left(\mathfrak{h}_1 \cap \mathfrak{h}_2\right)$. Therefore, $\left(\mathfrak{h}_1 + \mathfrak{h}_2\right)/\mathfrak{h}_2$ and $\mathfrak{h}_2$ are solvable Lie algebra. Consequently, $\mathfrak{h}_1 + \mathfrak{h}_2$ is a solvable ideal.

Theorem 3.7 (Lie's theorem) Let $\mathfrak{g}$ be a solvable Lie algebra over $\mathbb{C}$ and let $\pi : \mathfrak{g} \to \mathfrak{gl}(n, \mathbb{C})$ be a (complex) Lie algebra homomorphism. Then there exists a nonzero vector $v \in \mathbb{C}^n$ such that for every $X \in \mathfrak{g}$, $\pi(X)v = \lambda(X)v$ for some linear form λ defined on $\mathfrak{g}$.

Proof: We will prove this result by induction on the dimension of $\mathfrak{g}$. For the base case, if $\dim \mathfrak{g} = 1$, then $\dim \pi(\mathfrak{g}) \leq 1$ and choosing an eigenvector v for $\pi(X)$, gives the desired result. Suppose now that the result holds for all Lie algebras of dimension at most $n-1$. Let $\mathfrak{g}$ be a solvable Lie algebra such that $\dim \mathfrak{g} = n$ and let $\mathfrak{h}$ be a subalgebra of $\mathfrak{g}$ of dimension $n-1$ containing the ideal $D(\mathfrak{g}) = [\mathfrak{g}, \mathfrak{g}] = ad(\mathfrak{g})\mathfrak{g}$. Clearly, the algebra $\mathfrak{h}$ exists since $\mathfrak{g}$ is assumed to be solvable. In fact, it is easy to verify that $\mathfrak{h}$ must be an ideal since for arbitrary X in $\mathfrak{h}$ and $A \in \mathfrak{g}$, $[A, X] \in [\mathfrak{g}, \mathfrak{g}] \subseteq \mathfrak{h}$. Moreover, since $\mathfrak{g}/\mathfrak{h}$ is commutative, $\mathfrak{h}$ must be solvable. Let $\pi_{\mathfrak{h}}$ be the map obtained by restricting π to the ideal $\mathfrak{h}$. By the inductive hypothesis, there exists a nonzero vector $w \in \mathbb{C}^n$ such that $\pi_{\mathfrak{h}}(Y)w = \lambda(Y)w$ for all $Y \in \mathfrak{h}$ and a linear functional $\lambda \in \mathfrak{h}^*$. Next, let us write $\mathfrak{g} = \mathfrak{h} \oplus \mathbb{C}X$ for some nonzero element $X \in \mathfrak{g}$. For every natural number k, we define

$$w_k = \pi_{\mathfrak{h}}(X)^k w, \; w = w_0 \text{ and } w_{-1} = 0.$$

Let W be the complex span of the vectors $w_k, k \geq 0$. Clearly, W is invariant under the action of $\pi_{\mathfrak{h}}(X)$. Next, let v be an eigenvector for the endomorphism $\pi_{\mathfrak{h}}(X)$ in the vector space W. We aim to prove that v must be an eigenvector for each $\pi_{\mathfrak{h}}(Y)$ where $Y \in \mathfrak{h}$. First, we will prove by induction that

$$\pi_{\mathfrak{h}}(Y)w_k = \lambda(Y)w_k \left(\text{mod} \sum_{j=0}^{k-1} \mathbb{C}w_j\right).$$

That is, $\pi_{\mathfrak{h}}(Y)w_k = \lambda(Y)w_k + z$ for some vector $z \in \sum_{j=0}^{k-1} \mathbb{C}w_j$. This can be achieved by induction on k. The base case being clear, we shall only focus on the inductive step. To this end, verify that

$$\begin{aligned} \pi_{\mathfrak{h}}(Y)w_{k+1} &= \pi_{\mathfrak{h}}(Y)\pi_{\mathfrak{h}}(X)w_k \\ &= \pi_{\mathfrak{h}}(X)\pi_{\mathfrak{h}}(Y)w_k + \pi_{\mathfrak{h}}[Y, X]w_k. \end{aligned}$$

Now, since $[Y, X] \in \mathfrak{h}$, by the inductive hypothesis,

$$\pi_{\mathfrak{h}}(Y)w_{k+1} = \pi_{\mathfrak{h}}(X)\pi_{\mathfrak{h}}(Y)w_k + \lambda([Y, X])w_k \left(\text{mod} \sum_{j=0}^{k-1} \mathbb{C}w_j\right).$$

Appealing again to the inductive hypothesis, there must exist $z \in \text{mod} \sum_{j=0}^{k-1} \mathbb{C}w_j$ such that

$$\pi_{\mathfrak{h}}(Y)w_k = \lambda(Y)w_k + z$$

and as such,

$$\pi_{\mathfrak{h}}(X)\,\pi_{\mathfrak{h}}(Y)\,w_k = \pi_{\mathfrak{h}}(X)\left(\lambda(Y)\,w_k + z\right) = \lambda(Y)\,\pi_{\mathfrak{h}}(X)\,w_k + \pi_{\mathfrak{h}}(X)\,z.$$

Therefore,

$$\pi_{\mathfrak{h}}(Y)\,w_{k+1} = \lambda([Y,X])\,w_k \quad \left(\operatorname{mod} \sum_{j=0}^{k} \mathbb{C}w_j\right).$$

Next, we need to show that $\lambda([Y,X]) = 0$ for all $Y \in \mathfrak{h}$. Next, let $N = \dim W$ be the largest integer such that $\{w_1, \ldots, w_N\}$ is linearly independent. With respect to this ordered basis, the matrix representation of the endomorphism $\pi_{\mathfrak{h}}(Y)$ takes the form

$$\begin{bmatrix} \lambda(Y) & & * \\ & \ddots & \\ & & \lambda(Y) \end{bmatrix};$$

showing that the trace of $\pi_{\mathfrak{h}}(Y)$ is given by $\operatorname{Tr}\left(\pi_{\mathfrak{h}}(Y)\right) = (\dim W)\cdot\lambda(Y)$ and

$$\begin{aligned} \lambda([Y,X])\cdot \dim W &= \operatorname{Tr}\left(\pi_{\mathfrak{h}}([Y,X])\right) \\ &= \operatorname{Tr}\left(\left[\pi_{\mathfrak{h}}(Y), \pi_{\mathfrak{h}}(X)\right]\right) = 0. \end{aligned}$$

Secondly, we will prove that $\pi_{\mathfrak{h}}(Y)\,w_k = \lambda(Y)\,w_k$ for all $Y \in \mathfrak{h}$.

Again, we proceed by induction on k. For $p = 0$,

$$\pi_{\mathfrak{h}}(Y)\,w_0 = \pi_{\mathfrak{h}}(Y)\,w = \lambda(Y)\,w.$$

By the inductive hypothesis,

$$\begin{aligned} \pi_{\mathfrak{h}}(Y)\,w_{k+1} &= \pi_{\mathfrak{h}}(Y)\,\pi_{\mathfrak{h}}(X)\,w_k \\ &= \pi_{\mathfrak{h}}(X)\,\pi_{\mathfrak{h}}(Y)\,w_k + \pi_{\mathfrak{h}}[Y,X]\,w_k \\ &= \pi_{\mathfrak{h}}(X)\,\pi_{\mathfrak{h}}(Y)\,w_k + \lambda[Y,X]\,w_k \\ &= \pi_{\mathfrak{h}}(X)\,\pi_{\mathfrak{h}}(Y)\,w_k \\ &= \pi_{\mathfrak{h}}(X)\,\lambda(Y)\,w_k \\ &= \lambda(Y)\,\pi_{\mathfrak{h}}(X)\,w_k \\ &= \lambda(Y)\,w_{k+1}. \end{aligned}$$

Having established that $\pi_{\mathfrak{h}}(Y)\,z = \lambda(Y)\,z$ for all $Y \in \mathfrak{h}$ and all $z \in W$, it follows that $\pi_{\mathfrak{h}}(Y)\,v = \lambda(Y)\,v$. Therefore, v is an eigenvector for $\pi(X)$ which is an eigenvector for all $\pi(Y)$ with $Y \in \mathfrak{h}$.

We will now derive several useful applications of Lie's theorem. In particular, we will show that when the underlying field is complex (and thus algebraically closed) all elements of a linear solvable Lie algebra can be simultaneously upper-triangularized.

Corollary 3.1 Let $\mathfrak{g}$ be a solvable Lie algebra over the complex numbers and let $\pi : \mathfrak{g} \to \mathfrak{gl}(n, \mathbb{C})$ be a (complex) Lie algebra homomorphism.

1) There exists a basis with respect to every endomorphism such that $\pi(X)$ has a matrix representation of the form

$$\begin{bmatrix} \lambda(X) & & * \\ & \ddots & \\ & & \lambda(X) \end{bmatrix}$$

for some linear functional $\lambda \in \mathfrak{g}^*$.

2) There exist subalgebras $\mathfrak{g}_1, \mathfrak{g}_2, \dots, \mathfrak{g}_{l+1} = \{0\}$ such that (1) $\mathfrak{g}_{k+1}$ is an ideal of $\mathfrak{g}_k$ for $k = 1, \dots, l$, (2) $\dim\left(\mathfrak{g}_{k+1}/\mathfrak{g}_k\right) = 1$ for $k = 1, \dots, l$ and (3) each $\mathfrak{g}_k$ is an ideal in $\mathfrak{g}$.

Proof: For Part 1, we will show this result by induction on n. The base case clearly holds when $n = 1$, since all matrices of order 1 are automatically upper-triangular. By Theorem 3.7, select $v \in \mathbb{C}^n$ such that $\pi(X)v = \lambda(X)v$ for all $X \in \mathfrak{g}$. Then the vector space $\mathbb{C}v$ is invariant under the action of $\pi(X)$. Let $\{v_1 = v, v_2, \dots, v_n\}$ be a basis for V. Next, the homomorphism π induces an action of $\mathfrak{g}$ onto $V/\mathbb{C}v$ and by the inductive hypothesis, since $\dim(V/\mathbb{C}v) = n - 1$, it follows that

$$\pi(X)v_k = \lambda(X)v_k \mod \left(\sum_{j=1}^{k-1} \mathbb{C}v_j\right).$$

To prove Part 2, it suffices to apply Theorem 3.7 to the Lie algebra homomorphism $ad : \mathfrak{g} \to \mathrm{End}(\mathfrak{g})$.

Generally, if $\mathfrak{g}$ is a finite-dimensional Lie algebra defined over the field of real numbers, its maximal solvable ideal $\mathfrak{r}$ is called the radical of $\mathfrak{g}$. Any Lie algebra with a radical of dimension zero is called a semisimple Lie algebra. For instance, every simple Lie algebra is semisimple. But there exist semisimple Lie algebras which are not simple. Take, for instance, the Lie algebra spanned by matrices

$$\begin{bmatrix} 0 & -1 & 0 & 0 & 0 & 0 \\ 1 & 0 & 0 & 0 & 0 & 0 \\ 0 & 0 & 0 & 0 & 0 & 0 \\ 0 & 0 & 0 & 0 & 0 & 0 \\ 0 & 0 & 0 & 0 & 0 & 0 \\ 0 & 0 & 0 & 0 & 0 & 0 \end{bmatrix}, \begin{bmatrix} 0 & 0 & -1 & 0 & 0 & 0 \\ 0 & 0 & 0 & 0 & 0 & 0 \\ 1 & 0 & 0 & 0 & 0 & 0 \\ 0 & 0 & 0 & 0 & 0 & 0 \\ 0 & 0 & 0 & 0 & 0 & 0 \\ 0 & 0 & 0 & 0 & 0 & 0 \end{bmatrix}, \begin{bmatrix} 0 & 0 & 0 & 0 & 0 & 0 \\ 0 & 0 & -1 & 0 & 0 & 0 \\ 0 & 1 & 0 & 0 & 0 & 0 \\ 0 & 0 & 0 & 0 & 0 & 0 \\ 0 & 0 & 0 & 0 & 0 & 0 \\ 0 & 0 & 0 & 0 & 0 & 0 \end{bmatrix}$$

and

$$\begin{bmatrix} 0 & 0 & 0 & 0 & 0 & 0 \\ 0 & 0 & 0 & 0 & 0 & 0 \\ 0 & 0 & 0 & 0 & 0 & 0 \\ 0 & 0 & 0 & 0 & -1 & 0 \\ 0 & 0 & 0 & 1 & 0 & 0 \\ 0 & 0 & 0 & 0 & 0 & 0 \end{bmatrix}, \begin{bmatrix} 0 & 0 & 0 & 0 & 0 & 0 \\ 0 & 0 & 0 & 0 & 0 & 0 \\ 0 & 0 & 0 & 0 & 0 & 0 \\ 0 & 0 & 0 & 0 & 0 & -1 \\ 0 & 0 & 0 & 0 & 0 & 0 \\ 0 & 0 & 0 & 1 & 0 & 0 \end{bmatrix}, \begin{bmatrix} 0 & 0 & 0 & 0 & 0 & 0 \\ 0 & 0 & 0 & 0 & 0 & 0 \\ 0 & 0 & 0 & 0 & 0 & 0 \\ 0 & 0 & 0 & 0 & 0 & 0 \\ 0 & 0 & 0 & 0 & 0 & -1 \\ 0 & 0 & 0 & 0 & 1 & 0 \end{bmatrix}.$$

Clearly, such a Lie algebra has a nontrivial ideal but the dimension of its radical is trivial.

Finally, we say that a Lie group is solvable if it is the complexification of its Lie algebra is solvable, as defined above.

3.7 Construction of Spline-Type Partitions of Unity

Let G be a connected matrix Lie group and let U be a relatively compact open subset containing the identity in G. A family $\Psi = (\Psi_k)_{k\in I}$ of continuous and compactly supported functions is called a bounded uniform partition of unity of size U if the following conditions are satisfied:

(1) The range of Ψ_k is $[0, 1]$ for each fixed k in I.
(2) There is a well-spread family $(x_k)_{k\in I}$ in G such that each Ψ_k is supported in $x_k U$.
(3) $\sum_{k\in I} \Psi_k(x) \equiv 1$ for all $x \in G$.

Since bounded uniform partitions of unity are an important ingredient for the construction of frames and other frame-like objects [1–5], this section aims mainly to give an algorithmic procedure for the construction of a large class of smooth and bounded partitions of unity on connected Lie groups. Although our construction makes use of several technical Lie theoretic structures such as Levi's Decomposition and Iwasawa Decomposition, these concepts are not critical to the chapter's theme. Thus, to avoid digression, we will not attempt to either deeply motivate or prove their existence. Instead, we refer the interested reader to the textbook of Knapp [7, 8] for detailed treatment.

Let Q be a nontrivial Lie group. Then Q is said to be solvable if it is obtained by a finite sequence of extensions of abelian groups. Given subsets $A, B \subseteq Q$, denote by (A, B) the subgroup of G generated by $xyx^{-1}y^{-1}$ for $x \in A, y \in B$. Next, let $Q^{(1)} = (Q, Q)$ and define inductively $Q^{(k+1)} = \left(Q^{(k)}, Q^{(k)}\right)$. This gives a chain of normal subgroups

$$Q \rhd Q^{(1)} \rhd Q^{(2)} \cdots \rhd$$

and we say that G is solvable if this sequence has a finite length. In other words, there exists a nonnegative integer ℓ for which $Q^{(\ell)}$ is the trivial group containing the identity in Q.

A connected Lie group containing no nontrivial connected solvable normal subgroups is called a semisimple Lie group. In fact, a connected Lie group is semisimple if its Lie algebra is semisimple as well. This means that its Lie algebra contains no nontrivial abelian ideals. Next, let G be a connected Lie group with Lie algebra $\mathfrak{g}$. On the Lie algebra level, Levi states that any finite-dimensional real Lie algebra is the semidirect product of a maximal solvable ideal and a semisimple subalgebra. On the group level, Levi's Decomposition guarantees that it is always possible to decompose G into two factors: one solvable and the other semisimple. Precisely, there exists Q, a maximal solvable normal subgroup of G and S, a semisimple closed subgroup of G such that $G = QS$.

Furthermore, Iwasawa Decomposition is a procedure that finds inspiration in the well-known QR decomposition in linear algebra. This decomposition allows to factor a matrix A as a product of the QR such that Q is an orthogonal matrix and R is an upper triangular matrix. In precise terms, Iwasawa Decomposition makes the following valid. Every semisimple Lie group S admits a unique factorization of the form $S = KAN = NAK$ such that (see Knapp, Theorem 5.12) (1) K is a maximal compact subgroup of S, (2) NA is a connected solvable closed subgroup of S, (3) $K \cap NA$ is trivial, and (4) $(k, a, n) \mapsto nak$ defines a diffeomorphism from $K \times A \times N$ onto S.

Applying Levi's Decomposition together with Iwasawa Decomposition to a connected matrix Lie group G, we may write $G = QS$ such that Q is the maximal solvable part of G and S, is its semisimple part. S being semisimple, we may write $S = NAK$ and as such, $G = QNAK$, as summarized in the table below

$G = QNAK$	A connected Lie group	
Q	Maximal solvable Lie subgroup	
NAK	Semisimple part of G	(3.6)
QNA	A co-compact solvable subgroup	
K	A compact subgroup of the semisimple part	

Decomposition of a connected Lie group into its solvable and semisimple parts

Since QNA is a connected solvable Lie group, [8, Corollary 1.103] we write $QNA = C_1C_2 \cdots C_q$ such that each C_k is a one-parameter group (which may or may not be compact) and each subgroup $C_1C_2 \cdots C_k$ is normal in $C_1C_2 \cdots C_{k+1}$. This leads to a decomposition (see 3.6) of G into a product of the type $G = C_1C_2 \cdots C_qK$. Furthermore, for each index $k \in \{1, 2, \cdots, q\}$, we search for functions $f_k \in C_c(C_k)$ such that

$$\sum_{\gamma_k \in \Gamma_k} f_k(\gamma_k x) = 1$$

and the set Γ_k is some countable subset C_k for $1 \leq k < q$. Note that each component C_k of G, which is not compact, is necessarily simply connected. With this in mind, we first suppose that the factor C_k is simply connected. This gives a natural identification of C_k with $\mathbb{R}$, and with this identification, for any fixed natural number n, define

$$f_k(t) = \underbrace{1_{[-1/2,1/2)} * \cdots * 1_{[-1/2,1/2)}}_{n \text{ times}} \left(\frac{\epsilon n}{2} t\right)$$

and let $\Gamma_k = \frac{2}{\epsilon k}\mathbb{Z}$ be a lattice subgroup of $\mathbb{R}$. For compact factors C_k, C_k is necessarily a one-dimensional torus and we simply write:

$$C_k = \mathbb{T} = \left\{ e^{2\pi i x} : x \in \left[-\frac{1}{2}, \frac{1}{2}\right) \right\}.$$

In this case, the function f_k is just the indicator function of C_k and $\Gamma_k = \{e\}$ is the singleton containing the trivial element in C_k. We aim to establish the following. Let $G = QNAK$ and let $f \in C_c\left(C_1 C_2 \cdots C_q\right)$ such that $f\left(x_1 \cdots x_q\right) = f_1\left(x_1\right) \cdots f_q\left(x_q\right)$. Define $\varphi \in C_c(QNAK)$ such that $\varphi(dk) = f(d)$ for $d \in QNA$ and $k \in K$. Then

Theorem 3.8 $\sum_{\gamma \in \Gamma_1 \cdots \Gamma_q} \varphi(\gamma x) \equiv 1$ on G.

The proof of Theorem 3.8 requires a number of intermediate results. First (see Proposition 3.57), we prove that for any semidirect product of the form $G = NH$, it is possible to construct a bounded uniform partition of unity for G provided, we have access to bounded uniform partitions of unity on factors N and H. Secondly, (see Proposition 3.58) since any solvable Lie group has the structure for a semidirect product of the type $C_1 C_2 \cdots C_q$, a bounded partition of unity can be constructed iteratively from a family of bounded uniform partitions on each one-parameter factor C_k. Assuming that these claims are valid, then Theorem 3.8 must also hold since any connected Lie group is of the form $QNAK$ for some solvable Lie group QNA.

Proposition 3.57 Let G be a Lie group with the structure of a semidirect product such that $G = NH \simeq N \rtimes H$ and N, H are closed subgroups of G. For any function f on N and for g on H, the map $s : nh \mapsto s(nh) = f(n) \cdot g(h)$ is well-defined on G. Moreover, suppose that $f \in C_c(N)$ and $g \in C_c(H)$. Suppose additionally that Γ_N and Γ_H are some countable subsets of N and H, respectively, such that $\sum_{\gamma \in \Gamma_N} |f(\gamma x)| = 1$ for a.e. $x \in N$ and $\sum_{\gamma \in \Gamma_H} |g(\gamma x)| = 1$ for a.e. $x \in H$. Let $s \in C_c(G)$ such that $s(nh) = f(n) \cdot g(h)$. If $\Gamma = \Gamma_N \Gamma_H$ then $\sum_{\gamma \in \Gamma} |s(\gamma x)| = 1$ for a.e. $x \in G$.

Proposition 3.58 Let G be a connected solvable Lie group. Then G admits a connected normal Lie subgroup of codimension 1. Moreover, if Q is a connected solvable real Lie group, then there exists a finite family of one-parameter Lie subgroups C_k, $1 \leq k \leq q = \dim Q$ such that $Q = C_1 C_2 \cdots C_q$ and each element x in Q can be uniquely written as $x = x_1 \cdots x_q$ such that $x_k \in Q_k$. Additionally, if C_k is simply connected, then $C_k = \exp\left(\mathbb{R} A_k\right)$ for some vector A_k in the Lie algebra of Q. Otherwise, each C_k is Lie isomorphic to a one-dimensional torus.

Proof of Proposition 3.57: Let G be a semidirect product Lie group $G = NH$. For every $x \in G$, there exists a unique pair $(n, h) \in N \times H$ such that $x = nh$. Suppose that $x = nh = mk$ for $n, m \in N$ and $h, k \in H$. Then $m^{-1}n = kh^{-1} \in N \cap H = \{e\}$. Next, for any function f on N and for g on H, the map $s : nh \mapsto s(nh) = f(n) \cdot g(h)$ is well-defined on G. Moreover, for a fixed $x \in G$, there exists a unique pair $(n, h) \in N \times H$ such that $x = nh$ and

$$\sum_{\gamma \in \Gamma} |s(\gamma x)| = \sum_{\gamma_N \in \Gamma_N} \sum_{\gamma_H \in \Gamma_H} \left| s\left(\gamma_N \gamma_H nh\right)\right| = \left| s\left(\left(\gamma_N \gamma_H n \gamma_H^{-1}\right) \gamma_H h\right)\right|.$$

Since N is a normal subgroup of G, $\gamma_N \gamma_H n \gamma_H^{-1} \in N$ and by definition,

$$s\left(\gamma_N \gamma_H n \gamma_H^{-1} \gamma_H h\right) = f\left(\gamma_N \gamma_H n \gamma_H^{-1}\right) \cdot g\left(\gamma_H h\right).$$

This implies that

$$\sum_{\gamma \in \Gamma} |s(\gamma x)| = \left(\sum_{\gamma_H \in \Gamma_H} \left|g\left(\gamma_H h\right)\right|\right) \cdot \left(\sum_{\gamma_N \in \Gamma_N} \left|f\left(\gamma_N \left(\gamma_H n \gamma_H^{-1}\right)\right)\right|\right).$$

By assumption, for fixed $\gamma_H \in \Gamma_H$,

$$\sum_{\gamma_N \in \Gamma_N} \left|f\left(\gamma_N \left(\gamma_H n \gamma_H^{-1}\right)\right)\right| = 1 \quad \text{for a.e. } n \in N.$$

Finally,

$$\sum_{\gamma \in \Gamma} |s(\gamma x)| = \sum_{\gamma_H \in \Gamma_H} \left|g\left(\gamma_H h\right)\right| = 1 \quad \text{for a.e. } x \in G$$

as desired.

Proof of Proposition 3.58: Let H be the closure of the commutator of G. Since G is solvable, then H is necessarily a proper subgroup of G. Next, G/H is a non-trivial abelian Lie group and as such, it admits a connected Lie subgroup A of codimension one in G/H. Let $\pi : G \to G/H$ be the canonical quotient map. Then the inverse image of A under π is a subgroup of codimension one in G. The fact that such a subgroup is normal follows immediately from the solvability of G.

Proposition 3.59 Let $f_k \in C_c(C_k)$ such that $\sum_{\gamma_k \in \Gamma_k} f_k(\gamma_k x) = 1$, $\Gamma_k \subseteq C_k$ for $1 \le k < q$ and let $f \in C_c(Q)$ such that $f(x_1 \cdots x_q) = f_1(x_1) \cdots f_q(x_q)$. Then $\sum_{\gamma \in \Gamma_1 \cdots \Gamma_q} f(\gamma x) = 1$ for a.e. $x \in Q$.

Proposition 3.59 will be proved in several steps (see Claim 3.1 and Claim 3.2).

Claim 3.1 Let b_n be the function obtained by convolving the indicator function $1_{[-1/2,1/2)}$ with itself n times. Then $\forall t \in \mathbb{R}$, $\sum_{k \in \mathbb{Z}} b_n(t + k) = 1$.

Proof: By induction, since $\{[-1/2, 1/2) + k : k \in \mathbb{Z}\}$ is a partition of the real line, it is obvious that $\sum_{k\in\mathbb{Z}} 1_{[-1/2,1/2)}(t+k) = 1$ and the base case holds. Suppose next that $\sum_{k\in\mathbb{Z}} b_m(t+k) = 1$ for all natural numbers $m \leq n$. Then

$$\sum_{k\in\mathbb{Z}} b_{n+1}(t+k) = \sum_{k\in\mathbb{Z}} \left(b_n * 1_{[-1/2,1/2)}\right)(t+k)$$

and

$$\sum_{k\in\mathbb{Z}} b_{n+1}(t+k) = \sum_{k\in\mathbb{Z}} \int_{\mathbb{R}} b_n(x) \cdot 1_{[-1/2,1/2)}(t+k-x)\,dx.$$

The change of variable $s = t + k - x$ yields

$$\sum_{k\in\mathbb{Z}} b_{n+1}(t+k) = \int_{[-1/2,1/2)} \sum_{k\in\mathbb{Z}} b_n(k-s+t)\,ds$$

and by the inductive hypothesis, $\sum_{k\in\mathbb{Z}} b_n(k-s+t) = 1$. Therefore, $\sum_{k\in\mathbb{Z}} b_{n+1}(t+k) = \int_{[-1/2,1/2)} ds = 1$ for almost every real number t.

Claim 3.2 Let C be a one-parameter Lie group and let $\mathcal{O}$ be an open and connected neighborhood of the identity in C. Then there exists a function s supported on $\mathcal{O}$ and a countable set Γ of C such that $\sum_{\gamma\in\Gamma} |s(\gamma t)| = 1$ for a.e. $t \in C$. Moreover, s can be made continuous or as smooth as desired.

Proof: Two main cases are to be considered for the proof of this result: the simply connected case and the nonsimply connected case. For the first, assuming that C is simply connected, we may identify C with $\mathbb{R}$. Next, let ϵ be a positive real number and let n be a natural number. Let b_n be the function obtained by convolving the indicator function $1_{[-1/2,1/2)}$ with itself n times. Next, define $s_{n,\epsilon}(t) = b_n\left(\frac{\epsilon n}{2}t\right)$. Then $s_{n,\epsilon}$ is supported on the interval $\frac{2}{n\epsilon}\left[-\frac{n}{2}, \frac{n}{2}\right] = \left[-\frac{1}{\epsilon}, \frac{1}{\epsilon}\right]$. For every $t \in \mathbb{R}$,

$$\sum_{k\in\mathbb{Z}} \left|s_n\left(t + \frac{2}{\epsilon n}k\right)\right| = \sum_{k\in\mathbb{Z}} \left|b_n\left(\frac{\epsilon n}{2}\left(t + \frac{2}{\epsilon n}k\right)\right)\right| = \sum_{k\in\mathbb{Z}} b_n\left(k + \frac{nt\epsilon}{2}\right) = 1.$$

For the second case, if C is connected but not simply connected, we may identify C with the torus $\mathbb{T} = \left\{e^{2\pi i x} : x \in \left[-\frac{1}{2}, \frac{1}{2}\right)\right\}$. For a natural number L, define $E = \left\{e^{2\pi i x} : x \in \left(-\frac{1}{2L}, \frac{1}{2L}\right)\right\} \subset \mathbb{T}$. Let $f^{(n)}$ be the convolution (on the torus) of f with itself n times. We will prove by induction that

$$\sum_{k=0}^{L-1} \left(1_E\right)^{(n)} \left(e^{2\pi i(x+k)}\right) = \left(\frac{1}{L}\right)^{n-1} \Leftrightarrow \sum_{k=0}^{L-1} \left(L^{\frac{n-1}{n}} 1_E\right)^{(n)} \left(e^{2\pi i(x+k)}\right) = 1.$$

For the base case, it is easy to verify that $\sum_{k=0}^{L-1} 1_E\left(e^{2\pi i(x+k)}\right) = 1$ for almost every $x \in [0,1)$. Suppose next that $\sum_{k=0}^{L-1} \left(1_E\right)^{(n-1)}\left(e^{2\pi i(x+k)}\right) = \left(\frac{1}{L}\right)^{n-2}$. Then

$$\begin{aligned}\sum_{k=0}^{L-1}\left(1_E\right)^{(n+1)}\left(e^{2\pi i(x+k)}\right) &= \sum_{k=0}^{L-1}\left(\left(1_E\right)^{(n)} * 1_E\right)\left(e^{2\pi i(x+k)}\right)\\ &= \sum_{k=0}^{L-1}\int_0^1 \left(1_E\right)^{(n)}\left(e^{2\pi it}\right) 1_E\left(e^{2\pi i(x+k-t)}\right) dt\\ &= \sum_{k=0}^{L-1}\int_0^{1/L} \left(1_E\right)^{(n)}\left(e^{2\pi i(k-s+x)}\right) ds\end{aligned}$$

and

$$\begin{aligned}\sum_{k=0}^{L-1}\left(1_E\right)^{(n+1)}\left(e^{2\pi i(x+k)}\right) &= \int_0^{1/L}\sum_{k=0}^{L-1}\left(1_E\right)^{(n)}\left(e^{2\pi i(k-s+x)}\right) ds\\ &= \int_0^{1/L}\left(\frac{1}{L}\right)^{n-2} ds\\ &= \left(\frac{1}{L}\right)^{n-1}\end{aligned}$$

stepwise iteration of the procedure described above gives the desired result.

Finally, Theorem 3.8 is proved by analyzing the constructions above.

References

1 Lasse Borup and Morten Nielsen. Frame decomposition of decomposition spaces. *Journal of Fourier Analysis and Applications*, 13(1):39–70, 2007.

2 Hans G. Feichtinger and Peter Gröbner. Banach spaces of distributions defined by decomposition methods. I. *Mathematische Nachrichten*, 123:97–120, 1985.

3 Hans G. Feichtinger and K. H. Gröchenig. Banach spaces related to integrable group representations and their atomic decompositions. I. *Journal of Functional Analysis*, 86(2):307–340, 1989.

4 Hans G. Feichtinger and K. H. Gröchenig. Banach spaces related to integrable group representations and their atomic decompositions. II. *Monatshefte für Mathematik*, 108(2-3):129–148, 1989.

5 Hartmut Führ. Coorbit spaces and wavelet coefficient decay over general dilation groups. *Transactions of the American Mathematical Society*, 367(10):7373–7401, 2015.

6 Brian C. Hall. Lie groups, lie algebras, and representations. In *Quantum Theory for Mathematicians*, pages 333–366. Springer, 2013.

7 Anthony W. Knapp. *Representation theory of semisimple groups.* Princeton Landmarks in Mathematics. Princeton University Press, Princeton, NJ, 2001. An overview based on examples, Reprint of the 1986 original.

8 Anthony W. Knapp. *Lie groups beyond an introduction*, volume 140 of *Progress in Mathematics*. Birkhäuser, Boston, MA, second edition, 2002.

9 Anthony W. Knapp. *Basic real analysis.* Springer Science & Business Media, 2005.

10 John M. Lee. *Introduction to smooth manifolds*, volume 218 of *Graduate Texts in Mathematics*. Springer, New York, second edition, 2013.

4

Representation Theory

In this chapter, we will investigate the action of Lie groups on vector spaces, particularly function spaces over manifolds, and how this action can be understood through the lens of representation theory. Representation theory is a branch of mathematics that represents abstract groups as linear transformations on vector spaces. This theory was initiated by Sophus Lie in the late 19th century and later expanded upon by Hermann Weyl and Eugene Wigner in the early 20th century.

We will explain how to represent Lie groups as unitary operators on Hilbert spaces and how to use induced representations to create new representations from old ones. We will provide a basic introduction to these concepts, including invariant spaces, equivalent representations, irreducible representations, and Schur's Lemma. We will also discuss the concept of direct integral, which generalizes the notion of direct sums, and the concept of induced representation, along with various examples and applications to the theory of integrability of representations. For a more thorough understanding of these topics, we recommend consulting the texts by Folland [1] and Kaniuth and Taylor [3].

4.1 Representations of Lie Groups and Lie Algebras

Let V be an n-dimensional vector space (where n is possibly infinite) defined over the field of complex numbers, which is endowed with a topology induced by the Euclidean norm on V and let $GL(V)$ be the continuous invertible endomorphisms on V. In the strong topology on $GL(V)$, we say that a sequence of linear operators $\left(T_n\right)_{n\in\mathbb{N}}$ converges to T if and only if for each vector $v \in V$, one has that $T_n v$ converges to Tv. Next, the symbol $\mathfrak{gl}(V)$ denotes the vector space of all endomorphisms on V and $\mathcal{U}(V)$ is the group of all unitary operators acting on V.

A Bridge Between Lie Theory and Frame Theory: Applications of Lie Theory to Harmonic Analysis, First Edition. Vignon Oussa.

Definition 4.1 Let G be a Lie group. We say that $\pi : G \to GL(V)$ is a representation if π is a group homomorphism and is additionally, strongly continuous in the sense that the vector-valued function $x \mapsto \pi(x)v$ is continuous for all $v \in V$. If the range of π is a subset of $\mathcal{U}(V)$ and V is a Hilbert space, then we say that π is a unitary representation of G. The dimension of the representation π is defined to be the dimension of the vector space on which π is acting. In a similar fashion, if $\mathfrak{g}$ is a Lie algebra, then a representation π of $\mathfrak{g}$ is a Lie algebra homomorphism from $\mathfrak{g}$ into $\mathfrak{gl}(V)$ in the sense that not only π is an algebra homomorphism, but additionally, $\pi([X, Y]) = [\pi(X), \pi(Y)]$ for every $X, Y \in \mathfrak{g}$. In other words, π is Lie-bracket preserving.

Example 4.1 ***(The left regular representation)*** Let $\mathbb{K} = \mathbb{C}$, $V = L^2(\mathbb{R})$ and let G be the group of real numbers endowed with the usual addition on $\mathbb{R}$. For a real number x and for $f \in L^2(\mathbb{R})$, define $\left[\pi(x)f\right](t) = f(t-x)$. Then for arbitrary $x, y \in \mathbb{R}$, we have $\pi(x+y) = \pi(x)\pi(y)$ and $\pi(-x) = \pi(x)^{-1}$. This shows that π is a group homomorphism. Next, let $\left(x_n\right)_{n\in\mathbb{N}}$ be a convergent sequence of real numbers such that $\lim_{n\to\infty} x_n = x$ and let f be a continuous function of compact support. By the dominated convergence theorem,

$$\lim_{n\to\infty} \int_{\mathbb{R}} \left|f\left(t-x_n\right) - f(t)\right|^2 dt = \int_{\mathbb{R}} \lim_{n\to\infty} \left|f\left(t-x_n\right) - f(t)\right|^2 dt = 0.$$

The calculations above show that $\left(\pi\left(x_n\right)f\right)_{n\in\mathbb{N}}$ converges to $\pi(x)f$ in the $L^2(\mathbb{R})$ norm for all f belonging to a dense subset of $L^2(\mathbb{R})$. Next, let $f \in L^2(\mathbb{R})$ and let h be a continuous function of compact support such that $\|f - h\|_{L^2(\mathbb{R})} < \frac{\epsilon}{2}$ and ϵ is an arbitrary fixed positive real number. Since

$$\pi\left(x_n\right)f - \pi(x)f = \pi\left(x_n\right)f - \pi\left(x_n\right)h + \pi\left(x_n\right)h - \pi(x)h + \pi(x)h - \pi(x)f$$

appealing to triangle inequality, one obtains

$$\begin{aligned}
\left\|\pi\left(x_n\right)f - \pi(x)f\right\|_{L^2(\mathbb{R})} &\le \left\|\pi\left(x_n\right)f - \pi\left(x_n\right)h\right\|_{L^2(\mathbb{R})} \\
&\quad + \left\|\pi\left(x_n\right)h - \pi(x)h\right\|_{L^2(\mathbb{R})} + \|\pi(x)h - \pi(x)f\|_{L^2(\mathbb{R})} \\
&\le 2\|f - h\|_{L^2(\mathbb{R})} + \left\|\pi\left(x_n\right)h - \pi(x)h\right\|_{L^2(\mathbb{R})},
\end{aligned}$$

and

$$\begin{aligned}
&\operatorname{limsup}_{n\to\infty} \left\|\pi\left(x_n\right)f - \pi(x)f\right\|_{L^2(\mathbb{R})} \\
&\le 2\|f - h\|_{L^2(\mathbb{R})} + \lim_{n\to\infty} \left\|\pi\left(x_n\right)h - \pi(x)h\right\|_{L^2(\mathbb{R})} \le \epsilon.
\end{aligned}$$

This shows that $\lim_{n\to\infty} \left\|\pi\left(x_n\right)f - \pi(x)f\right\|_{L^2(\mathbb{R})} = 0$. Therefore, π is a unitary representation of $\mathbb{R}$ and more generally, if G is a Lie group and π is given by

$\pi(g)f(x) = f(g^{-1}x)$ where f is a square-integrable representation with respect to the Haar measure on G then π is a unitary representation of G, known as the regular representation of G.

Example 4.2 Let G be a matrix Lie group, and let $\sigma : G \to GL(V)$ be a representation of G acting in some complex and finite-dimensional vector space V. Then the action $(M, v) \mapsto Mv = \sigma(x)v$ is called a standard action of G and similarly, σ is called the standard representation of G. Furthermore, this action induces a natural representation π of G on the Hilbert space $L^2(V)$ defined as follows. For $f \in L^2(V)$, $[\pi(M)f](v) = f(M^{-1}v)$. Indeed, given $P, Q \in G$,

$$\begin{aligned}\pi(P)\pi(Q)f(v) &= \pi(Q)f(P^{-1}v) = f(Q^{-1}P^{-1}v)\\ &= f((PQ)^{-1}v)\\ &= \pi(PQ)f(v).\end{aligned}$$

For instance, let $G = SL(n, \mathbb{R})$ be the group of invertible matrices of order n with unit determinant. Given $f \in L^2(\mathbb{R}^n)$, let π be a representation of G acting in $L^2(\mathbb{R}^n)$ as follows: $\pi(M)f(x) = f(M^{-1}x)$. Since the Lebesgue measure on the n-Euclidean space is invariant under the action of the special linear group, it follows that π is a unitary representation of the special linear group.

Definition 4.2 Let $\pi : G \to GL(V)$ be a representation of a Lie group G. Next, let W be a subspace of V. We say that W is invariant under the action of π if $\pi(x)w \in W$ for all $w \in W$ and $x \in G$.

Definition 4.3 A representation $\pi : G \to GL(V)$ is said to be irreducible if the only closed invariant subspaces are the trivial ones: $\{0\}$ and W. The map $\pi_W : G \to GL(W)$ defined by $\pi_W(x) = \pi(x)|_W$ for all $x \in G$ is called a subrepresentation of π.

Definition 4.4 Let $\pi : \mathfrak{g} \to \mathfrak{gl}(V)$ be a representation of a Lie algebra $\mathfrak{g}$. Next, let W be a subspace of V. We say that W is invariant under the action of π if $\pi(X)w \in W$ for all $w \in W$ and $X \in \mathfrak{g}$. A representation is said to be irreducible if the only closed invariant subspaces are the trivial ones: $\{0\}$ and W. The map $\pi_W : \mathfrak{g} \to \mathfrak{gl}(W)$ defined by $\pi_W(X) = \pi(X)|_W$ for all $X \in \mathfrak{g}$ is called a subrepresentation of π.

Definition 4.5 A representation that is not irreducible is called reducible. If $V = \bigoplus_{k=1}^{n} V_k$ such that each V_k is invariant under the action of a representation π, we say that π is a direct sum of the representations π_{V_k} and we write $\pi = \bigoplus_{k=1}^{n} \pi_{V_k}$.

Definition 4.6 Let $\pi_V : G \to GL(V)$ and $\pi_W : G \to GL(W)$ be two representations of a Lie group G. Next, let $\pi_V \bigotimes \pi_W : G \to GL\left(V \bigotimes W\right)$ such that $\left(\left[\pi_V \bigotimes \pi_W\right](x)\right)\left(v \bigotimes w\right) = \pi_V(x) v \bigotimes \pi_W(x) w$. Then $\pi_V \bigotimes \pi_W$ is called a tensor representation of G.

Example 4.3 Fix $A = \begin{bmatrix} 0 & 1 & 1 \\ 0 & 0 & 1 \\ 0 & 0 & 0 \end{bmatrix}$ and let $\pi : \mathbb{R} \to GL\left(\mathbb{R}^3\right)$ such that

$$\pi(t) = \exp(tA) = \begin{bmatrix} 1 & t & \frac{t(t+2)}{2} \\ 0 & 1 & t \\ 0 & 0 & 1 \end{bmatrix}.$$

To verify that π is a group homomorphism, it suffices to check that for arbitrary real parameters t and s,

$$\pi(t+s) = \begin{bmatrix} 1 & s+t & \frac{1}{2}s^2 + st + s + \frac{1}{2}t^2 + t \\ 0 & 1 & s+t \\ 0 & 0 & 1 \end{bmatrix} = \begin{bmatrix} 1 & t & \frac{1}{2}t(t+2) \\ 0 & 1 & t \\ 0 & 0 & 1 \end{bmatrix} \begin{bmatrix} 1 & s & \frac{1}{2}s(s+2) \\ 0 & 1 & s \\ 0 & 0 & 1 \end{bmatrix}.$$

Since for any real number x,

$$\pi(t) \begin{bmatrix} x \\ 0 \\ 0 \end{bmatrix} = \begin{bmatrix} x \\ 0 \\ 0 \end{bmatrix},$$

it follows that π is a reducible representation of $\mathbb{R}$.

Example 4.4 Let $\mathcal{H}$ be the Hilbert space of square-integrable functions (with respect to Lebesgue measure) on $\mathbb{R}$, and let L be the left regular representation of $\mathbb{R}$ acting in $\mathcal{H}$. Next, let $\mathcal{K}$ be the Hilbert subspace of $\mathcal{H}$ consisting of vectors whose Fourier transforms are supported on $[0, 1]$. Since the Fourier transform intertwines translation operators with operators acting by multiplication by complex exponentials, it follows that $\mathcal{K}$ is a nontrivial Hilbert space that is invariant under the action of the regular representation. As a result, the left regular representation of the real line is a reducible unitary representation.

Definition 4.7 Let $\pi : G \to GL(V)$ and $\theta : G \to GL(W)$ be two representations of a Lie group G. Next, let $C(\pi, \sigma)$ be the collection of all bounded linear operators $T : V \to W$ such that $T \circ \pi(x) = \theta(x) \circ T$ for all $x \in G$. We say that π is equivalent (unitarily equivalent) to θ if $C(\pi, \sigma)$ contains an invertible operator (unitary) operator.

Definition 4.8 Let G be a Lie group and let V be a finite-dimensional vector space and let $\pi : G \to GL(V)$ be a representation of G. Then the differential $d\pi : \mathfrak{g} \to \mathfrak{gl}(V)$ of π at the identity of G is called the derived representation of π.

Example 4.5 Let $\pi : \mathbb{R} \to GL\left(\mathbb{R}^3\right)$ such that

$$\pi(t) = \exp\left(t\begin{bmatrix} 0 & 1 & 1 \\ 0 & 0 & 1 \\ 0 & 0 & 0 \end{bmatrix}\right) = \begin{bmatrix} 1 & t & \frac{1}{2}t(t+2) \\ 0 & 1 & t \\ 0 & 0 & 1 \end{bmatrix}.$$

Since

$$\lim_{h\to 0}\frac{\pi(t+h)-\pi(t)}{h} = \lim_{h\to 0}\begin{bmatrix} 0 & 1 & \frac{1}{2}h+t+1 \\ 0 & 0 & 1 \\ 0 & 0 & 0 \end{bmatrix} = \begin{bmatrix} 0 & 1 & t+1 \\ 0 & 0 & 1 \\ 0 & 0 & 0 \end{bmatrix}$$

then the derived representation of π is given by the linear map $d\pi : \mathbb{R} \to \mathfrak{gl}\left(\mathbb{R}^3\right)$ such that

$$d\pi(t) = \begin{bmatrix} 0 & t & t \\ 0 & 0 & t \\ 0 & 0 & 0 \end{bmatrix}.$$

Proposition 4.1 Let G be a connected Lie group and let $\pi : G \to GL(V)$ be a representation of G. Then π is irreducible if and only if $d\pi$ is irreducible.

Proof: Suppose that π is irreducible. Next, let W be a subspace of $d\pi$ which is invariant. Given $X \in \mathfrak{g}$ and $w \in W$, since W is a closed vector space, it follows that

$$e^{d\pi(X)}w = \lim_{k\to\infty}\left(\sum_{j=1}^{k}\frac{1}{j!}(d\pi(X))^j\right)w \in W.$$

Since G is connected, it is generated by a neighborhood of the identity. Thus, for arbitrary $g \in G$, we may write $g = \exp X_1 \cdots \exp X_r$ for some $X_1, \ldots, X_r \in \mathfrak{g}$. This shows that

$$\pi(g)w = \pi\left(\exp\left(X_1\right)\cdots\exp\left(X_r\right)\right)w = e^{d\pi(X_1)}\cdots e^{d\pi(X_r)}w.$$

Thus, W must be invariant under the action of π and since π is irreducible, it follows that its derived representation is irreducible as well. Next, let us suppose that $d\pi$ is irreducible and let W be a subspace of π which is invariant. Given $X \in \mathfrak{g}$, $w \in W$

$$\frac{\pi(\exp tX)w - w}{t} \in W,$$

and taking the limit of the above at $t = 0$, we obtain that

$$d\pi(X)w = \lim_{t \to 0} \frac{\pi(\exp tX)w - w}{t}$$

belongs to W. Thus, W must be trivial as well.

Proposition 4.2 ***(Schur's Lemma)***

1. Let $\pi : G \to GL(V)$ be an irreducible representation where V is a vector space defined over the complex numbers. If $T \in C(\pi, \pi)$, then T must be a multiple of the identity.
2. Let π and τ be two irreducible representations acting in complex vector spaces V, W. If $T \in C(\pi, \tau)$ and π and τ are not equivalent, then T must be the zero operator.

Proof: Let $T \in C(\pi, \pi)$. Then the null space of T and the image of T are invariant subspaces of V. By the irreducibility of π, such spaces must be trivial. As such, if $T \in C(\pi, \pi)$ is a nonzero operator, it must be invertible. Let λ be an eigenvalue of T. Then $T - \lambda I$ cannot be invertible and is also an element of $C(\pi, \pi)$. This shows that $T = \lambda I$. Suppose on the other hand that $T \in C(\pi, \tau)$ and that π and τ are not equivalent. Then T cannot be invertible, and as such, it must be the zero operator.

Example 4.6 Applying Schur's lemma, the following is immediate.

1. Let $\pi : GL(n, \mathbb{C}) \to GL(\mathbb{C}^n)$ be a representation of $GL(n, \mathbb{C})$. Then π is irreducible and letting $Z(GL(n, \mathbb{C}))$ be the center of $GL(n, \mathbb{C})$, it follows that $Z(GL(n, \mathbb{C}))$ consists only of scalar multiples of the identity matrix.
2. Let $\pi : SO(n, \mathbb{C}) \to GL(\mathbb{C}^n)$ be a representation of $SO(n, \mathbb{C})$. Then π is irreducible and letting $Z(SO(n, \mathbb{C}))$ be the center of $SO(n, \mathbb{C})$, it follows that $Z(SO(n, \mathbb{C}))$ consists of matrices of the form $\pm I$.
3. Any irreducible representation of a commutative Lie group is one-dimensional.

Proposition 4.3 Let $\pi : G \to \mathcal{U}(\mathcal{H})$ be a unitary representation of a Lie group G. Let $\mathcal{K}$ be a closed and invariant Hilbert subspace of $\mathcal{H}$. Put $\mathcal{H} = \mathcal{K} \oplus \mathcal{R}$. Then π is unitarily equivalent to the following direct sum $\pi_{\mathcal{K}} \oplus \pi_{\mathcal{R}}$.

Proof: Let $f \in \mathcal{K}$ and let $h \in \mathcal{R}$. Since $\pi(x)$ is a unitary map,

$$\langle \pi(x)h, f \rangle_{\mathcal{H}} = \left\langle h, \pi\left(x^{-1}\right) f \right\rangle_{\mathcal{H}} = 0$$

and since f is arbitrary, we can conclude that $\pi(x)h \in \mathcal{R}$. This shows that $\mathcal{R}$ is invariant under the action of π. Let $P_{\mathcal{K}}$ and $P_{\mathcal{R}}$ be the orthogonal projection on $\mathcal{K}$ and $\mathcal{R}$, respectively. Then $f \mapsto P_{\mathcal{K}}f \oplus P_{\mathcal{R}}f$ defines a unitary map in $C(\pi, \pi_{\mathcal{K}} \oplus \pi_{\mathcal{R}})$.

Definition 4.9 Let $\pi : G \to GL(V)$ be a representation of a Lie group G. We say that π is cyclic if there exists a vector $v \in V$ such that the linear span of $\pi(G)v$ is dense in V.

4.2 A Survey on the Theory of Direct Integrals

Let $G = \mathbb{Z}_n$ be the finite cyclic group endowed with the usual additive modular operation. Next, let L be the left regular representation of this group acting on the Hilbert space $l^2(\mathbb{Z}_n) = \mathbb{C}^n$. That is, given $\phi \in l^2(\mathbb{Z}_n)$ and $k, j \in \mathbb{Z}_n$,

$$[L(k)\phi](j) = \phi(j - k \bmod n) := \phi(j-k).$$

Next, we introduce the group Fourier transform on $\mathbb{Z}_n$ as a tool to decompose the representation L into a direct sum of its irreducible components. Indeed, the Fourier transform Λ is a unitary operator on $l^2(\mathbb{Z}_n)$ defined as follows: given $\phi \in l^2(\mathbb{Z}_n)$ and $\xi \in \mathbb{Z}_n$,

$$[\Lambda\phi](\xi) = \frac{1}{\sqrt{n}}\sum_{k=0}^{n-1}\phi(k)e^{-\frac{2\pi i k\xi}{n}}.$$

Next,

$$[\Lambda L(j)\phi](\xi) = \frac{1}{\sqrt{n}}\sum_{k=0}^{n-1}\phi(k-j)e^{-\frac{2\pi i k\xi}{n}},$$

and the change of variable $s = k - j$ gives

$$[\Lambda L(j)\phi](\xi) = e^{-\frac{2\pi i j\xi}{n}}\frac{1}{\sqrt{n}}\sum_{s=0}^{n-1}\phi(s)e^{-\frac{2\pi i s\xi}{n}} = e^{-\frac{2\pi i j\xi}{n}}[\Lambda\phi](\xi).$$

This shows that for every $j \in \mathbb{Z}_n$,

$$\Lambda L(j)\Lambda^{-1} = \bigoplus_{\xi \in \mathbb{Z}_n} e^{-\frac{2\pi i j\xi}{n}}.$$

In other words, letting χ_ξ be an irreducible representation of $\mathbb{Z}_n$ of dimension 1 over the complex numbers (a unitary character), it follows that the left regular representation of L is unitarily equivalent to the following direct sum of unitary characters: $\widehat{L} = \bigoplus_{\xi \in \mathbb{Z}_n} \chi_\xi$. The set $\mathbb{Z}_n$ is called the spectrum of the decomposition and this decomposition is canonical in the sense that any representation π of the cyclic group $\mathbb{Z}_n$ is unitarily equivalent to the left regular representation if its irreducible decomposition is of the form $\widehat{L}$ as described above.

Generally, it is not always possible to decompose a unitary representation of a group into a countable direct sum of irreducible representations as done for the left regular representation of the cyclic group of order n (see Lemma 3.10). In fact, it is known that there exist representations, which admit no irreducible subrepresentation (e.g., the regular representation of the real line) and the concept needed to generalize Proposition 3.37, is given to us by the theory of direct integrals.

Recall that a σ-algebra Σ on a set X is a collection of sets that includes X, is closed under taking complement, and countable unions as well. The pair (X, Σ) is called a Borel structure on X. Next, let X be a σ-compact and locally compact set and let μ be the completion of a Borel measure on X. A Hilbert bundle over X is a family of separable Hilbert spaces $\left(\mathcal{H}_x\right)_{x \in X}$ together with a family of functions $f_k : X \to \bigcup_{x \in X} \mathcal{H}_x$ such that $f_k(x)$ is a vector in $\mathcal{H}_x$ and the collection of nonzero elements in $\left(f_k(x)\right)_{k \in I}$ is an orthonormal basis for $\mathcal{H}_x$.

A section is a map

$$s : X \to \bigcup_{x \in X} \mathcal{H}_x$$

such that $s(x) \in \mathcal{H}_x$ for every $x \in X$. Next, a section is called measurable if for each index k, the mapping $x \mapsto \left\langle s(x), f_k(x) \right\rangle_{\mathcal{H}_x}$ is measurable and additionally, there exists a set I_s contained in I such that $x \mapsto \left\langle s(x), f_k(x) \right\rangle_{\mathcal{H}_x}$ is zero for every k not in I_s. A measurable section is called a μ-null section if it vanishes on a set whose complement has zero μ-measure. Moreover, the direct integral of Hilbert spaces $\mathcal{H} = \int_X^{\oplus} \mathcal{H}_x d\mu(x)$ is the set of all measurable sections s such that

$$\|s\|_{\mathcal{H}}^2 := \int_X \|s(x)\|_{\mathcal{H}_x}^2 \, d\mu(x) < \infty$$

modulo the collection of μ-null sections. From here on, we endow $\mathcal{H} = \int_X^{\oplus} \mathcal{H}_x d\mu(x)$ with the following inner product

$$\langle s, r \rangle_{\mathcal{H}} = \int_X \langle s(x), r(x) \rangle_{\mathcal{H}_x} d\mu(x).$$

Then it can be shown that $\mathcal{H}$ is a Hilbert space. Moreover, in the special case where $\mathcal{H}_x = \mathcal{K}$ for all $x \in X$, the direct integral $\mathcal{H}$ is just the Hilbert space $L^2(X, \mathcal{K})$ of $\mathcal{K}$-valued vectors defined on X. Next, for each $x \in X$, let $\pi : G \to \mathcal{U}\left(\mathcal{H}_x\right)$ be a unitary representation. We say that the family of representations $\left(\pi_x\right)_{x \in X}$

is measurable if for all $g \in G$, $x \mapsto \langle \pi_x(g)s(x), r(x)\rangle_{H_x}$ is Borel measurable. Define $\pi : G \to \mathcal{U}(\mathcal{H})$ such that given $s \in \mathcal{H}$

$$\left[\pi(g)s\right](x) = \pi_x(g)s(x). \tag{4.1}$$

Proposition 4.4 π is a strongly continuous representation of G.

Proof: It is easy to verify that for all $g, h \in G$, $\pi(gh) = \pi(g)\pi(h)$ and $\pi\left(g^{-1}\right) = \pi(g)$. Next, using the fact that each representation π_x is strongly continuous, together with the dominated convergence theorem, one verifies that π is strongly continuous as well.

The representation π defined in (4.1) is called a direct integral representation and is denoted by the symbol

$$\pi = \int_X^{\oplus} \pi_x d\mu(x).$$

Example 4.7 Let X be the set of natural numbers equipped with the counting measure μ. In this case,

$$\pi = \int_X^{\oplus} \pi_x d\mu(x) = \bigoplus_{k\in\mathbb{N}} \pi_k.$$

Example 4.8 Let L be the left regular representation of $\mathbb{R}$. Then L is unitarily equivalent to a direct integral of one-dimensional unitary representations of $\mathbb{R}$. Indeed, given any real number x, define $\chi_x : \mathbb{R} \to \mathcal{U}(\mathbb{C})$ such that $\chi_x(y) = e^{2\pi ixy}$. Then each χ_x is a one-dimensional unitary representation (unitary character) of $\mathbb{R}$ acting in $\mathbb{C}$. Next, let

$$\mathcal{F} : L^2(\mathbb{R}) \to \int_{\mathbb{R}}^{\oplus} \mathbb{C}dx = L^2(\mathbb{R})$$

be the Fourier transform defined as follows $\mathcal{F}f(x) = \widehat{f}(x) = \int_{\mathbb{R}} f(y)e^{-2\pi iyx}dy$. Then $\mathcal{F}$ is a unitary operator and additionally, $\mathcal{F} \in C\left(L, \int_{\mathbb{R}}^{\oplus} \chi_x dx\right)$. More precisely, the following holds true

$$\mathcal{F}(L(t)f)(y) = \int_{\mathbb{R}} (L(t)f)(y)e^{-2\pi iyx}dy = \int_{\mathbb{R}} f(y)e^{-2\pi i(y+t)x}dy$$

and

$$\mathcal{F}(L(t)f)(y) = e^{-2\pi itx} \cdot \int_{\mathbb{R}} f(y)e^{-2\pi iyx}dy = \chi_x(t)^{-1} \cdot \mathcal{F}f(x).$$

This shows that for every real number t, $\mathcal{F}L(t)\mathcal{F}^{-1} = \int_{\mathbb{R}}^{\oplus} \chi_x(t)^{-1}dx$.

4.3 Induced Representations

Consider a Lie group G, with a closed Lie subgroup H. The question at hand pertains to the development of a systematic and efficacious technique for constructing a unitary representation of G, given a representation of the subgroup H. In this section, we shall expound on a widely recognized method known as induction, which facilitates the generation of new representations from those of subgroups. Notable examples of representations derived through induction include the left regular representation of any group, obtained by inducing the trivial representation of the identity group, as well as representations pertinent to wavelet theory, Gabor analysis, and shearlet theory, which are derived by inducing unitary characters of appropriate normal subgroups [1, 3–5].

4.3.1 Quasi-invariant Measures on Cosets

Given $f \in C_c^\infty(G)$, the auxiliary function

$$f_H(x) = \int_H f(xh)\, d\mu_H(h), \tag{4.2}$$

where $d\mu_H$ is a left-invariant measure on H. Note that for $y \in H$,

$$f_H(xy) = \int_H f(xyh)\, d\mu_H(h).$$

Next, the change of variable $h \mapsto y^{-1}h$ gives

$$f_H(xy) = \int_H f(xh)\, d\mu_H\left(y^{-1}h\right).$$

Since $d\mu_H$ is a left-invariant measure on H, it follows that

$$f_H(xy) = \int_H f(xh)\, d\mu_H(h) = f_H(x).$$

This shows that f_H is constant on the left cosets of H. As a result, the following function:

$$G/H \ni xH \mapsto f_{G/H}(xH) = f_H(x)$$

is well defined.

Proposition 4.5 Given $\phi \in C_c(G/H)$, there exists a function $f \in C_c(G)$ such that $f_{G/H} = \phi$. Moreover, if ϕ is nonnegative, it is possible to choose a nonnegative function f in $C_c(G)$ satisfying the equation $f_{G/H} = \phi$.

Proof: Let us assume that K is a subset of G, which is compact and let us suppose additionally that the support of ϕ is contained in the image of K under the canonical quotient map π. Next, let τ be a nonnegative function compactly supported in G such that τ is positive on the compact set K. Then

$$\tau_{G/H}(xH) = \int_H \tau(xh)\, d\mu_H(h)$$

is strictly positive on the support of ϕ. Let ξ be a function defined on G/H such that

$$\xi(xH) = \frac{\phi(xH)}{\tau_{G/H}(xH)} \cdot 1_{\text{support}(\phi)}(xH).$$

Then ξ is continuous and compactly supported on G/H. Moreover given $x \in G$

$$f(x) = \xi(xH) \cdot \tau(x) = \frac{\phi(xH) \cdot 1_{\text{support}(\phi)}(xH) \cdot \tau(x)}{\int_H \tau(xh)\, d\mu_H(h)}.$$

Therefore, f is continuous and compactly supported on G and f is nonnegative whenever ϕ is nonnegative. Finally,

$$\begin{aligned} f_H(x) &= \int_H \xi(xH) \cdot \tau(xh)\, d\mu_H(h) \\ &= \xi(xH) \cdot \int_H \tau(xh)\, d\mu_H(h) \\ &= \xi(xH) \cdot \tau_{G/H}(xH). \end{aligned}$$

Using the definition of ξ, it follows that

$$\begin{aligned} \xi(xH) \cdot \tau_{G/H}(xH) &= \frac{\phi(xH) \cdot 1_{\text{support}(\phi)}(xH) \cdot \tau_{G/H}(xH)}{\tau_{G/H}(xH)} \\ &= \phi(xH) \cdot 1_{\text{support}(\phi)}(xH) \end{aligned}$$

for all $x \in G$.

Let G be a Lie group and H a closed subgroup of G. A rho-function for the pair (G, H) is a nonnegative function ρ defined on G, which is locally integrable and satisfies the following condition:

$$\rho(gh) = \frac{\Delta_H(h)}{\Delta_G(h)} \rho(g) \quad (g \in G, h \in H).$$

In other words,

$$\rho(gh) = \frac{\det\left(Ad_H(h)^{-1}\right)}{\det\left(Ad_G(h)^{-1}\right)} \rho(g).$$

Proposition 4.6 Let f be a smooth and nonnegative function which is compactly supported on G. Then

$$g \mapsto \int_H \left(\frac{\Delta_H(h)}{\Delta_G(h)}\right)^{-1} f(gh)\, d\mu_H(h)$$

is a continuous rho-function for the pair (G, H).

Proof: Put

$$\rho(g) = \int_H \left(\frac{\Delta_H(h)}{\Delta_G(h)}\right)^{-1} f(gh)\, d\mu_H(h).$$

Then given z in H,

$$\rho(gz) = \int_H \left(\frac{\Delta_H(h)}{\Delta_G(h)}\right)^{-1} f(gzh)\, d\mu_H(h).$$

Next, the change of variable $h \mapsto z^{-1}h$ yields

$$\rho(gz) = \int_H \left(\frac{\Delta_H(z^{-1}h)}{\Delta_G(z^{-1}h)}\right)^{-1} f(gh)\, d\mu_H(z^{-1}h).$$

Since $d\mu_H$ is left-invariant the equality

$$\left(\frac{\Delta_H(z^{-1}h)}{\Delta_G(z^{-1}h)}\right)^{-1} = \left(\frac{\Delta_H(z)}{\Delta_G(z)}\right) \cdot \left(\frac{\Delta_H(h)}{\Delta_G(h)}\right)^{-1}$$

gives

$$\rho(gz) = \left(\frac{\Delta_H(z)}{\Delta_G(z)}\right) \int_H \left(\frac{\Delta_H(h)}{\Delta_G(h)}\right)^{-1} f(gh)\, d\mu_H(h) = \left(\frac{\Delta_H(z)}{\Delta_G(z)}\right) \rho(g).$$

Proposition 4.7 Let ρ be a rho-function for the pair (G, H). Next, let f be a compactly supported function on G such that $f_{G/H}$ is identically zero on G/H. Then $\int_G f(g)\, \rho(g)\, d\mu_G(g) = 0$.

Proof: Let u be a continuous function which is compactly supported on G such that $u_{G/H}(xH) = 1$ for all x in the support of f. In other words, u_H is equal to 1 on the support of f. Recalling that $f_H(x) = \int_H f(xh)\, d\mu_H(h)$, since $d\mu_H(h^{-1}) = \Delta_H(h^{-1})\, d\mu_H(h^{-1})$, it follows that

$$\int_H f(xh)\, d\mu_H(h) = \int_H f(xh^{-1})\, d\mu_H(h^{-1}) = \int_H f(xh^{-1})\, \Delta_H(h^{-1})\, d\mu_H(h^{-1})$$

and as a result,

$$\begin{aligned} 0 &= \int_G \rho(x)\, u(x) \left(\int_H f(xh^{-1})\, \Delta_H(h^{-1})\, d\mu_H(h)\right) d\mu_G(x) \\ &= \int_H \int_G \rho(x)\, u(x) f(xh^{-1})\, d\mu_G(x)\, \Delta_H(h^{-1})\, d\mu_H(h). \end{aligned}$$

Replacing x with xh in the expression above, we obtain

$$0 = \int_H \int_G \rho(xh)\, u(xh) f(x)\, d\mu_G(xh)\, \Delta_H\left(h^{-1}\right) d\mu_H(h).$$

Since $d\mu_G(xh) = \Delta_G(h)\, d\mu_G(x)$, it follows that

$$0 = \int_H \Delta_G(h)\, \Delta_H\left(h^{-1}\right) \left(\int_G \rho(xh)\, u(xh) f(x)\, d\mu_G(x) \right) d\mu_H(h).$$

Thanks to Proposition 4.6, $\rho(xh) = \frac{\Delta_H(h)}{\Delta_G(h)} \rho(x)$, and it follows that

$$\begin{aligned} 0 &= \int_H \Delta_G(h)\, \Delta_H\left(h^{-1}\right) \frac{\Delta_H(h)}{\Delta_G(h)} \left(\int_G \rho(x)\, u(xh) f(x)\, d\mu_G(x) \right) d\mu_H(h) \\ &= \int_H \int_G \rho(x)\, u(xh) f(x)\, d\mu_G(x)\, d\mu_H(h) \\ &= \int_H f(x)\, \rho(x) \left(\int_G u(xh)\, d\mu_H(h) \right) d\mu_G(x). \end{aligned}$$

By construction, we have $u_H(x) = \int_H u(xh)\, d\mu_H(h) = 1$ and consequently, $0 = \int_H f(x)\, \rho(x)\, d\mu_G(x)$.

Proposition 4.8 Let ρ be a rho-function for the pair (G, H). Then there exists a regular Borel measure $d\mu_{(G/H,\rho)}$ such that for any function f, which is compactly supported and continuous on G,

$$\int_{G/H} f_{G/H}(x)\, d\mu_{(G/H,\rho)}(x) = \int_G f(x)\, \rho(x)\, d\mu_G(x).$$

Proof: The mapping $\ell : \phi_{G/H} \mapsto \ell\left(\phi_{G/H}\right) = \int_G \phi(g)\, \rho(g)\, d\mu_G(g)$ defines a positive linear functional on the Banach space $C_c(G/H)$. Appealing to Riesz representation theorem, there exists a regular Borel measure $d\mu_{(G/H,\rho)}$ such that $\int_{G/H} f_{G/H}(x)\, d\mu_{(G/H,\rho)}(x) = \int_G f(x)\, \rho(x)\, d\mu_G(x)$.

Proposition 4.9 Let G be a Lie group and let H be a closed subgroup of G. Then there exists a nonzero positive invariant regular Borel measure on G/H if and only if $\Delta_G(h) = \Delta_H(h)$ for all $h \in H$. In fact, such a measure is essentially unique. It other words, any invariant measure is unique up to multiplication by a constant.

Proof: First, let us assume that $\Delta_G(h) = \Delta_H(h)$ for all $h \in H$. Since $\rho(gh) = \frac{\Delta_H(h)}{\Delta_G(h)} \rho(g)$ for all $g \in G, h \in H$, it follows that $\rho(gh) = \rho(g)$ and the rho-function must be the constant function which is equal to 1 everywhere. Then there exists a regular Borel measure $d\mu_{(G/H,\rho)}$ on G/H such that

$$\int_{G/H} f_{G/H}(x)\, d\mu_{(G/H,\rho)}(x) = \int_G f(x)\, d\mu_G(x) \tag{4.3}$$

for all compactly supported continuous functions on G. Next, given $g \in G$, put $L_g f(x) = f(g^{-1}x)$. In a formal sense, let L_g be a translation operator. Then

$$\int_{G/H} f_{G/H}(g^{-1}x)\, d\mu_{(G/H,\rho)}(x) = \int_{G/H} (L_g f)_{G/H}(x)\, d\mu_{(G/H,\rho)}(x)$$

and by definition,

$$\int_{G/H} (L_g f)_{G/H}(x)\, d\mu_{(G/H,\rho)}(x) = \int_G (L_g f)(x)\, d\mu_G(x) = \int_G f(g^{-1}x)\, d\mu_G(x).$$

By the left-invariance of the Haar measure on G, we obtain

$$\int_G f(g^{-1}x)\, d\mu_G(x) = \int_G f(x)\, d\mu_G(x) = \int_{G/H} f_{G/H}(x)\, d\mu_{(G/H,\rho)}(x).$$

Suppose, on the other hand, that μ is a positive invariant measure on G/H. Next, let ℓ be a linear functional defined on the Banach space of compactly supported continuous functions on G such that

$$\ell : f \mapsto \ell(f) = \int_{G/H} f_{G/H}(x)\, d\mu(x).$$

Note ℓ is a left-invariant linear functional in the sense that $\ell(L_g f) = \ell(f)$ for all $g \in G$. Next, since up to normalization, the left-invariant Haar measure on G is unique, there must exist a positive constant C such that

$$\int_{G/H} f_{G/H}(x)\, d\mu(x) = C \int_G f(x)\, d\mu_G(x)$$

for every $f \in C_c(G)$.

Select a function $f \in C_c(G)$ such that $\int_G f(x)\, d\mu_G(x) = 1$. For arbitrary $h \in H$, define $s(x) = f(xh)\Delta_H(h)$ so that $s_{G/H} = f_{G/H}$ as shown here:

$$\begin{aligned}
s_{G/H}(xH) &= \int_H s(xz)\, d\mu_H(z) \\
&= \Delta_H(h) \cdot \int_H f(xzh)\, d\mu_H(z) \\
&= \Delta_H(h) \cdot \int_H f(xz)\, d\mu_H(zh^{-1}) \\
&= \int_H f(xz)\, d\mu_H(z).
\end{aligned}$$

Next,

$$\begin{aligned}
\frac{\Delta_H(h)}{\Delta_G(h)} - 1 &= \left(\frac{\Delta_H(h)}{\Delta_G(h)} - 1\right) \cdot \left(\int_G f(x)\, d\mu_G(x)\right) \\
&= \int_G \left(\frac{\Delta_H(h)}{\Delta_G(h)} f(x) - f(x)\right) d\mu_G(x)
\end{aligned}$$

and

$$\begin{aligned}\frac{\Delta_H(h)}{\Delta_G(h)} - 1 &= \int_G \left(\frac{\Delta_H(h)}{\Delta_G(h)} f(x)\right) d\mu_G(x) - \int_G f(x)\, d\mu_G(x) \\ &= \int_G \left(\frac{\Delta_H(h)}{\Delta_G(h)} f(xh)\right) d\mu_G(xh) - \int_G f(x)\, d\mu_G(x) \\ &= \int_G \left(\frac{\Delta_H(h)}{\Delta_G(h)} f(xh)\right) \Delta_G(h)\, d\mu_G(x) - \int_G f(x)\, d\mu_G(x) \\ &= \int_G f(xh)\,\Delta_H(h)\, d\mu_G(x) - \int_G f(x)\, d\mu_G(x).\end{aligned}$$

Since $s(x) = f(xh)\,\Delta_H(h)$, and $\int_{G/H} f_{G/H}(x)\, d\mu(x) = C\int_G f(x)\, d\mu_G(x)$, it follows that

$$\begin{aligned}\frac{\Delta_H(h)}{\Delta_G(h)} - 1 &= \int_G s(x)\, d\mu_G(x) - \int_G f(x)\, d\mu_G(x) \\ &= \frac{1}{C}\left(\int_{G/H} s_{G/H}(x)\, d\mu(x)\right) - \frac{1}{C}\left(\int_{G/H} f_{G/H}(x)\, d\mu(x)\right) \\ &= \frac{1}{C}\int_{G/H} \left(s_{G/H}(x) - f_{G/H}(x)\right) d\mu(x) = 0.\end{aligned}$$

This shows that $\frac{\Delta_H(h)}{\Delta_G(h)} = 1$. Therefore, $\Delta_G(h) = \Delta_H(h)$ for all $h \in H$.

Given a measure μ on G/H, we say that μ is quasi-invariant if for any $g \in G$ and $E \subseteq G/H$, the measure $d\mu_g$ defined by $d\mu_g(E) = \mu(gE)$ and μ are mutually absolutely continuous measures.

Proposition 4.10 Let ρ be a rho-function for the pair (G, H) and select a regular Borel measure $d\mu_{(G/H,\rho)}$ such that for any function f which is compactly supported and continuous on G,

$$\int_{G/H} f_{G/H}(x)\, d\mu_{(G/H,\rho)}(x) = \int_G f(x)\,\rho(x)\, d\mu_G(x).$$

Then

$$\int_G f(x)\,\rho(x)\, d\mu_G(x) = \int_{G/H}\int_H f(xh)\, d\mu_H(h)\, d\mu_{(G/H,\rho)}(x).$$

Proof: By definition, we have $f_{G/H}(xH) = \int_H f(xh)\, d\mu_H(h)$. By substituting $f_{G/H}(xH)$ for $\int_H f(xh)\, d\mu_H(h)$ in the following equation:

$$\int_{G/H} f_{G/H}(x)\, d\mu_{(G/H,\rho)}(x) = \int_G f(x)\,\rho(x)\, d\mu_G(x),$$

we obtain

$$\int_G f(x)\,\rho(x)\,d\mu_G(x) = \int_{G/H}\int_H f(xh)\,d\mu_H(h)\,d\mu_{(G/H,\rho)}(x)$$

as desired.

Remark 4.1 One could also establish Proposition 4.10 (via a differential geometric approach) by proving that G/H as a nontrivial G-invariant $(\dim G/H)$-form if and only if $\det\left(Ad_G(h)\right) = \det\left(Ad_H(h)\right)$ for all $h \in H$. For a complete presentation of these arguments, we refer the reader to Page 367 in [2].

4.3.2 Induced Unitary Characters

To motivate the concept of representations obtained by inducing unitary characters, we shall begin by exploring two examples.

Example 4.9 Let Γ be the Dihedral group of order 6. For instance, let us suppose that Γ is generated by matrices

$$R = \begin{bmatrix} 0 & 0 & 1 \\ 1 & 0 & 0 \\ 0 & 1 & 0 \end{bmatrix}, \quad S = \begin{bmatrix} 1 & 0 & 0 \\ 0 & 0 & 1 \\ 0 & 1 & 0 \end{bmatrix}$$

such that $R^3 = S^2 = I$ and $SRS^{-1} = R^{-1} = R^2$. Let $\langle R \rangle$ be the cyclic subgroup of order 3 generated by R. Given a unitary character χ of $\langle R \rangle$, consider the vector space

$$\mathcal{K}_\chi = \left\{ f : \Gamma \to \mathbb{C} : f\left(S^j R\right) = \overline{\chi(R)} \cdot f\left(S^j\right) \right\}.$$

Since there exists $\xi \in \mathbb{Z}_3$ such that $\chi(R) = e^{\frac{2\pi \xi i}{3}}$, if $f \in \mathcal{K}_\chi$ then $f\left(S^j R\right) = e^{-\frac{2\pi \xi i}{3}} \cdot f\left(S^j\right)$ and we may naturally identify $\mathcal{K}_\chi$ with $l^2\left(\mathbb{Z}_2\right)$. In other words, $\mathcal{K}_\chi \simeq \mathbb{C}^2$. Moreover, we claim that as a vector subspace of functions defined on Γ, $\mathcal{K}_\chi$ is Γ-invariant. Indeed, for arbitrary $f \in \mathcal{K}_\chi$, Γ acts on $\mathcal{K}_\chi$ by left translations as follows:

$$T_{R^k S^j} f\left(S^l\right) = f\left(\left(R^k S^j\right)^{-1} S^l\right).$$

Moreover,

$$T_{R^k S^j} f\left(S^l R\right) = f\left(S^{-j} R^{-k} S^l R\right) = f\left(S^{-j} S^l R^{3-k} R\right)$$

and

$$T_{R^k S^j} f\left(S^l R\right) = \overline{\chi(R)} \cdot f\left(S^{-j} R^{-k} S^l\right) = \overline{\chi(R)} \cdot T_{R^k S^j} f\left(S^l\right).$$

In fact, since

$$T_R f(I) = f\left(R^{-1}I\right) = f\left(IR^{-1}\right) = \chi(R) \cdot f(I),$$
$$T_R f(S) = f\left(R^{-1}S\right) = f(SR) = \overline{\chi(R)} \cdot f(S)$$

and

$$T_S f(I) = f(S),\ T_S f(S) = f(SS) = f(I);$$

it follows that

$$R \mapsto \begin{bmatrix} \chi(R) & 0 \\ 0 & \overline{\chi(R)} \end{bmatrix} \quad \text{and} \quad S \mapsto \begin{bmatrix} 0 & 1 \\ 1 & 0 \end{bmatrix}$$

defines a unitary representation of the Dihedral group which is associated with: (1) the unitary character χ of the subgroup $\langle R \rangle$ and (2) the left action of Γ on the set of cosets $\Gamma / \langle R \rangle$. Such a representation is called the induced representation of χ. In addition, since $\mathcal{K}_\chi$ has no trivial and invariant subspace, it is also true that this induced representation is irreducible as well.

Example 4.10 Let

$$G = O(2, \mathbb{R}) = \left\{ p(k,t) = \begin{bmatrix} -1 & 0 \\ 0 & 1 \end{bmatrix}^k \begin{bmatrix} \cos t & -\sin t \\ \sin t & \cos t \end{bmatrix} : k \in \mathbb{Z}_2 \text{ and } t \in \mathbb{R} \right\}$$

and let

$$H = SO(2, \mathbb{R}) = \left\{ \begin{bmatrix} \cos t & -\sin t \\ \sin t & \cos t \end{bmatrix} : t \in \mathbb{R} \right\}$$

be a closed connected Lie subgroup of G. Next, let χ be a unitary character of $SO(2, \mathbb{R})$ defined as

$$\chi\left(\begin{bmatrix} \cos t & -\sin t \\ \sin t & \cos t \end{bmatrix} \right) = e^{2\pi i t}$$

and let $\mathcal{K}_\chi$ be the space of continuous functions f defined on the orthogonal group such that

$$f(p(k,t)) = e^{-2\pi i t} \cdot f(p(k,0)).$$

Then $\mathcal{K}_\chi$ is invariant under left translations of elements of the orthogonal group and in fact, $\mathcal{K}_\chi$ can be identified with $l^2\left(\mathbb{Z}_2\right)$. For arbitrary $f \in \mathcal{K}_\chi$, verify that

$$f\left(p(0,t)^{-1} p(0,0)\right) = f\left(p(0,0) p(0,t)^{-1}\right) = e^{2\pi i t} \cdot f(p(0,0))$$

and

$$\begin{aligned} f\left(p(0,t)^{-1} p(1,0)\right) &= f\left(p(0,t)^{-1} p(1,0)\right) \\ &= f\left(p(1,0) p(1,0)^{-1} p(0,t)^{-1} p(1,0)\right) \\ &= f(p(1,0) p(0,t)). \end{aligned}$$

The preceding calculations show that the left action of $O(2,\mathbb{R})$ on $\mathcal{K}_\chi$ induces a two-dimensional unitary representation of $O(2,\mathbb{R})$ acting as:

$$\begin{bmatrix} -1 & 0 \\ 0 & 1 \end{bmatrix} \mapsto \begin{bmatrix} 0 & 1 \\ 1 & 0 \end{bmatrix}$$

and

$$\begin{bmatrix} \cos t & -\sin t \\ \sin t & \cos t \end{bmatrix} \mapsto \begin{bmatrix} e^{2\pi i t} & 0 \\ 0 & e^{-2\pi i t} \end{bmatrix}.$$

This representation is said be a representation of $O(2,\mathbb{R})$ induced by the unitary character χ and the action of $O(2,\mathbb{R})$ on the quotient group $O(2,\mathbb{R})/SO(2,\mathbb{R}) \simeq \mathbb{Z}_2$.

We will now generalize the constructions provided in the examples above. To this end, let G be a Lie group with a normal closed subgroup N and let χ be a unitary character of N. We aim to introduce a method for constructing a unitary representation π of G associated with the unitary character χ called an induced character. Consider the space $\mathcal{H}_\pi$ of all complex-valued functions f defined on G satisfying the following conditions: for all $x \in G, y \in N, f(xy) = \chi(y)^{-1} \cdot f(x)$ and

$$\int_{xN \in G/N} |f(x)|^2 d\mu(xN) < \infty.$$

We also equip $\mathcal{H}_\pi$ with the following inner product: given $f, h \in \mathcal{H}_\pi$,

$$\langle f, h\rangle_{\mathcal{H}_\pi} = \int_{xN \in G/N} f(x)\overline{h(x)} d\mu(xN).$$

It is an exercise for the interested reader to verify that $\mathcal{H}_\pi$ is indeed a Hilbert space, which is also G-invariant. Next, for each $x \in G$, let $\pi(x)$ be a unitary operator acting in $\mathcal{H}_\pi$ as follows:

$$\left[\pi(x)f\right](y) = f\left(x^{-1}y\right), y \in G.$$

Definition 4.10 The representation $\pi : G \to \mathcal{U}\left(\mathcal{H}_\pi\right)$ is called an induced representation of the unitary character χ and is denoted as $\pi = \mathrm{ind}_N^{NH}(\chi)$.

The interested reader is invited to refer to the following textbooks [1], for a complete introduction to the concept of induced representation. Fix a submanifold H of G such that $G = NH$, and additionally, the map $(n,h) \mapsto nh$ defines a diffeomorphism between $N \times H$ and NH. Suppose additionally that H is a Lie subgroup of G. Next, let $d\mu$ a left-invariant measure on H. Then for $n \in N, h, s \in H$, one has that

$$\pi(n)f(s) = \left[\mathrm{ind}_N^{NH}(\chi)(n)f\right](s) = f\left(n^{-1}s\right) = f\left(s\left(s^{-1}n^{-1}s\right)\right).$$

Since $s^{-1}n^{-1}s \in N$, by definition,

$$\left[\pi(n)f\right](s) = \chi\left(s^{-1}ns\right) \cdot f(s).$$

Also since H is a group, $h^{-1}s \in H$ and

$$\left[\pi(h)f\right](s) = \left[\mathrm{ind}_N^{NH}(\chi)(h)f\right](s) = f\left(h^{-1}s\right).$$

In the special case where χ is a trivial character, we call the induced representation π, a quasiregular representation associated with the pair (N, H). In light of the preceding observations, it is therefore natural to identify $\mathcal{H}_\pi$ with $L^2(H, d\mu)$ so that if $n \in N, h, s \in H$, then

$$\left[\pi(nh)f\right](s) = \chi\left(s^{-1}ns\right) \cdot f\left(h^{-1}s\right) = \left[s \star \chi\right](n) \cdot f\left(h^{-1}s\right).$$

4.4 Integrability of Induced Characters

Let G be a Lie group equipped with a left-invariant measure $d\mu_G$ and let π be a unitary representation of G acting in some Hilbert space $\mathcal{H}_\pi$. For fixed vectors $\varphi, \psi \in \mathcal{H}_\pi$, let

$$V_{\varphi,\psi}(x) = \langle \varphi, \pi(x)\psi \rangle.$$

In the special case where $\varphi = \psi$, simply write $V_{\varphi,\psi} = V_\varphi$.

Given a positive real number p, we say that π is p-integrable if there exists a nonzero vector f in $\mathcal{H}_\pi$ such that

$$\int_G |\langle f, \pi(x)f\rangle|^p d\mu_G(x) = \int_G \left|V_f(x)\right|^p d\mu_G(x) < \infty.$$

Any representation which is 1-integrable is called integrable, and any representation which is 2-integrable will be called square-integrable. Note also that since π is a continuous representation, then every matrix coefficient $V_{\varphi,\psi}$ is a continuous function defined on G, and as a result, any integrable representation is square-integrable as well.

Proposition 4.11 If $V_f \in L^p\left(G, d\mu_G\right)$ then for any $x \in G$, $V_{\pi(x)f} \in L^p\left(G, d\mu_G\right)$.

Proof: Let us assume that $V_f \in L^p\left(G, d\mu_G\right)$. For any fixed element $x \in G$,

$$\begin{aligned}\int_G |\langle \pi(x)f, \pi(y)\pi(x)f\rangle|^p d\mu_G(y) &= \int_G \left|\left\langle f, \pi(x)^{-1}\pi(y)\pi(x)f\right\rangle\right|^p d\mu_G(y)\\ &= \int_G \left|\left\langle f, \pi\left(x^{-1}yx\right)f\right\rangle\right|^p d\mu_G(y).\end{aligned}$$

Since $d\mu_G$ is left-invariant, it follows that

$$\begin{aligned}\int_G |\langle \pi(x)f, \pi(y)\pi(x)f\rangle|^p d\mu_G(y) &= \int_G |\langle f, \pi(yx)f\rangle|^p d\mu_G(y)\\ &= \int_G |\langle f, \pi(y)f\rangle|^p d\mu_G\left(yx^{-1}\right)\end{aligned}$$

and

$$\int_G |\langle \pi(x) f, \pi(y) \pi(x) f\rangle|^p d\mu_G(y) = \Delta_G\left(x^{-1}\right) \cdot \int_G |\langle f, \pi(y) f\rangle|^p d\mu_G(y) < \infty.$$

In this section, we will be primarily interested in the integrability of induced representations of the form $\mathrm{ind}_N^G(\chi)$, where $N = \exp \mathfrak{n}$ is a normal and simply connected closed subgroup of G, and χ is a unitary character of N.

Letting $d\mu_{G/N}$ be a Haar measure on G/N, we may realize the induced representation $\pi = \mathrm{ind}_N^G(\chi)$ as acting in the Hilbert space $L^2\left(G/N, d\mu_{G/N}\right)$ as follows. Suppose that $H \subseteq G$ is a submanifold of G which is a smooth transversal of G/N. If $n \in N, h, s \in H$ and $f \in L^2\left(G/N, d\mu_{G/N}\right)$, then

$$\left[\pi(nh) f\right](sN) = \chi\left(s^{-1} ns\right) \cdot f\left(h^{-1} sN\right).$$

Since by assumption, χ is a unitary character of N, there exists a linear functional $\lambda \in \mathfrak{n}^*$ such that

$$\chi(n) = \chi_\lambda(n) = e^{2\pi i \langle \lambda, \log n\rangle}.$$

To see this, it suffices to compute the derived representation of χ as a linear representation of the Lie algebra of N. Moreover,

$$\left[\pi(nh) f\right](sN) = e^{2\pi i \left\langle \lambda, \log\left(s^{-1} ns\right)\right\rangle} f\left(h^{-1} sN\right).$$

Letting $n = \exp X$,

$$\log\left(s^{-1} \exp(X) s\right) = \log\left(\exp\left(Ad(s)^{-1} X\right)\right) = Ad(s)^{-1} X.$$

That is,

$$\log\left(s^{-1} ns\right) = Ad(s)^{-1} \log(n)$$

and as a result,

$$\begin{aligned} \left[\pi(nh) f\right](sN) &= e^{2\pi i \left\langle \left(Ad(s)^{-1}\right)^* \lambda, \log(n)\right\rangle} \cdot f\left(h^{-1} sN\right) \\ &= \chi_{\left(Ad(s)^{-1}\right)^* \lambda}(n) \cdot f\left(h^{-1} sN\right). \end{aligned}$$

For $g \in G$, put

$$Ad^\star(g) = \left(Ad(g)^{-1}\right)^*.$$

Then

$$\left[\pi(nh) f\right](sN) = \chi_{Ad^\star(s)\lambda}(n) \cdot f\left(h^{-1} sN\right).$$

Next, $V_f(nh) = \langle f, \pi(nh) f\rangle$ and letting $n = \exp X$ for some $X \in \mathfrak{n}$,

$$\begin{aligned} V_f(\exp(X) h) &= \langle f, \pi(\exp(X) h) f\rangle \\ &= \int_{G/N} e^{-2\pi i \left\langle Ad^\star(s)\lambda, X\right\rangle} \left(f(sN) \cdot \overline{f\left(h^{-1} sN\right)}\right) d\mu_{G/N}(sN). \end{aligned}$$

Defining $F_h(s) = f(sN) \cdot \overline{f(h^{-1}sN)}$ and letting $\beta(s) = Ad^{\star}(s)\lambda$, it is clear that $\beta : H \to \mathfrak{n}^*$ defines a smooth map between manifolds H and $\mathfrak{n}^*$. Consequently,

$$\begin{aligned} V_f(\exp(X)h) &= \int_{G/N} e^{-2\pi i \langle Ad^{\star}(s)\lambda, X\rangle} F_h(s)\, d\mu_{G/N}(sN) \\ &= \int_{G/N} e^{-2\pi i \langle \beta(s), X\rangle} F_h(s)\, d\mu_{G/N}(sN). \end{aligned}$$

Example 4.11 Let $G = \mathbb{R}^2 \rtimes \mathbb{R}$ be the semidirect product group equipped with the operation $(x,t)(y,s) = (x + t \bullet y, t+s)$ where

$$t \bullet y = \exp \begin{bmatrix} t & t \\ 0 & t \end{bmatrix} y = \begin{bmatrix} e^t & te^t \\ 0 & e^t \end{bmatrix} \begin{bmatrix} y_1 \\ y_2 \end{bmatrix} = \begin{bmatrix} e^t(y_1 + ty_2) \\ y_2 e^t \end{bmatrix}.$$

Next, let $\pi = \mathrm{ind}_{\mathbb{R}^2}^G(\chi)$ where χ is a unitary character of $\mathbb{R}^2$ defined as follows: $\chi(x) = e^{2\pi i (x_1 + x_2)}$. Then π can be realized as acting in $L^2(\mathbb{R})$ such that for $f \in L^2(\mathbb{R})$ and $(x,y) \in G$,

$$\begin{aligned} \pi(x,y)f(t) &= e^{2\pi i \left\langle \begin{bmatrix} e^{-t} \\ -e^{-t}(t-1) \end{bmatrix}, \begin{bmatrix} x_1 \\ x_2 \end{bmatrix} \right\rangle} f(t-y) \\ &= e^{2\pi i (e^{-t}x_1 - e^{-t}(t-1)x_2)} f(t-y). \end{aligned}$$

Thus, a matrix coefficient for the representation π, associated with $f \in L^2(\mathbb{R})$, is a complex-valued continuous map defined on $\mathbb{R}^3 \rtimes \mathbb{R}$ by means of integration as follows:

$$\begin{aligned} V_f(x,y) &= \int_{\mathbb{R}} e^{-2\pi i (e^{-t}x_1 - e^{-t}(t-1)x_2)} f(t) \overline{f(t-y)} dt \\ &= \int_{\mathbb{R}} e^{-2\pi i (e^{-t}x_1)} e^{2\pi i (e^{-t}(t-1)x_2)} f(t) \overline{f(t-y)} dt. \end{aligned}$$

The change of variable $\xi = e^{-t}$ gives

$$V_f(x,y) = \int_{\mathbb{R}} e^{-2\pi i (\xi x_1)} \frac{e^{2\pi i (\xi(\ln(\xi^{-1})-1)x_2)} f\left(\ln\left(\xi^{-1}\right)\right) \overline{f(\ln(\xi^{-1}) - y)}}{\xi} d\xi.$$

Now, let

$$\Theta_{y,x_2}(\xi) = \frac{e^{2\pi i (\xi(\ln(\xi^{-1})-1)x_2)} f\left(\ln\left(\xi^{-1}\right)\right) \overline{f(\ln(\xi^{-1}) - y)}}{\xi}$$

and let $\widehat{\Theta_{y,x_2}}$ be its Fourier transform with respect to the argument ξ. Then

$$V_f(x,y) = \int_{\mathbb{R}} e^{-2\pi i (\xi x_1)} \Theta_{y,x_2}(\xi)\, d\xi = \widehat{\Theta_{y,x_2}}(x_1).$$

A straightforward application of Plancherel's theorem gives

$$\begin{aligned}\int_{\mathbb{R}}\left|V_f\left(x_1,x_2,y\right)\right|^2 dx_1 &= \int_{\mathbb{R}}\left|\widehat{\Theta_{y,x_2}}\left(x_1\right)\right|^2 dx_1\\ &= \int_{\mathbb{R}}\left|\Theta_{y,x_2}\left(\xi\right)\right|^2 d\xi\\ &= \int_{\mathbb{R}}\left|\frac{e^{2\pi i(\xi(\ln(\xi^{-1})-1)x_2)}f\left(\ln\left(\xi^{-1}\right)\right)\overline{f\left(\ln\left(\xi^{-1}\right)-y\right)}}{\xi}\right|^2 d\xi\\ &= \int_{\mathbb{R}}\left|\frac{f\left(\ln\left(\xi^{-1}\right)\right)\overline{f\left(\ln\left(\xi^{-1}\right)-y\right)}}{\xi}\right|^2 d\xi.\end{aligned}$$

Changing variable by letting $\xi = e^{-t}$, we obtain

$$\int_{\mathbb{R}}\left|V_f\left(x_1,x_2,y\right)\right|^2 dx_1 = \int_{\mathbb{R}}\left|\frac{f(t)f(t-y)}{e^{-\frac{t}{2}}}\right|^2 dt.$$

Since

$$\begin{aligned}\int_{\mathbb{R}}\int_{\mathbb{R}}\left|V_f\left(x_1,x_2,y\right)\right|^2 dx_1 dx_2 &= \int_{\mathbb{R}}\int_{\mathbb{R}}\left|e^{\frac{t}{2}}f(t)f(t-y)\right|^2 dt dx_2\\ &= \left(\int_{\mathbb{R}} dx_2\right)\cdot\left(\int_{\mathbb{R}}\left|e^{\frac{t}{2}}f(t)f(t-y)\right|^2 dt\right)\\ &= \infty,\end{aligned}$$

it follows that the representation π is not a square-integrable representation of G.

Proposition 4.7 gives a sufficient geometric condition under which, the induced representation $\mathrm{ind}_N^G(\chi)$ is not square-integrable.

Proposition 4.12 If $\dim N > \dim G/N$ then the induced representation $\pi = \mathrm{ind}_N^G(\chi)$ cannot be square-integrable.

Proof: Having already established that for any fixed vector $f \in L^2\left(G/N, d\mu_{G/N}\right)$,

$$V_f(\exp(X)h) = \int_{G/N} e^{-2\pi i\langle\beta(s),X\rangle}F_h(s)\,d\mu_{G/N}(sN),$$

where $F_h(s) = f(sN)\cdot\overline{f(h^{-1}sN)}$, since $\dim \mathfrak{n}^* > \dim H$, then β cannot be a submersion. In fact, by computing the differential of the map $Ad^\star(s)\lambda$ at the identity element, one verifies that there exists a linear projection map $P^* : \mathfrak{n}^* \to \mathfrak{n}_1^*$ such that $\dim \mathfrak{n}_1^* = \dim \mathfrak{h}$, $P^* \circ \beta : H \to \mathfrak{n}_1^*$, and additionally, the span of the range of $P^* \circ \beta$ coincides with the vector space $\mathfrak{n}_1^*$ and $P^* \circ \beta$ is a submersion almost everywhere on H. Noting that $\mathfrak{n}_1^*$ is a proper vector subspace of $\mathfrak{n}^*$ and letting P be the adjoint of the linear map P^*,

$$\begin{aligned} V_f(\exp(X)h) &= V_f(\exp(PX + (X - PX))h) \\ &= \int_{G/N} e^{-2\pi i\langle \beta(s), PX + (X-PX)\rangle} F_h(s)\, d\mu_{G/N}(sN) \\ &= \int_{G/N} e^{-2\pi i\langle \beta(s), PX\rangle} \left(e^{-2\pi i\langle \beta(s), X-PX\rangle} F_h(s)\right) d\mu_{G/N}(sN). \end{aligned}$$

Since P is a projection, $\langle \beta(s), PX\rangle = \left\langle \beta(s), P^2X\right\rangle = \langle P^*\beta(s), PX\rangle$, and as a result,

$$V_f(\exp(X)h) = \int_{G/N} e^{-2\pi i\langle P^* \circ \beta(s), PX\rangle} e^{-2\pi i\langle \beta(s), X-PX\rangle} F_h(s)\, d\mu_{G/N}(sN).$$

In fact, letting $Y = X - PX$,

$$V_f(\exp(PX + Y)h) = \int_{G/N} e^{-2\pi i\langle P^* \circ \beta(s), PX\rangle} e^{-2\pi i\langle \beta(s), Y\rangle} F_h(s)\, d\mu_{G/N}(sN).$$

Next, let

$$d\mu_F(sN) = e^{-2\pi i\langle \beta(s), Y\rangle} F_h(s)\, d\mu_{G/N}(sN)$$

be a complex-valued measure defined on the quotient group G/N and let $d\omega_F$ be the measure on $\mathfrak{n}_1^*$ given by the push-forward of $d\mu_F$ via $P^* \circ \beta$. Then

$$V_f(\exp(PX + Y)h) = \int_{G/N} e^{-2\pi i\langle P^* \circ \beta(s), PX\rangle} d\mu_F(sN) = \int_{\mathfrak{n}_1^*} e^{-2\pi i\langle \xi, PX\rangle} d\omega_F(\xi).$$

Letting $d\omega_F(\xi) = R(\xi)\, d\xi$, where $d\xi$ is the Lebesgue measure on $\mathfrak{n}_1^*$, it follows that for $Z = PX$,

$$V_f(\exp(Z + Y)h) = \int_{\mathfrak{n}_1^*} e^{-2\pi i\langle \xi, Z\rangle} R(\xi)\, d\xi.$$

In other words, for each fixed Y, $Z \mapsto V_f(\exp(Z + Y)h)$ is the Euclidean Fourier transform of R. Next, appealing to Parseval's equality,

$$\int_{\mathfrak{n}_1} \left|V_f(\exp(Z + Y)h)\right|^2 dZ = \int_{\mathfrak{n}_1^*} \left|e^{-2\pi i\langle \xi, Z\rangle} R(\xi)\right|^2 d\xi = \int_{\mathfrak{n}_1^*} |R(\xi)|^2 d\xi.$$

Since $Y \mapsto \int_{\mathfrak{n}_1} \left|V_f(\exp(Z + Y)h)\right|^2 dZ$ is constant, as proved in the equality above, it follows that no matrix coefficient of the representation π can indeed be square-integrable.

Proposition 4.13 Let μ_N be a fixed left-invariant measure on N and let $\delta : H \to (0, \infty)$ be a positive and smooth function uniquely determined as follows $d\mu_N\left(znz^{-1}\right) = \delta(z)^{-1} d\mu_N(n)$. Then $\delta(h)\, d\mu_{G/N}(hN)\, d\mu_N(n)$ is a Haar measure on the Lie group G.

Proof: For a positive function u, which is compactly supported on G, and for $z \in H$ and $m \in N$, let

$$\begin{aligned} I(mz) &= \int_N \int_{G/N} u\left((mz)^{-1} nh\right) \delta(h)\, d\mu_{G/N}(hN)\, d\mu_N(n) \\ &= \int_N \int_{G/N} u\left(z^{-1}\left(m^{-1}n\right) h\right) \delta(h)\, d\mu_{G/N}(hN)\, d\mu_N(n). \end{aligned}$$

The change of variable $n \mapsto mn$ gives

$$\begin{aligned} I(mz) &= \int_N \int_{G/N} u\left(z^{-1} nh\right) \delta(h)\, d\mu_{G/N}(hN)\, d\mu_N(n) \\ &= \int_N \int_{G/N} u\left(\left(z^{-1}nz\right)\left(z^{-1}h\right)\right) \delta(h)\, d\mu_{G/N}(hN)\, d\mu_N(n). \end{aligned}$$

Moreover, making the change of variable: $h \mapsto zh$, we obtain

$$\begin{aligned} I(mz) &= \int_N \int_{G/N} u(nh)\, \delta(zh)\, d\mu_{G/N}(hN)\, d\mu_N\left(znz^{-1}\right) \\ &= \int_N \int_{G/N} u(nh)\, \delta(zh)\, \delta(z)^{-1} d\mu_{G/N}(hN)\, d\mu_N(n) \\ &= \int_N \int_{G/N} u(nh)\, \delta(h)\, d\mu_{G/N}(hN)\, d\mu_N(n) \\ &= I(e). \end{aligned}$$

Example 4.12 Let G be the semidirect product Lie group $\mathbb{R} \rtimes \mathbb{R}$ with group operation $(x, a)(y, b) = (x + e^a y, a + b)$. Let χ be a unitary character of $\mathbb{R} \rtimes \{0\}$ given by $\chi(x, 0) = e^{2\pi i x}$ and let $\pi = \mathrm{ind}_{\mathbb{R} \rtimes \{0\}}^{\mathbb{R} \rtimes \mathbb{R}}(\chi)$ be a representation of G acting in $L^2(\mathbb{R})$ such that $\pi(x, a) f(t) = e^{2\pi i e^{-t} x} f(t - a)$. We aim to show that there exists a nonzero vector $g \in L^2(\mathbb{R})$ such that

$$\langle g, \pi(x, a) g \rangle = \int_{\mathbb{R}} \left[g(t) \cdot g(t - a)\right] e^{-2\pi i e^{-t} x} dt$$

is p-integrable for every $p \geq 1$ with respect to the left Haar measure on G. Let $g(t) = e^{-t} \varphi(t)$ such that φ is smooth positive and $\varphi(\ln(1/s))$ is compactly supported and smooth as well. The change of variable $s = e^{-t}$ in $\int_{\mathbb{R}} \left[g(t) g(t - a)\right] e^{-2\pi i e^{-t} x} dt$ gives $t = \ln\left(s^{-1}\right)$ and $dt = \frac{ds}{s}$. Consequently, letting $\mathcal{F}$ be the Euclidean Fourier transform defined on $L^2(\mathbb{R})$, we obtain

$$\langle g, \pi(x, a) g \rangle = e^a \cdot \mathcal{F}\left(s \mapsto s \cdot \varphi\left(\ln\left(s^{-1}\right)\right) \varphi\left(\ln\left(s^{-1}\right) - a\right) \cdot 1_{(0,\infty)}(s)\right)(x).$$

Computing the integral $I(g) = \int_{\mathbb{R}} \int_{\mathbb{R}} |\langle g, \pi(x, a) g \rangle| e^{-a} dx da$,

$$I(g) = \int_{\mathbb{R}} e^a \left\| \mathcal{F}\left(s \mapsto s \cdot \varphi\left(\ln\left(s^{-1}\right)\right) \varphi\left(\ln\left(s^{-1}\right) - a\right)\right) \right\|_{L^1} e^{-a} da.$$

Note that the function

$$a \mapsto \left\| \mathcal{F}\left(s \mapsto s \cdot \varphi\left(\ln\left(s^{-1}\right)\right) \varphi\left(\ln\left(s^{-1}\right) - a\right)\right) \right\|_{L^1(\mathbb{R})}$$

is continuous and compactly supported. Therefore, the integral

$$\int_{\mathbb{R}} \left\| \mathcal{F}\left(s \mapsto s \cdot \varphi\left(\ln\left(s^{-1}\right)\right) \cdot \varphi\left(\ln\left(s^{-1}\right) - a\right)\right) \right\|_{L^1(\mathbb{R})} da$$

is finite. This shows that π is p-integrable for every $p \geq 1$.

Example 4.13 Let G be the three-dimensional Heisenberg Lie group with Lie algebra $\mathfrak{g}$ spanned by vectors X, Y, Z satisfying the nontrivial Lie brackets $[X, Y] = Z$. Let $\{Z^*, Y^*, X^*\}$ be the canonical dual basis for the vector space $\mathfrak{g}^*$ of linear functionals on $\mathfrak{g}$. Put

$$\Omega = \{\lambda Z^* + \upsilon Y^* + \omega X^* \in \mathfrak{g}^* : \lambda, \upsilon, \omega \in \mathbb{R} \text{ and } \lambda \neq 0\}.$$

Each irreducible representation

$$\pi_{\lambda Z^* + \upsilon Y^* + \omega X^*} = \operatorname{ind}_{\exp \mathbb{R}Z + \mathbb{R}Y}^{G} \left(\chi_{\lambda Z^* + \upsilon Y^* + \omega X^*}\right)$$

associated with a linear functional $\ell = \lambda Z^* + \upsilon Y^* + \omega X^*$ in the coadjoint orbit of ℓ can be realized as acting in the Hilbert space $L^2(\mathbb{R})$ as follows:

$$\pi_\ell (\exp zZ \exp yY \exp xX) f(t) = e^{2\pi i \left\langle \begin{bmatrix} \lambda \\ \upsilon - t\lambda \end{bmatrix}, \begin{bmatrix} z \\ y \end{bmatrix} \right\rangle} f(t - x).$$

Given $\varphi, \psi \in C_c^\infty(\mathbb{R}) \setminus \{0\}$, define

$$V_{\varphi,\psi}^{(\ell)}(y, x) = \left\langle \varphi, \pi_\ell (\exp yY \exp xX) \psi \right\rangle = e^{-2\pi i \upsilon y} \int_{\mathbb{R}} \varphi(t) \overline{\psi(t - x) e^{2\pi i (t\lambda) y}} dt.$$

Given $\varphi, \psi \in C_c^\infty(\mathbb{R}) \setminus \{0\}$, since φ, ψ are Schwartz,

$$(x, y) \mapsto \varphi(t) \cdot \overline{\psi(t - x) e^{2\pi i (t\lambda) y}}$$

is a Schwartz function defined on $\mathbb{R}^2$ and consequently,

$$0 < \int_{\mathbb{R}} \int_{\mathbb{R}} \left| V_{\varphi,\psi}^{(\lambda Z^* + \upsilon Y^* + \omega X^*)}(y, x) \right| dy dx < \infty.$$

Observe that the kernel of the representation $\pi_{\lambda Z^* + \upsilon Y^* + \omega X^*}$ is a discrete subgroup contained in the center of the Heisenberg Lie group. Next, let $\overline{\pi}_{\lambda Z^* + \upsilon Y^* + \omega X^*}$ be an irreducible representation of $G/\ker\left(\pi_{\lambda Z^* + \upsilon Y^* + \omega X^*}\right)$ given by $\overline{\pi}_{\lambda Z^* + \upsilon Y^* + \omega X^*} = \pi_{\lambda Z^* + \upsilon Y^* + \omega X^*} \circ q$ where q is the canonical quotient map from G onto $G/\ker\left(\pi_{\lambda Z^* + \upsilon Y^* + \omega X^*}\right)$. Since the center of $G/\ker\left(\pi_{\lambda Z^* + \upsilon Y^* + \omega X^*}\right)$ is compact, it follows that $\overline{\pi}_{\lambda Z^* + \upsilon Y^* + \omega X^*}$ is p-integrable for every $p \geq 1$.

Next, describe a sufficient differential geometric condition under which π is an integrable representation.

Theorem 4.1 Let $\beta : H \to \mathfrak{n}^*$ be the smooth function defined by $\beta(h) = Ad^\star(h)\lambda$. If β is a local diffeomorphism at the identity in H then the induced representation π is integrable.

Proof: Since β is an immersion at the identity, it is a smooth function which is also locally injective on some open subset O around the identity element in H. Let $f \in C_c^\infty(G/N)$ and let us assume additionally that f is nonzero and that f is compactly supported inside OH. Then

$$\langle f, \pi(nh)f\rangle_{L^2(G/N, d\mu_{G/N})} = \int_{OH} f(s)\overline{f(h^{-1}s)}e^{-2\pi i\langle \beta(s), \log n\rangle} d\mu_{G/N}(sN).$$

Making the change of variable $\xi = \beta(s)$ and letting

$$\omega(\xi) = \frac{d\mu_{G/N}\left(\beta^{-1}(\xi)N\right)}{d\xi}$$

be the Radon Nikodym where $d\xi$ is the standard Lebesgue measure defined on $\mathfrak{n}^*$, we obtain that

$$\langle f, \pi(nh)f\rangle_{L^2(G/N, d\mu_{G/N})}$$
$$= \int_{\beta(OH)} e^{-2\pi i\langle \xi, \log n\rangle} \left(f\left(\beta^{-1}(\xi)\right) \cdot \overline{f(h^{-1}\beta^{-1}(\xi))} \cdot \omega(\xi)\right) d\xi.$$

Let

$$\Phi_h(\xi) = f\left(\beta^{-1}(\xi)\right) \cdot \overline{f(h^{-1}\beta^{-1}(\xi))} \cdot \omega(\xi)$$

and additionally, let $\widehat{\Phi_h}$ be the Fourier transform of Φ_h. Then

$$\langle f, \pi(nh)f\rangle_{L^2(G/N, d\mu_{G/N})} = \langle f, \pi(\exp(X)h)f\rangle_{L^2(G/N, d\mu_{G/N})} = \widehat{\Phi}_h(X).$$

This implies that, for fixed $h \in H$, since each $\widehat{\Phi}_h$ is the Fourier transform of a smooth function which is compactly supported, for each fixed $h \in H$, each $\widehat{\Phi}_h$ is a Schwartz function defined over $\mathfrak{n}$. Consequently,

$$\int_H \int_N \left|\langle f, \pi(nh)f\rangle_{L^2(G/N, d\mu_{G/N})}\right| \delta(h)\, d\mu_N(n)\, d\mu_H(h)$$
$$= \int_H \delta(h) \cdot \int_N \left|\langle f, \pi(nh)f\rangle_{L^2(G/N, d\mu_{G/N})}\right| d\mu_N(n)\, d\mu_H(h)$$
$$= \int_H \delta(h) \cdot \int_N \left|\widehat{\Phi}_h(\log n)\right| d\mu_N(n)\, d\mu_H(h)$$
$$= \int_H \delta(h) \cdot \left\|\widehat{\Phi}_h\right\|_{L^1(\mathfrak{n})} d\mu_H(h).$$

For a fixed basis $\{X_1, \dots, X_{\dim N}\}$ of $\mathfrak{n}$ and for $\varphi \in C^\infty(\mathfrak{n})$, put

$$|\alpha| = |\alpha_1 + \cdots + \alpha_{\dim N}|, |\beta| = |\beta_1 + \cdots + \beta_{\dim N}|,$$
$$X^\alpha = X_1^{\alpha_1} \cdots X_1^{\alpha_{\dim N}},$$
$$D^\beta = \frac{\partial}{\partial X_1}^{\beta_1} \circ \cdots \circ \frac{\partial}{\partial X_{\dim N}}^{\dim N},$$

$X(x) = \sum_{k=1}^{\dim N} x_k X_k$, and let

$$\|\varphi\|_{(s)} = \sup_{|\alpha|,|\beta| \le s} \sup_{X \in \mathfrak{n}} \left| X^\alpha \cdot D^\beta (\varphi(X)) \right|.$$

Moreover, let $S(\mathfrak{n})$ be the set of all functions $\varphi \in C^\infty(\mathfrak{n})$ such that $\|\varphi\|_{(s)} < \infty$ for all $s \ge 0$. That is, $S(\mathfrak{n})$ is the set of a Schwartz functions defined on $\mathfrak{n}$. Observe that

$$\begin{aligned}\left\|\widehat{\Phi}_h\right\|_{L^1(\mathfrak{n})} &= \int_{\mathfrak{n}} \left|\widehat{\Phi}_h(X)\right| dX \\ &\le \int_{\mathfrak{n}} \sup_{X \in \mathfrak{n}} \left((1 + \|X\|)^{(\dim N+1)} \cdot \left|\widehat{\Phi}_h(X)\right|\right) \cdot (1 + \|X\|)^{-(\dim N+1)} dX \\ &\le \sup_{X \in \mathfrak{n}} \left((1 + \|X\|)^{(\dim N+1)} \cdot \left|\widehat{\Phi}_h(X)\right|\right) \cdot \left(\int_{\mathfrak{n}} \frac{1}{(1 + \|X\|)^{(\dim N+1)}} dX\right).\end{aligned}$$

Letting $C_1 = \int_{\mathfrak{n}} \frac{1}{(1+\|X\|)^{(\dim N+1)}} dX$, there exists $\ell \ge 0$ such that

$$\left\|\widehat{\Phi}_h\right\|_{L^1(\mathfrak{n})} \le C_1 \cdot \left\|\widehat{\Phi}_h\right\|_{(\ell)}.$$

This shows that

$$\int_H \delta(h) \left\|\widehat{\Phi}_h\right\|_{L^1(\mathfrak{n})} d\mu_H(h) \le C_1 \int_H \delta(h) \cdot \left\|\widehat{\Phi}_h\right\|_{(\ell)} d\mu_H(h).$$

Since $\left\|\widehat{\Phi}_h\right\|_{(\ell)}$ is a Schwartz seminorm, then there exist (a) a positive constant C_2 and (b) $\ell' \ge 0$ such that $\left\|\widehat{\Phi}_h\right\|_{(\ell)} \le C_2 \cdot \|\Phi_h\|_{(\ell')}$. As a result,

$$\begin{aligned}\int_H \delta(h) \left\|\widehat{\Phi}_h\right\|_{L^1(\mathfrak{n})} d\mu_H(h) &\le C_1 \cdot C_2 \int_H \delta(h) \cdot \|\Phi_h\|_{(\ell')} d\mu_H(h) \\ &\le C_1 \cdot C_2 \int_H \delta(h) \\ &\quad \cdot \sup_{|\alpha|,|\beta| \le \ell'} \sup_{\xi \in \mathfrak{n}} \left|\xi^\alpha \cdot D^\beta\left(\Phi_h(\xi)\right)\right| d\mu_H(h).\end{aligned}$$

Using the assumption that

$$(\xi, h) \mapsto \Phi_h(\xi) = f\left(\beta^{-1}(\xi)\right) \cdot \overline{f(h^{-1}\beta^{-1}(\xi))} \cdot \omega(\xi)$$

is smooth and compactly supported, it is also a Schwartz function, and as a result,

$$\Psi : h \mapsto \delta(h) \cdot \sup_{|\alpha|,|\beta| \le \ell'} \sup_{\xi \in \mathfrak{n}} \left|\xi^\alpha \cdot D^\beta\left(\Phi_h(\xi)\right)\right|$$

is a continuous assignment on H. Finally,

$$\begin{aligned}\int_H \delta(h) \left\|\widehat{\Phi}_h\right\|_{L^1(\mathfrak{n})} d\mu_H(h) &\leq C_1 \cdot C_2 \int_H \Psi(h)\, d\mu_H(h) \\ &\leq C_1 \cdot C_2 \cdot \sup_{h\in H}(\Psi(h)) \cdot \mu_H(\text{support}(\Psi)) \\ &< \infty.\end{aligned}$$

Proposition 4.14 Let Σ be a submanifold of $\mathfrak{n}^*$ and let $\pi = \int_\Sigma^\oplus \mathrm{ind}_N^G(\chi_\lambda)\, d\lambda$ be a direct integral representation of G where $d\lambda$ is the Lebesgue measure on Σ. Suppose that the function $b : \Sigma \times H \to \mathfrak{n}^*$ defined by $b(\lambda, s) = Ad^\star(s)\lambda$ is a local diffeomorphism at some point $p \in \Sigma \times H$. Then π is an integrable representation of G.

Proof: Since by assumption, $b : \Sigma \times H \to \mathfrak{n}$ is a local diffeomorphism, then there exists an open and relatively compact subset $O_1 \times O_2$ of $\Sigma \times H$ such that the restriction of b to $O_1 \times O_2$ defines a diffeomorphism between $O_1 \times O_2$ and $b(O_1 \times O_2)$. Let $f \in L^2(\Sigma \times G/N, d\mu_{G/N} d\lambda)$. Suppose additionally that f is smooth and compactly supported inside $O_1 \times O_2/N \subseteq \Sigma \times G/N$. Then a matrix coefficient of π associated to f takes the form:

$$V_f(\exp(X)h) = \int_\Sigma \int_{G/N} e^{-2\pi i\langle b(\lambda,s),X\rangle} F_h(\lambda, s)\, d\mu_{G/N}(sN)\, d\lambda,$$

where $F_h(\lambda, s) = f(\lambda, sN) \cdot \overline{f(\lambda, h^{-1}sN)}$ and $b(\lambda, s) = Ad^\star(s)\lambda$. Next, letting $d\mu_H$ be a measure on H defined by the equality $\int_{G/N} d\mu_{G/N}(sN) = \int_H d\mu_H(s)$,

$$V_f(\exp(X)h) = \int_\Sigma \int_H e^{-2\pi i\langle b(\lambda,s),X\rangle} F_h(\lambda, s)\, d\mu_H(s)\, d\lambda$$

Put $d\psi(\lambda, s) = d\mu_H(s)\, d\lambda$. Then

$$\begin{aligned}V_f(\exp(X)h) &= \int_{\Sigma\times H} e^{-2\pi i\langle b(\lambda,s),X\rangle} F_h(\lambda, s)\, d\psi(\lambda, s) \\ &= \int_{O_1\times O_2} e^{-2\pi i\langle b(\lambda,s),X\rangle} F_h(\lambda, s)\, d\psi(\lambda, s)\end{aligned}$$

and the change of variable $\xi = b(\lambda, s)$ yields

$$V_f(\exp(X)h) = \int_{b(O_1\times O_2)} e^{-2\pi i\langle \xi,X\rangle} F_h\left(b^{-1}(\xi)\right) d\psi\left(b^{-1}(\xi)\right).$$

Next, letting $d\psi\left(b^{-1}(\xi)\right) = W(\xi)\, d\xi$, where $d\xi$ is the Lebesgue measure on $\mathfrak{n}^*$, we derive

$$V_f(\exp(X)h) = \int_{\mathfrak{n}^*} e^{-2\pi i\langle \xi,X\rangle} F_h\left(b^{-1}(\xi)\right) W(\xi)\, d\xi.$$

Furthermore, letting $\Theta_h(\xi) = F_h\left(b^{-1}(\xi)\right) W(\xi)$, and $\widehat{\Theta_h}$ be its Fourier transform, it follows that

$$V_f(\exp(X)h) = \int_{\mathfrak{n}^*} e^{-2\pi i\langle\xi,X\rangle}\Theta_h(\xi)\,d\xi = \widehat{\Theta_h}(X).$$

Since

$$\int_H\int_{\mathfrak{n}}\left|V_f(\exp(X)h)\right|\delta(h)\,dX d\mu_H(h) = \int_H \delta(h)\cdot\left\|\widehat{\Theta_h}\right\|_{L^1(\mathfrak{n})} d\mu_H(h),$$

there exists a positive constant C_1 such that $\left\|\widehat{\Theta_h}\right\|_{L^1(\mathfrak{n})} \leq C_1\cdot\left\|\widehat{\Theta_h}\right\|_{(\ell)}$, where

$$\left\|\widehat{\Theta_h}\right\|_{(\ell)} = \sup_{|\alpha|,|\beta|\leq\ell}\ \sup_{X\in\mathfrak{n}}\left|X^\alpha\cdot D^\beta\left(\widehat{\Theta_h}(X)\right)\right|.$$

This shows that

$$\int_H \delta(h)\cdot\left\|\widehat{\Theta_h}\right\|_{L^1(\mathfrak{n})} d\mu_H(h) \leq C_1\int_H \delta(h)\cdot\left\|\widehat{\Theta_h}\right\|_{(\ell)} d\mu_H(h).$$

Since $\left\|\widehat{\Theta_h}\right\|_{(\ell)}$ is a Schwartz seminorm, then there exist a positive constant C_2 and $\ell'\geq 0$ such that $\left\|\widehat{\Theta_h}\right\|_{(\ell)} \leq C_2\cdot\left\|\Theta_h\right\|_{(\ell')}$ and

$$\begin{aligned}
&\int_H \delta(h)\cdot\left\|\widehat{\Theta_h}\right\|_{L^1(\mathfrak{n})} d\mu_H(h)\\
&\quad\leq C_1\cdot C_2\int_H \delta(h)\cdot\left\|\Theta_h\right\|_{(\ell')} d\mu_H(h)\\
&\quad\leq C_1\cdot C_2\int_H \delta(h)\cdot\sup_{|\alpha|,|\beta|\leq\ell'}\ \sup_{\xi\in\mathfrak{n}}\left|\xi^\alpha\cdot D^\beta\left(F_h\left(b^{-1}(\xi)\right)W(\xi)\right)\right| d\mu_H(h).
\end{aligned}$$

Put

$$\Psi(h) = C_1\cdot C_2\cdot\delta(h)\sup_{|\alpha|,|\beta|\leq\ell'}\ \sup_{\xi\in\mathfrak{n}}\left|\xi^\alpha\cdot D^\beta\left(F_h\left(b^{-1}(\xi)\right)W(\xi)\right)\right|.$$

Then

$$\int_H \delta(h)\cdot\left\|\widehat{\Theta_h}\right\|_{L^1(\mathfrak{n})} d\mu_H(h) \leq \int_H \Psi(h)\,d\mu_H(h).$$

Since Ψ is continuous and compactly supported on H, it follows that

$$\int_H\int_{\mathfrak{n}}\left|V_f(\exp(X)h)\right|\delta(h)\,dX d\mu_H(h) \leq \int_H \Psi(h)\,d\mu_H(h) < \infty$$

and as a result, π is an integrable representation of G.

Example 4.14 Equip $G = \mathbb{R}^3 \rtimes \mathbb{R}$ with group operation $(v,t)(w,s) = (v + e^{tA}w, t+s)$ where

$$A = \begin{bmatrix} 1 & 1 & 0\\ 0 & 1 & 1\\ 0 & 0 & 1\end{bmatrix}.$$

Let $N = \mathbb{R}^3 \rtimes \{0\}$ be a maximal normal and abelian Lie subgroup of G. Suppose next that Σ is a unit sphere in $\mathbb{R}^3$ parametrized as follows:

$$\Sigma = \left\{ \lambda_{(\theta,\phi)} = \begin{bmatrix} \sin\phi\cos\theta \\ \sin\phi\sin\theta \\ \cos\phi \end{bmatrix} : (\theta,\phi) \in [0,2\pi) \times [0,\pi] \right\}.$$

For each $\lambda_{(\theta,\phi)} \in \Sigma$, let $\chi_{\lambda_{(\theta,\phi)}}$ be a unitary character of N defined as follows: $\chi_{\lambda_{(\theta,\phi)}}(v) = e^{2\pi i \langle \lambda_{(\theta,\phi)}, v\rangle}$. Then

$$b(\phi,\theta,s) = e^{-sA^T}\lambda_{(\theta,\phi)} = \begin{bmatrix} (\cos\theta\sin\phi)\,e^{-s} \\ (\sin\phi)\,e^{-s}(\sin\theta - s\cos\theta) \\ \frac{1}{2}e^{-s}(2\cos\phi + s\sin\phi(s\cos\theta - 2\sin\theta)) \end{bmatrix}.$$

Computing the determinant of the differential of b at $\left(\frac{\pi}{2},\frac{\pi}{2},0\right)$, verify that

$$\det\begin{bmatrix} 0 & -1 & 0 \\ -1 & 0 & 0 \\ -1 & 0 & -1 \end{bmatrix} = 1.$$

Therefore, b is indeed a local diffeomorphism at $\left(0,\frac{\pi}{2},\frac{\pi}{2}\right)$ and as a result,

$$\int_{[0,2\pi)\times[0,\pi]}^{\oplus} \operatorname{ind}_N^G\left(\chi_{\lambda_{(\theta,\phi)}}\right) d(\theta,\phi)$$

is an integrable representation of G.

Example 4.15 Let $G = \mathbb{R}^4 \rtimes \mathbb{R}$ be $(v,t)(w,s) = \left(v + e^{tA}w, t+s\right)$, where

$$A = \begin{bmatrix} 0 & 1 & 0 & 0 \\ -1 & 0 & 0 & 0 \\ 0 & 0 & 0 & \sqrt{2} \\ 0 & 0 & -\sqrt{2} & 0 \end{bmatrix}.$$

Next, let $N = \mathbb{R}^4 \rtimes \{0\}$ be a maximal normal and abelian Lie subgroup of G. For any unitary character χ of N, there exists a linear functional $\lambda \in \mathbb{R}^4$ such that $\chi(v,0) = \chi_\lambda(v,0) = e^{2\pi i\langle\lambda,v\rangle}$. For fixed $\lambda \in \mathbb{R}^4$, we compute b as follows:

$$b(\lambda,s) = e^{tA}\lambda = \begin{bmatrix} \lambda_1\cos s + \lambda_2\sin s \\ \lambda_2\cos s - \lambda_1\sin s \\ \lambda_4\cos\sqrt{2}s + \lambda_5\sin\sqrt{2}s \\ \lambda_5\cos\sqrt{2}s - \lambda_4\sin\sqrt{2}s \end{bmatrix}.$$

Suppose next that Σ is a three-dimensional manifold described as follows:

$$\Sigma = \left\{ \lambda_{(\alpha,\beta,\gamma)} = \begin{bmatrix} \alpha \\ 0 \\ \beta \\ \gamma \end{bmatrix} : (\alpha, \beta, \gamma) \in \mathbb{R}^3 \right\}.$$

For each $\lambda_{(\alpha,\beta,\gamma)}$ in Σ, let $\chi_{\lambda_{(\alpha,\beta,\gamma)}}$ be a unitary character of N defined as follows: $\chi_{\lambda_{(\alpha,\beta,\gamma)}}(v) = e^{2\pi i \langle \lambda_{(\alpha,\beta,\gamma)}, v \rangle}$. Then

$$b(\alpha, \beta, \gamma, s) = e^{-sA^T} \lambda_{(\alpha,\beta,\gamma)} = \begin{bmatrix} \alpha \cos s \\ -\alpha \sin s \\ \beta \cos\left(\sqrt{2}s\right) + \gamma \sin\left(\sqrt{2}s\right) \\ \gamma \cos\left(\sqrt{2}s\right) - \beta \sin\left(\sqrt{2}s\right) \end{bmatrix}.$$

Computing the determinant of the differential of b at $(1, 1, 1, 0)$, we obtain

$$\det \begin{bmatrix} 1 & 0 & 0 & 0 \\ 0 & 0 & 0 & -1 \\ 0 & 1 & 0 & \sqrt{2} \\ 0 & 0 & 1 & -\sqrt{2} \end{bmatrix} = -1.$$

Therefore, b is indeed a local diffeomorphism at $(1, 1, 1, 0)$, and as a result,

$$\int_{\mathbb{R}^3}^{\oplus} \mathrm{ind}_N^G \left(\chi_{\lambda_{(\alpha,\beta,\gamma)}} \right) d(\alpha, \beta, \gamma)$$

is an integrable unitary representation of G.

Example 4.16 Let $G = \mathbb{R}^3 \rtimes \mathbb{R}^2$ be $(v, t)(w, s) = \left(v + e^{t_1 A_1 + t_2 A_2} w, t + s\right)$, where

$$A_1 = \begin{bmatrix} 1 & 0 & 0 \\ 0 & 0 & 0 \\ 0 & 0 & 1 \end{bmatrix}, \quad A_2 = \begin{bmatrix} 0 & 0 & 0 \\ 0 & 1 & 0 \\ 0 & 0 & 1 \end{bmatrix}.$$

Let $N = \mathbb{R}^3 \rtimes \{0\}$ be a maximal normal and abelian Lie subgroup of G. For any unitary character χ of N, there exists a linear functional $\lambda \in \mathbb{R}^3$ such that $\chi(v, 0) = \chi_\lambda(v, 0) = e^{2\pi i \langle \lambda, v \rangle}$. Suppose Σ is a curve in $\mathbb{R}^3$ defined as follows:

$$\Sigma = \left\{ \lambda_\alpha = \begin{bmatrix} \alpha \\ \alpha \\ \alpha \end{bmatrix} : \alpha \in \mathbb{R} \right\}.$$

For each $\lambda_\alpha \in \Sigma$, let χ_{λ_α} be a unitary character of N defined as follows: $\chi_{\lambda_\alpha}(v) = e^{2\pi i \langle \lambda_\alpha, v\rangle}$. Then

$$b(\alpha, s) = e^{-sA^T}\lambda_\alpha = \begin{bmatrix} e^{-s_1}\alpha \\ e^{-s_2}\alpha \\ e^{-s_1-s_2}\alpha \end{bmatrix}$$

and the Jacobian of b at (α, s) is

$$Jb_{(\alpha,s)} = \begin{bmatrix} e^{-s_1} & -e^{-s_1}\alpha & 0 \\ e^{-s_2} & 0 & -e^{-s_2}\alpha \\ e^{-s_1-s_2} & -e^{-s_1-s_2}\alpha & -e^{-s_1-s_2}\alpha \end{bmatrix}.$$

Thus, whenever α is not equal to zero, $Jb_{(\alpha,s)}$ is invertible and as a result, b is a local diffeomorphism at (α_0, s) as long as $\alpha_0 \neq 0$. This shows that the direct integral representation $\int_{\mathbb{R}}^{\oplus} \mathrm{ind}_N^G\left(\chi_{\lambda_\alpha}\right) d\alpha$ is an integrable representation of G.

It is worth noting that the local diffeomorphism condition imposed in Theorem 4.1 is not always necessary. For example, the left regular representation of any Lie group is unitarily equivalent to a representation obtained by inducing the trivial character from the singleton subgroup containing the identity. Letting λ be the zero linear functional associated with this trivial representation, it is clear that in this context, $\beta : G \to \mathfrak{n}^*$ is the zero map sending every element in $\mathfrak{g}$ to the zero functional. Clearly such a map is nowhere an immersion, yet as shown in Theorem 4.1, the left regular representation is integrable.

Proposition 4.15 The left regular representation of any Lie group G is p-integrable for any positive real number p.

Proof: Given $f \in C_c(G)$ put

$$V_f(x) = \langle f, L(x)f\rangle = \int_G f(y)\overline{f(x^{-1}y)}d\mu_G(y).$$

Let S be the support of f and suppose additionally that S is symmetric, in the sense that $S = S^{-1}$. Then by assumption, S is a compact subset of G and additionally, $S \cap xS$ is nonempty if and only if $x \in SS^{-1} = S^2$. In other words,

$$V_f(x) = \langle f, L(x)f\rangle = \begin{cases} \int_{S\cap xS} f(y)\overline{f(x^{-1}y)}d\mu_G(y) & \text{if } x \in S^2 \\ 0 \quad \text{if } x \notin S^2 \end{cases}.$$

Since the mapping $\sigma : (x, y) \mapsto xy$ is a continuous function from $G \times G$ into G, it follows that $\sigma(S \times S) = S^2$ is a compact subset of G. This shows that V_f is a function which is compactly supported on G. Appealing to Cauchy–Schwartz inequality,

$$\begin{aligned}
\int_G \left|V_f(x)\right|^p d\mu_G(x) &= \int_{S^2} \left|V_f(x)\right|^p d\mu_G(x) \\
&\leq \int_{S^2} \left| \|f\|_{L^2(G,d\mu_G)} \cdot \|L(x) f\|_{L^2(G,d\mu_G)} \right|^p d\mu_G(x) \\
&= \|f\|^{2p}_{L^2(G,d\mu_G)} \cdot \left(\int_{S^2} d\mu_G(x) \right) \\
&\leq \|f\|^{2p}_{L^2(G,d\mu_G)} \cdot \mu_G\left(S^2\right) < \infty.
\end{aligned}$$

References

1 Gerald B. Folland. *A course in abstract harmonic analysis*. Studies in Advanced Mathematics. CRC Press, Boca Raton, FL, 1995. ISBN 0-8493-8490-7.

2 Sigurdur Helgason. *Geometric analysis on symmetric spaces*, volume 39 of *Mathematical Surveys and Monographs*. American Mathematical Society, Providence, RI, second edition, 2008.

3 Eberhard Kaniuth and Keith F. Taylor. *Induced representations of locally compact groups*, volume 197. Cambridge University Press, 2013.

4 Vignon Oussa. Frames arising from irreducible solvable actions I. *Journal of Functional Analysis*, 274(4):1202–1254, 2018.

5 Vignon Oussa. Compactly supported bounded frames on lie groups. *Journal of Functional Analysis*, 277(6):1718–1762, 2019.

5

Frame Theory

In a Hilbert space, a frame is a set of vectors that can be used to represent any vector in the space. In finite-dimensional vector spaces, a frame is simply a spanning set, a set of vectors that can be used to represent all the vectors in the space. However, in infinite-dimensional vector spaces, the situation is more complicated because the convergence of infinite sums (representations of a vector as a linear combination of an infinite set of vectors) is not guaranteed. In general, spanning sets and frames do not coincide in infinite-dimensional vector spaces.

Before we delve into the concept of frames, we will review the notion of an orthonormal basis. We will then generalize orthonormal bases to Riesz bases, which are a generalization of bases that do not necessarily possess the property of orthonormality. We will also examine Parseval's equality, which is a property that is typically associated with orthonormal bases, but can be extended to other systems that do not necessarily have the properties of bases or orthonormality. This will lead us naturally to the concept of frames.

For readers who wish to learn more about frames, we recommend the textbooks of Christensen and Heil, which specialize in these topics [2, 4].

5.1 Series Expansions in Hilbert Spaces

Let $\mathcal{H}$ be a separable Hilbert space and let $(f_k)_{k\in J}$ be a countable subset of $\mathcal{H}$ such that the closure of the span of $(f_k)_{k\in J}$ coincides with $\mathcal{H}$. Then, there exists a subfamily $(f_k)_{k=1}^{\infty}$ such that (a) f_{n+1} does not belong to the span of $(f_k)_{k=1}^{n}$ and (b) the span of $(f_k)_{k=1}^{\infty}$ is dense in $\mathcal{H}$. Appealing to Gram–Schmidt procedure, it is well known [6] that it is possible to iteratively construct a countable family of vectors $(e_k)_{k\in I}$ in $\mathcal{H}$, which is an orthonormal basis for $\mathcal{H}$. In other words,

$$\langle e_k, e_j\rangle_{\mathcal{H}} = \begin{cases} 0 & \text{if } k \neq j \\ 1 & \text{if } k = j \end{cases}$$

A Bridge Between Lie Theory and Frame Theory: Applications of Lie Theory to Harmonic Analysis, First Edition. Vignon Oussa.

and if $\langle f, e_j \rangle_{\mathcal{H}} = 0$ for all $j \in I$ then f must be the zero vector in $\mathcal{H}$. This shows that every separable Hilbert space has a countable orthonormal basis.

For a fixed index $k \in I$, and $f \in \mathcal{H}$, the scalar $\langle f, e_k \rangle_{\mathcal{H}}$ is called a Fourier coefficient of f with respect to the sequence $(e_k)_{k \in I}$ and additionally, the series $\sum_{k \in I} \langle f, e_k \rangle e_k$ is often called the Fourier series of f with respect to the orthonormal basis $(e_k)_{k \in I}$.

Proposition 5.1 ***(Generalized Pythagorean theorem)*** If $(e_k)_{k=1}^{n}$ is a finite orthonormal sequence then

$$\left\| \sum_{k=1}^{n} e_k \right\|_{\mathcal{H}}^2 = \sum_{k=1}^{n} \|e_k\|_{\mathcal{H}}^2 .$$

Proof: Since $(e_k)_{k=1}^{n}$ is an orthonormal sequence, it is easy to verify that

$$\left\| \sum_{k=1}^{n} e_k \right\|_{\mathcal{H}}^2 = \left\langle \sum_{k=1}^{n} e_k, \sum_{j=1}^{n} e_j \right\rangle_{\mathcal{H}} = \sum_{k=1}^{n} \sum_{j=1}^{n} \langle e_k, e_j \rangle_{\mathcal{H}},$$

and

$$\left\| \sum_{k=1}^{n} e_k \right\|_{\mathcal{H}}^2 = \sum_{k=1}^{n} \langle e_k, e_k \rangle_{\mathcal{H}} = \sum_{k=1}^{n} \|e_k\|_{\mathcal{H}}^2 .$$

Proposition 5.2 Let $(e_k)_{k=1}^{\infty}$ be a complete orthonormal sequence in an infinite-dimensional Hilbert space $\mathcal{H}$. For any vector $f \in \mathcal{H}$, $f = \sum_{k=1}^{\infty} \langle f, e_k \rangle_{\mathcal{H}} e_k$ and additionally, Parseval's identity holds. That is, $\|f\|_{\mathcal{H}}^2 = \sum_{k=1}^{\infty} \left| \langle f, e_k \rangle_{\mathcal{H}} \right|^2$.

Proof: For a fixed natural number n, put $f_n = \sum_{k=1}^{n} \langle f, e_k \rangle_{\mathcal{H}} e_k$. By linearity of inner product,

$$\begin{aligned} \|f - f_n\|_{\mathcal{H}}^2 = \langle f - f_n, f - f_n \rangle_{\mathcal{H}} &= \langle f, f \rangle_{\mathcal{H}} - \langle f_n, f \rangle_{\mathcal{H}} - \langle f, f_n \rangle_{\mathcal{H}} \\ &\quad + \langle f_n, f_n \rangle_{\mathcal{H}}. \end{aligned}$$

Since $\langle f_n, f_n \rangle_{\mathcal{H}} = \left\| \sum_{k=1}^{n} \langle f, e_k \rangle_{\mathcal{H}} e_k \right\|_{\mathcal{H}}^2$,

$$\begin{aligned} \|f - f_n\|_{\mathcal{H}}^2 &= \|f\|_{\mathcal{H}}^2 + \left\| \sum_{k=1}^{n} \langle f, e_k \rangle_{\mathcal{H}} e_k \right\|_{\mathcal{H}}^2 - \sum_{k=1}^{n} \langle f, e_k \rangle_{\mathcal{H}} \langle e_k, f \rangle_{\mathcal{H}} \\ &\quad - \sum_{k=1}^{n} \overline{\langle f, e_k \rangle_{\mathcal{H}}} \langle f, e_k \rangle_{\mathcal{H}} \\ &= \|f\|_{\mathcal{H}}^2 + \sum_{k=1}^{n} \left| \langle f, e_k \rangle_{\mathcal{H}} \right|^2 - \sum_{k=1}^{n} \left| \langle f, e_k \rangle_{\mathcal{H}} \right|^2 - \sum_{k=1}^{n} \left| \langle f, e_k \rangle_{\mathcal{H}} \right|^2 \end{aligned}$$

and $\|f - f_n\|_{\mathcal{H}}^2 = \|f\|_{\mathcal{H}}^2 - \sum_{k=1}^{n} \left|\langle f, e_k\rangle_{\mathcal{H}}\right|^2$. As a result, for any natural number n,

$$\sum_{k=1}^{n} \left|\langle f, e_k\rangle_{\mathcal{H}}\right|^2 = \|f\|_{\mathcal{H}}^2 - \|f - f_n\|_{\mathcal{H}}^2 \le \|f\|_{\mathcal{H}}^2 .$$

Taking the limit on each side of the inequality above, we obtain

$$\sum_{k=1}^{\infty} \left|\langle f, e_k\rangle_{\mathcal{H}}\right|^2 = \lim_{n\to\infty} \sum_{k=1}^{n} \left|\langle f, e_k\rangle_{\mathcal{H}}\right|^2 \le \lim_{n\to\infty} \|f\|_{\mathcal{H}}^2 = \|f\|_{\mathcal{H}}^2 .$$

This shows that for every orthonormal sequence $(e_k)_{k=1}^{\infty}$ (complete or incomplete) and for every vector $f \in \mathcal{H}$, $\sum_{k=1}^{\infty} \left|\langle f, e_k\rangle_{\mathcal{H}}\right|^2 \le \|f\|_{\mathcal{H}}^2$. This inequality is also known as Bessel's inequality.

We aim now to show that whenever the series $\sum_{k=1}^{\infty} \left|\langle f, e_k\rangle_{\mathcal{H}}\right|^2$ is convergent then $\sum_{k=1}^{\infty} \langle f, e_k\rangle_{\mathcal{H}} e_k$ must necessarily be an element of the Hilbert space $\mathcal{H}$. To this end, put $f_n = \sum_{k=1}^{n} \langle f, e_k\rangle_{\mathcal{H}} e_k$ for some fixed natural number n. Appealing to the generalized Pythagorean theorem, for arbitrary natural numbers l, m,

$$\begin{aligned}
\|f_{m+l} - f_m\|_{\mathcal{H}}^2 &= \left\| \sum_{k=1}^{m+l} \langle f, e_k\rangle_{\mathcal{H}} e_k - \sum_{k=1}^{m} \langle f, e_k\rangle_{\mathcal{H}} e_k \right\|_{\mathcal{H}}^2 \\
&= \left\| \sum_{k=m+1}^{m+l} \langle f, e_k\rangle_{\mathcal{H}} e_k \right\|_{\mathcal{H}}^2 \\
&= \sum_{k=m+1}^{m+l} \left|\langle f, e_k\rangle_{\mathcal{H}}\right|^2 .
\end{aligned}$$

Taking the limit of each side of the equality $\|f_{m+l} - f_m\|_{\mathcal{H}}^2 = \sum_{k=m+1}^{m+l} \left|\langle f, e_k\rangle_{\mathcal{H}}\right|^2$,

$$\lim_{m\to\infty} \|f_{m+l} - f_m\|_{\mathcal{H}}^2 = \lim_{m\to\infty} \sum_{k=m+1}^{m+l} \left|\langle f, e_k\rangle_{\mathcal{H}}\right|^2 = 0.$$

This shows that $(f_n)_{n=1}^{\infty}$ is a Cauchy sequence in $\mathcal{H}$. Since $\mathcal{H}$ is a complete normed space, it follows that

$$\lim_{n\to\infty} \sum_{k=1}^{n} \langle f, e_k\rangle_{\mathcal{H}} e_k = \sum_{k=1}^{\infty} \langle f, e_k\rangle_{\mathcal{H}} e_k \in \mathcal{H}.$$

Now, put $g = f - \sum_{k=1}^{\infty} \langle f, e_k\rangle_{\mathcal{H}} e_k$. Then, for every natural number j,

$$\begin{aligned}
\langle g, e_j\rangle_{\mathcal{H}} &= \left\langle f - \sum_{k=1}^{\infty} \langle f, e_k\rangle_{\mathcal{H}} e_k, e_j \right\rangle_{\mathcal{H}} \\
&= \langle f, e_j\rangle_{\mathcal{H}} - \sum_{k=1}^{\infty} \langle f, e_k\rangle_{\mathcal{H}} \langle e_k, e_j\rangle_{\mathcal{H}} \\
&= \langle f, e_j\rangle_{\mathcal{H}} - \langle f, e_j\rangle_{\mathcal{H}} = 0.
\end{aligned}$$

Since $(e_k)_{k=1}^{\infty}$ is assumed to be complete, it follows that $f - \sum_{k=1}^{\infty} \langle f, e_k \rangle_{\mathcal{H}} e_k = 0$ and as a result, $f = \sum_{k=1}^{\infty} \langle f, e_k \rangle_{\mathcal{H}} e_k$. Finally,

$$\left\| \sum_{k=1}^{n} \langle f, e_k \rangle_{\mathcal{H}} e_k \right\|_{\mathcal{H}}^2 = \left\langle \sum_{k=1}^{n} \langle f, e_k \rangle_{\mathcal{H}} e_k, \sum_{j=1}^{n} \langle f, e_j \rangle_{\mathcal{H}} e_j \right\rangle$$
$$= \sum_{k=1}^{n} \sum_{j=1}^{n} \langle f, e_k \rangle_{\mathcal{H}} \overline{\langle f, e_j \rangle_{\mathcal{H}}} \langle e_k, e_j \rangle_{\mathcal{H}}$$

and

$$\left\| \sum_{k=1}^{n} \langle f, e_k \rangle_{\mathcal{H}} e_k \right\|_{\mathcal{H}}^2 = \sum_{k=1}^{n} \langle f, e_k \rangle_{\mathcal{H}} \overline{\langle f, e_k \rangle_{\mathcal{H}}} = \sum_{k=1}^{n} \left| \langle f, e_k \rangle_{\mathcal{H}} \right|^2.$$

Next,

$$\lim_{n \to \infty} \left\| \sum_{k=1}^{n} \langle f, e_k \rangle_{\mathcal{H}} e_k \right\|_{\mathcal{H}}^2 = \lim_{n \to \infty} \sum_{k=1}^{n} \left| \langle f, e_k \rangle_{\mathcal{H}} \right|^2$$

and since $f \mapsto \|f\|_{\mathcal{H}}$ is a continuous function, it follows that

$$\|f\|_{\mathcal{H}}^2 = \left\| \sum_{k=1}^{\infty} \langle f, e_k \rangle_{\mathcal{H}} e_k \right\|_{\mathcal{H}}^2 = \sum_{k=1}^{\infty} \left| \langle f, e_k \rangle_{\mathcal{H}} \right|^2.$$

Let $(e_k)_{k \in I}$ be an orthonormal basis for $\mathcal{H}$ and let $\varphi : I \to I$ be a bijection. Since, $(e_k)_{k \in \varphi(I)}$ is also an orthonormal basis for $\mathcal{H}$, it follows that for any vector $f \in \mathcal{H}$, $f = \sum_{k \in \varphi(I)} \langle f, e_k \rangle_{\mathcal{H}} e_k$ and additionally, $\|f\|_{\mathcal{H}}^2 = \sum_{k \in \varphi(I)} \left| \langle f, e_k \rangle_{\mathcal{H}} \right|^2$. Therefore, $\sum_{k \in I} \langle f, e_k \rangle_{\mathcal{H}} e_k$ is an unconditional convergent series expansion of f.

5.2 Riesz Bases

Definition 5.1 A Riesz basis for a separable Hilbert space $\mathcal{H}$ is a family of vectors of the form $(J e_k)_{k=1}^{\infty}$ where $(e_k)_{k=1}^{\infty}$ is an orthonormal basis and $J : \mathcal{H} \to \mathcal{H}$ is a bounded linear operator which is also a bijection.

Proposition 5.3 If $(f_k)_{k=1}^{\infty}$ is a Riesz basis for $\mathcal{H}$ then there exists a unique sequence $(\phi_k)_{k=1}^{\infty}$ in $\mathcal{H}$ such that for every vector $f \in \mathcal{H}$, $f = \sum_{k=1}^{\infty} \langle f, \phi_k \rangle_{\mathcal{H}} f_k$. This series converges unconditionally in $\mathcal{H}$ and additionally, the sequence $(\phi_k)_{k=1}^{\infty}$ is also a Riesz basis for $\mathcal{H}$ satisfying the biorthogonality condition:

$$\langle f_k, \phi_j \rangle_{\mathcal{H}} = \begin{cases} 0 & \text{if } k \neq j \\ 1 & \text{if } k = j \end{cases}.$$

Proof: Let $J : \mathcal{H} \to \mathcal{H}$ be a bounded and bijective operator such that $f_k = J e_k$ for every natural number k where $(e_k)_{k=1}^{\infty}$ is some fixed orthonormal basis for $\mathcal{H}$. Let Λ be the inverse of the operator J with adjoint Λ^*. Then

$$\Lambda f = \sum_{k=1}^{\infty} \langle \Lambda f, e_k \rangle_{\mathcal{H}} e_k = \sum_{k=1}^{\infty} \langle f, \Lambda^* e_k \rangle_{\mathcal{H}} e_k.$$

Put $\phi_k = \Lambda^* e_k$. Then

$$f = J\Lambda f = J\left(\sum_{k=1}^{\infty} \langle f, \Lambda^* e_k \rangle_{\mathcal{H}} e_k \right) = \sum_{k=1}^{\infty} \langle f, \Lambda^* e_k \rangle_{\mathcal{H}} J e_k = \sum_{k=1}^{\infty} \langle f, \phi_k \rangle_{\mathcal{H}} f_k.$$

By construction, $\Lambda^* : \mathcal{H} \to \mathcal{H}$ is a bounded linear operator which is also a bijection. This shows that the sequence of vectors $(\phi_k)_{k=1}^{\infty}$ is also a Riesz basis for $\mathcal{H}$. Finally,

$$\langle f_k, \phi_j \rangle_{\mathcal{H}} = \langle J e_k, \Lambda^* e_j \rangle_{\mathcal{H}} = \langle e_k, J^* \Lambda^* e_j \rangle_{\mathcal{H}} = \langle e_k, e_j \rangle_{\mathcal{H}}.$$

Definition 5.2 The pair $\left((f_k)_{k=1}^{\infty}, (\phi_k)_{k=1}^{\infty} \right)$ is called a pair of dual Riesz bases.

Proposition 5.4 Let $(f_k)_{k=1}^{\infty}$ be a Riesz basis for $\mathcal{H}$. Then there exist positive constants A, B such that for any fixed $f \in \mathcal{H}$,

$$A \cdot \|f\|_{\mathcal{H}}^2 \le \sum_{k=1}^{\infty} \left| \langle f, f_k \rangle_{\mathcal{H}} \right|^2 \le B \cdot \|f\|_{\mathcal{H}}^2 .$$

Proof: Let $J : \mathcal{H} \to \mathcal{H}$ be a bounded bijective operator such that $f_k = J e_k$ for every natural number k, where $(e_k)_{k=1}^{\infty}$ is some orthonormal basis for $\mathcal{H}$. Then

$$\sum_{k=1}^{\infty} \left| \langle f, f_k \rangle_{\mathcal{H}} \right|^2 = \sum_{k=1}^{\infty} \left| \langle f, J e_k \rangle_{\mathcal{H}} \right|^2 = \sum_{k=1}^{\infty} \left| \langle J^* f, e_k \rangle_{\mathcal{H}} \right|^2 .$$

By Parseval's equality (see Proposition 5.2) $\sum_{k=1}^{\infty} \left| \langle J^* f, e_k \rangle_{\mathcal{H}} \right|^2 = \|J^* f\|_{\mathcal{H}}^2$ and

$$\sum_{k=1}^{\infty} \left| \langle f, f_k \rangle_{\mathcal{H}} \right|^2 = \|J^* f\|_{\mathcal{H}}^2 \le \|J^*\|_{\mathcal{H}\to\mathcal{H}}^2 \cdot \|f\|_{\mathcal{H}}^2 .$$

Next, since $f = (J^*)^{-1} J^* f$,

$$\|f\|_{\mathcal{H}} = \left\| (J^*)^{-1} J^* f \right\|_{\mathcal{H}} \le \left\| (J^*)^{-1} \right\|_{\mathcal{H}\to\mathcal{H}} \cdot \|J^* f\|_{\mathcal{H}}.$$

Therefore,

$$\|f\|_{\mathcal{H}}^2 \le \left\| (J^*)^{-1} \right\|_{\mathcal{H}\to\mathcal{H}}^2 \cdot \|J^* f\|_{\mathcal{H}}^2 = \left(\left\| (J^*)^{-1} \right\|_{\mathcal{H}\to\mathcal{H}}^2 \right) \cdot \left(\sum_{k=1}^{\infty} \left| \langle J^* f, e_k \rangle_{\mathcal{H}} \right|^2 \right)$$

and dividing the above by $\left\|(J^*)^{-1}\right\|^2_{\mathcal{H}\to\mathcal{H}}$,

$$\frac{1}{\left\|(J^*)^{-1}\right\|^2_{\mathcal{H}\to\mathcal{H}}}\|f\|^2_{\mathcal{H}} \le \sum_{k=1}^{\infty}\left|\langle J^*f, e_k\rangle_{\mathcal{H}}\right|^2 \le \|J^*\|^2_{\mathcal{H}\to\mathcal{H}}\cdot\|f\|^2_{\mathcal{H}}.$$

Proposition 5.5 The following are equivalent statements:

1. $(f_k)_{k=1}^{\infty}$ is a Riesz basis for $\mathcal{H}$.
2. $(f_k)_{k=1}^{\infty}$ is a complete set and additionally, there exist positive constants A, B such that for any sequence $(\alpha_k)_{k=1}^{\infty}$ of complex numbers which has finitely many nonzero terms, the following holds:

$$A\cdot\left(\sum_{k=1}^{\infty}|\alpha_k|^2\right) \le \left\|\sum_{k=1}^{\infty}\alpha_k f_k\right\|^2_{\mathcal{H}} \le B\cdot\left(\sum_{k=1}^{\infty}|\alpha_k|^2\right).$$

Proof: Let $(f_k)_{k=1}^{\infty}$ be a Riesz basis for $\mathcal{H}$ and let $J : \mathcal{H} \to \mathcal{H}$ be a bounded bijective operator such that $f_k = Je_k$ for every natural number k, where $(e_k)_{k=1}^{\infty}$ is an orthonormal basis for $\mathcal{H}$. Then $(f_k)_{k=1}^{\infty}$ is a complete set, since if $\langle f, f_k\rangle_{\mathcal{H}} = 0$ for every natural number k, then $\langle J^*f, e_k\rangle_{\mathcal{H}} = 0$ for every $k \in \mathbb{N}$ and consequently, $J^*f = f = 0$. By assumption,

$$\left\|\sum_{k=1}^{\infty}\alpha_k f_k\right\|^2_{\mathcal{H}} = \left\|J\left(\sum_{k=1}^{\infty}\alpha_k e_k\right)\right\|^2_{\mathcal{H}} \le \|J\|^2_{\mathcal{H}\to\mathcal{H}}\cdot\left\|\sum_{k=1}^{\infty}\alpha_k e_k\right\|^2_{\mathcal{H}}.$$

Moreover,

$$\sum_{k=1}^{\infty}|\alpha_k|^2 = \left\|\sum_{k=1}^{\infty}\alpha_k e_k\right\|^2_{\mathcal{H}} = \left\|J^{-1}J\left(\sum_{k=1}^{\infty}\alpha_k e_k\right)\right\|^2_{\mathcal{H}}$$

$$\le \|J^{-1}\|^2\cdot\left\|J\left(\sum_{k=1}^{\infty}\alpha_k e_k\right)\right\|^2_{\mathcal{H}}.$$

This shows that

$$\sum_{k=1}^{\infty}|\alpha_k|^2 \le \left(\|J^{-1}\|^2_{\mathcal{H}\to\mathcal{H}}\right)\cdot\left\|\sum_{k=1}^{\infty}\alpha_k f_k\right\|^2_{\mathcal{H}}$$

and

$$\frac{1}{\|J^{-1}\|^2_{\mathcal{H}\to\mathcal{H}}}\sum_{k=1}^{\infty}|\alpha_k|^2 \le \left\|\sum_{k=1}^{\infty}\alpha_k f_k\right\|^2_{\mathcal{H}} \le \left(\|J\|^2_{\mathcal{H}\to\mathcal{H}}\right)\cdot\sum_{k=1}^{\infty}|\alpha_k|^2.$$

Conversely, let us assume that (2) holds. Let $(e_k)_{k=1}^{\infty}$ be an orthonormal basis for $\mathcal{H}$. Extend the mapping $e_k \mapsto f_k$ to a bounded operator $J : \mathcal{H} \to \mathcal{H}$ such that $Je_k = f_k$.

The fact that J is bounded follows from the following calculations. Let $(\alpha_k)_{k=1}^{n}$ be a finite sequence of complex numbers. Then

$$\left\| J\left(\sum_{k=1}^{n}\alpha_k e_k\right)\right\|_{\mathcal{H}}^2 = \left\|\sum_{k=1}^{n}\alpha_k J e_k\right\|_{\mathcal{H}}^2$$

and

$$\left\|\sum_{k=1}^{n}\alpha_k J e_k\right\|_{\mathcal{H}}^2 = \left\|\sum_{k=1}^{n}\alpha_k f_k\right\|_{\mathcal{H}}^2 \leq B\cdot\left(\sum_{k=1}^{n}|\alpha_k|^2\right) = B\cdot\left\|\sum_{k=1}^{n}\alpha_k e_k\right\|_{\mathcal{H}}^2 .$$

Similarly, we extend the mapping $f_k \mapsto e_k$ to a bounded operator $\Lambda : \mathcal{H} \to \mathcal{H}$ such that $\Lambda f_k = e_k$. Since $\Lambda J = J\Lambda$ is the identity operator, it follows that $(f_k)_{k=1}^{\infty}$ is a Riesz basis for $\mathcal{H}$.

5.3 Frames

We have seen earlier that every separable Hilbert space $\mathcal{H}$ has a countable orthonormal basis $(e_k)_{k\in I}$. Moreover, any such fixed orthonormal basis satisfies Parseval's identity: for every $f \in \mathcal{H}$, $\sum_{k\in I}\left|\langle f, e_k\rangle\right|^2 = \|f\|_{\mathcal{H}}^2$.

Observe that it is also possible for a sequence to satisfy Parseval's identity without having the properties of an orthonormal basis. For example, let $(b_k)_{k=1}^{\infty}$ be an orthonormal basis for some Hilbert space $\mathcal{H}$. Then for any vector $f \in \mathcal{H}$,

$$\|f\|^2 = \sum_{k=1}^{\infty}\left|\langle f, b_k\rangle\right|^2 = \sum_{k=1}^{\infty}2\left|\left\langle f, \frac{1}{\sqrt{2}}b_k\right\rangle\right|^2 .$$

This shows that the following sequence of vectors:

$$\underbrace{\frac{b_1}{\sqrt{2}}, \frac{b_1}{\sqrt{2}}}_{2}, \underbrace{\frac{b_2}{\sqrt{2}}, \frac{b_2}{\sqrt{2}}}_{2}, \underbrace{\frac{b_3}{\sqrt{2}}, \frac{b_3}{\sqrt{2}}}_{2}, \ldots$$

satisfies Parseval's equality for $\mathcal{H}$, demonstrating that the concept of Parseval's equality does not require orthogonality. In fact, we may further relax this notion of Parseval's identity as follows:

Definition 5.3 A countable set $(f_k)_{k\in I}$ in a Hilbert space $\mathcal{H}$ is a frame for $\mathcal{H}$ if there exist constants $A, B > 0$ such that the pseudo-Parseval's identity holds:

$$A\|f\|_{\mathcal{H}}^2 \leq \sum_{k\in I}\left|\langle f, f_k\rangle_{\mathcal{H}}\right|^2 \leq B\|f\|_{\mathcal{H}}^2$$

for all $f \in \mathcal{H}$.

The constants A, B are called frame bounds, and clearly, any fixed frame has an uncountable number of pairs of frame bounds. However, the largest possible lower frame bound is called an optimal lower frame bound and the smallest possible upper frame bound is called an optimal upper frame bound.

Duffin and Schaeffer introduced the concept of frame in 1952 to study problems in nonharmonic Fourier series. The theory was later revisited by Young revisited in 1980. However, the work of Duffin, Schaeffer, and Young did not generate much interest until the emergence of the seminal paper of Daubechies et al. [3], in which they show how indispensable frames are for topics such as signal analysis.

Intuitively, the first inequality

$$A\|f\|^2 \leq \sum_{k\in I} \left|\langle f, f_k\rangle\right|^2$$

means that if $\left(\langle f, f_k\rangle\right)_{k\in I}$ has small $l^2(I)$-norm, then f has small $\mathcal{H}$-norm. Similarly, the second inequality

$$\sum_{k\in I} \left|\langle f, f_k\rangle\right|^2 \leq B\|f\|^2$$

implies that if f has small $\mathcal{H}$-norm then $\left(\langle f, f_k\rangle\right)_{k\in I}$ has small $l^2(I)$-norm as well. Thus, a frame gives an equivalence between the norm of $\mathcal{H}$ and the usual norm of $l^2(I)$.

Moreover, $(f_k)_{k\in I}$ is called a tight frame if there exists a positive real number A such that

$$\sum_{k\in I} \left|\langle f, f_k\rangle_{\mathcal{H}}\right|^2 = A \cdot \|f\|^2_{\mathcal{H}}$$

for all $x \in \mathcal{H}$.

Furthermore $(f_k)_{k\in I}$ is said to be Parseval frame if it is a tight frame with frame bound $A = 1$. Note that any tight frame can be suitably normalized to construct a Parseval frame since

$$\sum_{k\in I} \left|\langle f, f_k\rangle_{\mathcal{H}}\right|^2 = A\,\|f\|^2_{\mathcal{H}} \Leftrightarrow \sum_{k\in I} \left|\langle f, A^{-1/2} f_k\rangle_{\mathcal{H}}\right|^2 = \|f\|^2_{\mathcal{H}}.$$

We aim to show that the concept of frame is a natural generalization of that of a basis, in the sense that any vector in the Hilbert space admits stable series expansions in terms of the elements of a fixed frame. To this end, we will first need to introduce the concept of a frame operator.

Let $(f_k)_{k\in I}$ be a frame for $\mathcal{H}$ with frame bounds A, B. Formally define the frame operator $S : \mathcal{H} \to \mathcal{H}$ associated with $(f_k)_{k\in I}$ as follows:

$$Sf = \sum_{k\in I} \langle f, f_k\rangle_{\mathcal{H}} f_k.$$

Proposition 5.6 The following statements hold true:

1. S is a bounded
2. S is self-adjoint
3. S is positive
4. S is invertible
5. $(S^{-1} f_k)_{k\in I}$ is a frame called the canonical dual frame of $(f_k)_{k\in I}$ with frame bounds B^{-1}, A^{-1} with frame operator S^{-1}.

Proof: For the first part, let $V : \mathcal{H} \to l^2(I)$ be a linear operator given by $Vf = (\langle f, f_k\rangle_{\mathcal{H}})_{k\in I}$. V is also known as the analysis operator associated with frame $(f_k)_{k\in I}$. Next, V is a bounded linear operator and its adjoint $V^* : l^2(I) \to \mathcal{H}$ called the synthesis operator, is defined as follows:

$$V^*(a) = \sum_{k\in I} a_k f_k.$$

Moreover, composing the analysis operator with the synthesis operator,

$$V^*(Vf) = V^*\left(\left(\langle f, f_k\rangle_{\mathcal{H}}\right)_{k\in I}\right) = \sum_{k\in I} \langle f, f_k\rangle_{\mathcal{H}} f_k.$$

This shows that $S = V^* V$ is a bounded linear operator, since

$$\|S\|_{\mathcal{H}\to\mathcal{H}} = \|V^* V\|_{\mathcal{H}\to\mathcal{H}} \le \|V^*\|_{\mathcal{H}\to\mathcal{H}} \cdot \|V\|_{\mathcal{H}\to\mathcal{H}} = \|V\|^2_{\mathcal{H}\to\mathcal{H}}.$$

Part (2) is clearly true, since $S^* = (V^* V)^* = V^* V = S$.
Part (3) follows from the following calculations:

$$\langle Sf, f\rangle = \left\langle \sum_{k\in I} \langle f, f_k\rangle_{\mathcal{H}} f_k, f \right\rangle_{\mathcal{H}} = \sum_{k\in I} \langle f, f_k\rangle_{\mathcal{H}} \cdot \langle f_k, f\rangle_{\mathcal{H}}$$
$$= \sum_{k\in I} \left|\langle f, f_k\rangle_{\mathcal{H}}\right|^2 \ge 0.$$

To prove (4), observe that

$$\|I_{\mathcal{H}} - B^{-1}S\|_{\mathcal{H}\to\mathcal{H}} = \sup\left\{\left\langle (I_{\mathcal{H}} - B^{-1}S)f, f\right\rangle_{\mathcal{H}} : \|f\|_{\mathcal{H}} = 1\right\}.$$

Since

$$\left\langle (I_{\mathcal{H}} - B^{-1}S)f, f\right\rangle_{\mathcal{H}} = \langle f, f\rangle_{\mathcal{H}} - B^{-1}\langle Sf, f\rangle_{\mathcal{H}}$$
$$= \|f\|^2_{\mathcal{H}} - B^{-1}\sum_{k\in I}\left|\langle f, f_k\rangle_{\mathcal{H}}\right|^2$$
$$\le \|f\|^2_{\mathcal{H}} - \frac{A\|f\|^2_{\mathcal{H}}}{B} = \left(1 - \frac{A}{B}\right) \cdot \|f\|^2_{\mathcal{H}},$$

it follows that

$$\|I_{\mathcal{H}} - B^{-1}S\|_{\mathcal{H}\to\mathcal{H}} \leq \sup\left\{\left(1-\frac{A}{B}\right)\cdot\|f\|_{\mathcal{H}}^2 : \|f\|_{\mathcal{H}} = 1\right\} \leq 1-\frac{A}{B} < 1.$$

Consequently, $B^{-1}S$ is an invertible operator and so is B.

Finally, to establish (5), observe that

$$\sum_{k\in I}\left|\langle f, S^{-1}f_k\rangle_{\mathcal{H}}\right|^2 = \sum_{k\in I}\left|\langle S^{-1}f, f_k\rangle_{\mathcal{H}}\right|^2 \leq B\cdot\|S^{-1}f\|_{\mathcal{H}}^2$$
$$\leq \left(B\cdot\|S^{-1}\|_{\mathcal{H}\to\mathcal{H}}^2\right)\cdot\|f\|_{\mathcal{H}}^2.$$

Letting

$$Rf = \sum_{k\in I}\langle f, S^{-1}f_k\rangle_{\mathcal{H}}\, S^{-1}f_k,$$

it follows that R is a bounded operator defined on $\mathcal{H}$. Additionally,

$$\sum_{k\in I}\left|\langle f, S^{-1}f_k\rangle_{\mathcal{H}}\right|^2 = \sum_{k\in I}\langle f, S^{-1}f_k\rangle_{\mathcal{H}}\langle S^{-1}f_k, f\rangle_{\mathcal{H}}$$
$$= \left\langle \sum_{k\in I}\langle f, S^{-1}f_k\rangle_{\mathcal{H}}\, S^{-1}f_k, f\right\rangle_{\mathcal{H}} = \langle Rf, f\rangle_{\mathcal{H}}$$

and

$$Rf = S^{-1}\left(\sum_{k\in I}\langle f, S^{-1}f_k\rangle_{\mathcal{H}} f_k\right) = S^{-1}\left(\sum_{k\in I}\langle S^{-1}f, f_k\rangle_{\mathcal{H}} f_k\right)$$
$$= S^{-1}SS^{-1}f = S^{-1}f.$$

Therefore, since $AI_{\mathcal{H}} \leq S \leq BI_{\mathcal{H}}$, we derive that $B^{-1}I_{\mathcal{H}} \leq S^{-1} \leq A^{-1}I_{\mathcal{H}}$ and

$$\sum_{k\in I}\left|\langle f, S^{-1}f_k\rangle_{\mathcal{H}}\right|^2 = \langle S^{-1}f, f\rangle_{\mathcal{H}} \geq B^{-1}\cdot\|f\|_{\mathcal{H}}^2.$$

Proposition 5.7 Let $(f_k)_{k\in I}$ be a frame for $\mathcal{H}$ with frame bounds A, B. Then for any vector $f \in \mathcal{H}$,

$$f = \sum_{k\in I}\langle f, S^{-1}f_k\rangle_{\mathcal{H}} f_k = \sum_{k\in I}\langle f, f_k\rangle_{\mathcal{H}}\, S^{-1}f_k.$$

Proof: Factoring the identity, we obtain the following series expansions with unconditional convergence:

$$f = S\left(S^{-1}f\right) = \sum_{k\in I}\langle S^{-1}f, f_k\rangle_{\mathcal{H}} f_k.$$

Since S^{-1} is self-adjoint, $f = \sum_{k\in I} \langle f, S^{-1} f_k \rangle_{\mathcal{H}} f_k$. Similarly,

$$
\begin{aligned}
f = S^{-1}(Sf) &= \sum_{k\in I} \langle Sf, S^{-1} f_k \rangle_{\mathcal{H}} S^{-1} f_k \\
&= \sum_{k\in I} \langle f, SS^{-1} f_k \rangle_{\mathcal{H}} S^{-1} f_k \\
&= \sum_{k\in I} \langle f, f_k \rangle_{\mathcal{H}} S^{-1} f_k.
\end{aligned}
$$

Example 5.1 Let $\mathcal{H} = L^2[0,1]$ and let $\mathcal{E}(\Lambda) = \{x \mapsto e^{2\pi i\lambda x} : \lambda \in \Lambda\}$ be a countable set of exponential functions and $\Lambda \subset \mathbb{R}$. Then $\mathcal{E}$ is a frame for $\mathcal{H}$ if there exist positive constants A, B such that for any $f \in \mathcal{H}$,

$$
A\int_{[0,1]} |f(t)|^2 dt \le \sum_{\lambda\in\Lambda} \left| \int_{[0,1]} f(t) e^{-2\pi i\lambda x} dt \right|^2 \le B\int_{[0,1]} |f(t)|^2 dt.
$$

Let $\lambda \mapsto \int_{[0,1]} f(t) e^{-2\pi i\lambda x} dt = \mathcal{F}f(\lambda)$ be the Fourier transform of f. Then $\mathcal{E}(\Lambda)$ is a frame for $L^2[0,1]$ if there exist positive constants A, B such that for any $f \in \mathcal{H}$,

$$
A \cdot \int_{[0,1]} |f(t)|^2 dt \le \sum_{\lambda\in\Lambda} |\mathcal{F}f(\lambda)|^2 \le B \cdot \int_{[0,1]} |f(t)|^2 dt.
$$

Noting that the inequality above implies that $\sum_{\lambda\in\Lambda} |\mathcal{F}f(\lambda)|^2$ must be a convergent series, it is clear for instance that if Λ has at least one accumulation point as a subset of the support of f, then $\mathcal{E}(\Lambda)$ cannot be a frame.

Proposition 5.8 Let $(f_k)_{k\in I}$ be a frame for a Hilbert space $\mathcal{H}$ and let T be a nonzero, bounded and surjective operator defined on $\mathcal{H}$. Then $(Tf_k)_{k\in I}$ is a frame for $\mathcal{H}$.

Proof: First, suppose that $0 < A \le B$ are frame bounds for $(f_k)_{k\in I}$ and let f be a vector in $\mathcal{H}$. Then

$$
\sum_{k\in I} \left| \langle f, Tf_k \rangle_{\mathcal{H}} \right|^2 = \sum_{k\in I} \left| \langle T^* f, f_k \rangle_{\mathcal{H}} \right|^2 \le B\|T^* f\|^2 \le B \cdot \|T^*\|^2_{\mathcal{H}\to\mathcal{H}} \cdot \|f\|^2_{\mathcal{H}}.
$$

This shows that the positive constant $B \cdot \|T^*\|^2_{\mathcal{H}\to\mathcal{H}}$ is an upper frame bound for $(Tf_k)_{k\in I}$. Secondly, let φ be a vector in $\mathcal{H}$. Since $TT^*(TT^*)^{-1}$ is an orthogonal projection onto $\mathcal{H}$, letting $\varphi = Tf$,

$$
\varphi = TT^*(TT^*)^{-1}\varphi = \left(TT^*(TT^*)^{-1}\right)^* Tf = \left(T^*(TT^*)^{-1}\right)^* T^*Tf.
$$

Computing the $\mathcal{H}$-norm of φ, we obtain

$$
\|\varphi\|^2_{\mathcal{H}} \le \frac{\left\| \left(T^*(TT^*)^{-1}\right)^* \right\|^2_{\mathcal{H}\to\mathcal{H}}}{A} \sum_{k\in I} \left| \langle T^*Tf, f_k \rangle_{\mathcal{H}} \right|^2,
$$

showing that

$$\|\varphi\|_{\mathcal{H}}^2 \leq \frac{\left\|\left(T^*(TT^*)^{-1}\right)^*\right\|_{\mathcal{H}\to\mathcal{H}}^2}{A} \sum_{k\in I} \left|\left\langle \varphi, Tf_k \right\rangle_{\mathcal{H}}\right|^2.$$

Therefore, $\frac{\|(T^*(TT^*)^{-1})^*\|_{\mathcal{H}\to\mathcal{H}}^2}{A}$ is a positive lower frame bound for $\left(Tf_k\right)_{k\in I}$.

Proposition 5.9 If $\left(f_k\right)_{k\in I}$ is a Parseval frame then its associated frame operator is the identity map.

Proof: Recall that $S : f \mapsto \sum_{k\in I} \langle f, f_k \rangle_{\mathcal{H}} f_k$ is a bounded, self-adjoint, and invertible operator acting on $\mathcal{H}$. Suppose that $\left(f_k\right)_{k\in I}$ is a Parseval frame. Then for any vector $f \in \mathcal{H}$,

$$\langle Sf, f \rangle_{\mathcal{H}} = \left\langle \sum_{k\in I} \langle f, f_k \rangle_{\mathcal{H}} f_k, f \right\rangle_{\mathcal{H}} = \sum_{k\in I} \langle f, f_k \rangle_{\mathcal{H}} \cdot \langle f_k, f \rangle_{\mathcal{H}}$$

and

$$\langle Sf, f \rangle_{\mathcal{H}} = \sum_{k\in I} \left|\langle f, f_k \rangle_{\mathcal{H}}\right|^2 = \|f\|_{\mathcal{H}}^2 .$$

This shows that both S and its inverse coincide with the identity operator on $\mathcal{H}$.

If $\left(f_k\right)_{k\in I}$ is a Parseval frame, then every vector in $\mathcal{H}$ admits a series expansion (in terms of elements of the sequence $\left(f_k\right)_{k\in I}$), which is very similar to that of an orthonormal basis. Indeed, appealing to Proposition 5.7, for any vector $f \in \mathcal{H}$,

$$f = \sum_{k\in I} \langle f, S^{-1} f_k \rangle_{\mathcal{H}} f_k = \sum_{k\in I} \langle f, f_k \rangle_{\mathcal{H}} f_k.$$

Generally, if $\left(f_k\right)_{k\in I}$ is a frame that is not a Parseval frame, more work is needed to compute the inverse of the frame operator. For instance, assuming that $\left(f_k\right)_{k\in I}$ is a frame with frame bounds $A, B > 0$, since $\frac{A}{B} I_{\mathcal{H}} \leq \frac{1}{B} S \leq I_{\mathcal{H}}$, it is clear that

$$\left\| I_{\mathcal{H}} - \frac{1}{B} S \right\|_{\mathcal{H}\to\mathcal{H}} \leq \left| 1 - \frac{A}{B} \right| < 1.$$

Using Neumann series [5]

$$BS^{-1} = \left(I_{\mathcal{H}} - \left(I_{\mathcal{H}} - \frac{1}{B} S\right)\right)^{-1} = \sum_{k=0}^{\infty} \left(I_{\mathcal{H}} - \frac{1}{B} S\right)^k.$$

As a result, the inverse of the frame operator is a series expansion in the vector space spanned by S and the identity operator, and it is explicitly given by:

$$S^{-1} = \frac{1}{B} \sum_{k=0}^{\infty} \left(I_{\mathcal{H}} - \frac{1}{B} S\right)^k. \tag{5.1}$$

We will now turn our attention to the following fundamental question: How does one construct a frame for a Hilbert space? A reasonable approach is to fix a vector in the given space and then use this vector to generate a countable family of vectors in the same space. There are, of course, many ways to achieve this. For instance, if the Hilbert space of interest is acted upon unitarily by a group, a family of operators would naturally arise from a unitary representation afforded by the group action. The range of this representation offers an extensive reservoir of unitary operators acting on the space of interest. If the group at hand is a connected Lie group, the collection of vectors arising by applying these operators to a fixed vector may be too large. To reduce redundancy, one could then ask if it is possible to choose inside this orbit, a countable set of vectors that could serve as a frame for the Hilbert space.

Several questions naturally emerge from this procedure. How does one choose the window vector giving rise to a particular frame? Which group actions in the Hilbert space lead to the successful construction of frames? Is it possible to construct frames enjoying some prescribed properties that could potentially be useful for specific applications?

We will formalize this research agenda by stating "The Discretization Problem" (see Problem 5.1) of unitary representation of a connected Lie group to construct systems such as frames, Parseval frames, orthonormal, and Riesz bases. We will also discuss how this problem naturally arises in wavelet and time-frequency analyses. Furthermore, we will develop an approach that allows us to view these subjects as specific occurrences of a much deeper phenomenon. In fact, using the geometric tools developed in the preceding chapters, we will present a unified, systematic procedure for handling a large class of representations.

Problem 5.1 ***The Discretization Problem*** Let G be a Lie group. Can we describe explicitly, triples of the type (f, π, Γ), where π is a strongly continuous unitary representation of G acting in some Hilbert space $\mathcal{H}_\pi$, f is a fixed vector in $\mathcal{H}_\pi$, and $\Gamma \subset G$ is a countable set such that $\mathfrak{F}(f, \pi, \Gamma) = \{\pi(\gamma)f \; ; \; \gamma \in \Gamma\}$ is a frame (or Parseval frame, or Riesz, or an orthonormal basis) for $\mathcal{H}_\pi$?

The Discretization Problem aims to answer the following question: to which extent, is it possible to analyze a vector φ in the Hilbert space $\mathcal{H}_\pi$ in terms of some subset of an orbit of a fixed vector f in the same Hilbert space. Precisely, we are interested in series expansions of the form

$$\varphi = \sum_{\gamma \in \Gamma} c_\gamma \pi(\gamma) f,$$

where $c_\gamma \in \mathbb{C}$ and Γ is some countable subset of G.

We will mostly be concerned with discretizing unitary representations, acting smoothly on some function space (which is also a Hilbert space) defined over some smooth manifold. We will also pay a special attention to sufficient conditions under which nonpathological frames exist. From our viewpoint, a nonpathological frame will often mean a type of frame enjoying at least one of the following properties: boundedness, compact support, continuity, fast decay, and differentiability.

For other related, but significantly broader approach, we also refer the interested reader to the theory of coorbit, which was developed by Feichtinger and Gröchenig, and its generalizations due to G. Olafsson, J. Christensen, and others [1].

Suppose that $\mathfrak{F}(f, \pi, \Gamma)$ is a frame for $\mathcal{H}_\pi$ with frame bounds $A, B > 0$ and associated frame operator S. Then for every $\varphi \in \mathcal{H}_\pi$, thanks to Proposition 5.7,

$$\varphi = \sum_{\gamma\in\Gamma} \left\langle \varphi, S^{-1}\pi(\gamma)f \right\rangle_{\mathcal{H}_\pi} \pi(\gamma)f$$

and the sequence of complex numbers $\left(\left\langle \varphi, S^{-1}\pi(\gamma)f \right\rangle_{\mathcal{H}_\pi}\right)_{\gamma\in\Gamma}$ is square-integrable over Γ and

$$A\,\|\varphi\|^2_{\mathcal{H}_\pi} \le \sum_{\gamma\in\Gamma} \left|\langle \varphi, \pi(\gamma)f\rangle_{\mathcal{H}_\pi}\right|^2 \le B\,\|\varphi\|^2_{\mathcal{H}_\pi}$$

for all $\varphi \in \mathcal{H}_\pi$. Note also that if $\mathfrak{F}(f, \pi, \Gamma)$ is a frame for $\mathcal{H}_\pi$ and if additionally, J is a bounded and invertible operator commuting with every unitary operator in the set $\pi(\Gamma)$ then $\mathfrak{F}(Af, \pi, \Gamma)$ is also a frame for $\mathcal{H}_\pi$.

References

1 Eirik Berge. A primer on coorbit theory. *Journal of Fourier Analysis and Applications*, 28(1):2, 2022.

2 Ole Christensen. *An introduction to frames and Riesz bases*. Applied and Numerical Harmonic Analysis. Birkhäuser, Boston, MA, 2003.

3 Ingrid Daubechies, Alex Grossmann, and Yves Meyer. Painless nonorthogonal expansions. *Journal of Mathematical Physics*, 27(2):1271–1283, 1986.

4 Christopher Heil. *A basis theory primer: expanded edition*. Springer Science & Business Media, 2010.

5 Rainer Kress. *Linear integral equations*. Springer, 1989.

6 Steven J. Leon, Åke Björck, and Walter Gander. Gram-Schmidt orthogonalization: 100 years and more. *Numerical Linear Algebra with Applications*, 20(3):492–532, 2013.

6

Frames on Euclidean Spaces

In this chapter, we delve into the study of frames in the Hilbert space $L^2(M)$, where M is a simply connected Euclidean manifold. Our focus will be on the discretization of two classes of induced representations associated with the affine Lie group, commonly referred to as the *ax+b* Lie group [1, 3–5]. Additionally, we will revisit the well-established Gabor wavelets, which have been widely studied in the literature [1, 3–5]. These examples will serve as a foundation for the construction of a more general class of frames in subsequent chapters.

6.1 Wavelets and the *ax+b* Group

Let $G = \mathbb{R} \rtimes \mathbb{R}$ be the semidirect product Lie group equipped with group operation: $(x,t)(y,s) = \left(x + e^t y, t+s\right)$ so that $N = \mathbb{R} \rtimes \{0\}$ is a normal commutative subgroup of G. Fix a nonzero real number ξ and let χ_ξ be a unitary character of N defined as $\chi_\xi(x,0) = e^{2\pi i x\xi}$. Next, let $\pi_\xi = \mathrm{ind}_N^G\left(\chi_\xi\right)$ be an induced representation of G acting in $L^2(\mathbb{R})$ as follows:

$$\left[\pi_\xi(x,t)f\right](s) = e^{2\pi i e^{-s} x\xi} \cdot f(s-t) \quad (x,t) \in \mathbb{R}^2.$$

On the one hand, the family of operators $\left\{\pi_\xi(0,t) : t \in \mathbb{R}\right\}$ acts on $L^2(\mathbb{R})$ by translations, and as shown in Frames of Translates Chapter 9, it is known that the left regular representation of $\mathbb{R}$ cannot be discretized to construct frames for $L^2(\mathbb{R})$. On the other hand, the set $\left\{\pi_\xi(x,0) : x \in \mathbb{R}\right\}$ consists of norm-preserving operators acting by complex exponentials multiplication, and it is an exercise for the reader to verify that this action cannot on its own generate frames for $L^2(\mathbb{R})$ either.

The discretization problem for the induced representation π_ξ can be reformulated in the following terms. Describe (if possible) triples of the form $\left(f, \pi_\xi, \Gamma\right)$ such that $f \in L^2(\mathbb{R})$ and $\Gamma \subset \mathbb{R} \rtimes \mathbb{R}$ is a countable set for which the corresponding countable subset $\mathfrak{F}(f,\pi,\Gamma) = \{\pi(\gamma)f : \gamma \in \Gamma\}$ of an orbit of f is a frame for $L^2(\mathbb{R})$.

A Bridge Between Lie Theory and Frame Theory: Applications of Lie Theory to Harmonic Analysis, First Edition. Vignon Oussa.

Suppose that f is a continuous function that is compactly supported on the real line. Our first result shows that there exists a countable subset $\Gamma \subset G$ such that $\pi_\xi(\Gamma)f$ is a frame for $L^2(\mathbb{R})$. Additionally, our method produces precise bounds for the constructed frames.

Theorem 6.1 Let $\Gamma(a,b) = \Gamma_1(a)\Gamma_2(b)$ such that

$$\Gamma_1(a) = \{0\} \rtimes a\mathbb{Z} \quad \text{and} \quad \Gamma_2(b) = b\mathbb{Z} \rtimes \{0\}$$

where a and b are some positive real numbers. Let $f \in C_c(\mathbb{R})$ and put

$$\epsilon \geq \max\left\{\left|\inf\left(\xi e^{-\operatorname{supp}(f)}\right)\right|, \quad \sup\left|\left(\xi e^{-\operatorname{supp}(f)}\right)\right|\right\} > 0.$$

Next, let r be a positive real number such that

$$0 < \operatorname{essinf}\left(s \mapsto \sum_{k\in\mathbb{Z}} \frac{|f(s-rk)|^2}{|\xi e^{-(s-rk)}|}\right) \leq \operatorname{esssup}\left(s \mapsto \sum_{k\in\mathbb{Z}} \frac{|f(s-rk)|^2}{|\xi e^{-(s-rk)}|}\right) < \infty.$$

Then $\pi_\xi\left(\Gamma\left(r, \frac{1}{2\epsilon}\right)\right)f$ is a frame for $L^2(\mathbb{R})$ with frame bounds

$$0 < \operatorname{essinf}\left(s \mapsto \sum_{k\in\mathbb{Z}} \frac{2\epsilon \cdot |f(s-rk)|^2}{|\xi e^{-(s-rk)}|}\right)$$
$$\leq \operatorname{esssup}\left(s \mapsto \sum_{k\in\mathbb{Z}} \frac{2\epsilon \cdot |f(s-rk)|^2}{|\xi e^{-(s-rk)}|}\right) < \infty.$$

Proof: Let supp (f) and supp (h) be the support of f and h, respectively. Suppose additionally that f and h are continuous and compactly supported functions. Given integers k and j, verify that

$$\left\langle h, \pi_\xi(0,ak)\pi_\xi(bj,0)f\right\rangle_{L^2(\mathbb{R})} = \left\langle \pi_\xi(0,-ak)h, \pi_\xi(bj,0)f\right\rangle_{L^2(\mathbb{R})}$$

and

$$\left\langle h, \pi_\xi(0,ak)\pi_\xi(bj,0)f\right\rangle_{L^2(\mathbb{R})}$$
$$= \int_{\operatorname{supp}(f)} e^{-2\pi i(\xi e^{-s})bj}\left(\left[\pi_\xi(0,-ak)h\right](s)\cdot\overline{f(s)}\right)ds. \tag{6.1}$$

Thus, the matrix coefficient $\left\langle h, \pi_\xi(0,ak)\pi_\xi(bj,0)f\right\rangle_{L^2(\mathbb{R})}$ is obtained by integrating

$$s \mapsto \left[\pi_\xi(0,-ak)h\right](s)\cdot\overline{f(s)}\cdot 1_{\operatorname{supp}(f)}(s)$$

against the complex exponential $s \mapsto e^{-2\pi i(\xi e^{-s})bj}$. Noting that $s \mapsto e^{-2\pi i(\xi e^{-s})bj}$ is obtained by composing $s \mapsto e^{-2\pi isbj}$ with $s \mapsto \xi e^{-s}$, the change of variable $x = \xi e^{-s}$ yields

$$s = \ln\left(\frac{\xi}{x}\right) \quad \text{and} \quad \frac{d}{dx}\left(\ln\left(\frac{\xi}{x}\right)\right) = \frac{1}{s},$$

and

$$\begin{aligned}
&\left\langle h, \pi_{\xi}(0, ak)\,\pi_{\xi}(bj, 0) f\right\rangle_{L^2(\mathbb{R})} \\
&= \int_{\xi e^{-\operatorname{supp}(f)}} e^{-2\pi ix(bj)} \frac{\left[\pi_{\xi}(0, -ak) h\right]\left(\ln\left(\frac{\xi}{x}\right)\right)\overline{f\left(\ln\left(\frac{\xi}{x}\right)\right)}}{|x|} dx.
\end{aligned}$$

In other words, for fixed k, the function $bj \mapsto \left\langle h, \pi_{\xi}(0, ak)\,\pi_{\xi}(bj, 0) f\right\rangle_{L^2(\mathbb{R})}$ is just the Fourier transform of

$$x \mapsto \frac{\left[\pi_{\xi}(0, -ak) h\right]\left(\ln\left(\frac{\xi}{x}\right)\right)\overline{f\left(\ln\left(\frac{\xi}{x}\right)\right)}}{|x|} \cdot 1_{\xi e^{-\operatorname{supp}(f)}}(x)$$

evaluated at bj. Next, let ϵ be a positive real number such that

$$\xi e^{-\operatorname{supp}(f)} \subseteq [-\epsilon, \epsilon]$$

where

$$\epsilon \geq \max\left\{\left|\inf\left(\xi e^{-\operatorname{supp}(f)}\right)\right|, \sup\left|\left(\xi e^{-\operatorname{supp}(f)}\right)\right|\right\}.$$

Letting $b = \frac{1}{2\epsilon}$, note that

$$\int_{\mathbb{R}} \left| e^{-2\pi ix\frac{j}{2\epsilon}} \cdot (2\epsilon)^{-1/2} \cdot 1_{[-\epsilon,\epsilon]}(x)\right|^2 dx = \frac{1}{2\epsilon}\int_{[-\epsilon,\epsilon]} dx = 1$$

and

$$\begin{aligned}
&\int_{\mathbb{R}} \left(e^{-2\pi ix\frac{j}{2\epsilon}} \cdot (2\epsilon)^{-1/2} \cdot 1_{[-\epsilon,\epsilon]}(x)\right)\left(e^{2\pi ix\frac{k}{2\epsilon}} \cdot (2\epsilon)^{-1/2} \cdot 1_{[-\epsilon,\epsilon]}(x)\right) dx \\
&= (2\epsilon)^{-1}\int_{[-\epsilon,\epsilon]} \left(e^{-2\pi ix\frac{j}{2\epsilon}}\right)\left(e^{-2\pi ix\frac{k}{2\epsilon}}\right) dx \\
&= (2\epsilon)^{-1}\int_{[-\epsilon,\epsilon]} \left(e^{-2\pi ix\left(\frac{j+k}{2\epsilon}\right)}\right) dx \\
&= \int_{[-\epsilon,\epsilon]} \left(e^{-2\pi ix(j+k)}\right) dx.
\end{aligned}$$

Next, the trigonometric system of complex exponentials

$$\left\{x \mapsto e^{-2\pi i x \frac{j}{2\epsilon}} \cdot (2\epsilon)^{-1/2} \cdot 1_{[-\epsilon,\epsilon]}(x) : j \in \mathbb{Z}\right\}$$

is a complete countable subset of $L^2([-\epsilon,\epsilon])$ and in light of the calculations above, it follows that this trigonometric system is also an orthonormal basis for $L^2([-\epsilon,\epsilon])$. As a result, letting $e^{-2\pi i x(bj)} = e^{-2\pi i x \frac{j}{2\epsilon}} \cdot (2\epsilon)^{-1/2} \cdot 2\epsilon$, it is clear that

$$\sum_{j\in\mathbb{Z}} \left| \left\langle h, \pi_\xi(0, ak)\pi_\xi\left(\frac{j}{2\epsilon}, 0\right) f \right\rangle_{L^2(\mathbb{R})} \right|^2$$

$$= 2\epsilon \cdot \int_{\xi e^{-\mathrm{supp}(f)}} \left| \frac{\left[\pi_\xi(0,-ak)h\right]\left(\ln\left(\frac{\xi}{x}\right)\right)\overline{f\left(\ln\left(\frac{\xi}{x}\right)\right)}}{|x|} \right|^2 dx$$

and

$$\sum_{j\in\mathbb{Z}} \left| \left\langle h, \pi_\xi(0, ak)\pi_\xi\left(\frac{j}{2\epsilon}, 0\right) f \right\rangle_{L^2(\mathbb{R})} \right|^2$$

$$= 2\epsilon \cdot \int_{\xi e^{-\mathrm{supp}(f)}} \left| \frac{\left[\pi_\xi(0,-ak)h\right]\left(\ln\left(\frac{\xi}{x}\right)\right)\overline{f\left(\ln\left(\frac{\xi}{x}\right)\right)}}{|x|^{1/2}} \right|^2 \frac{1}{|x|} dx.$$

Changing variable again by letting $s = \ln\left(\frac{\xi}{x}\right)$,

$$\sum_{j\in\mathbb{Z}} \left| \left\langle h, \pi_\xi(0, ak)\pi_\xi\left(\frac{j}{2\epsilon}, 0\right) f \right\rangle_{L^2(\mathbb{R})} \right|^2$$

$$= 2\epsilon \cdot \int_{\mathrm{supp}(f)} \left| \frac{\left[\pi_\xi(0,-ak)h\right](s)\overline{f(s)}}{|\xi e^{-s}|^{1/2}} \right|^2 ds$$

$$= 2\epsilon \cdot \int_{\mathrm{supp}(f)} \left| \frac{h(s+ak)\overline{f(s)}}{|\xi e^{-s}|^{1/2}} \right|^2 ds$$

$$= 2\epsilon \cdot \int_{\mathbb{R}} |h(s)|^2 \frac{|f(s-ak)|^2}{|\xi e^{-(s-ak)}|} ds.$$

This gives

$$\sum_{k\in\mathbb{Z}}\sum_{j\in\mathbb{Z}} \left| \left\langle h, \pi_\xi(0, ak)\pi_\xi\left(\frac{j}{2\epsilon}, 0\right) f \right\rangle_{L^2(\mathbb{R})} \right|^2$$

$$= \int_{\mathbb{R}} |h(s)|^2 \cdot \left(\sum_{k\in\mathbb{Z}} \frac{2\epsilon \cdot |f(s-ak)|^2}{|\xi e^{-(s-ak)}|} \right) ds.$$

Put

$$m_{a,\epsilon}(f) = \operatorname{essinf}\left(s \mapsto \sum_{k\in\mathbb{Z}} \frac{2\epsilon \cdot |f(s-ak)|^2}{|\xi e^{-(s-ak)}|}\right)$$

and

$$M_{a,\epsilon}(f) = \operatorname{esssup}\left(s \mapsto \sum_{k\in\mathbb{Z}} \frac{2\epsilon \cdot |f(s-ak)|^2}{|\xi e^{-(s-ak)}|}\right).$$

Since $f \in C_c(\mathbb{R})$, we may choose $a > 0$ to be small enough to ensure that $0 < m_{a,\epsilon}(f) \le M_{a,\epsilon}(f) < \infty$. Finally,

$$\begin{aligned}
m_{a,\epsilon}(f) \cdot \|h\|^2_{L^2(\mathbb{R})} &= m_{a,\epsilon}(f) \cdot \left(\int_{\mathbb{R}} |h(s)|^2 ds\right) \\
&\le \sum_{k\in\mathbb{Z}} \sum_{j\in\mathbb{Z}} \left| \left\langle h, \pi_\xi(0, ak)\, \pi_\xi\left(\frac{j}{2\epsilon}, 0\right) f \right\rangle_{L^2(\mathbb{R})} \right|^2 \\
&= \int_{\mathbb{R}} |h(s)|^2 \cdot \left(\sum_{k\in\mathbb{Z}} \frac{2\epsilon \cdot |f(s-ak)|^2}{|\xi e^{-(s-ak)}|} \right) ds \\
&\le M_{a,\epsilon}(f) \cdot \int_{\mathbb{R}} |h(s)|^2 ds \\
&= M_{a,\epsilon}(f) \cdot \|h\|^2_{L^2(\mathbb{R})}.
\end{aligned}$$

Since the space of compactly supported continuous functions is dense in $L^2(\mathbb{R})$, it is safe to assert that the system of vectors $\pi_\xi\left(\Gamma\left(a, (2\epsilon)^{-1/2}\right)\right) f$ is a frame for $L^2(\mathbb{R})$ with frame bounds $0 < m_{a,\epsilon}(f) \le M_{a,\epsilon}(f)$ (Figure 6.1).

Example 6.1 To construct a triple of the type (f, π_ξ, Γ) such that $f \in L^2(\mathbb{R})$ and $\Gamma \subset \mathbb{R} \rtimes \mathbb{R}$ is a countable set such that the system $\mathfrak{F}(f, \pi_\xi, \Gamma) = \{\pi_\xi(\gamma) f : \gamma \in \Gamma\}$ is a frame, let

$$f(s) = |\xi|^{\frac{1}{2}} \cdot e^{-\frac{s}{2}} \cdot \left(1_{\left[-\frac{1}{2},\frac{1}{2}\right]} * 1_{\left[-\frac{1}{2},\frac{1}{2}\right]} * 1_{\left[-\frac{1}{2},\frac{1}{2}\right]}\right)(s)$$

where $*$ stands for the usual convolution on the real line

Explicitly,

$$f(s) = \begin{cases} |\xi|^{\frac{1}{2}} \cdot e^{-\frac{s}{2}} \left(\frac{3}{4} - s^2\right) & \text{if } s \in \left[-\frac{1}{2}, \frac{1}{2}\right] \\ |\xi|^{\frac{1}{2}} \cdot e^{-\frac{s}{2}} \frac{(3-2s)^2}{8} & \text{if } s \in \left(\frac{1}{2}, \frac{3}{2}\right) \\ |\xi|^{\frac{1}{2}} \cdot e^{-\frac{s}{2}} \frac{(3+2s)^2}{8} & \text{if } s \in \left(-\frac{3}{2}, -\frac{1}{2}\right) \\ 0 & \text{otherwise} \end{cases}$$

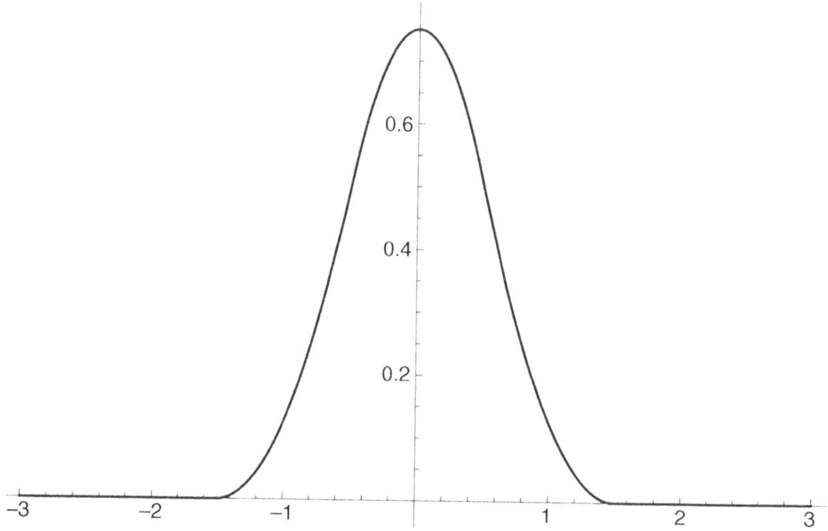

Figure 6.1 A window function generating a frame.

and an orbit of f is the set of vectors

$$\left\{ s \mapsto e^{2\pi i e^{-s}x} \cdot |\xi|^{\frac{1}{2}} \cdot e^{-\frac{s}{2}} \cdot \left(1_{\left[-\frac{1}{2},\frac{1}{2}\right]} * 1_{\left[-\frac{1}{2},\frac{1}{2}\right]} * 1_{\left[-\frac{1}{2},\frac{1}{2}\right]} \right)(s-t) \, : \, t, x \in \mathbb{R} \right\}.$$

Since for any positive real number r,

$$\Theta(s) = \sum_{k \in \mathbb{Z}} \frac{|f(s-rk)|^2}{\left|\xi e^{-(s-rk)}\right|} = \sum_{k \in \mathbb{Z}} \left| \left(1_{\left[-\frac{1}{2},\frac{1}{2}\right]} * 1_{\left[-\frac{1}{2},\frac{1}{2}\right]} * 1_{\left[-\frac{1}{2},\frac{1}{2}\right]} \right)(s-rk) \right|^2$$

and since $\mathrm{supp}(f) = \left[-\frac{3}{2}, \frac{3}{2}\right]$, assuming additionally that r is a positive real number, which is strictly less that 3, it is clear that Θ is bounded on the real line. Furthermore, since $r < 3$, the essential infimum of Θ is also a positive constant. Finally, fix a positive real number ϵ such that $\epsilon > \frac{3}{2}$ and define a countable subset $\Gamma\left(r, \frac{1}{2\epsilon}\right) = \Gamma_1(r)\Gamma_2\left(\frac{1}{2\epsilon}\right)$ of the ax+b Lie group consisting of points $\left(\frac{e^{rj}k}{2\epsilon}, rj\right)$, where $k, j \in \mathbb{Z}$ (Figure 6.2).

Then

$$\pi_\xi\left(\frac{e^{rj}k}{2\epsilon}, rj\right) f(s) = e^{2\pi i \frac{e^{-s}e^{rj}k}{2\epsilon}} \cdot f(s-rj)$$

and Theorem 6.1, guarantees that the system of vectors $\pi_\xi\left(\Gamma\left(r, \frac{1}{2\epsilon}\right)\right) f$ is a frame for $L^2(\mathbb{R})$ (Figure 6.3).

Figure 6.2 A sampling set.

Remark 6.1 It may also be worth noting that it is possible to construct a frame of the type $\pi_\xi(\Gamma)f$ without requiring f to be of compact support. Suppose that the set $e^{-\text{supp}(f)}$ given by the image of the support of f under the exponential function $x \mapsto e^{-x}$ is a compact subset of $\mathbb{R}$. Suppose additionally that the restriction of $t \mapsto \frac{f(t)}{\sqrt{|\xi e^{-t}|}}$ to the countable set $s - r\mathbb{Z}$ belongs to the Hilbert space $l^2(s - r\mathbb{Z})$ of square-summable sequences for almost every real number s. This in a sense measures how quickly f decays as t grows on the positive reals without boundary. Indeed, with very little adjustment, the steps presented in the proof of Theorem 6.1 are easily modified to construct a countable set $\Gamma \subset G$ such that the system $\pi_\xi(\Gamma)f$ is a frame for $L^2(\mathbb{R})$.

Using an approach similar to the proof of Theorem 6.1, we present next a solution of the discretization problem for the direct sum of two nonequivalent irreducible representations of the $ax+b$ Lie group.

Theorem 6.2 Let $\tau = \pi_1 \oplus \pi_{-1}$ be a reducible multiplicity-free representation of G acting in $L^2(\mathbb{R}) \oplus L^2(\mathbb{R})$. If $f_1 + f_{-1} \in L^2(\mathbb{R}) \oplus L^2(\mathbb{R})$ such that f_1, f_{-1} are nonzero continuous and compactly supported function such that $-e^{-\text{supp}(f_1)}$ and $-e^{-\text{supp}(f_{-1})}$ are disjoint subsets of $\mathbb{R}$ of positive Lebesgue measure, then

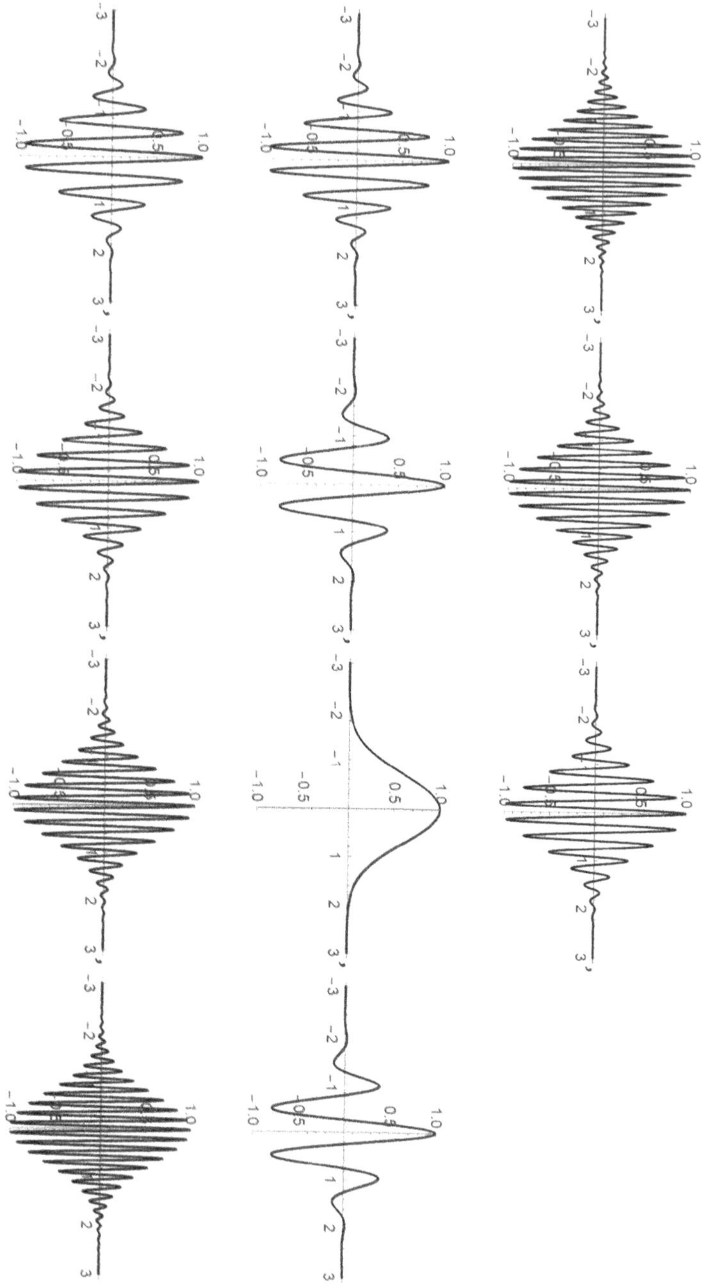

Figure 6.3 Some elements of a frame generated by the *ax*+*b* Lie group.

the following holds. If (a) there exists a positive real number ϵ such that $E = -\left(e^{-\operatorname{supp}(f_1)} \cup e^{-\operatorname{supp}(f_{-1})}\right) \subseteq [-\epsilon, \epsilon]$ and (b) there exists a positive real number a small enough to ensure that

$$0 < m_{a,\epsilon}(f) = \min_{l \in \{-1,1\}} \left(\operatorname{essinf}\left(s \mapsto \sum_{k\in\mathbb{Z}} \frac{2\epsilon \cdot \left|f_l(s-ak)\right|^2}{\left|e^{-(s-ak)}\right|} \right)\right)$$

and

$$M_{a,\epsilon}(f) = \max_{l \in \{-1,1\}} \left(\operatorname{esssup}\left(s \mapsto \sum_{k\in\mathbb{Z}} \frac{2\epsilon \cdot |f(s-ak)|^2}{\left|e^{-(s-ak)}\right|} \right)\right) < \infty,$$

then $\tau\left(\Gamma\left(a, (2\epsilon)^{-1/2}\right)\right)\left(f_1 + f_{-1}\right)$ is a frame for $L^2(\mathbb{R}) \oplus L^2(\mathbb{R})$ with associated frame bounds $0 < m_{a,\epsilon}(f) \leq M_{a,\epsilon}(f)$.

Proof: Let $f_1, f_{-1}, h_1, h_{-1} \in C_c(\mathbb{R})$. Put $h = h_1 + h_{-1}$ and define

$$\begin{aligned} s(k,j) &= \left\langle h_1 + h_{-1}, \tau(0, ak)\,\tau(bj, 0)\left(f_1 + f_{-1}\right)\right\rangle_{L^2(\mathbb{R})\oplus L^2(\mathbb{R})} \\ &= \sum_{\xi\in\{-1,1\}} \int_{\operatorname{supp}(f)} e^{-2\pi i \xi e^{-s} bj} \left(\left[\pi_\xi(0, -ak)\, h\right](s) \cdot \overline{f(s)} \right) ds. \end{aligned}$$

The change of variable $x = \xi e^{-s}$ yields

$$\begin{aligned} s(k,j) = &\int_{e^{-\operatorname{supp}(f_1)}} e^{-2\pi i x b j} \frac{\left[\pi_1(0,-ak)\,h_1\right]\left(\ln\left(\frac{1}{x}\right)\right) \cdot \overline{f_1\left(\ln\left(\frac{1}{x}\right)\right)}}{|x|} dx \\ &+ \int_{-e^{-\operatorname{supp}(f_{-1})}} e^{2\pi i x b j} \frac{\left[\pi_{-1}(0,-ak)\,h_{-1}\right]\left(\ln\left(\frac{-1}{x}\right)\right) \cdot \overline{f_{-1}\left(\ln\left(\frac{-1}{x}\right)\right)}}{|x|} dx. \end{aligned}$$

Replacing x with $-x$ in the first integral above gives

$$\begin{aligned} s(k,j) = &\int_{-e^{-\operatorname{supp}(f_1)}} e^{2\pi i x b j} \frac{\left[\pi_1(0,-ak)\,h_1\right]\left(\ln\left(-\frac{1}{x}\right)\right) \cdot \overline{f_1\left(\ln\left(-\frac{1}{x}\right)\right)}}{|x|} dx \\ &+ \int_{-e^{-\operatorname{supp}(f_{-1})}} e^{2\pi i x b j} \frac{\left[\pi_{-1}(0,-ak)\,h_{-1}\right]\left(\ln\left(-\frac{1}{x}\right)\right) \cdot \overline{f_{-1}\left(\ln\left(-\frac{1}{x}\right)\right)}}{|x|} dx. \end{aligned}$$

The map $x \mapsto -x$ adjusts the sign in the exponential function in the first integral above, so that we may view $s(k,j)$ as a single integral. Thus, for each index k, define

$$\begin{aligned} F_k(x) = &\frac{\left[\pi_1(0,-ak)\,h_1\right]\left(\ln\left(-\frac{\xi}{x}\right)\right) \cdot \overline{f_1\left(\ln\left(-\frac{\xi}{x}\right)\right)}}{|x|} 1_{-e^{-\operatorname{supp}(f_1)}}(x) \\ &+ \frac{\left[\pi_{-1}(0,-ak)\,h_{-1}\right]\left(\ln\left(\frac{\xi}{x}\right)\right) \cdot \overline{f_{-1}\left(\ln\left(\frac{\xi}{x}\right)\right)}}{|x|} 1_{-e^{-\operatorname{supp}(f_{-1})}}(x) \end{aligned}$$

and let $E = -\left(e^{-\mathrm{supp}(f_1)} \cup e^{-\mathrm{supp}(f_{-1})}\right)$. Then $s(k,j) = \int_E e^{2\pi i x b j} \cdot F_k(x)\,dx$ is merely the Fourier transform of F_k evaluated at bj. Since E is a subset of $[-\epsilon, \epsilon]$, a straightforward application of Plancherel's theorem gives

$$\sum_{j\in\mathbb{Z}} |s(k,j)|^2 = 2\epsilon \cdot \sum_{j\in\mathbb{Z}} \left| \int_E e^{-2\pi i x \frac{j}{2\epsilon}} \cdot (2\epsilon)^{-1/2} F_k(x)\,dx \right|^2 = 2\epsilon \cdot \int_E |F_k(x)|^2 dx.$$

Explicitly,

$$\sum_{j\in\mathbb{Z}} |s(k,j)|^2 = 2\epsilon \int_{-e^{-\mathrm{supp}(f_1)}} \left| \frac{\left[\pi_1(0,-ak)h_1\right]\left(\ln\left(-\frac{\xi}{x}\right)\right)\overline{f_1\left(\ln\left(-\frac{\xi}{x}\right)\right)}}{|x|} \right|^2 dx$$
$$+ 2\epsilon \int_{-e^{-\mathrm{supp}(f_{-1})}} \left| \frac{\left[\pi_{-1}(0,-ak)h_{-1}\right]\left(\ln\left(\frac{\xi}{x}\right)\right)\overline{f_{-1}\left(\ln\left(\frac{\xi}{x}\right)\right)}}{|x|} \right|^2 dx$$

and additionally, the change of variable $x \mapsto -x$ in the first integral, followed by the change of variable $s = \ln\left(\frac{\xi}{x}\right)$ gives:

$$\sum_{j\in\mathbb{Z}} |s(k,j)|^2 = 2\epsilon \int_{\mathrm{supp}(f_1)} \left| \frac{\left[\pi_1(0,-ak)h_1\right](s)\overline{f_1(s)}}{e^{-s/2}} \right|^2 ds$$
$$+ 2\epsilon \int_{\mathrm{supp}(f_{-1})} \left| \frac{\left[\pi_{-1}(0,-ak)h_{-1}\right](s)\overline{f_{-1}(s)}}{e^{-s/2}} \right|^2 ds.$$

Summing $|s(k,j)|^2$ over $(k,j) \in \mathbb{Z}\times\mathbb{Z}$, derive the following:

$$\sum_{k\in\mathbb{Z}}\sum_{j\in\mathbb{Z}} |s(k,j)|^2 = \int_{\mathbb{R}} |h_1(s)|^2 \left(\sum_{k\in\mathbb{Z}} \frac{2\epsilon \cdot |f_1(s-ak)|^2}{|e^{-(s-ak)}|} \right) ds$$
$$+ \int_{\mathbb{R}} |h_{-1}(s)|^2 \left(\sum_{k\in\mathbb{Z}} \frac{2\epsilon \cdot |f_{-1}(s-ak)|^2}{|e^{-(s-ak)}|} \right) ds.$$

Since $f_1, f_{-1} \in C_c(\mathbb{R})$, we may indeed, choose $a > 0$ to be small enough so that $0 < m_{a,\epsilon}(f) \le M_{a,\epsilon}(f) < \infty$ and

$$m_{a,\epsilon}(f) \cdot \|h\|^2_{L^2(\mathbb{R})} = m_{a,\epsilon}(f) \cdot \left(\int_{\mathbb{R}} |h(s)|^2 ds \right).$$

Therefore,

$$m_{a,\epsilon}(f) \cdot \|h\|^2_{L^2(\mathbb{R})} \le \sum_{k\in\mathbb{Z}}\sum_{j\in\mathbb{Z}} \left| \left\langle h_1 + h_{-1}, \tau(0,ak)\,\tau(bj,0)\left(f_1 + f_{-1}\right) \right\rangle_{L^2(\mathbb{R})\oplus L^2(\mathbb{R})} \right|^2$$
$$\le M_{a,\epsilon}(f) \cdot \|h\|^2_{L^2(\mathbb{R})}.$$

Since the space of functions of the type $h_1 + h_{-1}$ where h_1, h_{-1} are compactly supported continuous functions is dense in $L^2(\mathbb{R}) \oplus L^2(\mathbb{R})$, it follows that $\tau\left(\Gamma\left(a,(2\epsilon)^{-1/2}\right)\right)\left(f_1 + f_{-1}\right)$ is a frame for $L^2(\mathbb{R}) \oplus L^2(\mathbb{R})$ with frame bounds $0 < m_{a,\epsilon}(f) \leq M_{a,\epsilon}(f)$.

6.1.1 The Wavelet Representation

Recall that $G = \mathbb{R} \rtimes \mathbb{R}$ is a noncommutative semidirect product group equipped with group operation $(x,t)(y,s) = \left(x + e^t y, t + s\right)$. Let

$$H = \{(0,s) : s \in \mathbb{R}\}$$

and let π be a representation $\mathrm{ind}_H^G(1)$ obtained by inducing the trivial character from the closed subgroup H of G. Then π is called a quasiregular representation of G. The "quasi" part of this term suggests explicitly that in a sense which we will make precise, π behaves almost like a left regular representation. In fact, the action of π is induced by a left and differentiable action of G on the homogeneous space G/H. To further elaborate this, put $N = \{(t,0) : t \in \mathbb{R}\}$. Then N is a closed submanifold of G, which is also a smooth transversal for cosets in G/H. Additionally, for fixed $(0,s) \in H$,

$$(0,s)(t,0) = \left(e^s t, s\right) = \left(e^s t, 0\right)(0,s) \Rightarrow (0,s)(t,0)H = \left(e^s t, 0\right)H$$

and given $(x,0) \in N$,

$$(x,0)(t,0) = (t+x,0) \Rightarrow (x,0)(t,0)H = (t+x,0)H.$$

The conormal action of H on G induces a dilation action on the function space $L^2(\mathbb{R})$ and the normal part N acts on this space by translation. Finally, the left action $(g,hH) \mapsto ghH$ of G on G/H induces a unitary (but reducible) representation on the Hilbert space of square-integrable functions on G/H, which we describe explicitly in terms of the generators of G as follows. For arbitrarily fixed $f \in L^2(\mathbb{R})$,

$$\left[\pi(x,0)f\right](t) = f(t-x) \text{ and } \left[\pi(0,s)f\right](t) = e^{-s/2} \cdot f\left(e^{-s}t\right). \tag{6.2}$$

While each operator $\pi(x,0)$ acts by translation, each $\pi(0,s)$ acts unitarily on $L^2(\mathbb{R})$ by dilation, and we are interested in sampling the orbit of a function to construct a frame for $L^2(\mathbb{R})$.

Definition 6.1 A frame (Parseval frame, orthonormal basis, and Riesz basis) wavelet is a frame (Parseval frame, orthonormal basis, and Riesz basis) of the type $\pi(\Gamma)f$, where Γ is a countable subset of G and $f \in L^2(\mathbb{R})$. Moreover, the representation π is a called a standard wavelet representation.

To state our next results, we fix notation as follows. Let $f_1 + f_2 \in L^2(\mathbb{R}) \oplus L^2(\mathbb{R})$ such that f_1, f_2 are nonzero continuous and compactly supported function where $-e^{-\mathrm{supp}(f_1)}$ and $-e^{-\mathrm{supp}(f_2)}$ are disjoint subsets of $\mathbb{R}$ of positive Lebesgue measure. Choose $\epsilon, a > 0$, such that $E = \left(-e^{-\mathrm{supp}(f_1)}\right) \cup \left(-e^{-\mathrm{supp}(f_{-1})}\right) \subseteq [-\epsilon, \epsilon]$,

$$0 < m_{a,\epsilon}(f) = \min_{l \in \{-1,1\}} \left(\operatorname{essinf}\left(s \mapsto \sum_{k \in \mathbb{Z}} \frac{2\epsilon \cdot \left| f_l(s - ak) \right|^2}{\left| e^{-(s-ak)} \right|} \right) \right)$$

and

$$M_{a,\epsilon}(f) = \max_{l \in \{-1,1\}} \left(\operatorname{esssup}\left(s \mapsto \sum_{k \in \mathbb{Z}} \frac{2\epsilon \cdot \left| f_l(s - ak) \right|^2}{\left| e^{-(s-ak)} \right|} \right) \right) < \infty.$$

Let φ_1, φ_2 be the Fourier transform of f restricted to $(0, \infty)$ and $(-\infty, 0)$, respectively, where each

$$f_1 : t \mapsto e^{-\frac{t}{2}} \varphi_k\left(\pm e^t\right) \quad \text{and} \quad f_{-1} : t \mapsto e^{-\frac{t}{2}} \varphi_k\left(-e^t\right)$$

and $f_1, f_{-1} \in L^2(\mathbb{R})$. Define $J : L^2(\mathbb{R}) \to L^2(\mathbb{R}) \oplus L^2(\mathbb{R})$ such that

$$Ju = \left(t \mapsto e^{-\frac{t}{2}} \mathcal{F}u\left(e^{-t}\right)\right) + \left(t \mapsto e^{-\frac{t}{2}} \mathcal{F}u\left(-e^{-t}\right)\right).$$

Proposition 6.1 The following holds true:

1. The wavelet representation π (6.2) is unitarily equivalent to the direct sum representation $\tau = \pi_1 \oplus \pi_{-1}$.
2. $\pi\left(\Gamma\left(a, (2\epsilon)^{-1/2}\right)\right)\left(f_1 + f_2\right)$ is a frame for $L^2(\mathbb{R})$ with frame bounds $0 < m_{a,\epsilon}(f) \leq M_{a,\epsilon}(f)$.

Proof: **Proof of Part (1)**: Let $J : L^2(\mathbb{R}) \to L^2(\mathbb{R}) \oplus L^2(\mathbb{R})$ such that

$$Jf = \left(t \mapsto e^{-\frac{t}{2}} \cdot \mathcal{F}f\left(e^{-t}\right)\right) + \left(t \mapsto e^{-\frac{t}{2}} \cdot \mathcal{F}f\left(-e^{-t}\right)\right)$$

where $\mathcal{F}$ is the Fourier transform defined as follows: $\mathcal{F}f(t) = \int_{\mathbb{R}} f(x) e^{-2\pi i x t} dx$. The operator J takes the restriction of any square-integrable function f onto the positive reals and negative reals and outputs a pair obtained by dilating each restriction while preserving their respective norm.

Step 1: We will first prove that J is an isometry. Computing the norm of Jf,

$$\begin{aligned} \|Jf\|^2_{L^2(\mathbb{R}) \oplus L^2(\mathbb{R})} &= \int_{\mathbb{R}} \left| e^{-\frac{t}{2}} \mathcal{F}f\left(e^{-t}\right) \right|^2 dt + \int_{\mathbb{R}} \left| e^{-\frac{t}{2}} \mathcal{F}f\left(-e^{-t}\right) \right|^2 dt \\ &= \int_{\mathbb{R}} \left| \mathcal{F}f\left(e^{-t}\right) \right|^2 e^{-t} dt + \int_{\mathbb{R}} \left| \mathcal{F}f\left(-e^{-t}\right) \right|^2 e^{-t} dt. \end{aligned}$$

The change of variable $\xi = e^{-t}$ gives

$$\|Jf\|^2_{L^2(\mathbb{R})\oplus L^2(\mathbb{R})} = \int_0^\infty |\mathcal{F}f(\xi)|^2 d\xi + \int_0^\infty |\mathcal{F}f(-\xi)|^2 d\xi$$
$$= \int_0^\infty |\mathcal{F}f(\xi)|^2 d\xi + \int_{-\infty}^0 |\mathcal{F}f(\xi)|^2 d\xi$$
$$= \|f\|^2_{L^2(\mathbb{R})}.$$

Secondly, we aim to show that J is surjective. To this end, fix $g_1 + g_2 \in L^2(\mathbb{R}) \oplus L^2(\mathbb{R})$. Let $f \in L^2(\mathbb{R})$ and let φ_1, φ_2 be the Fourier transform of f restricted to $(0,\infty)$ and $(-\infty, 0)$, where each $g_k : t \mapsto e^{-\frac{t}{2}} \cdot \varphi_k\left(\pm e^t\right) \in L^2(\mathbb{R})$. Then $Jf = g_1 + g_2$.

Step 2: We will now show that J intertwines π and $\pi_{-1} \oplus \pi_1$. Note that for every real number x and s,

$$J\pi(x,0)f = \left(t \mapsto e^{-\frac{t}{2}}\left[\mathcal{F}\pi(x,0)f\right]\left(e^{-t}\right)\right)$$
$$+ \left(t \mapsto e^{-\frac{t}{2}}\left[\mathcal{F}\pi(x,0)f\right]\left(-e^{-t}\right)\right)$$
$$J\pi(0,s)f = \left(t \mapsto e^{-\frac{t}{2}}\left[\mathcal{F}\pi(0,s)f\right]\left(e^{-t}\right)\right)$$
$$+ \left(t \mapsto e^{-\frac{t}{2}}\left[\mathcal{F}\pi(0,s)f\right]\left(-e^{-t}\right)\right).$$

Since

$$\left[\mathcal{F}\pi(x,0)f\right](t) = \int_{\mathbb{R}} \left[\pi(x,0)f\right](s)\, e^{2\pi ist} ds = e^{2\pi ixt}\mathcal{F}f(t)$$

and

$$\left[\mathcal{F}\pi(0,s)f\right](t) = \int_{\mathbb{R}} \left[\pi(0,s)f\right](r)\, e^{-2\pi irt} dr = \int_{\mathbb{R}} e^{-s/2} f\left(e^{-s}r\right) e^{-2\pi irt} dr$$

it follows that

$$\left[\mathcal{F}\pi(0,s)f\right](t) = e^{-s/2} \cdot \int_{\mathbb{R}} f\left(e^{-s}r\right) e^{-2\pi irt} dr$$

Moreover,

$$\left[\mathcal{F}\pi(0,s)f\right](t) = e^{-s/2}e^{s} \cdot \int_{\mathbb{R}} f(r)\, e^{-2\pi ie^{s}rt} dr = e^{\frac{s}{2}} \left[\mathcal{F}f\right]\left(e^{s}t\right).$$

Therefore,

$$J\pi(x,0)f = \left(t \mapsto e^{-\frac{t}{2}} e^{2\pi ixe^{-t}} \mathcal{F}f\left(e^{-t}\right)\right) + \left(t \mapsto e^{-\frac{t}{2}} e^{-2\pi ixe^{-t}} \mathcal{F}f\left(-e^{-t}\right)\right)$$
$$= \left(t \mapsto e^{2\pi ixe^{-t}}\left(e^{-\frac{t}{2}}\mathcal{F}f\left(e^{-t}\right)\right)\right) + \left(t \mapsto e^{-2\pi ixe^{-t}}\left(e^{-\frac{t}{2}}\mathcal{F}f\left(-e^{-t}\right)\right)\right)$$
$$= \left[\pi_1 \oplus \pi_{-1}\right](x,0)\left(t \mapsto e^{-\frac{t}{2}}\mathcal{F}f\left(e^{-t}\right)\right) + \left(t \mapsto e^{-\frac{t}{2}}\mathcal{F}f\left(-e^{-t}\right)\right)$$
$$= \left[\pi_1 \oplus \pi_{-1}\right](x,0)Jf.$$

Similarly,

$$\begin{aligned} J\pi(0,s)f &= \left(t \mapsto e^{\frac{s-t}{2}} \mathcal{F}f\left(e^{s-t}\right)\right) + \left(t \mapsto e^{\frac{s-t}{2}} \mathcal{F}f\left(-e^{s-t}\right)\right) \\ &= \left[\pi_1 \oplus \pi_{-1}\right](s,0)\left(t \mapsto e^{-\frac{t}{2}} \mathcal{F}f\left(e^{-t}\right)\right) + \left(t \mapsto e^{-\frac{t}{2}} \mathcal{F}f\left(-e^{-t}\right)\right) \\ &= \left[\pi_1 \oplus \pi_{-1}\right](s,0)Jf. \end{aligned}$$

Thus, J is a unitary operator intertwining π with $\pi_1 \oplus \pi_{-1}$.

Proof of Part (2): Since

$$J\pi\left(\Gamma\left(a,(2\epsilon)^{-1/2}\right)\right)f = \left[\pi_1 \oplus \pi_{-1}\right]\left(\Gamma\left(a,(2\epsilon)^{-1/2}\right)\right)\left(f_1 + f_2\right)$$

and since $\left[\pi_1 \oplus \pi_{-1}\right]\left(\Gamma\left(a,(2\epsilon)^{-1/2}\right)\right)\left(f_1 + f_2\right)$ is a frame with frame bounds $0 < m_{a,\epsilon}(f) \leq M_{a,\epsilon}(f)$, it follows that $\pi\left(\Gamma\left(a,(2\epsilon)^{-1/2}\right)\right)f$ is a frame for $L^2(\mathbb{R})$ with frame bounds $0 < m_{a,\epsilon}(f) \leq M_{a,\epsilon}(f)$.

6.2 Gabor Systems and the Heisenberg Group

Let G be the three-dimensional Heisenberg Lie group with Lie algebra $\mathfrak{g}$ spanned by vectors X, Y, Z satisfying nontrivial Lie bracket $[X, Y] = Z$. Let

$$\pi_{\lambda Z^* + \upsilon Y^* + \omega X^*} = \mathrm{ind}_{\exp \mathbb{R}Z + \mathbb{R}Y}^{G}\left(\chi_{\lambda Z^* + \upsilon Y^* + \omega X^*}\right)$$

be a unitary representation of the Heisenberg Lie group, where $\ell = \lambda Z^* + \upsilon Y^* + \omega X^*$ is a linear functional in $\mathfrak{g}^*$ and λ is a nonzero real number. Moreover, each π_ℓ is realized as acting in $L^2(\mathbb{R})$ as follows. For $\exp zZ \exp yY \exp xX \in G$ and $f \in L^2(\mathbb{R})$,

$$\left[\pi_\ell\left(\exp zZ \exp yY \exp xX\right)f\right](t) = e^{2\pi i\left\langle \begin{bmatrix} \lambda \\ \upsilon - t\lambda \end{bmatrix}, \begin{bmatrix} z \\ y \end{bmatrix} \right\rangle} f(t-x).$$

On the one hand, the operators $\left\{\pi_\ell(\exp xX) : x \in \mathbb{R}\right\}$ act on $L^2(\mathbb{R})$ by translations, and as observed earlier, this action alone cannot be discretized to construct frames for the entire Hilbert space $L^2(\mathbb{R})$. The operators in $\left\{\pi_\ell(\exp yY) : y \in \mathbb{R}\right\}$ act as frequency shifts on $L^2(\mathbb{R})$.

Definition 6.2 Let Γ be a countable subset of G. Any frame of the form $\pi_{\lambda Z^* + \upsilon Y^* + \omega X^*}(\Gamma)f$ is called a Gabor frame.

The following is worth recording. First, since $\pi_{Z^*}(\exp \mathbb{Z}Y)1_{[0,1)}$ is an orthonormal basis for $L^2([0,1))$ and since $\{[0,1) + k : k \in \mathbb{Z}\}$ is a partition of the real line, the family of vectors $\pi_{Z^*}(\exp \mathbb{Z}Y \exp \mathbb{Z}X)1_{[0,1)}$ generated by the indicator function of the unit interval $[0,1)$ is an orthonormal basis for $L^2(\mathbb{R})$. Moreover, the set $\pi_{Z^*}(\exp \mathbb{Z}Y \exp \mathbb{Z}X)$ is a commutative group of operators.

Secondly, suppose that $\pi_\ell(\exp \mathbb{Z}Y \exp \mathbb{Z}X) f$ is an orthonormal basis for $L^2(\mathbb{R})$. Then $\pi_\ell(\exp \mathbb{Z}Y \exp \mathbb{Z}X) \varphi$ is an orthonormal basis for $L^2(\mathbb{R})$ if and only if there exists a unitary operator J such that $J\pi_\ell(\exp kY \exp jX) = \pi_\ell(\exp kY \exp jX) J$ for all integers k, j and $\varphi = Jf$. Indeed, suppose that J is a unitary operator such that $J\pi_\ell(\exp kY \exp jX) = \pi_\ell(\exp kY \exp jX) J$ for all integers k, j. Since the image of an orthonormal basis under a unitary operator is an orthonormal basis, it is clear that

$$J\pi_\ell(\exp kY \exp jX) f = \pi_\ell(\exp kY \exp jX) Jf$$

and as a result, $\pi_\ell(\exp \mathbb{Z}Y \exp \mathbb{Z}X) \varphi$ must be an orthonormal basis. Conversely, let us assume that $\pi_\ell(\exp \mathbb{Z}Y \exp \mathbb{Z}X) \varphi$ is an orthonormal basis for $L^2(\mathbb{R})$. Define a linear operator such that $J : L^2(\mathbb{R}) \to L^2(\mathbb{R})$ and $J\pi_\ell(\exp kY \exp jX) f = \pi_\ell(\exp kY \exp jX) \varphi$ for every $(k, j) \in \mathbb{Z}^2$. By assumption, $J\pi_\ell(\exp 0Y \exp 0X) f = Jf = \varphi$ and

$$J\pi_\ell(\exp kY \exp jX) f = \pi_\ell(\exp kY \exp jX) \varphi = \pi_\ell(\exp kY \exp jX) Jf$$

for every integer k, j. Therefore, $J\pi_\ell(\exp kY \exp jX) = \pi_\ell(\exp kY \exp jX) J$ for all integers k, j.

Similarly to the irreducible representation of the $ax+b$ Lie group, we will see that whenever f is continuous and compactly supported, there exists a sufficiently dense and discrete subset Γ of the Heisenberg group, such that $\pi(\Gamma) f$ is a frame for $L^2(\mathbb{R})$. In fact, Γ is a function of (1) the support of f and (2) the family of l^2-norms obtained by restricting f to $t + a\mathbb{Z}$ for some suitable positive real number a where $t \in \mathbb{R}$.

For any positive real numbers a, b, let $\Gamma_{(a,b)} = \exp(b\mathbb{Z}Y)\exp(a\mathbb{Z}X)$. Let $f \in C_c(\mathbb{R})$ and suppose that $b^{-1} > 0$ is large enough to ensure that $\lambda \cdot \operatorname{supp}(f)$ $(\lambda Z^* + \upsilon Y^* + \omega X^* \in \mathfrak{g}^*)$ is a subset of the half-open interval $\left[-\frac{b^{-1}}{2}, \frac{b^{-1}}{2}\right)$ and let a be a positive real number. For a fixed continuous and compactly supported function f, let $\Theta_{(f,a)}$ be a (formally defined) real-valued function associated with f and a, and with explicit formula:

$$\Theta_{(f,a)}(t) = \sum_{j \in \mathbb{Z}} |f(t - aj)|^2.$$

Such $\Theta_{(f,a)}$ is in a formal sense, a-periodic almost everywhere, and long as f is continuous and has compact support, as we shall assume here, then for any positive real number a, $\Theta_{(f,a)}$ is a bounded function. But of course its essential lower bound of $\Theta_{(f,a)}$ can possibly be equal to zero.

Theorem 6.3 Let $\pi = \pi_{\lambda Z^* + \upsilon Y^* + \omega X^*}$. The following holds true

1. If $0 < \operatorname{essinf}\left(\Theta_{(f,a)}\right) \leq \operatorname{esssup}\left(\Theta_{(f,a)}\right) < \infty$ then $\pi\left(\Gamma_{(a,b)}\right) f$ is a frame for $L^2(\mathbb{R})$ with frame bounds

$$0 < \operatorname{essinf}\left((b|\lambda|)^{-1}\Theta_{(f,a)}\right) \leq \operatorname{esssup}\left((b|\lambda|)^{-1}\Theta_{(f,a)}\right) < \infty.$$

2. Suppose that $\pi\left(\Gamma_{(a,b)}\right) f$ is a frame for $L^2(\mathbb{R})$. If $J : L^2(\mathbb{R}) \to L^2(\mathbb{R})$ is a bounded and invertible linear operator such that $J\pi(\gamma)J^{-1} = \pi_\ell(\gamma)$ for every $\gamma \in \Gamma_{(a,b)}$, then $\pi\left(\Gamma_{(a,b)}\right) Jf$ is a frame for $L^2(\mathbb{R})$.

Proof: **Proof of Part (1)**: Given $h, f \in C_c(\mathbb{R})$, let

$$V_{h,f}(x,y,z) = \langle h, \pi(\exp zZ \exp yY \exp xX) f\rangle_{L^2(\mathbb{R})}.$$

Then

$$V_{h,f}(x,y,z) = e^{-2\pi i\lambda z} \cdot \langle h, \pi(\exp yY \exp xX) f\rangle = e^{-2\pi i\lambda z} V_{h,f}(x,y,0)$$

and for any real numbers x, y, z, $\left|V_{h,f}(x,y,z)\right| = \left|V_{h,f}(x,y,0)\right|$. Next,

$$\begin{aligned} V_{h,f}(x,y,0) &= \int_{\mathbb{R}} h(t)\overline{f(t-x)}e^{-2\pi i(v-t\lambda)y}dt \\ &= e^{-2\pi i v y}\int_{\text{supp}(f)} h(t+x)\overline{f(t)}e^{2\pi i t\lambda y}dt. \end{aligned} \tag{6.3}$$

Digressing for a short moment, it is instructive to pause for a second, to contrast (6.3) with (6.1). We draw the reader's attention to the fact that $t \mapsto e^{2\pi i t\lambda y}$ is obtained by composition the exponential function with a function only linear in the t-variable, while in Formula (6.3), one is integrating against a function given by composing an exponential, with a second exponential. These disparities, which, at first may deceptively appear to be describing mild differences in features, turn out to be at the root of the spectacular differences between frames generated by the actions of the Heisenberg and $ax+b$ Lie groups as recorded in the literature [2, 3]. The change of variable $s = t\lambda$, yields

$$V_{h,f}(x,y,0) = \frac{e^{-2\pi i v y}}{|\lambda|}\int_{\lambda\text{supp}(f)}\left(h\left(\frac{s}{\lambda}+x\right)\cdot\overline{f\left(\frac{s}{\lambda}\right)}\right)\cdot e^{2\pi i s y}ds.$$

Assuming that $y = bk$, where $b > 0$ and k is an integer, suppose that b^{-1} is large enough to ensure that $\lambda \cdot \text{supp}(f)$ is a subset of $\left[-\frac{b^{-1}}{2}, \frac{b^{-1}}{2}\right)$. Then

$$V_{h,f}(x,bk,0) = \frac{e^{-2\pi i v bk}}{b^{1/2}|\lambda|}\int_{\left[-\frac{b^{-1}}{2},\frac{b^{-1}}{2}\right)} h\left(\frac{s}{\lambda}+x\right)\cdot\overline{f\left(\frac{s}{\lambda}\right)}b^{1/2}e^{2\pi i s bk}ds.$$

Applying first an absolute value to each side of the equation above, and applying Parseval's equality next, we derive the following:

$$\begin{aligned} \sum_{k\in\mathbb{Z}}\left|V_{h,f}(x,bk,0)\right|^2 &= \frac{1}{b|\lambda|^2}\sum_{k\in\mathbb{Z}}\left|\int_{\left[-\frac{b^{-1}}{2},\frac{b^{-1}}{2}\right)} h\left(\frac{s}{\lambda}+x\right)\overline{f\left(\frac{s}{\lambda}\right)}b^{1/2}e^{2\pi i s bk}ds\right|^2 \\ &= \frac{1}{b|\lambda|^2}\int_{\left[-\frac{b^{-1}}{2},\frac{b^{-1}}{2}\right)}\left|h\left(\frac{s}{\lambda}+x\right)f\left(\frac{s}{\lambda}\right)\right|^2 ds \\ &= \frac{1}{b|\lambda|}\int_{\mathbb{R}}|h(t+x)f(t)|^2dt. \end{aligned}$$

This, clearly leads to the following conclusion:

$$\sum_{k\in\mathbb{Z}}\left|V_{h,f}(x,bk,0)\right|^2=\int_{\mathbb{R}}|h(t)|^2\frac{|f(t-x)|^2}{b|\lambda|}dt.$$

Replacing x with $aj\in a\mathbb{Z}$,

$$\sum_{j\in\mathbb{Z}}\sum_{k\in\mathbb{Z}}\left|V_{h,f}(aj,bk,0)\right|^2=\int_{\mathbb{R}}|h(t)|^2\sum_{j\in\mathbb{Z}}\frac{|f(t-aj)|^2}{b|\lambda|}dt.$$

Additionally, assuming that $a>0$ is small enough to ensure that

$$\begin{aligned}0<m_{a,b}(f)&=\operatorname{essinf}\left((b|\lambda|)^{-1}\Theta_{(f,a)}\right)\le M_{a,b}(f)\\&=\operatorname{esssup}\left((b|\lambda|)^{-1}\Theta_{(f,a)}\right)<\infty,\end{aligned}$$

we finally derive

$$m_{a,b}(f)\,\|h\|^2_{L^2(\mathbb{R})}\le\sum_{j\in\mathbb{Z}}\sum_{k\in\mathbb{Z}}\left|V_{h,f}(aj,bk,0)\right|^2\le M_{a,b}(f)\,\|h\|^2_{L^2(\mathbb{R})}.$$

Since $C_c(\mathbb{R})$ is dense in $L^2(\mathbb{R})$, it follows that $\pi_{\lambda Z^*+\upsilon Y^*+\omega X^*}(\exp(b\mathbb{Z}Y)\exp(a\mathbb{Z}X))$ is a frame for $L^2(\mathbb{R})$ with frame bounds $m_{a,b}(f), M_{a,b}(f)$.

Proof of Part (2): This result follows from the fact that the image of a frame under a bounded and invertible linear operator is a frame as well.

The following examples illustrate the results in Theorem 6.3.

Example 6.2 Let $f(s)=\left(1_{\left[-\frac{1}{2},\frac{1}{2}\right]}*1_{\left[-\frac{1}{2},\frac{1}{2}\right]}*1_{\left[-\frac{1}{2},\frac{1}{2}\right]}\right)(s)$ where $*$ stands for the usual convolution in $L^1(\mathbb{R})$. Then

$$f(s)=\begin{cases}\left(\frac{3}{4}-s^2\right) & \text{if } s\in\left[-\frac{1}{2},\frac{1}{2}\right]\\ \frac{(3-2s)^2}{8} & \text{if } s\in\left(\frac{1}{2},\frac{3}{2}\right)\\ \frac{(3+2s)^2}{8} & \text{if } s\in\left(-\frac{3}{2},-\frac{1}{2}\right)\\ 0 & \text{otherwise}\end{cases}.$$

Letting $\Theta(s)=\sum_{j\in\mathbb{Z}}|f(s-j)|^2$, one verifies that Θ is a 1-periodic function, continuous on the real line,

$$\inf_{s\in\mathbb{R}}(\Theta(s))=\Theta\left(\frac{1}{2}\right)>0\quad\text{and}\quad\sup_{s\in\mathbb{R}}(\Theta(s))=\Theta(0)<\infty.$$

Put $\Gamma_{\left(1,\frac{1}{3}\right)}=\exp\left(\frac{1}{3}\mathbb{Z}Y\right)\exp(a\mathbb{Z}X)$. Then $\pi_{Z^*}\left(\Gamma_{\left(1,\frac{1}{3}\right)}\right)f$ is a frame for $L^2(\mathbb{R})$ with frame bounds $3\Theta\left(\frac{1}{2}\right)>1$ and $3\Theta(0)<1.3$.

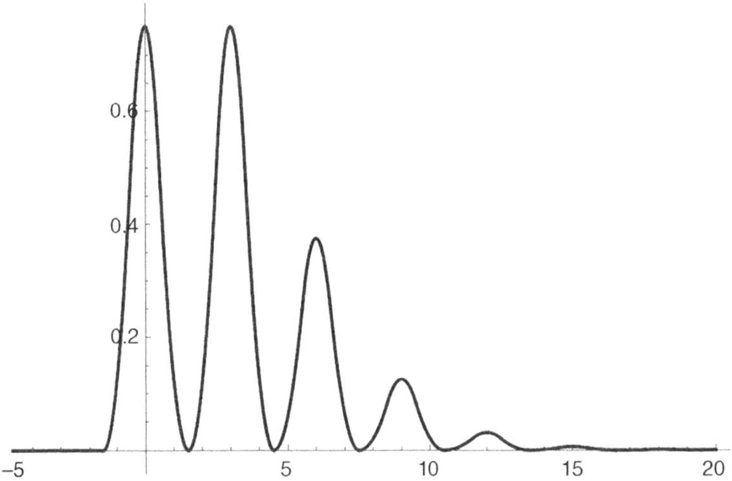

Figure 6.4 A generator of a frame induced by an irreducible action of the Heisenberg Lie group

To construct new frames from $\pi_{Z^*}\left(\Gamma_{\left(1,\frac{1}{3}\right)}\right)f$, we suggest the following approach. Fix $A = \pi_{Z^*}(\exp 3X)$ as a unitary operator acting on $L^2(\mathbb{R})$. Note that A commutes with every element in the set $\pi_{Z^*}\left(\Gamma_{\left(1,\frac{1}{3}\right)}\right)$, since π is a homomorphism and additionally,

$$\begin{bmatrix} 1 & j & 0 \\ 0 & 1 & \frac{k}{3} \\ 0 & 0 & 1 \end{bmatrix}\begin{bmatrix} 1 & 3 & 0 \\ 0 & 1 & 0 \\ 0 & 0 & 1 \end{bmatrix}\begin{bmatrix} 1 & j & 0 \\ 0 & 1 & \frac{k}{3} \\ 0 & 0 & 1 \end{bmatrix}^{-1} = \begin{bmatrix} 1 & 3 & -k \\ 0 & 1 & 0 \\ 0 & 0 & 1 \end{bmatrix}.$$

Next, $A\pi_{Z^*}\left(\Gamma_{\left(1,\frac{1}{3}\right)}\right)f = \pi_{Z^*}\left(\Gamma_{\left(1,\frac{1}{3}\right)}\right)Af$ and

$$e^A f(t) = e^{\pi_{Z^*}(\exp 3X)} f(t) = \sum_{k=0}^{\infty} \frac{f(t-3k)}{k!}$$

as illustrated below (Figure 6.4. This shows that $\pi_{Z^*}\left(\Gamma_{\left(1,\frac{1}{3}\right)}\right)e^A f$ is a frame for $L^2(\mathbb{R})$.

References

1 Ole Christensen. *An introduction to frames and Riesz bases*. Applied and Numerical Harmonic Analysis. Birkhäuser, Boston, MA, 2003.

2 Ole Christensen. *An introduction to frames and Riesz bases*, pages 67–108. Springer, 2016.
3 Christopher Heil. *A basis theory primer: expanded edition*. Springer Science & Business Media, 2010.
4 Vignon Oussa. Frames arising from irreducible solvable actions I. *Journal of Functional Analysis*, 274(4):1202–1254, 2018.
5 Vignon Oussa. Compactly supported bounded frames on lie groups. *Journal of Functional Analysis*, 277(6):1718–1762, 2019.

7

Frames on Lie Groups

The initial construction of classical wavelets in the literature provides only an implicit link to the discretization problem of the affine group on the real line. However, subsequent contributors have fortified this connection, precipitating extensive mathematical endeavors to broaden these constructions to other groups, such as the shearlet group [5]. Hartmut Führ led one of the most pivotal expansions of wavelet construction to a larger class of groups, building on theories such as coorbit theory [1], which Feichtinger and Gröchenig first introduced in the nineties. The particularly inspirational publication [4] demonstrated the significant advantages of using more abstract tools in representation theory in generalizing these concepts. A considerable obstacle in this process is the heavy reliance on many constructions, algorithms, and results on integrability conditions. Checking whether a representation is integrable generally poses a challenge. To address this, I chose to study the action of the representation more intensively, focusing on its differential geometric attributes. As I began to compute examples explicitly, I found that for a broad class of irreducible representations, the integrability condition could be circumvented if one were prepared to explore a more innovative approach to sampling the orbits of a representation. In this chapter, I aim to present my ideas in a manner that is accessible, providing insights into the construction of frames and wavelets from a differential geometric standpoint.

Treating the frames constructed in the preceding chapter (by discretizing the $ax+b$ and Heisenberg Lie groups) as prototypes of a much more general phenomenon, this chapter aims to develop a far-reaching and unified theory capable of handling a large class of induced representations. As a byproduct, we produce an algorithmic procedure that outputs frames for function spaces over any linear Lie group. Additionally, the frames constructed with our methods satisfy several desired regular conditions that make them suitable for applications in signal analysis, time-frequency, generalized wavelet, shearlets, and operator theories.

A Bridge Between Lie Theory and Frame Theory: Applications of Lie Theory to Harmonic Analysis, First Edition. Vignon Oussa.

7.1 Discretization of Induced Characters

7.1.1 Connection to Wavelet Theory and Time-Frequency Analysis

Let H be a Lie group endowed with a fixed left-invariant measure $d\mu_H$. We will construct frames of the form $\pi(\Gamma)f$, where π is a unitary representation of some Lie group G acting in the Hilbert space $\mathfrak{H} = L^2(H, d\mu_H)$ such that the window function f can be chosen to satisfy at least one (if not all) of following conditions: boundedness, compactly support, continuity, and smoothness. Any frame possessing at least one of these properties will be called a "nice" frame.

Recall the three-dimensional Heisenberg group, realized as a semidirect product group $\mathbb{R}^2 \rtimes \mathbb{R}$ and equipped with the noncommutative operation

$$(v,t)(w,s) = \left(v + \begin{bmatrix} 1 & t \\ 0 & 1 \end{bmatrix} w, t+s\right)$$

where $(v,t), (w,s) \in \mathbb{R}^2 \times \mathbb{R}$. The infinite-dimensional irreducible representations [2] of the three-dimensional Heisenberg Lie group act in $L^2(\mathbb{R})$ as follows:

$$\left[\pi_\lambda(v,t)f\right](s) = e^{2\pi i \left\langle \begin{bmatrix} 1 & t \\ -t & 1 \end{bmatrix} \begin{bmatrix} \lambda \\ 0 \end{bmatrix}, \begin{bmatrix} v_1 \\ v_2 \end{bmatrix} \right\rangle} f(s-t).$$

Three classes of operators: scalar multiplications, modulation operators (phase translations), and translations, defined by

$$\pi_\lambda\left(\begin{bmatrix} v_1 \\ 0 \end{bmatrix}, 0\right), \pi_\lambda\left(\begin{bmatrix} 0 \\ v_2 \end{bmatrix}, 0\right) \text{ and } \pi_\lambda(0,t)$$

respectively, emerge from the action π. For our purpose, the role played by the operators acting by scalar multiplication is negligible. However, we have seen that the combined actions of modulations and translations can be discretized to construct frames and orthonormal bases for $L^2(\mathbb{R})$ (Figure 7.1).

Furthermore, we have proved that it is possible to construct nice frames of the form $\pi_\lambda(\Gamma)f$, where Γ is a countable subgroup of the Heisenberg Lie group. Similarly, let G be the $ax+b$ Lie group: the semidirect product group $\mathbb{R} \rtimes \mathbb{R}$ with group operation $(x,a)(y,b) = (x + e^a y, a+b)$. Up to unitary equivalence, G admits two infinite-dimensional irreducible representations [2], realized as acting in the space of square integrable functions on the real line (Figure 7.2).

$$\pi_\pm(x,a)f(t) = e^{\pm 2\pi i e^{-t} x} f(t-a). \tag{7.1}$$

The action of $\pi_\pm$ gives rise to two main classes of operators: frequency shifts acting by multiplication by exponentials with nonlinear phase and translations determined by $\pi_\pm(x,0)$ and $\pi_\pm(0,a)$ respectively.

As discussed in the previous chapter, it is always possible to discretize each representation $\pi_\pm$ to construct frames generated by a "nice" vector. It is also

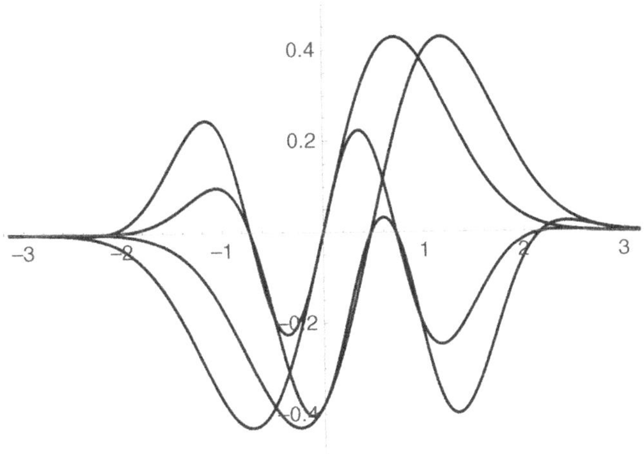

Figure 7.1 Showing a sample of orbit points in the orbit of the Heisenberg group.

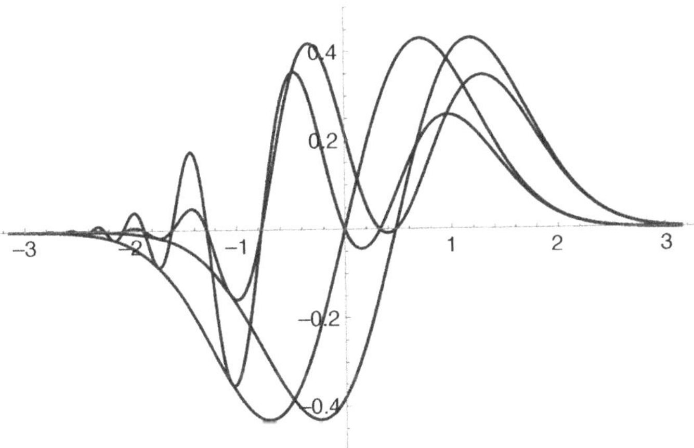

Figure 7.2 Showing a sample of orbit points in the orbit of the $ax+b$ group.

worth highlighting that the two representations of interest are monomial as well. This means that they are induced by some unitary characters of some closed Lie subgroups. In fact, in both cases, the irreducible representations of interest are obtained by inducing a fixed character from some closed normal subgroup.

To extend these constructions, we will describe a class of semidirect product groups, which we regard as a natural generalization of our prototypical groups ($ax+b$ and Heisenberg Lie groups). To this end, let G be a Lie group such that $G = NH, H, N$ are connected closed subgroups and N is simply connected, solvable, normal closed in G. Additionally, the exponential map defines a diffeomorphism between the Lie algebra $\mathfrak{n}$ and N. We also assume that π is a unitary representation of G obtained by inducing a unitary character χ of N. This requires that the commutator subgroup of N is contained in the kernel of this character. Moreover, letting μ_H be a fixed left-invariant measure on H, we realize π as acting in the function space $L^2\left(H, d\mu_H\right)$. By assumption, the unitary representation χ of N is associated to a linear functional $\lambda \in \mathfrak{n}^*$ such that $\chi\left(\exp X\right) = e^{2\pi i\langle\lambda,X\rangle}$ for every $X \in \mathfrak{n}$. Next, on the level of the Lie algebra, we may decompose $\mathfrak{g} = \mathfrak{n} \oplus \mathfrak{h}$ where $\mathfrak{n}$ and $\mathfrak{h}$ is spanned by $\left\{X_1, \ldots, X_n\right\}$ and $\left\{A_1, \ldots, A_r\right\}$ respectively. Letting $\left\{X_1^*, X_2^*, \ldots, X_n^*\right\}$ be a dual basis for the vector space $\mathfrak{n}^*$ satisfying $\left\langle X_j^*, X_k\right\rangle = \delta_{j,k}$ for $1 \leq j, k \leq n$, there exist real numbers $\lambda_k, 1 \leq k \leq n$ such that $\lambda = \sum_{k=1}^n \lambda_k X_k^*$. Note also that since χ is assumed to be a unitary character of N, the derived ideal $[\mathfrak{n}, \mathfrak{n}] = \mathbb{R}\text{-span}\{[X, Y] : X, Y \in \mathfrak{n}\}$ of $\mathfrak{n}$ must necessarily be contained in the kernel of the linear functional λ.

Let

$$\pi = \mathrm{ind}_N^{NH}\left(\chi_\lambda\right) \tag{7.2}$$

be a unitary representation of G acting in Hilbert space $\mathfrak{H} = L^2\left(H, d\mu_H\right)$ as follows. For arbitrary $f \in \mathfrak{H}, x \in N$, and $z \in H$,

$$[\pi\left(x\right)]f\left(h\right) = e^{2\pi i\left\langle Ad\left(h^{-1}\right)^*\lambda, \log(x)\right\rangle} \cdot f\left(h\right) \text{ and } [\pi\left(z\right)]f\left(h\right) = f\left(z^{-1}h\right). \tag{7.3}$$

On the one hand, remark that given $x \in N$, the operator $\pi\left(x\right)$ acts in a way that also resembles a modulation action. In the case of the Heisenberg group, $h \mapsto Ad\left(h^{-1}\right)^*\lambda$ is just a linear polynomial in the coordinates of h and when G is the ax+b Lie group, $h \mapsto Ad\left(h^{-1}\right)^*\lambda$ can be viewed as an exponential function defined on the real line. Generally, $h \mapsto Ad\left(h^{-1}\right)^*\lambda$ will be a smooth and analytic function in the coordinates of h. We call the action induced by $\pi\left(N\right)$ a generalized modulation action. On the other hand, each element in the family of operators $\left(\pi\left(z\right)\right)_{z\in H}$ acts on the Hilbert space $\mathfrak{H}$ by left translations. This generalizes the translation action induced by the conormal part of the Heisenberg and the ax+b groups.

We will prove that under a mild geometric and easily verified condition imposed on $h \mapsto Ad\left(h^{-1}\right)^*\lambda$, π_λ can be discretized in a manner which is analogous to the examples of the Heisenberg and ax+b group for the construction of nice frames.

The following are condensed versions of our main results.

Theorem 7.1 If $h \mapsto Ad(h^{-1})^*\lambda$ is an immersion at the identity of H, then there exist a countable set Γ and a continuous (or a smooth) function f compactly supported on H such that $\pi\left(\Gamma\right)f$ is a frame for $L^2\left(H, d\mu_H\right)$.

Theorem 7.2 If $h \mapsto Ad(h^{-1})^*\lambda$ is an immersion at the identity of H and if H is a simply connected and solvable Lie group for which the exponential map defines a bijection between $\mathfrak{h}$ and H, then there exist a countable set Γ and a continuous (or a smooth) function f compactly supported on H such that $\pi(\Gamma)f$ is a Parseval frame for $L^2(H, d\mu_H)$.

The following is a detailed and technical reformulation of Theorems 7.1 and 7.2.

Theorem 7.3 Suppose that $h \mapsto Ad(h^{-1})^* \lambda$ is an immersion at the identity of H.

1) Then there exists a relatively compact neighborhood $\mathcal{O}$ of the identity in H such that if f is a continuous function supported on $\mathcal{O}$, then there exists a countable set $\Gamma \subset G$ such that $\pi(\Gamma)f$ is a frame for $L^2(H, d\mu_H)$ Moreover, f can be chosen to be in $C^k(H)$ for any arbitrary natural number k (and even infinitely differentiable).
2) If H is a solvable Lie group and $X \mapsto \exp X$ defines a diffeomorphism between $\mathfrak{h}$ and H then there exist a relatively compact neighborhood $\mathcal{O}$ of the identity in H, a continuous function f supported on $\mathcal{O}$, and a countable subset Γ of G such that $\pi(\Gamma)f$ is a Parseval frame. In the specific cases where f can be selected to be of unit norm, then $\pi(\Gamma)f$ is necessarily an orthonormal basis for $L^2(H, d\mu_H)$.

As a direct consequence of the results above, we derive the following. Let $\pi_\lambda = \text{ind}_N^{NH}(\chi_\lambda)$ where χ_λ is a unitary character of N and additionally, let us suppose that $h \mapsto Ad(h^{-1})^* \lambda$ is an immersion at the identity of H. Then there exist countable sets $\Gamma_N \subset N$ and $\Gamma_H \subset H$ and a nice vector $f \in L^2(H, d\mu_H)$ such that $\pi_\lambda(\Gamma_H \Gamma_N) f$ is a frame for $L^2(H, d\mu_H)$.

Let R be the right regular representation of H acting in $L^2(H, d\mu_H)$ as follows: $R_k f(h) = \Delta(k)^{1/2} f(hk)$. Then every operator R_k, commutes with all operators $\pi_\lambda(h), h \in H$, since the restriction of π_λ to H coincides with the left regular representation of the group. Moreover, for $x \in N$,

$$R_k\left[\pi_\lambda(x) f\right](h) = \Delta(k)^{1/2} e^{2\pi i \langle \beta_\lambda(hk), \log x \rangle} f(hk).$$

Letting

$$H_{(\lambda, \Gamma_N)} = \left\{k \in H : \left\langle \beta_\lambda(hk), \log \gamma_N \right\rangle - \left\langle \beta_\lambda(h), \log \gamma_N \right\rangle \in \mathbb{Z}, \forall \gamma_N \in \Gamma_N, h \in H\right\},$$

it follows that for any $\gamma_N \in \Gamma_N$

$$\begin{aligned} k \in H_{(\lambda, \Gamma_N)} \Rightarrow R_k\left[\pi_\lambda(\gamma_N) f\right](h) &= \Delta(k)^{1/2} e^{2\pi i \langle \beta_\lambda(h), \log x \rangle} f(hs) \\ &= \pi_\lambda(\gamma_N) R_k f(h) \end{aligned}$$

and as a result, the unitary operator R_s commutes with every operator in the set $\pi_\lambda(\Gamma_H \Gamma_N)$ and Corollary 7.1 is thus, just an easy derivation of Theorem 7.3. In fact, Corollary 7.1, may be genuinely understood as a statement that gives a simple sufficient condition under which, it is possible to perturb the frames constructed in Theorem 7.3. This result also suggests that the larger $\Gamma_H^{\star}$ is as a set, the higher the number of possible ways to perturb $\pi_\lambda(\Gamma) f$ into a frame of the type $\pi_\lambda(\Gamma) A f$, where A is a bounded and invertible operator.

Corollary 7.1 Let $\mathcal{A}_G$ be the set of all bounded and invertible operators generated by the unitary operators contained in $\pi_\lambda\left(H_{(\lambda,\Gamma_N)}\right)$. Put $\Gamma = \Gamma_H \Gamma_N$. If $\pi_\lambda(\Gamma) f$ is a frame for $L^2(H, d\mu_H)$ and $A \in \mathcal{A}_G$ then $\pi_\lambda(\Gamma) A f$ is a frame for $L^2(H, d\mu_H)$ as well.

Remark 7.1 For the class of groups discussed here, it is very instructive to view Corollary 7.1) through the lenses of the Heisenberg and *ax+b* Lie groups.

1) Let $\mathbb{H}$ be the Heisenberg Lie group with Lie algebra spanned by

$$Z = \begin{bmatrix} 0 & 0 & 1 \\ 0 & 0 & 0 \\ 0 & 0 & 0 \end{bmatrix}, \; Y = \begin{bmatrix} 0 & 0 & 0 \\ 0 & 0 & 1 \\ 0 & 0 & 0 \end{bmatrix}, \; X = \begin{bmatrix} 0 & 1 & 0 \\ 0 & 0 & 0 \\ 0 & 0 & 0 \end{bmatrix}$$

such that $[X, Y] = Z$. Here, $N = \exp \mathbb{R}Z \exp \mathbb{R}Y$ and $H = \exp \mathbb{R}X$. In addition, since H is commutative, it is unimodular and the left and right regular representations necessarily coincide. Letting $\lambda = Z^*$, it is then easy to verify that

$$\beta_\lambda(t) = \begin{bmatrix} 1 & 0 \\ -t & 1 \end{bmatrix} \begin{bmatrix} 1 \\ 0 \end{bmatrix} = \begin{bmatrix} 1 \\ -t \end{bmatrix}.$$

Suppose next that $\pi(\exp \mathbb{Z}aX \exp \mathbb{Z}bY) f$ is a frame for $L^2(\mathbb{R})$. In local coordinates, $t \mapsto \beta_\lambda(t)$ is the linear function mapping every real number to its additive opposite, and it is not hard to verify that

$$H_{(Z^*, \exp \mathbb{Z}bY)} = \{\exp xX \in \mathbb{R} : -bkx \in \mathbb{Z} \text{ for all } k \in \mathbb{Z}\} = \exp\left(\mathbb{Z}\frac{X}{b}\right).$$

Therefore, for the Heisenberg group, $\mathcal{A}_{\mathbb{H}}$ is the set of all bounded and invertible operators generated by the unitary operators in $\pi_{Z^*}\left(\exp\left(\mathbb{Z}\frac{X}{b}\right)\right)$. For the general case, where $G = NH$, in principle, the closer the map $t \mapsto \beta_\lambda(t)$ is to being linear (for example, when β_λ is a polynomial function), the larger one should expect $H_{(\lambda,\Gamma_N)}$ to be.

2) Let us contrast the example of the *ax+b* group with that of the Heisenberg Lie group. To this end, let $\mathbb{A}$ be *ax+b* Lie group with Lie algebra spanned by

$$X = \begin{bmatrix} 0 & 1 \\ 0 & 0 \end{bmatrix}, \; A = \begin{bmatrix} 1 & 0 \\ 0 & 0 \end{bmatrix}$$

such that $[A, X] = X$. Here, $N = \exp \mathbb{R}X$ and $H = \exp \mathbb{R}A$. Since H is commutative, it is also unimodular. Letting $\lambda = X^*$, then $\beta_\lambda(t) = e^{-t}$and β_λ is highly nonlinear when compared to $t \mapsto -t$. We should not therefore (at least intuitively) expect $\mathcal{A}_{\mathbb{A}}$ to be small. Indeed, suppose next that $\pi\left(\exp r\mathbb{Z}A \exp \alpha\mathbb{Z}X\right)f$ is a frame for $L^2(\mathbb{R})$. Then, it is not hard to check that

$$H_{(X^*,\exp\alpha\mathbb{Z}X)} = \left\{\exp aA \in \mathbb{R} : k\alpha e^{-t} \cdot \left(e^{-a} - 1\right) \in \mathbb{Z} \text{ for all } k \in \mathbb{Z} \text{and } t \in \mathbb{R}\right\}.$$

Thus, $H_{(X^*,\exp\alpha\mathbb{Z}X)} = \{0\}$ and as such, $\mathcal{A}_{\mathbb{A}}$ is trivial. This is perhaps another subtle evidence for the stark disparities as observed in the literature between time-frequency analysis and wavelet theory. Since we are, however, primarily interested in exploiting geometric properties of Lie groups giving rise to frames, we will make no attempt, in this chapter to push these observations any further.

Theorem 7.3 Part (1) follows from a direct application of Propositions 7.1, 7.2, 7.3, 7.4 and Theorem 7.3 Part (2) is a byproduct of Lemma 7.1, and Theorems 7.1 and 7.2. follow from Theorem 7.3.

7.1.2 A Toy Example

We will present a toy example that captures the central ideas in the proof of Theorem 7.3. This example is meant to prepare the reader for the abstraction required for the general case.

Let $G = NH$ be a simply connected and connected Lie group with Lie algebra spanned by $\left\{X_1, X_2, A_1\right\}$ such that $N = \exp\left(\mathbb{R}X_1 + \mathbb{R}X_2\right)$ and $H = \exp \mathbb{R}A_1$ are commutative closed subgroups of G satisfying nontrivial Lie brackets: $\left[A_1, X_2\right] = X_1, \left[A_1, X_2\right] = X_2 + X_1$. Let A be a matrix representation of the endomorphism $ad\left(aA_1\right)$ restricted to the ideal spanned by X_1, X_2. Then

$$A = \begin{bmatrix} a & a \\ 0 & a \end{bmatrix} \text{ and } e^A = \begin{bmatrix} e^a & ae^a \\ 0 & e^a \end{bmatrix}. \tag{7.4}$$

In order to simplify presentation, we abuse notation by identifying G with the semidirect product group $\mathbb{R}^2 \rtimes \mathbb{R}$ with group operation

$$\left(x_1, x_2, a\right)\left(t_1, t_2, b\right) = \left(x_1 + e^a\left(t_1 + at_2\right), x_2 + e^a t_2, a + b\right).$$

As a result, G has the usual topology of $\mathbb{R}^3$ and we may realize the induced representation π_λ as acting in $L^2(\mathbb{R})$ as follows:

$$\left[\pi_\lambda\left(x_1, x_2, a\right)f\right](t) = \exp\left(2\pi i\left\langle\begin{bmatrix} e^{-t} \\ -te^{-t} \end{bmatrix}, \begin{bmatrix} x_1 \\ x_2 \end{bmatrix}\right\rangle\right) f(t-a). \tag{7.5}$$

Observe that

$$t \mapsto \begin{bmatrix} e^{-t} \\ -te^{-t} \end{bmatrix}$$

is a smooth injective function defined on $\mathbb{R}$.

For a generic vector f which is square-integrable on the real line, the orbit $\pi_\lambda(G)f$ of f is parametrized by the three-dimensional manifold $\pi_\lambda(G)$. Contrasting this example with that of the $ax+b$ Lie group (see (7.1), one may realize π_λ and (7.1), as distinct group representations acting in the same space $L^2(\mathbb{R})$). However, in terms of its dimensions, G is larger than the $ax+b$ Lie group. Since π_λ acts on the real line, in principle, two parameters should provide enough sampling sets for the existence of a nice frame. The fact that G is a three-parameter group coupled with π_λ being faithful, indicates that as a whole, π_λ provides more operators than needed. One should then look for a way to shave off the part of the group which is not useful. Letting,

$$S = \begin{bmatrix} 0 & 1 \\ 0 & 0 \end{bmatrix}, \quad A = \begin{bmatrix} 1 & 0 \\ 0 & 0 \end{bmatrix},$$

and

$$Z = \begin{bmatrix} 0 & 0 & 1 \\ 0 & 0 & 0 \\ 0 & 0 & 0 \end{bmatrix}, \quad Y = \begin{bmatrix} 0 & 0 & 0 \\ 0 & 0 & 1 \\ 0 & 0 & 0 \end{bmatrix}, \quad X = \begin{bmatrix} 0 & 1 & 0 \\ 0 & 0 & 0 \\ 0 & 0 & 0 \end{bmatrix}$$

and

$$X_1 = \begin{bmatrix} 0 & 0 & 1 \\ 0 & 0 & 0 \\ 0 & 0 & 0 \end{bmatrix}, \quad X_2 = \begin{bmatrix} 0 & 0 & 0 \\ 0 & 0 & 1 \\ 0 & 0 & 0 \end{bmatrix}, \quad X_3 = \begin{bmatrix} 1 & 1 & 0 \\ 0 & 1 & 0 \\ 0 & 0 & 0 \end{bmatrix},$$

we summarize the observations above in this table.

	Lie algebra	**Sampling manifold**	**Not useful part**	**Sampling set**
$ax+b$	S, A	$\pi(\exp \mathbb{R}A \exp \mathbb{R}S)$	None	$\pi(\exp \mathbb{Z}aA \exp \mathbb{Z}bS)$
Heisenberg	Z, Y, X	$\pi(\exp \mathbb{R}Y \exp \mathbb{R}X)$	$\exp \mathbb{R}Z$	$\pi(\exp \mathbb{Z}aX \exp \mathbb{Z}bY)$
G	X_1, X_2, X_3	To be determined	To be determined	To be determined

(7.6)

The following graphics illustrate a manifold parametrizing an orbit together with a sampling set for the $ax+b$ and Heisenberg Lie groups (Figure 7.3).

We will now attempt to turn these intuitive ideas into mathematically precise statements that can be substantiated. As a starter (Figure 7.4), let

$$P = \begin{bmatrix} 0 & 0 \\ 0 & 1 \end{bmatrix}$$

be a rank-1 projection. Considering the action of the normal part, restricted to the range of P, one observes that

$$\left\langle \begin{bmatrix} e^{-t} \\ -te^{-t} \end{bmatrix}, \begin{bmatrix} 0 & 0 \\ 0 & 1 \end{bmatrix} \begin{bmatrix} x_1 \\ x_2 \end{bmatrix} \right\rangle = \left\langle \begin{bmatrix} e^{-t} \\ -te^{-t} \end{bmatrix}, \begin{bmatrix} 0 \\ x_2 \end{bmatrix} \right\rangle = -te^{-t}x_2.$$

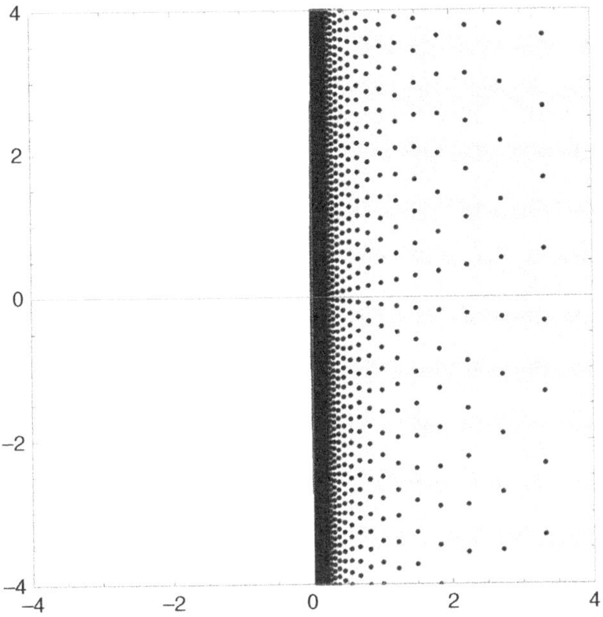

Figure 7.3 Orbit and sampling set on the $ax+b$ group.

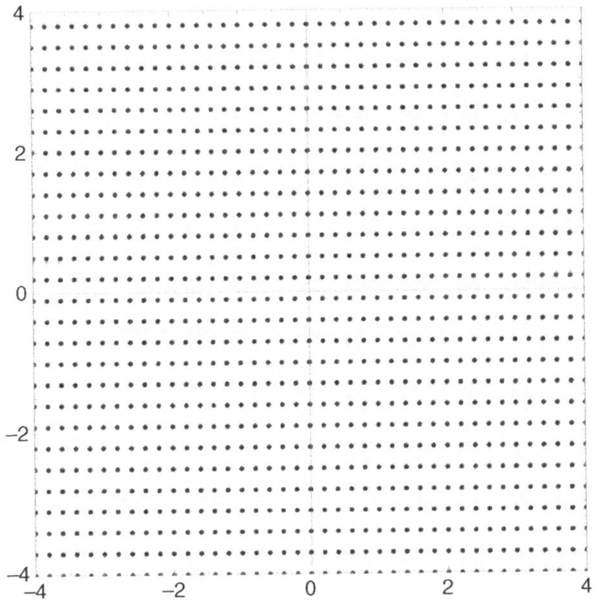

Figure 7.4 Orbit and sampling set on the Heisenberg group.

Put simply, $\left[\pi_\lambda\left(0,x_2,a\right)f\right](t) = e^{-2\pi\left(te^{-t}x_2\right)i}f(t-a)$. So even though the collection of operators $\pi_\lambda\left(0,x_2,a\right)$ does not have a group structure, at this point, we have managed to reduce the dimension of the set parametrizing the orbit of f from 3 to 2. (So how indispensable is an underlying group structure for the existence of nice frames?) It remains however, to check if it is possible at all, to construct a nice frame by taking a sample inside the family of operators $\pi_\lambda\left(0,x_2,a\right)$.

Let $\beta(t) = -te^{-t}$. Then β is smooth and noninjective as depicted below

However, since $\beta'(t) = e^{-t}(t-1) \Rightarrow \beta'(0) = -1$, β must be an immersion at zero. In other words, β is locally injective at zero, and letting β be the restriction of $t \mapsto -te^{-t}$ to $\left(-\frac{1}{2},\frac{1}{2}\right)$, one easily verifies that β is an injective function (Figure 7.5).

Next, let f and h be two smooth functions defined on $\mathcal{O} = \left(-\frac{1}{2},\frac{1}{2}\right)$. Put $E = \left(-\frac{1}{2\sqrt{e}},\frac{\sqrt{e}}{2}\right)$. Then

$$\left\langle h,\pi_\lambda\left(0,x_2,0\right)f\right\rangle_{L^2(\mathcal{O})} = \int_{\mathcal{O}}\left(g(t)\cdot\overline{f(t)}\right)\cdot e^{2\pi i\beta(t)x_2}dt$$

and the change of variable $\xi = \beta(t)$ gives

$$\left\langle h,\pi_\lambda\left(0,x_2,0\right)f\right\rangle_{L^2(\mathcal{O})} = \int_{E}\left(g\left(\beta^{-1}(\xi)\right)\cdot\overline{f\left(\beta^{-1}(\xi)\right)}\right)e^{2\pi i\xi x_2}d\left(\beta^{-1}(\xi)\right).$$

Since

$$d\left(\beta^{-1}(\xi)\right) = \left|\frac{\beta^{-1}(\xi)}{\xi+\xi\beta^{-1}(\xi)}\right|d\xi$$

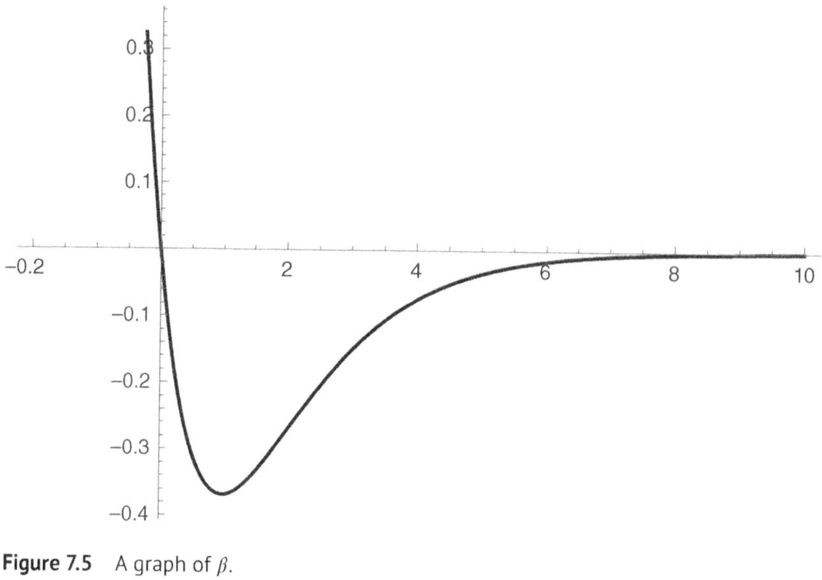

Figure 7.5 A graph of β.

define $W(\xi) = \left|\frac{\beta^{-1}(\xi)}{\xi+\xi\beta^{-1}(\xi)}\right|$. Then $W(\xi)$ is smooth and positive on E and additionally, $W(0) = 1$. Moreover, the matrix coefficient $\left\langle h, \pi_\lambda(0, x_2, 0) f\right\rangle_{L^2(\mathcal{O})}$ is the Fourier transform of

$$\xi \mapsto W(\xi) \cdot \left(g\left(\beta^{-1}(\xi)\right) \cdot \overline{f\left(\beta^{-1}(\xi)\right)}\right) \cdot 1_E(\xi)$$

evaluated at x_2. In other words,

$$\left\langle h, \pi_\lambda(0, x_2, 0) f\right\rangle_{L^2(\mathcal{O})} = \int_E \left(W(\xi) \cdot g\left(\beta^{-1}(\xi)\right) \cdot \overline{f\left(\beta^{-1}(\xi)\right)}\right) e^{2\pi i \xi x_2} d\xi.$$

Since E is an interval of length $\rho = \frac{1+e}{2\sqrt{e}}$, E is up to a null set a fundamental domain of the lattice $\frac{1+e}{2\sqrt{e}}\mathbb{Z} = \mathcal{L}$. Thus, the trigonometric system $\left\{\xi \mapsto \frac{e^{2\pi i \xi k}}{\rho^{1/2}} : k \in \mathcal{L}\right\}$ is an orthonormal basis for $L^2(E, d\xi)$. Then

$$\begin{aligned}\sum_{x_2\in\mathcal{L}} &\left|\left\langle h, \pi_\lambda(0, x_2, 0) f\right\rangle_{L^2(\mathcal{O})}\right|^2 \\ &= \rho \sum_{x_2\in\mathcal{L}} \left|\int_E \left(W(\xi) \cdot g\left(\beta^{-1}(\xi)\right) \cdot \overline{f\left(\beta^{-1}(\xi)\right)}\right) \frac{e^{2\pi i \xi x_2}}{\rho^{1/2}} d\xi\right|^2\end{aligned}$$

and as such,

$$\begin{aligned}\sum_{x_2\in\mathcal{L}} &\left|\left\langle h, \pi_\lambda(0, x_2, 0) f\right\rangle_{L^2(\mathcal{O})}\right|^2 \\ &= \rho \int_E \left|\left(g\left(\beta^{-1}(\xi)\right) \cdot \sqrt{W(\xi)} f\left(\beta^{-1}(\xi)\right)\right)\right|^2 W(\xi)\, d\xi.\end{aligned}$$

Suppose now that $\sqrt{W(\xi)} f\left(\beta^{-1}(\xi)\right) = 1_E(\xi)$. This is the case if and only if

$$f(t) = \left|\frac{e^t}{t+1}\right|^{-1/2} \cdot 1_{\mathcal{O}}(t).$$

Therefore

$$\sum_{x_2\in\mathcal{L}} \left|\left\langle h, \pi_\lambda(0, x_2, 0) f\right\rangle_{L^2(\mathcal{O})}\right|^2 = \rho \int_E \left|g\left(\beta^{-1}(\xi)\right)\right|^2 W(\xi)\, d\xi$$

and changing variables back to t,

$$\sum_{x_2\in\mathcal{L}} \left|\left\langle h, \pi_\lambda(0, x_2, 0) f\right\rangle_{L^2(\mathcal{O})}\right|^2 = \rho \int_{\mathcal{O}} |h(t)|^2 dt.$$

Since $\left\{\left[\frac{1}{2} - a, \frac{1}{2} - a\right) : a \in \mathbb{Z}\right\}$ is a tiling of the real line,

$$\sum_{x_2\in\mathcal{L}} \left|\left\langle \pi_\lambda(0, 0, -a) h, \pi_\lambda(0, x_2, 0) f\right\rangle_{L^2(\mathcal{O})}\right|^2 = \rho \int_{\mathcal{O}} |h(t+a)|^2 dt$$

and

$$\sum_{a\in\mathbb{Z}}\sum_{x_2\in\mathcal{L}}\left|\left\langle h,\pi_\lambda(0,0,a)\,\pi_\lambda\left(0,x_2,0\right)f\right\rangle_{L^2(\mathcal{O})}\right|^2=\rho\|h\|^2.$$

Since the set of all smooth functions which are compactly supported is dense in $L^2(\mathbb{R})$, we conclude that $\pi_\lambda\left(\exp\left(\mathbb{Z}A_1\right)\exp(\mathcal{L}\mathbb{Z})\right)\rho^{-1/2}f$ is a nice Parseval frame for $L^2(\mathbb{R})$.

We are now ready to complete Equation 7.6:

	Lie algebra	Sampling manifold	Not useful part	Sampling set
G	X_1, X_2, X_3	$\exp\mathbb{R}X_3\exp\mathbb{R}X_2$	$\exp\mathbb{R}X_1$	$\exp\mathbb{Z}aX_3\exp\mathbb{Z}bX_2$
		$\exp\mathbb{R}X_3\exp\mathbb{R}X_1$	$\exp\mathbb{R}X_2$	$\exp\mathbb{Z}aX_3\exp\mathbb{Z}bX_1$

Additionally, the following illustrates $\exp\mathbb{Z}aX_3\exp\mathbb{Z}bX_2$ as a subset of $\exp\mathbb{R}X_3\exp\mathbb{R}X_2$ regarded as a closed submanifold of G

We conclude by noting that the sampling manifolds $\exp\mathbb{R}X_3\exp\mathbb{R}X_2$ and $\exp\mathbb{R}X_3\exp\mathbb{R}X_1$ do not form an exhaustive list. On the contrary, with very little creativity, other sampling manifolds can easily be carved out of G.

7.1.3 Proofs of Main Results

Let

$$\beta_\lambda : H\to\beta_\lambda(H)\subseteq\mathfrak{n}^* \tag{7.7}$$

such that

$$\beta_\lambda(h)=Ad\left(h^{-1}\right)^*\lambda. \tag{7.8}$$

By assumption, β_λ is smooth and defines an immersion at the neutral element in H. In other words, the differential of β_λ at the identity of the Lie group H is an injective linear operator taking $\mathfrak{h}$ into $\mathfrak{n}^*$. Thus, $n=\dim N$ is greater than or equal to $r=\dim H$. Note also that no assumption is made on the irreducibility of π. Letting $r=\dim(H)$, there is a smooth chart

$$(\mathcal{O},\varphi) \tag{7.9}$$

centered at the identity in H such that $\varphi:\mathcal{O}\to\varphi(\mathcal{O})\subseteq\mathfrak{h}$ is a diffeomorphism taking the identity in H to the zero element in its Lie algebra, and $\beta_\lambda\circ\varphi^{-1}$ has a constant rank equal to r. Since the exponential map defines a local diffeomorphism at the neutral element of the Lie algebra $\mathfrak{h}$, it is clear that φ can be taken to be the logarithmic function, and in local coordinates. Let D be the differential of $\beta_\lambda\circ\varphi^{-1}$ at the zero element in the Lie algebra and put $\varphi=\log$.

Proposition 7.1 The following holds true

1) If $s, h \in H$, then $\beta_\lambda(sh) = \beta_{\beta_\lambda(h)}(s)$.
2) β_λ is an immersion at the identity of H if and only if $\det\left(D^T D\right)$ is nonzero.
3) Let $\omega \in \mathfrak{n}^*$ and let $\{X_1, X_2, \ldots, X_n\}$, $\{A_1, A_2, \ldots, A_r\}$ be a basis for $\mathfrak{n}$, $\mathfrak{h}$, respectively. Then

$$D_\omega = \begin{bmatrix} \langle \omega, [X_1, A_1] \rangle & \cdots & \langle \omega, [X_1, A_r] \rangle \\ \vdots & \ddots & \vdots \\ \langle \omega, [X_n, A_1] \rangle & \cdots & \langle \omega, [X_n, A_r] \rangle \end{bmatrix}$$

is a matrix representation of the differential of $\beta_\omega \circ \varphi^{-1}$ at 0 with respect to the fixed bases.

Proof: **Proof of Part (1):** Since Ad^* is a representation of G acting on $\mathfrak{n}^*$,

$$\beta_\lambda(sh) = Ad\left(h^{-1}s^{-1}\right)^* \lambda = \left[Ad\left(h^{-1}\right) Ad\left(s^{-1}\right)\right]^* \lambda.$$

Computing the adjoint of the product $Ad\left(h^{-1}\right) Ad\left(s^{-1}\right)$, $\beta_\lambda(sh) = \beta_{Ad(h^{-1})\lambda}(s) = \beta_{\beta_\lambda(h)}(s)$ as desired.

Proof of Part (2): Recall that a linear operator is injective if and only if its Grammian matrix is invertible. Therefore, β_λ is an immersion at the identity of H if and only if the null space of D is trivial.

Proof of Part (3): For arbitrary $t \in \mathbb{R}, A \in \mathfrak{h}$ and $X \in \mathfrak{n}$, computing the differential of $t \mapsto \left[Ad(\exp(-tA))\right]^* \omega$ at zero, we obtain

$$\left\langle \lim_{t \to 0} \frac{\left[Ad(\exp(-tA))\right]^* \omega - \omega}{t}, X \right\rangle = \lim_{t \to 0} \frac{\left\langle \omega, \sum_{k=1}^{\infty} \frac{ad(-tA)^k}{k!} X \right\rangle}{t}.$$

Next, we split $\sum_{k=1}^{\infty} \frac{ad(-tA)^k}{k!} X$ into two terms:

$$\sum_{k=1}^{\infty} \frac{ad(-tA)^k}{k!} X = ad(-tA)X + \sum_{k=2}^{\infty} \frac{ad(-tA)^k}{k!} X.$$

Since $\lim_{t \to 0} \sum_{k=2}^{\infty} \frac{t^{k-1} ad(-A)^k}{k!} X = 0$,

$$\left\langle \lim_{t \to 0} \frac{\left[Ad(\exp(-tA))\right]^* \omega - \omega}{t}, X \right\rangle = -\langle \omega, [A, X] \rangle = \langle \omega, [X, A] \rangle.$$

Since $\beta_\lambda \circ \exp$ is assumed to be an immersion at the identity, its differential $\left(\beta_\lambda \circ \exp\right)_{*_0}$ at the zero element in the Lie algebra is an injective map. For a fixed subset J of cardinality $\dim H = r$, let $\left[\left(\beta_\lambda \circ \exp\right)_{*_0}\right]_J$ be the submatrix obtained by retaining all rows of the matrix representation of $(\beta_\lambda \circ \exp)_{*_0}$ parametrized by the set J. In other words, suppose that in matrix form $(\beta_\lambda \circ \exp)_{*_0}$ is given by

$$\left[(\beta_\lambda \circ \exp)_{*_0}\right] = \begin{bmatrix} \mathrm{row}_1 \left[(\beta_\lambda \circ \exp)_{*_0}\right] \\ \vdots \\ \mathrm{row}_{\dim N} \left[(\beta_\lambda \circ \exp)_{*_0}\right] \end{bmatrix}.$$

Next, let J be a subset of $\{1, \dots, r\}$ such that $J = \{j_1, \cdots, j_r\}$. Then

$$\left[(\beta_\lambda \circ \exp)_{*_0}\right]_J = \begin{bmatrix} \mathrm{row}_{j_1} \left[(\beta_\lambda \circ \exp)_{*_0}\right] \\ \vdots \\ \mathrm{row}_{j_r} \left[(\beta_\lambda \circ \exp)_{*_0}\right] \end{bmatrix}.$$

By the inverse function theorem (Theorem 2.2), there is a linear projection

$$P = P_J \, : \, \mathfrak{n} \to \mathfrak{n} \tag{7.10}$$

of rank r (depending on the set J) given by

$$P\left(X_k\right) = \begin{cases} X_k & \text{if } k \in J \\ 0 & \text{if } k \notin J \end{cases} \tag{7.11}$$

such that

$$\Theta_{J,\lambda} = \Theta_\lambda = P^* \beta_\lambda \exp \, : \, \log(\mathcal{O}) \to \Theta_\lambda\left(\log(\mathcal{O})\right)$$

is a local diffeomorphism at the zero element in $\mathfrak{h}$.

Example 7.1 Let $G = \mathbb{R}^4 \rtimes H$ and suppose that $\mathfrak{h}$ is spanned by

$$A_1 = \begin{bmatrix} 1 & -1 & 0 & 0 \\ 1 & 1 & 0 & 0 \\ 0 & 0 & 0 & 0 \\ 0 & 0 & 0 & 1 \end{bmatrix}, \; A_2 = \begin{bmatrix} 0 & 0 & 0 & 0 \\ 0 & 0 & 0 & 0 \\ 0 & 0 & 0 & 1 \\ 0 & 0 & 0 & 0 \end{bmatrix}, \; A_3 = \begin{bmatrix} 0 & 0 & 0 & 0 \\ 0 & 0 & 0 & 0 \\ 0 & 0 & -1 & 0 \\ 0 & 0 & 0 & 1 \end{bmatrix}$$

with nontrivial Lie brackets given by $[A_1, A_2] = -A_2$, $[A_2, A_3] = 2A_2$. Then H is a connected solvable Lie group and every element in H can be uniquely factored as follows:

$$\begin{bmatrix} 1 & 0 & 0 & 0 \\ 0 & 1 & 0 & 0 \\ 0 & 0 & 1 & x \\ 0 & 0 & 0 & 1 \end{bmatrix} \begin{bmatrix} 1 & 0 & 0 & 0 \\ 0 & 1 & 0 & 0 \\ 0 & 0 & e^{-y} & 0 \\ 0 & 0 & 0 & e^{y} \end{bmatrix} \begin{bmatrix} e^a \cos a & -e^a \sin a & 0 & 0 \\ e^a \sin a & e^a \cos a & 0 & 0 \\ 0 & 0 & 1 & 0 \\ 0 & 0 & 0 & e^a \end{bmatrix}.$$

Given a linear functional $\lambda \in (\mathbb{R}^4)^*$, the Jacobian of β_λ at the neutral element of H with respect to the coordinates described above is given by

$$D = (\beta_\lambda \circ \exp)_{*_0} = \begin{bmatrix} -\lambda_1 - \lambda_2 & 0 & 0 \\ \lambda_1 - \lambda_2 & 0 & 0 \\ 0 & 0 & \lambda_3 \\ \lambda_4 & -\lambda_3 & -\lambda_4 \end{bmatrix}.$$

Thus, D has full rank if and only if the determinant of its Grammian $D^T D$ is nonzero. Since

$$\det\left(D^T D\right) = 2\lambda_3^4\left(\lambda_1^2 + \lambda_2^2\right) \neq 0,$$

it follows that as long as the coordinates of the linear functional λ satisfy the condition $2\lambda_3^4\left(\lambda_1^2 + \lambda_2^2\right) \neq 0$, the map β_λ defines an immersion at the identity of H. The table below summarizes which submatrices of D are invertible

J	*Submatrix*	*Determinant of submatrix*
$\{1,2,3\}$	$\begin{bmatrix} -\lambda_1 - \lambda_2 & 0 & 0 \\ \lambda_1 - \lambda_2 & 0 & 0 \\ 0 & 0 & \lambda_3 \end{bmatrix}$	0
$\{1,2,4\}$	$\begin{bmatrix} -\lambda_1 - \lambda_2 & 0 & 0 \\ \lambda_1 - \lambda_2 & 0 & 0 \\ \lambda_4 & -\lambda_3 & -\lambda_4 \end{bmatrix}$	0
$\{1,3,4\}$	$\begin{bmatrix} -\lambda_1 - \lambda_2 & 0 & 0 \\ 0 & 0 & \lambda_3 \\ \lambda_4 & -\lambda_3 & -\lambda_4 \end{bmatrix}$	$-\lambda_3^2\left(\lambda_1 + \lambda_2\right)$
$\{2,3,4\}$	$\begin{bmatrix} \lambda_1 - \lambda_2 & 0 & 0 \\ 0 & 0 & \lambda_3 \\ \lambda_4 & -\lambda_3 & -\lambda_4 \end{bmatrix}$	$\lambda_3^2\left(\lambda_1 - \lambda_2\right)$

Example 7.2 Given Lie algebras

$$\mathfrak{n} = \left\{ \begin{bmatrix} 0 & 0 & 0 & x_1 \\ 0 & 0 & 0 & x_2 \\ 0 & 0 & 0 & x_3 \\ 0 & 0 & 0 & 0 \end{bmatrix} : x_1, x_2, x_3 \in \mathbb{R} \right\} \text{ and } \mathfrak{h} = \left\{ \begin{bmatrix} a & -a & 0 & 0 \\ a & a & 0 & 0 \\ 0 & 0 & a & 0 \\ 0 & 0 & 0 & 0 \end{bmatrix} : a \in \mathbb{R} \right\},$$

let G be a simply connected four-dimensional Lie group with Lie algebra $\mathfrak{n} \oplus \mathfrak{h}$. Then G is a connected solvable Lie group, and every element in $G = NH$ can be uniquely factored as follows:

$$\begin{bmatrix} 1 & 0 & 0 & x_1 \\ 0 & 1 & 0 & x_2 \\ 0 & 0 & 1 & x_3 \\ 0 & 0 & 0 & 1 \end{bmatrix} \begin{bmatrix} e^a \cos a & -e^a \sin a & 0 & 0 \\ e^a \sin a & e^a \cos a & 0 & 0 \\ 0 & 0 & e^a & 0 \\ 0 & 0 & 0 & 1 \end{bmatrix}.$$

Moreover, given a linear functional $\lambda = \lambda_1 X_1^* + \lambda_2 X_2^* + \lambda_3 X_3^* \in \mathfrak{n}$, it is not hard to verify that

$$\beta_\lambda (a) = \left(e^{-a} \left(\lambda_1 \cos a - \lambda_2 \sin a\right), e^{-a} \left(\lambda_1 \cos a + \lambda_2 \sin a\right), e^{-a} \lambda_3, \lambda_4\right)$$

and

$$\left[\left(\beta_\lambda \circ \exp\right)_{*_0}\right] = \begin{bmatrix} -\lambda_1 - \lambda_2 \\ \lambda_2 - \lambda_1 \\ -\lambda_3 \\ 0 \end{bmatrix}.$$

Furthermore, $\left\|\left[\left(\beta_\lambda \circ \exp\right)_{*_0}\right]\right\|^2 = 2\left(\lambda_1^2 + \lambda_2^2\right) + \lambda_3^2$, and as long as $2\left(\lambda_1^2 + \lambda_2^2\right) + \lambda_3^2$ is nonzero, then β_λ is an immersion at the identity of H. There are many possible choices for J. One may fix J to be equal to any of the following three singletons: $\{1\}, \{2\}, \{3\}$.

From here on, fix a set J as described in Proposition 7.3. Next, let $\mathcal{O}$ be a relatively compact subset around the identity element in H such that the restriction

$$\Theta_\lambda|_{\log(\mathcal{O})} : \log(\mathcal{O}) \to \Theta_\lambda\left(\log(\mathcal{O})\right)$$

of Θ_λ to $\log(\mathcal{O})$ is a diffeomorphism. As a result, $\Theta_\lambda\left(\log(\mathcal{O})\right)$ is a relatively compact subset of $P^*(\mathfrak{n}^*)$.

Example 7.3 Let H be a matrix group acting by automorphisms on a simply connected commutative group N equipped with the structure of a vector space, and let G be the corresponding semidirect product group. The table below shows that Proposition 7.1 is quite useful, since it provides simple necessary and sufficient condition for the immersion of β_λ. Let

$$\mathfrak{h}_1 = \mathbb{R}A_1 + \mathbb{R}A_2 + \mathbb{R}A_3,\ A_1 = \begin{bmatrix} 0 & 0 & 1 \\ 0 & 0 & 0 \\ 0 & 0 & 0 \end{bmatrix},\ A_2 = \begin{bmatrix} 0 & 0 & 0 \\ 0 & 0 & 1 \\ 0 & 0 & 0 \end{bmatrix},\ A_3 = \begin{bmatrix} 0 & 1 & 0 \\ 0 & 0 & 0 \\ 0 & 0 & 0 \end{bmatrix}$$

$$\mathfrak{h}_2 = \mathbb{R}A_1 + \mathbb{R}A_2 + \mathbb{R}A_3,\ A_1 = \begin{bmatrix} 1 & 0 & 0 \\ 0 & 1 & 0 \\ 0 & 0 & 1 \end{bmatrix},\ A_2 = \begin{bmatrix} 0 & 1 & 0 \\ 0 & 0 & 1 \\ 0 & 0 & 0 \end{bmatrix},\ A_3 = \begin{bmatrix} 0 & 0 & 1 \\ 0 & 0 & 0 \\ 0 & 0 & 0 \end{bmatrix}$$

and

$$\mathfrak{h}_3 = \mathbb{R}A_1 + \mathbb{R}A_2 \quad \text{with } A_1 = \begin{bmatrix} 0 & 1 \\ 0 & 0 \end{bmatrix},\ A_2 = \begin{bmatrix} 1 & 0 \\ 0 & 0 \end{bmatrix}.$$

One easily verifies the following:

	$-D_\lambda$	$D_\lambda^T D_\lambda$	$\det\left(D_\lambda^T D_\lambda\right)$
$\mathfrak{h}_1$	$\begin{bmatrix} 0 & 0 & 0 \\ 0 & 0 & \lambda_1 \\ \lambda_1 & \lambda_2 & 0 \end{bmatrix}$	$\begin{bmatrix} \lambda_1^2 & \lambda_1\lambda_2 & 0 \\ \lambda_1\lambda_2 & \lambda_2^2 & 0 \\ 0 & 0 & \lambda_1^2 \end{bmatrix}$	0
$\mathfrak{h}_2$	$\begin{bmatrix} \lambda_1 & 0 & 0 \\ \lambda_2 & \lambda_1 & 0 \\ \lambda_3 & \lambda_2 & \lambda_1 \end{bmatrix}$	$\begin{bmatrix} \lambda_1^2+\lambda_2^2+\lambda_3^2 & \lambda_2\left(\lambda_1+\lambda_3\right) & \lambda_1\lambda_3 \\ \lambda_2\left(\lambda_1+\lambda_3\right) & \lambda_1^2+\lambda_2^2 & \lambda_1\lambda_2 \\ \lambda_1\lambda_3 & \lambda_1\lambda_2 & \lambda_1^2 \end{bmatrix}$	λ_1^6
$\mathfrak{h}_3$	$\begin{bmatrix} 0 & \lambda_1 \\ \lambda_1 & 0 \end{bmatrix}$	$\begin{bmatrix} \lambda_1^2 & 0 \\ 0 & \lambda_1^2 \end{bmatrix}$	λ_1^4

Remark 7.2 Suppose that $H = SL(2,\mathbb{R})$ is the special linear acting by automorphism on a simply connected commutative group N equipped with the structure of a two-dimensional vector space. Let G be the semidirect product group associated with this action of the Special Linear Group on $\mathbb{R}^2$. Fixing a vector $\lambda \in \mathbb{R}^2$, define $\beta_\lambda(h) = h^{-T}\lambda$. Since $\dim H = 3 > 2$, β_λ is nowhere an immersion, and the methods developed in this chapter do not apply (although we will later see that the action of π_λ restricted to H can be discretized to construct frames which unlike the frames constructed here are quite pathological. They are not "nice" frames!)

Proposition 7.2 offers a general necessary and sufficient condition under which β_λ is an immersion.

Proposition 7.2 β_λ is an immersion at the identity of H if and only if

$$H_\lambda = \left\{h \in H : Ad\left(h^{-1}\right)^* \lambda = \lambda\right\}$$

is a discrete subgroup of H.

Proof: On the one hand, suppose that $h \mapsto Ad\left(h^{-1}\right)^{*}\lambda$ is an immersion at the identity of H. Then $r = \dim\left(H/H_{\lambda}\right) = \dim \beta_{\lambda}(H) = \dim H$. As a result, H_{λ} must be a discrete subgroup of H. Suppose on the other hand that H_{λ} is a discrete subgroup of H. Then there exists an open set U containing the identity of H such that $H_{\lambda} \cap U$ is a singleton. Finally, the restriction of β_{λ} to U, is an immersion at the identity of H.

Let Γ_H, Γ_N be a countable subset of H, N, respectively. For arbitrary vectors g and f in $L^2\left(H, d\mu_H\right)$, define

$$s\left(\Gamma_H, \Gamma_N, g, f\right) = \sum_{\ell \in \Gamma_H} \sum_{\exp PX \in \Gamma_N} \left|\left\langle g, \pi\left(\ell^{-1}\right)\pi\left(\exp(PX)\right)f\right\rangle_{\mathfrak{H}}\right|^2. \tag{7.12}$$

The sequence $\left(\left\langle g, \pi\left(\ell^{-1}\right)\pi\left(\exp(PX)\right)f\right\rangle_{\mathfrak{H}}\right)_{(\ell, \exp PX) \in \Gamma_H \times \Gamma_N}$ corresponds to a sampling of a matrix coefficient of π and the sum above is just a formal l^2-norm associated with this sequence.

Let f be a continuous function on H which is also supported in $\mathcal{O}$. For the time being, glossing over all convergence issues related to the definition of the infinite sum $s\left(\Gamma_H, \Gamma_N, g, f\right)$, since for $X \in \mathfrak{n}$ and $\ell \in H$,

$$\left[\pi\left(\exp(X)\right)\right]f(h) = e^{2\pi i\left\langle Ad\left(h^{-1}\right)^{*}\lambda, X\right\rangle} \cdot f(h) \text{ and } \left[\pi(\ell)\right]f(h) = f\left(\ell^{-1}h\right),$$

it is clear (at least formally) that $I_{g,f,\ell,X} = \left\langle g, \pi\left(\ell^{-1}\right)\pi\left(\exp(PX)\right)f\right\rangle_{\mathfrak{H}}$. Since

$$\left\langle g, \pi\left(\ell^{-1}\right)\pi\left(\exp(PX)\right)f\right\rangle_{\mathfrak{H}} = \left\langle \pi(\ell)g, \pi\left(\exp(PX)\right)f\right\rangle_{\mathfrak{H}},$$

$$I_{g,f,\ell,X} = \int_{\mathcal{O}} \left(g\left(\ell^{-1}h\right) \cdot \overline{f(h)}\right) \cdot e^{-2\pi i\left\langle Ad\left(h^{-1}\right)^{*}\lambda, PX\right\rangle} d\mu_H(h).$$

We have now arrived at a place where the assumption that $h \mapsto Ad\left(h^{-1}\right)^{*}\lambda$ is a local immersion is crucial. Since a local immersion is necessarily locally invertible and the inverse of $h \mapsto Ad\left(h^{-1}\right)^{*}\lambda$ defines a differentiable function between a suitable open subset of H and its image, the change of variable $h = \exp(A)$ allows to view the formal integral $I_{g,f,\ell,X}$ as a family of Euclidean Fourier transforms. Armed with this interpretation of $I_{g,f,\ell,X}$, we will be able to appeal to powerful methods of Euclidean Fourier analysis (such as Plancherel's equality) for a more transparent description of $I_{g,f,\ell,X}$.

Letting $h = \exp(A)$, and $Q_{\ell}(A) = g\left(\ell^{-1}\exp A\right) \cdot \overline{f(\exp A)}$,

$$I_{g,f,\ell,X} = \int_{\varphi(\mathcal{O})} Q_{\ell} \cdot e^{-2\pi i\left\langle \beta_{\lambda}(\exp A), PX\right\rangle} d\mu_H(\exp A).$$

Note that P is assumed to be a linear projection, and as such, its adjoint acts on the dual vector space $\mathfrak{n}^{*}$. Additionally, since $P = P^2 = P^{*}P$,

$$\left\langle \beta_{\lambda}(\exp A), PX\right\rangle = \left\langle P^{*}\beta_{\lambda}(\exp A), PX\right\rangle.$$

Recalling that $\Theta_\lambda = P^* \beta_\lambda \exp : \log(\mathcal{O}) \to \Theta_\lambda(\log(\mathcal{O}))$,

$$I_{g,f,\ell,X} = \int_{\varphi(\mathcal{O})} Q_\ell(A) \cdot e^{-2\pi i \langle \Theta_\lambda(A), PX \rangle} d\mu_H(\exp A).$$

The change of variable $\xi = \Theta_\lambda(A)$ gives

$$I_{g,f,\ell,X} = \int_{\Theta_\lambda(\varphi(\mathcal{O}))} Q_\ell(A) \cdot e^{-2\pi i \langle \xi, PX \rangle} d\mu_H\left(\exp\left(\Theta_\lambda^{-1}(\xi)\right)\right).$$

Put

$$F_{g,f}(\xi) = g\left(\ell^{-1} \exp\left(\Theta_\lambda^{-1}(\xi)\right)\right) \cdot \overline{f\left(\exp\left(\Theta_\lambda^{-1}(\xi)\right)\right)}.$$

Then

$$I_{g,f,\ell,X} = \int_{\Theta_\lambda(\log(\mathcal{O}))} F_{g,f}(\xi) \cdot e^{-2\pi i \langle \xi, PX \rangle} d\mu_H\left(\exp\left(\Theta_\lambda^{-1}(\xi)\right)\right).$$

Define $W_\lambda : \Theta_\lambda(\log(\mathcal{O})) \to \mathbb{R}$ such that

$$W_\lambda(\xi) = \frac{d\mu_H\left(\exp\left(\Theta_\lambda^{-1}(\xi)\right)\right)}{d\xi}$$

is the Radon–Nikodym derivative of the Haar measure on H with respect to the Lebesgue measure on $\Theta_\lambda(\varphi(\mathcal{O}))$.

Proposition 7.3 The following holds true.

1) If s is a continuous compactly supported function on $\mathcal{O}$, then $h \mapsto s(h) \sqrt{W_\lambda\left(\Theta_\lambda\left(\log(h)\right)\right)}$ is also continuous and compactly supported on $\mathcal{O}$.
2) For g and f in $L^2\left(H, d\mu_H\right)$ such that f is a continuous function compactly supported in $\mathcal{O}$,

$$s\left(\Gamma_H, \Gamma_N, g, f\right) = \sum_{\ell \in \Gamma_H} \sum_{PX \in \log \Gamma_N} \left| \int_{\Theta_\lambda(\varphi(\mathcal{O}))} \left[F_{g,f}(\xi) \cdot W_\lambda(\xi)\right] \cdot e^{-2\pi i \langle \xi, PX \rangle} d\xi \right|^2.$$

3) Let f be a nonzero positive function that is compactly supported on H. Then there exists a countable set $\Gamma_H \subset H$ such that $A \leq \sum_{\gamma \in \Gamma_H} f(\gamma x) \leq B$ hold for all $x \in H$ with $0 < A \leq B < \infty$ independent of x.

Proof: **Proof of Part (1)**: We show that W_λ is a strictly positive function defined on $\Theta_\lambda\left(\log(\mathcal{O})\right)$. To this end, let u be a positive function which is compactly supported on $\mathcal{O}$. For arbitrary $h \in \mathcal{O}$, let $A \in \log(\mathcal{O}) \subset \mathfrak{h}$ such that $h = \exp(A)$. Then

$$\int_{\mathcal{O}} u(h)\, d\mu_H(h) = \int_{\varphi(\mathcal{O})} u(\exp(A)) \cdot d\mu_H(\exp(A)).$$

Letting $A = \Theta_\lambda^{-1}(\xi)$,

$$\int_{\mathcal{O}} u(h)\, d\mu_H(h) = \int_{\Theta_\lambda(\log(\mathcal{O}))} u\left(\exp\left(\Theta_\lambda^{-1}(\xi)\right)\right) d\mu_H\left(\exp\left(\Theta_\lambda^{-1}(\xi)\right)\right).$$

Since Θ_λ is a local diffeomorphism, $d\mu_H\left(\exp\left(\Theta_\lambda^{-1}(\xi)\right)\right) = W_\lambda(\xi)\,d\xi$ and additionally W_λ is a strictly positive and continuous function on the set $\Theta_\lambda\left(\log(\mathcal{O})\right)$. As a result,

$$h \mapsto s(h) \cdot \sqrt{W_\lambda\left(\Theta_\lambda\left(\log(h)\right)\right)}$$

is also continuous and supported on $\mathcal{O}$.

Proof of Part (2): Part (2) is a straightforward application of Part (1).

Proof of Part (3): We start by fixing a set $\Gamma \subset H$ which is U-separated, for some open set U around the neutral element of H, select an open and symmetric set V such that $V^2 \subset U$. Recall that a countable set $\Gamma \subset H$ is U-separated if $\gamma U \cap \gamma' U$ is empty for $\gamma \neq \gamma'$ in Γ. Moreover, a countable set $\Gamma \subset H$ is called V-dense if $H = \Gamma V$. Let S be the support of f. Since f is compact, $S = \bigcup_{i=1}^{k} x_i V$ where $x_i \in \Gamma$. For fixed $x_i \in \Gamma$, we claim that each $x_i V$ contains no more than one element from Γ. Indeed, suppose by contradiction that this is not the case. Then for arbitrary $\kappa, \gamma \in \Gamma$ in $x_i V$, one has $\kappa^{-1}\gamma \in V^2 \subset U$ and this contradicts the assumption that Γ is a U-separated countable set. Therefore,

$$\sum_{\gamma\in\Gamma} f(\gamma x) \leq k \cdot \left(\sup\left\{f(x) : x \in H\right\}\right) < \infty.$$

Next, let E be an open set contained in H such that $f(x) > \epsilon$ for all $x \in E$ and for some positive real number $\epsilon > 0$. Without loss of generality, we may assume that E is an open set around the neutral element in H. Choose a symmetric open set V around the identity in H such that $V^2 \subset E$. Then every V-dense subset Γ must intersect E. Indeed, letting $y \in V$ and $\gamma \in \Gamma$ such that $y \in \gamma V, \gamma \in yV^{-1} \subset V^2 \subset E$ and $\sum_{\gamma\in\Gamma} f(\gamma x) \geq \epsilon > 0$. Finally, let Q be a subset of V such that Q is symmetric, Q^2 is contained in V. Next, choose a Q-separated subset, which is maximal with respect to inclusion. Then such a set is V-dense and Z-separated as well.

Let $w(h) = W_\lambda\left(\Theta_\lambda\left(\log(h)\right)\right)$. Appealing to Proposition 7.3 Part (3), the following is immediate. Let f be a continuous or smooth function which is supported in $\mathcal{O}$. Then $f\sqrt{w}$ is continuous and supported in $\mathcal{O}$ and there exists a discrete subset Γ_H of H such that given

$$m_{Hf} = \operatorname{essinf}_{h\in H}\left(\sum_{\ell\in\Gamma_H}\left|f(\ell h)\cdot\sqrt{w(\ell h)}\right|^2\right) \tag{7.13}$$

and

$$M_{Hf} = \operatorname{esssup}_{h\in H}\left(\sum_{\ell\in\Gamma_H}\left|f(\ell h)\cdot\sqrt{w(\ell h)}\right|^2\right),$$

we have $0 < m_{Hf} \leq M_{Hf} < \infty$.

Next, the closure of the set $\Theta_\lambda(\varphi(\mathcal{O}))$ is assumed to be compact in $P^*(\mathfrak{n}^*)$ and $\Theta_\lambda(\varphi(\mathcal{O}))$ has a positive Lebesgue measure in

$$\mathfrak{p}^* = P^*(\mathfrak{n}^*) \simeq \mathbb{R}^r. \tag{7.14}$$

Given $C \subset \mathfrak{p}^*$, and Γ_N contained in $\exp(\mathfrak{p})$, define

$$\mathcal{E}(C, \Gamma_N) = \left\{ \frac{e^{2\pi i \langle \xi, Y \rangle}}{|C|^{1/2}} : Y \in \log(\Gamma_N) \right\} \tag{7.15}$$

and put

$$\mathfrak{p} = P_J(\mathfrak{n}) = P(\mathfrak{n}). \tag{7.16}$$

Clearly, there exists a compact subset C of $\mathfrak{p}^*$ such that $\Theta_\lambda(\log(\mathcal{O})) \subseteq C$ and additionally, there exists a countable subset Γ_N of $\exp(\mathfrak{p})$ such that the trigonometric system $\mathcal{E}(C, \Gamma_N)$ is an orthonormal basis for $L^2(C, d\xi)$. Indeed, let

$$\vartheta = \left\{ \left\| \Theta_\lambda(\log(h)) \right\|_{\max} : h \in \mathcal{O} \right\} > 0$$

and order J such that $J = \{j_1 < \cdots < j_r\} = \{1, 2, \ldots, n\}$. Put

$$C = \sum_{k=1}^{r} [-\vartheta, \vartheta) X_{j_k}^* \subset \mathfrak{p}^*$$

and define $\Gamma_N = \exp\left(\sum_{k=1}^{r} \frac{1}{2\vartheta} \mathbb{Z} X_{j_k} \right) \subset N$. Clearly,

$$\left\{ C + k : k \in \sum_{k=1}^{r} 2\vartheta \mathbb{Z} X_{j_k}^* \right\}$$

is a measurable partition of $\mathfrak{p}^*$and as a result, the system $\mathcal{E}(C, \Gamma_N)$ is an orthonormal basis for $L^2(C, d\xi)$.

Proposition 7.4 Let f be a continuous (or a smooth) function which is compactly supported on $\mathcal{O}$. For any countable set $\Gamma \subset G$, define $\mathcal{S}(f, \Gamma) = \{\pi(x) f : x \in \Gamma\}$.

1) Suppose additionally that Γ_H and Γ_N are countable sets satisfying $0 < m_{Hf} \leq M_{Hf} < \infty$. Moreover, C is a compact subset of $\mathfrak{p}^*$ such that $\Theta_\lambda(\varphi(\mathcal{O})) \subseteq C$ and Γ_N is a discrete subset of $\exp(\mathfrak{p})$ such that $\mathcal{E}(C, \Gamma_N)$ is an orthonormal basis for $L^2(C, d\xi)$. Then $\mathcal{S}(f, \Gamma_H^{-1}\Gamma_N)$ is a frame for $\mathfrak{H}$ with frame bounds $0 < m_{Hf} \cdot |C| \leq M_{Hf} \cdot |C| < \infty$. Additionally, (1) If $m_{Hf} = M_{Hf}$ holds then $\mathcal{S}(f, \Gamma_H^{-1}\Gamma_N)$ is a tight frame for $\mathfrak{H}$ and (2) if $m_{Hf} = M_{Hf} = |C|^{-1}$ and $\int_H |f(h)|^2 d\mu_H(h) = 1$ then $\mathcal{S}(f, \Gamma_H^{-1}\Gamma_N)$ is an orthonormal basis for $\mathfrak{H}$.
2) Choose a countable set $\Gamma_N \subset \exp(\mathfrak{p})$, where C is a compact subset of $\mathfrak{p}^*$ such that $\Theta_\lambda(\varphi(\mathcal{O})) \subseteq C$ and $\mathcal{E}(C, \Gamma_N)$ is an orthonormal basis for $L^2(C, d\xi)$. Given a

discrete set Λ of H, the following holds true. If $S\left(f, \Lambda^{-1}\Gamma_N\right)$ is a frame for $\mathfrak{H}$ with frame bounds A, B, then for almost every $h \in H$,

$$A|C|^{-1} \leq \sum_{\ell \in \Lambda} \left| f(\ell h) \sqrt{w(\ell h)} \right|^2 \leq B|C|^{-1}.$$

In particular, f must be bounded. Moreover if $S\left(f, \Lambda^{-1}\Gamma_N\right)$ is a Parseval frame for $\mathfrak{H}$ and suppose that $\left\{\ell^{-1}\Omega : \ell \in \Lambda\right\}$ is a partition of H for some relatively compact subset Ω of H. Then $\frac{\mu_H(\Omega)}{|C|} < 1$.

3) Let Γ_H be a countable subset of H such that $0 < m_{Hf} \leq M_{Hf} < \infty$ as defined in (7.13). Next, let C be a compact subset of $\mathfrak{p}^*$ where $\Theta_\lambda(\varphi(\mathcal{O})) \subseteq C$ and Γ_N is a discrete subset of $\exp(\mathfrak{p})$ such that the trigonometric system $\mathcal{E}\left(C, \Gamma_N\right)$ is an orthonormal basis for $L^2(C, d\xi)$. Put $F(x) = \sum_{\ell \in \Gamma_H} \left| f(\ell x) w(\ell x)^{1/2} \right|^2$. Then the frame operator S corresponding to $S\left(f, \Gamma_H^{-1}\Gamma_N\right)$ and its inverse are given by $S\varphi = (|C| \cdot F)\varphi$ and $S^{-1}\varphi = (|C| \cdot F)^{-1}\varphi$, respectively, for all $\varphi \in \mathfrak{H}$.

Proof: **Proof of Part (1)**: Let g be a continuous function, which is compactly supported in H. Put

$$\mathrm{s}\left(\Gamma_H, \Gamma_N, g, f\right) = \sum_{\ell \in \Gamma_H} \sum_{PX \in \log \Gamma_N} \left| \left\langle g, \pi\left(\ell^{-1}\right) \pi(\exp(PX)) f \right\rangle_{\mathfrak{H}} \right|^2. \tag{7.17}$$

According to Proposition 7.3,

$$\begin{aligned} &\mathrm{s}\left(\Gamma_H, \Gamma_N, g, f\right) \\ &= \sum_{\ell \in \Gamma_H} \sum_{PX \in \log \Gamma_N} \left| \int_{\Theta_\lambda(\varphi(\mathcal{O}))} \left[g\left(\ell^{-1}\varphi^{-1}\left(\Theta_\lambda^{-1}(\xi)\right)\right) \cdot \overline{f\left(\varphi^{-1}\left(\Theta_\lambda^{-1}(\xi)\right)\right)} \cdot W_\lambda(\xi) \right] \right. \\ &\qquad \left. \times e^{-2\pi i \langle \xi, PX \rangle} d\xi \right|^2. \end{aligned}$$

To avoid cluster of notation, let $u(\xi) = \varphi^{-1}\left(\Theta_\lambda^{-1}(\xi)\right) \in H$. Then

$$\begin{aligned} &\mathrm{s}\left(\Gamma_H, \Gamma_N, g, f\right) \\ &= \sum_{\ell \in \Gamma_H} \sum_{PX \in \log \Gamma_N} \left| \int_{\Theta_\lambda(\varphi(\mathcal{O}))} g\left(\ell^{-1}u(\xi)\right) \cdot \overline{f(u(\xi))} \cdot W_\lambda(\xi) e^{-2\pi i \langle \xi, PX \rangle} d\xi \right|^2. \end{aligned}$$

Letting $\mathfrak{F}_C$ be the Fourier series defined on $L^2(C)$, it follows that

$$\begin{aligned} &\int_{\Theta_\lambda(\varphi(\mathcal{O}))} g\left(\ell^{-1}u(\xi)\right) \cdot \overline{f(u(\xi))} \cdot W_\lambda(\xi) e^{-2\pi i \langle \xi, PX \rangle} d\xi \\ &\quad = \mathfrak{F}_C\left(\xi \mapsto g\left(\ell^{-1}u(\xi)\right) \cdot \overline{f(u(\xi))} \cdot W_\lambda(\xi) \cdot 1_{\Theta_\lambda(\varphi(\mathcal{O}))}(\xi)\right)(PX). \end{aligned}$$

Next, Parseval's equality gives

$$s(\Gamma_H, \Gamma_N, g, f) = |C| \cdot \sum_{\ell \in \Gamma_H} \int_{\Theta_\lambda(\varphi(\mathcal{O}))} \left| g\left(\ell^{-1} u(\xi)\right) \cdot \overline{f(u(\xi))} \cdot W_\lambda(\xi)^{1/2} \right|^2 W_\lambda(\xi)\, d\xi.$$

Making the change of variable $h = u(\xi) = \varphi^{-1}\left(\Theta_\lambda^{-1}(\xi)\right)$, and applying Tonelli's theorem,

$$s(\Gamma_H, \Gamma_N, g, f) = \sum_{\ell \in \Gamma_H} \int_{\ell^{-1}\mathcal{O}} |g(h)|^2 \cdot \left| f(\ell h) w(\ell h)^{1/2} \right|^2 d\mu_H(h) \tag{7.18}$$

$$= \sum_{\ell \in \Gamma_H} \int_H |g(h)|^2 \cdot \left| f(\ell h) w(\ell h)^{1/2} \right|^2 d\mu_H(h) \tag{7.19}$$

$$= |C| \cdot \int_H |g(h)|^2 \sum_{\ell \in \Gamma_H} \left| f(\ell h) w(\ell h)^{1/2} \right|^2 d\mu_H(h) \tag{7.20}$$

Furthermore,

$$s(\Gamma_H, \Gamma_N, g, f) \leq |C| \cdot M_{Hf} \left(\int_H |g(h)|^2 d\mu_H(h) \right) = |C| \cdot M_{Hf} \|g\|_{\mathfrak{H}}^2$$

and

$$s(\Gamma_H, \Gamma_N, g, f) \geq |C| \cdot m_{Hf} \left(\int_H |g(h)|^2 d\mu_H(h) \right) = |C| \cdot m_{Hf} \|g\|_{\mathfrak{H}}^2 .$$

Since the set of all continuous and compactly supported functions is dense in $\mathfrak{H}$, it follows that $\left\{\pi\left(\ell^{-1}\kappa\right) f : (\ell, \kappa) \in \Gamma_H \times \Gamma_N\right\}$ is a frame for the Hilbert space $\mathfrak{H}$ with frame bounds $|C| \cdot m_{Hf} \leq |C| \cdot M_{Hf}$. This proves the first part of the result. To prove (1), let us assume that $m_{Hf} = M_{Hf}$. In other words, the lower and upper frame bound coincide and as a result, $S\left(f, \Gamma_H^{-1}\Gamma_N\right)$ is a tight frame for $\mathfrak{H}$. To prove (2), let us assume that $|C| \cdot m_{Hf} = |C| \cdot M_{Hf} = \int_H |f(h)|^2 d\mu_H(h) = 1$. Then $S\left(f, \Gamma_H^{-1}\Gamma_N\right)$ is a unit norm Parseval frame and therefore, it must necessarily be an orthonormal basis.

Proof of Part (2): Define

$$s(\Lambda, \Gamma_N, g, f) = \sum_{\ell \in \Lambda} \sum_{\exp(PX) \in \Gamma_N} \left| \left\langle g, \pi\left(\ell^{-1}\right) \pi(\exp(PX)) f \right\rangle_{\mathfrak{H}} \right|^2 .$$

For arbitrary $g \in \mathfrak{H}$, we have

$$s(\Lambda, \Gamma_N, g, f) = \int_H |g(h)|^2 \left(|C| \sum_{\ell \in \Lambda} \left| f(\ell h) w(\ell h)^{1/2} \right|^2 \right) d\mu_H(h)$$

$$\geq \int_H A |g(h)|^2 d\mu_H(h).$$

As such,

$$\int_H |g(h)|^2 \left(|C| \sum_{\ell \in \Lambda} \left| f(\ell h) w(\ell h)^{1/2} \right|^2 - A \right) d\mu_H(h) \geq 0.$$

Suppose now that $|C| \sum_{\ell\in\Gamma_H} \left|f(\ell h) w(\ell h)^{1/2}\right|^2 < A$ on some subset $\Omega \subseteq H$ of positive and finite Haar measure. Letting g be the indicator function of Ω, we obtain

$$\begin{aligned} s(\Lambda, \Gamma_N, g, f) &= \int_\Omega |C| \cdot \sum_{\ell\in\Lambda} \left|f(\ell h) w(\ell h)^{1/2}\right|^2 d\mu_H(h) \\ &< A \cdot \int_\Omega d\mu_H(h) = A\|g\|^2. \end{aligned}$$

Therefore, $A\|g\|^2 > s(\Lambda, \Gamma_N, g, f)$. This, however, contradicts the assumption that $S(f, \Lambda^{-1}\Gamma_N)$ is a frame for $\mathfrak{H}$ with frame bounds A, B. Similarly, suppose that $|C| \cdot \sum_{\ell\in\Lambda} \left|f(\ell h) w(\ell h)^{1/2}\right|^2 > B$ on some subset Ω' of positive and finite Haar measure. Let g be the indicator function of Ω'. Then

$$\sum_{\ell\in\Lambda} \sum_{\exp(PX)\in\Gamma_N} \left|\left\langle g, \pi\left(\ell^{-1}\right) \pi(\exp(PX)) f\right\rangle_{\mathfrak{H}}\right|^2 > B\|g\|^2$$

and this contradicts that $S(f, \Lambda^{-1}\Gamma_N)$ is a frame for $\mathfrak{H}$ with upper frame bound B. For the second part, if $S(f, \Lambda^{-1}\Gamma_N)$ is a Parseval frame for $\mathfrak{H}$ then for almost every $h \in H$,

$$\sum_{\ell\in\Lambda} \left|f(\ell h) \sqrt{w(\ell h)}\right|^2 = |C|^{-1}.$$

Suppose also that $\{\ell^{-1}\Omega : \ell \in \Lambda\}$ is a partition of H for some relatively compact subset Ω of H. Then

$$\|f\|^2 = \int_H |f(h)|^2 d\mu_H(h) = \int_\Omega \sum_{\ell\in\Lambda} \left|f(\ell h) \sqrt{w(\ell h)}\right|^2 d\mu_H(h)$$

and the assumption that $S(f, \Lambda^{-1}\Gamma_N)$ is a Parseval frame for $\mathfrak{H}$ implies that

$$\int_\Omega |C|^{-1} d\mu_H(h) = \frac{\mu_H(\Omega)}{|C|} < 1.$$

Proof of Part (3): Let us assume that

$$S\left(f, \Gamma_H^{-1}\Gamma_N\right) = \left\{\pi\left(\ell^{-1}\kappa\right) f : (\ell, \kappa) \in \Gamma_H \times \Gamma_N\right\}$$

is a frame for $\mathfrak{H}$ with frame operator $S : \mathfrak{H} \to \mathfrak{H}$ given by

$$S\varphi = \sum_{(\ell,\kappa)\in\Gamma_H\times\Gamma_N} \left\langle \varphi, \pi\left(\ell^{-1}\kappa\right) f\right\rangle_{\mathfrak{H}} \pi\left(\ell^{-1}\kappa\right) f \tag{7.21}$$

for $\varphi \in \mathfrak{H}$. Assuming that φ has compact support,

$$\begin{aligned} \langle S\varphi, \varphi\rangle_{\mathfrak{H}} &= \left\langle \sum_{(\ell,\kappa)\in\Gamma_H\times\Gamma_N} \left\langle \varphi, \pi\left(\ell^{-1}\kappa\right) f\right\rangle_{\mathfrak{H}} \pi\left(\ell^{-1}\kappa\right) f, \varphi \right\rangle_{\mathfrak{H}} \\ &= \sum_{(\ell,\kappa)\in\Gamma_H\times\Gamma_N} \left\langle \varphi, \pi\left(\ell^{-1}\kappa\right) f\right\rangle_{\mathfrak{H}} \left\langle \pi\left(\ell^{-1}\kappa\right) f, \varphi\right\rangle_{\mathfrak{H}}. \end{aligned}$$

Thus,

$$\begin{aligned}\langle S\varphi,\varphi\rangle_{\mathfrak{H}} &= \sum_{(\ell,\kappa)\in\Gamma_H\times\Gamma_N}\left\langle\varphi,\pi\left(\ell^{-1}\kappa\right)f\right\rangle_{\mathfrak{H}}\overline{\langle\varphi,\pi\left(\ell^{-1}\kappa\right)f\rangle_{\mathfrak{H}}}\\ &= \sum_{(\ell,\kappa)\in\Gamma_H\times\Gamma_N}\left|\left\langle\varphi,\pi\left(\ell^{-1}\kappa\right)f\right\rangle_{\mathfrak{H}}\right|^2\end{aligned}$$

and

$$\langle S\varphi,\varphi\rangle_{\mathfrak{H}} = |C|\cdot\int_H|\varphi(x)|^2\left(\sum_{\ell\in\Gamma_H}\left|f(\ell x)w(\ell x)^{1/2}\right|^2\right)d\mu_H(x).$$

Since the set of all continuous functions which are compactly supported is dense in $\mathfrak{H}$, and since S is a bounded linear operator, it follows that for all $\varphi\in\mathfrak{H}$,

$$\langle S\varphi,\varphi\rangle_{\mathfrak{H}} = |C|\cdot\int_H|\varphi(x)|^2F(x)\,d\mu_H(x) = \int_H(|C|\cdot\varphi(x)F(x))\overline{\varphi(x)}d\mu_H(x)$$

and $\langle S\varphi,\varphi\rangle_{\mathfrak{H}} = \langle|C|\cdot F\varphi,\varphi\rangle_{\mathfrak{H}}$. These calculations show that the action of the operator S on a vector φ in the Hilbert space $\mathfrak{H}$ is given by $Sh = |C|\cdot F\varphi$. In a similar fashion, the operator S is can be inverted as follows $S^{-1}\varphi = \frac{1}{|C|\cdot F}\varphi$.

The following theorem offers concrete conditions for the existence of Parseval frames and orthonormal bases of the type $\mathcal{S}\left(s\cdot 1_A,\Gamma_H^{-1}\Gamma_N\right)$ for some measurable function s defined on $\mathcal{O}$ and a $d\mu_H$-measurable subset A of $\mathcal{O}$.

Proposition 7.5 Let f be a vector in $L^2\left(\mathcal{O},d\mu_H\right)$ such that

$$f\left(\varphi^{-1}\left(\Theta_\lambda^{-1}(\xi)\right)\right) = W_\lambda(\xi)^{-1/2}\cdot 1_{\Theta_\lambda(\varphi(\mathcal{O}))}(\xi).$$

(a) C is a compact subset of $\mathfrak{p}^*$ such that $\Theta_\lambda(\varphi(\mathcal{O}))\subseteq C$ and Γ_N is a discrete subset of $\exp(\mathfrak{p})$ such that $\mathcal{E}\left(C,\Gamma_N\right)$ is an orthonormal basis for $L^2(C,d\xi)$ and (b) there is a discrete subset Γ_H of H such that $\left\{\ell^{-1}\mathcal{O}:\ell\in\Gamma_H\right\}$ is a tiling of H.

1) Then $\mathcal{S}\left(|C|^{-1/2}f,\Gamma_H^{-1}\Gamma_N\right)$ is a Parseval frame for $\mathfrak{H}$.
2) If additionally, $\frac{\|f\|_{\mathfrak{H}}}{\sqrt{|C|}} = 1$ then $\mathcal{S}\left(|C|^{-1/2}f,\Gamma_H^{-1}\Gamma_N\right)$ is an orthonormal basis for $\mathfrak{H}$.

Proof: To prove (1), note that

$$\int_{\mathcal{O}}|f(h)|^2d\mu_H(h) = \int_{\varphi(\mathcal{O})}\left|f\left(\varphi^{-1}(A)\right)\right|^2d\mu_H\left(\varphi^{-1}(A)\right).$$

By a suitable change of variables, it follows that,

$$\int_{\mathcal{O}}|f(h)|^2d\mu_H(h) = \int_{\Theta_\lambda(\varphi(\mathcal{O}))}\left|W_\lambda(\xi)^{1/2}\cdot f\left(\varphi^{-1}\left(\Theta_\lambda^{-1}(\xi)\right)\right)\right|^2d\xi$$

and $\int_{\mathcal{O}} |f(h)|^2 d\mu_H(h)$ is equal to the Lebesgue measure of $\Theta_\lambda(\varphi(\mathcal{O}))$. Given $X \in \mathfrak{p}$, and letting g be a continuous function defined on $\mathcal{O}$,

$$I = \langle g, \pi(\exp X) f\rangle_{\mathfrak{H}} = \int_{\mathcal{O}} g(h) e^{2\pi i \langle Ad(h^{-1})^* \lambda, X\rangle} \overline{f(h)} d\mu_H(h).$$

Recalling that $\beta_\lambda\left(\varphi^{-1}(A)\right) = Ad\left(h^{-1}\right)^* \lambda$

$$I = \int_{\varphi(\mathcal{O})} \left(g\left(\varphi^{-1}(A)\right)\overline{f\left(\varphi^{-1}(A)\right)}\right) \cdot e^{2\pi i \langle \beta_\lambda(\varphi^{-1}(A)), X\rangle} d\mu_H\left(\varphi^{-1}(A)\right).$$

Since $X = P^* X \in \mathfrak{p}$, and $\Theta_\lambda(A) = P^* \beta_\lambda \varphi^{-1}(A)$

$$I = \int_{\varphi(\mathcal{O})} g\left(\varphi^{-1}(A)\right) e^{2\pi i \langle \Theta_\lambda(A), X\rangle} \overline{f\left(\varphi^{-1}(A)\right)} d\mu_H\left(\varphi^{-1}(A)\right).$$

Making the change of variable $\xi = \Theta_\lambda(A)$,

$$I = \int_{\Theta_\lambda(\varphi(\mathcal{O}))} e^{2\pi i \langle \xi, X\rangle} \left(g\left(\varphi^{-1}\left(\Theta_\lambda^{-1}(\xi)\right)\right) \overline{f\left(\varphi^{-1}\left(\Theta_\lambda^{-1}(\xi)\right)\right)} W_\lambda(\xi)\right) d\xi.$$

Next, since $\mathcal{E}\left(C, \Gamma_N\right)$ is an orthonormal basis for $L^2(C, d\xi)$, the trigonometric system

$$\left\{\frac{e^{2\pi i \langle \xi, X\rangle} \cdot 1_{\Theta_\lambda(\varphi(\mathcal{O}))}(\xi)}{|C|^{1/2}} : \exp(X) \in \Gamma_N\right\} \tag{7.22}$$

is a Parseval frame for $L^2\left(\Theta_\lambda(\varphi(\mathcal{O}))\right)$and as a result,

$$\begin{aligned} I_1 &= \sum_{\exp X \in \Gamma_N} \left|\langle g, \pi(\exp X) f\rangle_{\mathfrak{H}}\right|^2 \\ &= \sum_{\exp X \in \Gamma_N} \left|\int_{\Theta_\lambda(\varphi(\mathcal{O}))} \frac{e^{2\pi i \langle \xi, X\rangle}}{|C|^{1/2}} g\left(\varphi^{-1}\left(\Theta_\lambda^{-1}(\xi)\right)\right) |C|^{1/2} W_\lambda(\xi)^{1/2} d\xi\right|^2. \end{aligned}$$

Set $u(\xi) = \varphi^{-1}\left(\Theta_\lambda^{-1}(\xi)\right)$. Then

$$\left||C|^{1/2} g(u(\xi)) W_\lambda(\xi)^{1/2}\right|^2 = |C| \cdot |g(u(\xi))|^2 W_\lambda(\xi)$$

and

$$I_1 = \int_{\Theta_\lambda(\varphi(\mathcal{O}))} \left||C|^{1/2} g(u(\xi)) W_\lambda(\xi)^{1/2}\right|^2 d\xi = |C| \cdot \int_{\mathcal{O}} |g(h)|^2 d\mu_H(h).$$

Since $\Gamma_H \subset H$ is assumed to be a countable subset such that $\left\{\ell^{-1}\mathcal{O} : \ell \in \Gamma_H\right\}$ is a tiling of H,

$$\begin{aligned} &\sum_{\ell \in \Gamma_H} \sum_{\exp X \in \Gamma_N} \left|\left\langle g, \pi\left(\ell^{-1}\right) \pi(\exp X) f\right\rangle_{\mathfrak{H}}\right|^2 \\ &= \sum_{\ell \in \Gamma_H} \sum_{\exp X \in \Gamma_N} \left|\left\langle \pi(\ell) g \cdot 1_{\mathcal{O}}, 1_{\mathcal{O}} \cdot \pi(\exp X) f\right\rangle_{\mathfrak{H}}\right|^2 \end{aligned}$$

and

$$\sum_{\ell \in \Gamma_H} \sum_{\exp X \in \Gamma_N} \left| \left\langle g, \pi\left(\ell^{-1}\right) \pi(\exp X) f \right\rangle_{\mathfrak{H}} \right|^2 = |C| \cdot \sum_{\ell \in \Gamma_H} \left\| \pi(\ell)\left(g \cdot 1_{\mathcal{O}}\right) \right\|_{\mathfrak{H}}^2 .$$

Using the fact that $\pi(\ell)$ is a unitary operator,

$$\sum_{\ell \in \Gamma_H} \sum_{\exp X \in \Gamma_N} \left| \left\langle g, \pi\left(\ell^{-1}\right) \pi(\exp X) f \right\rangle_{\mathfrak{H}} \right|^2 = |C| \cdot \sum_{\ell \in \Gamma_H} \int_{\ell^{-1}\mathcal{O}} |g(h)|^2 d\mu_H(h)$$

and

$$\sum_{\ell \in \Gamma_H} \sum_{\exp X \in \Gamma_N} \left| \left\langle g, \pi\left(\ell^{-1}\right) \pi(\exp X) f \right\rangle_{\mathfrak{H}} \right|^2 = |C| \cdot \int_H |g(h)|^2 d\mu_H(h).$$

Therefore, the countable system $\left\{\pi\left(\ell^{-1}\kappa\right) f : \kappa \in \Gamma_N, \ell \in \Gamma_H\right\}$ is a tight frame for $L^2\left(H, d\mu_H(h)\right)$ with frame bound $|C|$ and by normalizing, it follows that $\left\{\pi\left(\ell^{-1}\kappa\right) |C|^{-1/2} f : \kappa \in \Gamma_N, \ell \in \Gamma_H\right\}$ is a Parseval frame. For (2), suppose additionally that

$$|C|^{-1/2} \cdot \|f\|_{\mathfrak{H}} = |C|^{-1/2} \cdot \left|\Theta_\lambda(\varphi(\mathcal{O}))\right|^{1/2} = 1. \tag{7.23}$$

Then

$$\left\| |C|^{-1/2} f \right\|_{\mathfrak{H}} = \left(|C|^{-1} \int_{\mathcal{O}} |f(x)|^2 d\mu_H(x) \right)^{1/2}$$

and

$$\left\| |C|^{-1/2} f \right\|_{\mathfrak{H}} = \left(|C|^{-1} \int_{\Theta_\lambda(\varphi(\mathcal{O}))} \left| f\left(\varphi^{-1}\left(\Theta_\lambda^{-1}(\xi)\right)\right) \right|^2 d\mu_H\left(\varphi^{-1}\left(\Theta_\lambda^{-1}(\xi)\right)\right) \right)^{1/2} .$$

This shows that

$$\left\| |C|^{-1/2} f \right\|_{\mathfrak{H}} = \left(|C|^{-1} \int_{\Theta_\lambda(\varphi(\mathcal{O}))} \left| f\left(\varphi^{-1}\left(\Theta_\lambda^{-1}(\xi)\right)\right) W_\lambda(\xi)^{1/2} \right|^2 d\xi \right)^{1/2}$$

and consequently,

$$\begin{aligned} \left\| |C|^{-1/2} f \right\|_{\mathfrak{H}} &= \left(|C|^{-1} \int_{\Theta_\lambda(\varphi(\mathcal{O}))} \frac{1_{\Theta_\lambda(\varphi(\mathcal{O}))}(\xi)}{\left(W_\lambda(\xi)\right)^{1/2}} \left(W_\lambda(\xi)\right)^{1/2} d\xi \right)^{1/2} \\ &= \left(|C|^{-1} \int_{\Theta_\lambda(\varphi(\mathcal{O}))} d\xi \right)^{1/2} . \end{aligned}$$

Referring back to (7.23),

$$\left\| |C|^{-1/2} f \right\|_{\mathfrak{H}} = |C|^{-1/2} \cdot \left|\Theta_\lambda(\varphi(\mathcal{O}))\right|^{1/2} = 1.$$

Therefore, $\left\{\pi\left(\ell^{-1}\kappa\right)\left(|C|^{-1/2} f\right) : \kappa \in \Gamma_N, \ell \in \Gamma_H\right\}$ is a unit-norm Parseval frame, and thus necessarily an orthonormal basis for $\mathfrak{H}$.

To prove Theorem 7.3 Part (2), it suffices to verify the following statement. Let $E\left(C,\Gamma_N\right)$ be as defined in (7.15).

Lemma 7.1 Let s be a continuous (or a smooth) function which is compactly supported on $\mathcal{O}$. Suppose that Γ_H is a discrete subset of H such that $\sum_{\ell\in\Gamma_H}|s(\ell h)|^2 = 1$ for almost every $h \in H$. Let C be a compact subset of $\mathfrak{p}^*$ such that $\Theta_\lambda(\varphi(\mathcal{O})) \subseteq C$ and Γ_N is a countable subset of $\exp(\mathfrak{p})$, where the trigonometric system $\mathcal{E}\left(C,\Gamma_N\right)$ is an orthonormal basis for $L^2(C, d\xi)$. Then the system

$$\left\{\left[\pi\left(\ell^{-1}\kappa\right)\right]\left(|C|^{-1/2} s w^{-1/2}\right) : (\ell,\kappa) \in \Gamma_H \times \Gamma_N\right\}$$

is a Parseval frame generated by the compactly supported and continuous function $sw^{-1/2}|C|^{-1/2}$.

Proof: Letting $f = sw^{-1/2}$,

$$\sum_{\kappa\in\Gamma_H}\left|f(\kappa h)\sqrt{w(\kappa h)}\right|^2 = \sum_{\kappa\in\Gamma_H}\left|\frac{s(\kappa h)}{\sqrt{w(\kappa h)}}\sqrt{w(\kappa h)}\right|^2 = \sum_{\kappa\in\Gamma_H}|s(kh)|^2 = 1.$$

Suppose that H is a connected exponential solvable Lie group. We will prove by induction on the dimension of H that there exists a continuous function s compactly supported in $\mathcal{O}$ such that $\sum_{\ell\in\Gamma_H}|s(\ell h)|^2 = 1$ for every element h in H. For the base case, let us suppose that the dimension of H is equal to 1. Moreover, let us suppose that $\mathcal{O} = \left(-\epsilon^{-1}, \epsilon^{-1}\right)$ for some positive real number ϵ.

Put $s(t) = \left(1_{[-1/2,1/2]} * 1_{[-1/2,1/2]}\right)^{\frac{1}{2}}(\epsilon t)$, where $*$ is the usual convolution product: $(f * g)(y) = \int_{\mathbb{R}} f(x)g(y-x)\,dx$. Since

$$\sum_{k\in\mathbb{Z}}\left(1_{[-1/2,1/2]} * 1_{[-1/2,1/2]}\right)(t+k) = 1 \text{ for all } t \in \mathbb{R},$$

it follows that

$$\sum_{k\in\mathbb{Z}}\left|s\left(t+\epsilon^{-1}k\right)\right|^2 = \sum_{k\in\mathbb{Z}}\left(1_{[-1/2,1/2]} * 1_{[-1/2,1/2]}\right)(t\epsilon + k) = 1$$

holds for every real number t. Moreover, the smoothness of s can be improved by replacing w with a spline-type function of higher order. To this end, let b_n be the function obtained by convolving the function $1_{[-1/2,1/2]}$ with itself n-many times and let $s_n(t) = \left(b_n\right)^{\frac{1}{2}}\left(\frac{\epsilon n}{2}t\right)$. Then

$$\sum_{k\in\mathbb{Z}}\left|s_n\left(t+\frac{2}{\epsilon n}k\right)\right|^2 = \sum_{k\in\mathbb{Z}}\left|\left(b_n\right)^{\frac{1}{2}}\left(\frac{\epsilon n}{2}\left(t+\frac{2}{\epsilon n}k\right)\right)\right|^2 = \sum_{k\in\mathbb{Z}} b_n\left(k+\frac{nt\epsilon}{2}\right)$$

and

$$\sum_{k\in\mathbb{Z}}\left|s_n\left(t+\frac{2}{\epsilon n}k\right)\right|^2 = 1$$

for every $t \in \mathbb{R}$ as desired. By the inductive hypothesis, we may assume that the stated result holds true whenever, $\dim H \leq r$ for some natural number $r \geq 1$. Next, let H be a solvable Lie group such that $\dim(H) = r + 1$. Since the exponential map is assumed to define a diffeomorphism between $\mathfrak{h}$ and H, we may write $H = H_{\mathrm{n}} H_{\mathrm{c}}$ such that H_{n} is normal in H and H_{c} is a one-parameter subgroup of H. Without loss of generality, we may also assume that $\mathcal{O} = \mathcal{O}_{\mathrm{n}} \mathcal{O}_{\mathrm{c}}$ such that $\mathcal{O}_{\mathrm{n}} \subset H_{\mathrm{n}}$ and $\mathcal{O}_{\mathrm{c}} \subset H_{\mathrm{c}}$ such that the following holds. There exist $\Gamma_{\mathrm{n}} \subset H_{\mathrm{n}}$ and $\Gamma_{\mathrm{c}} \subset H_{\mathrm{c}}$ and continuous functions (or smooth functions) s_{n} and s_{c} such that $\sum_{\ell_{\mathrm{n}} \in \Gamma_{\mathrm{n}}} \left| s_{\mathrm{n}} \left(\ell_{\mathrm{n}} h' \right) \right|^2 = 1$ for almost every $h' \in H_{\mathrm{n}}$ and $\sum_{\ell_{\mathrm{c}} \in \Gamma_{\mathrm{c}}} \left| s_{\mathrm{c}} \left(\ell_{\mathrm{c}} h'' \right) \right|^2 = 1$ for almost every $h'' \in H_{\mathrm{c}}$. Define a continuous function s on H as follows. Given $h \in H$, since $h = h_{\mathrm{n}} h_{\mathrm{c}}$ with $h_{\mathrm{n}} \in H_{\mathrm{n}}$ and $h_{\mathrm{c}} \in H_{\mathrm{c}}$, we define $s(h) = s\left(h_{\mathrm{n}} h_{\mathrm{c}}\right) = s_{\mathrm{n}}\left(h_{\mathrm{n}}\right) \cdot s_{\mathrm{c}}\left(h_{\mathrm{c}}\right)$ for all h in H. Thus, for $h \in H$,

$$\sum_{\ell_{\mathrm{c}} \in \Gamma_{\mathrm{c}}} \sum_{\ell_{\mathrm{n}} \in \Gamma_{\mathrm{n}}} \left| s\left(\ell_{\mathrm{n}} \ell_{\mathrm{c}} h\right) \right|^2 = \sum_{\ell_{\mathrm{c}} \in \Gamma_{\mathrm{c}}} \sum_{\ell_{\mathrm{n}} \in \Gamma_{\mathrm{n}}} \left| s\left(\ell_{\mathrm{n}} \ell_{\mathrm{c}} h_{\mathrm{n}} h_{\mathrm{c}}\right) \right|^2$$

and

$$\sum_{\ell_{\mathrm{c}} \in \Gamma_{\mathrm{c}}} \sum_{\ell_{\mathrm{n}} \in \Gamma_{\mathrm{n}}} \left| s\left(\ell_{\mathrm{n}} \ell_{\mathrm{c}} h\right) \right|^2 = \sum_{\ell_{\mathrm{c}} \in \Gamma_{\mathrm{c}}} \sum_{\ell_{\mathrm{n}} \in \Gamma_{\mathrm{n}}} \left| s\left(\ell_{\mathrm{n}} \ell_{\mathrm{c}} h_{\mathrm{n}} \left(\ell_{\mathrm{c}}^{-1} \ell_{\mathrm{c}}\right) h_{\mathrm{c}}\right) \right|^2.$$

Since H_{n} is normal in H, $\ell_{\mathrm{n}} \ell_{\mathrm{c}} h_{\mathrm{n}} \ell_{\mathrm{c}}^{-1} \in H_{\mathrm{n}}$ and

$$\sum_{\ell_{\mathrm{c}} \in \Gamma_{\mathrm{c}}} \sum_{\ell_{\mathrm{n}} \in \Gamma_{\mathrm{n}}} \left| s\left(\left(\ell_{\mathrm{n}} \ell_{\mathrm{c}} h_{\mathrm{n}} \ell_{\mathrm{c}}^{-1}\right) \left(\ell_{\mathrm{c}} h_{\mathrm{c}}\right)\right) \right|^2$$

$$= \sum_{\ell_{\mathrm{c}} \in \Gamma_{\mathrm{c}}} \sum_{\ell_{\mathrm{n}} \in \Gamma_{\mathrm{n}}} \left| s_{\mathrm{n}}\left(\ell_{\mathrm{n}} \ell_{\mathrm{c}} h_{\mathrm{n}} \ell_{\mathrm{c}}^{-1}\right) \right|^2 \cdot \left| s_{\mathrm{c}}\left(\ell_{\mathrm{c}} h_{\mathrm{c}}\right) \right|^2.$$

By assumption, for any fixed $\ell_{\mathrm{c}} \in \Gamma_{\mathrm{c}}$, since $\ell_{\mathrm{c}} h_{\mathrm{n}} \ell_{\mathrm{c}}^{-1} \in H_{\mathrm{n}}$,

$$\sum_{\ell_{\mathrm{n}} \in \Gamma_{\mathrm{n}}} \left| s_{\mathrm{n}}\left(\ell_{\mathrm{n}} \ell_{\mathrm{c}} h_{\mathrm{n}} \ell_{\mathrm{c}}^{-1}\right) \right|^2 = 1.$$

Therefore,

$$\sum_{\ell_{\mathrm{c}} \in \Gamma_{\mathrm{c}}} \sum_{\ell_{\mathrm{n}} \in \Gamma_{\mathrm{n}}} \left| s_{\mathrm{n}}\left(\ell_{\mathrm{n}} \ell_{\mathrm{c}} h_{\mathrm{n}} \ell_{\mathrm{c}}^{-1}\right) \right|^2 \cdot \left| s_{\mathrm{c}}\left(\ell_{\mathrm{c}} h_{\mathrm{c}}\right) \right|^2 = \sum_{\ell_{\mathrm{c}} \in \Gamma_{\mathrm{c}}} \left| s_{\mathrm{c}}\left(\ell_{\mathrm{c}} h_{\mathrm{c}}\right) \right|^2 = 1,$$

and we can now conclude that $\sum_{\ell_{\mathrm{c}} \in \Gamma_{\mathrm{c}}} \sum_{\ell_{\mathrm{n}} \in \Gamma_{\mathrm{n}}} \left| s\left(\ell_{\mathrm{n}} \ell_{\mathrm{c}} h\right) \right|^2 = 1$ for all h in H.

7.2 Localized Frames on Matrix Lie Groups

In this section, we apply Theorem 7.3 to construct "nice" frames on any closed matrix subgroup of a general linear group. Precisely, we will prove the following general result.

Theorem 7.4 Let H be a matrix Lie group. Then there exists a group of the type $G = NH$ such that N is normal and G. Moreover, there exists a sufficiently small neighborhood $\mathcal{O}_e$ of the identity in H such that if f is a continuous and compactly supported on $\mathcal{O}_e$, then there exists a countable set Γ of G, a unitary character χ of N, and an irreducible representation $\pi = \mathrm{ind}_N^{NH}(\chi)$ of G such that $\pi(\Gamma)f$ is a frame for $L^2(H, d\mu_H)$.

Remark 7.3 Note that whenever $\pi(\Gamma)f$ is a frame for $L^2(H, d\mu_H)$ with frame operator

$$S = \sum_{\gamma \in \Gamma} \pi(\gamma) f \otimes \pi(\gamma) f$$

(recall that S is an invertible self-adjoint operator acting on $L^2(H, d\mu_H)$) then $S^{-1/2}\pi(\Gamma)f$ must be Parseval frame. If additionally, $S^{-1/2}$ happens to commute with all operators in $\pi(\Gamma)$ then $\pi(\Gamma)S^{-1/2}f$ is a Parseval frame for $L^2(H, d\mu_H)$. This is the case, for instance whenever Γ is a subgroup of G. In addition, as observed earlier if $\pi(\Gamma)f$ is a frame for $L^2(H, d\mu_H)$ and if A is a bounded and invertible operator commuting with $\pi(\Gamma)$ then $\pi(\Gamma)Af$ is also a frame for $L^2(H, d\mu_H)$. Thus, a full description of the commutant of $\pi(\Gamma)$ in the algebra of bounded operators on $L^2(H, d\mu_H)$ provides access to a wealth of new frames of the type $\pi(\Gamma)\varphi$.

The class of group G presented in this section follows a construction introduced by J. Wolf to study the representation theory of maximal parabolic subgroups of classical Lie groups [6].

Let

$$H = \left\{ [0, h] = \begin{bmatrix} h & 0_n \\ 0_n & \mathrm{Id}_n \end{bmatrix} \in \mathrm{GL}(2n, \mathbb{R}) : h \in K \right\}$$

and define

$$N = \left\{ [x, e] = \begin{bmatrix} \mathrm{Id}_n & x \\ 0_n & \mathrm{Id}_n \end{bmatrix} \in \mathrm{GL}(2n, \mathbb{R}) : x \in \mathfrak{gl}(n, \mathbb{R}) \right\}$$

where K is a closed and connected matrix subgroup of $GL(n, \mathbb{R})$.

H and N are embedded closed subgroups of the general linear $\mathrm{GL}(2n, \mathbb{R})$. Moreover, N is a normal closed subgroup of G. To see this, it suffices to verify that for arbitrary $h \in K$ and $x \in \mathfrak{gl}(n, \mathbb{R})$, the following matrix $[0, h]\,[x, e]\,[0, h]^{-1} = [hx, e]$ remains an element of N. Since N is a commutative group, it is clear that N is invariant under the conjugation action of G. Next, put $G = NH$ so that

$$G = \left\{ [x, h] = [x, e]\,[0, h] \in \mathrm{GL}(2n, \mathbb{R}) : h \in K \text{and} x \in \mathfrak{gl}(n, \mathbb{R}) \right\}.$$

Then the map $h \mapsto [0, h]$ defines a Lie group isomorphism between K and H, and it is quite convenient to identify K with H via this isomorphism. By assumption, the dimension of N must be greater than or equal to that of H, and the Lie algebra of N is just the matrix space

$$\mathfrak{n} = \left\{ \log[x, e] = \begin{bmatrix} 0_n & x \\ 0_n & 0_n \end{bmatrix} : x \in \mathfrak{gl}(n, \mathbb{R}) \right\}.$$

Identify $\mathfrak{n}$ with its linear dual $\mathfrak{n}^*$ such that for $\lambda \in \mathfrak{n}^*$ and $X \in \mathfrak{n}$, the pairing $\langle \lambda, X \rangle$ is given by the trace of the product $\lambda^T X$. Put

$$\lambda = \begin{bmatrix} 0_n & \mathrm{Id}_n \\ 0_n & 0_n \end{bmatrix} \in \mathfrak{n}^* \text{ and } X(x) = \begin{bmatrix} 0_n & x \\ 0_n & 0_n \end{bmatrix} \in \mathfrak{n}. \tag{7.24}$$

Then the linear functional λ determines a unitary character of the normal subgroup of N defined as follows:

$$\chi_\lambda(\exp X(x)) = e^{2\pi i \langle \lambda, x \rangle} = e^{2\pi i \cdot \mathrm{Tr}\left(\lambda^T X(x)\right)}.$$

Consider the induced representation [3, Chapter 6] $\pi = \mathrm{ind}_N^{NH}(\chi_\lambda)$ of G realized as acting on the Hilbert space $\mathfrak{H} = L^2(H, d\mu_H)$ as described subsequently. Given $x \in G$, define the endomorphism $Ad(s)$ as the differential of the map $y \mapsto sys^{-1}$ at the neutral element of G. For $f \in \mathfrak{H}$, $x \in N$, and $z \in H$, the action of π is determined by the equation:

$$[\pi(xz)]f(h) = e^{2\pi i \left\langle Ad_{\mathfrak{n}}(h^{-1})^* \lambda, \log(x) \right\rangle} f\left(z^{-1}h\right) \tag{7.25}$$

where $Ad_{\mathfrak{n}}\left(h^{-1}\right)$ is the restriction of the endomorphism $Ad\left(h^{-1}\right) : \mathfrak{g} \to \mathfrak{g}$ to the ideal $\mathfrak{n}$. Let $[0, h] \in H$ and $X(x) \in \mathfrak{n}$. Then $Ad(h)X = X(hx)$. Indeed, let $c : \mathbb{R} \to \mathrm{GL}(2n, \mathbb{R})$ be a curve defined as $c(t) = [0, h]\,[tx, e]\,[0, h]^{-1}$. Then $\frac{c(r)-c(0)}{r} = X(hx)$ and consequently $\lim_{r \to 0} \frac{c(r)-c(0)}{r} = X(hx)$.

Remark 7.4 Let $\lambda_\omega = \begin{bmatrix} 0_n & \omega \\ 0_n & 0_n \end{bmatrix} \in \mathfrak{n}^*$ such that $\omega \in \mathfrak{gl}(n, \mathbb{R})$ with $\det(\omega) \neq 0$. Define $\beta_{\lambda_\omega} : H \to \beta_\lambda(H) \subseteq \mathfrak{n}^*$ such that $\beta_{\lambda_\omega}(h) = Ad\left(h^{-1}\right)^* \lambda_\omega$. Note that if ω is a matrix of order n of full rank, β_{λ_ω} is an immersion at the identity of H satisfying $\beta_{\lambda_\omega}([0, h]) = \lambda_{(h^{-1})^T \omega}$ for $h \in K$. To see this, we argue as follows. Given $X(x)$ where $x \in \mathfrak{gl}(n, \mathbb{R})$, one verifies that

$$\left\langle \lambda_\omega, Ad\left(h^{-1}\right) X \right\rangle = \mathrm{Tr}\left(\begin{bmatrix} 0_n & 0_n \\ \omega^T h^{-1} & 0_n \end{bmatrix} \begin{bmatrix} 0_n & x \\ 0_n & 0_n \end{bmatrix} \right).$$

As a result, $Ad\left(h^{-1}\right)^* \lambda_\omega = \lambda_{(h^{-1})^T \omega}$. The fact that ω is a matrix of order n of full rank implies that β_{λ_ω} is an immersion at the identity of H. Note that $\beta_\lambda(h) = \lambda$ if and only if h is the identity matrix in H.

Finally, appealing to Theorem 7.3, Theorem 7.4 follows immediately, and a detailed version of Theorem 7.4 may be formulated as shown below:

Theorem 7.5 Let f be a continuous (or a smooth) function that is compactly supported on $\mathcal{O}$. If Γ_H and Λ are discrete subsets of H and $\exp(\mathfrak{p}) \subseteq N$ respectively such that (a) $0 < m_{\Gamma_H f} \leq M_{\Gamma_H f} < \infty$, (b) C is a compact subset of $\mathfrak{p}^*$ with $\Theta_\lambda(\varphi(\mathcal{O})) \subseteq C$ and the trigonometric system $\mathcal{E}(C, \Lambda)$ is an orthonormal basis for $L^2(C, d\xi)$ then the collection $S\left(f, \Gamma_H^{-1}\Lambda\right)$ is a frame for $\mathfrak{H}$ with frame bounds $0 < m_{\Gamma_H f} \cdot |C| \leq M_{\Gamma_H f} \cdot |C| < \infty$. Furthermore, if additionally, $m_{\Gamma_H f} = M_{\Gamma_H f} = |C|^{-1}$ and $\|f\|_{\mathfrak{H}} = 1$ then $S\left(f, \Gamma_H^{-1}\Lambda\right)$ is an orthonormal basis for $\mathfrak{H}$.

7.3 A Generalization

Our main purpose in this chapter is to generalize Theorem 7.3 to handle representations obtained by inducing conormally representations which are not necessarily characters.

Suppose that $G = NH$ such that N is a closed and normal subgroup of G that H is a set which may or may not have a group structure. Let π be a unitary representation of G obtained by inducing a representation τ of N acting in $\mathcal{H}_\tau$. Since N is assumed to be normal, the homogeneous space G/N admits a G-invariant measure denoted by $d(sN)$. Realize π as acting by left translations in the space of vector-valued functions $f : G \to \mathcal{H}_\tau$ such that (a) $f(hn) = \tau(n)^* f(h)$ and (b) $\int_{G/N} \|f(g)\|^2_{\mathcal{H}_\tau} d(gN) < \infty$. This means that the values of any function in this space are essentially determined by their restriction to the subgroup H via the representation τ.

Since $\pi = \mathrm{ind}_N^{NH}(\tau)$, G acts on this space $\mathcal{H}_\pi$ as follows. For $n \in N, h \in H$,

$$\pi(n) f(h) = f\left(n^{-1}h\right) = f\left(hh^{-1}n^{-1}h\right) = \tau\left(h^{-1}nh\right) f(h).$$

Moreover, letting $s, h \in H$, $\pi(s) f(h) = f\left(s^{-1}h\right)$. It is thus convenient to realize the action of π in $L^2\left(G/N, \mathcal{H}_\tau\right)$

For a fixed countable set $\Gamma_N \subset N$ and a vector $f \in L^2\left(G/N, \mathcal{H}_\tau\right)$, formally,

$$\sum_{\gamma\in\Gamma} \left|\langle \varphi, \pi(\gamma) f\rangle_{L^2(G/N,\mathcal{H}_\tau)}\right|^2 = \sum_{\gamma\in\Gamma} \left|\int_{G/N} \left\langle \varphi(s), \tau\left(s^{-1}\gamma s\right) f(s)\right\rangle_{L^2(\mathcal{H}_\tau)} d(sN)\right|^2.$$

Let τ_s be a unitary representation of N given by $\tau_s(n) = \tau\left(s^{-1}ns\right)$. Then

$$\sum_{\gamma\in\Gamma} \left|\langle \varphi, \pi(\gamma) f\rangle_{L^2(G/N,\mathcal{H}_\tau)}\right|^2 = \sum_{\gamma\in\Gamma} \left|\int_{G/N} \left\langle \varphi(s), \tau_s(n) f(s)\right\rangle_{L^2(\mathcal{H}_\tau)} d(sN)\right|^2.$$

We make the following assumption

Condition 7.1 There exist a set a countable set $\Gamma \subset N$ and $f \in L^2\left(G/N, \mathcal{H}_\tau\right)$ such that

$$\sum_{\gamma \in \Gamma} \left| \langle \varphi, \pi(\gamma) f \rangle_{L^2(G/N, \mathcal{H}_\tau)} \right|^2 = \int_{G/N} \|\varphi(s)\|^2_{L^2(\mathcal{H}_\tau)} \|d(s) \cdot f(s)\|^2_{L^2(\mathcal{H}_\tau)} \, d(sN)$$

for all $\varphi \in L^2\left(G/N, \mathcal{H}_\tau\right)$ and some continuous function d defined on G.

Then for every $\kappa \in G$,

$$\sum_{\gamma \in \Gamma} \left| \left\langle \pi\left(\kappa^{-1}\right) \varphi, \pi(\gamma) f \right\rangle_{L^2(G/N, \mathcal{H}_\tau)} \right|^2$$
$$= \int_{G/N} \|\varphi(\kappa s)\|^2_{L^2(\mathcal{H}_\tau)} \cdot \|d(s) f(s)\|^2_{L^2(\mathcal{H}_\tau)} \, d(sN).$$

and changing variable $g = \kappa s$,

$$\sum_{\gamma \in \Gamma} \left| \left\langle \pi\left(\kappa^{-1}\right) \varphi, \pi(\gamma) f \right\rangle_{L^2(G/N, \mathcal{H}_\tau)} \right|^2$$
$$= \int_{G/N} \|\varphi(g)\|^2_{L^2(\mathcal{H}_\tau)} \cdot \left\| d\left(\kappa^{-1} g\right) f\left(\kappa^{-1} g\right) \right\|^2_{L^2(\mathcal{H}_\tau)} d(gN).$$

Next, letting κ run through some countable set Λ,

$$\sum_{\kappa \in \Lambda} \sum_{\gamma \in \Gamma} \left| \langle \varphi, \pi(\kappa\gamma) f \rangle_{L^2(G/N, \mathcal{H}_\tau)} \right|^2$$
$$= \sum_{\kappa \in \Lambda} \int_{G/N} \|\varphi(g)\|^2_{L^2(\mathcal{H}_\tau)} \left\| d\left(\kappa^{-1} g\right) f\left(\kappa^{-1} g\right) \right\|^2_{L^2(\mathcal{H}_\tau)} d(gN)$$

Next, we make the following assumption.

Condition 7.2 There exists a countable subset Λ of G such that

$$\sum_{\kappa \in \Lambda} \int_{G/N} \|\varphi(g)\|^2_{L^2(\mathcal{H}_\tau)} \left\| d\left(\kappa^{-1} g\right) f\left(\kappa^{-1} g\right) \right\|^2_{L^2(\mathcal{H}_\tau)} d(gN)$$
$$= \int_{G/N} \|\varphi(g)\|^2_{L^2(\mathcal{H}_\tau)} \sum_{\kappa \in \Lambda} \left\| d\left(\kappa^{-1} g\right) f\left(\kappa^{-1} g\right) \right\|^2_{L^2(\mathcal{H}_\tau)} d(gN)$$

for every $\varphi \in L^2\left(G/N, \mathcal{H}_\tau\right)$ and additionally, there exist positive real numbers $m_{\Lambda f} \leq M_{\Lambda f}$ such that

$$\operatorname{esssup}_{g \in G} \left(\sum_{\kappa \in \Lambda} \left\| d\left(\kappa^{-1} g\right) f\left(\kappa^{-1} g\right) \right\|^2_{L^2(\mathcal{H}_\tau)} \right) = m_{\Lambda f}$$

$$\operatorname{esssup}_{g \in G} \left(\sum_{\kappa \in \Lambda} \left\| d\left(\kappa^{-1} g\right) f\left(\kappa^{-1} g\right) \right\|^2_{L^2(\mathcal{H}_\tau)} \right) = M_{\Lambda f}.$$

Then

$$m_{\Lambda,f} \cdot \|\varphi\|^2 = m_{\Lambda,f} \cdot \int_{G/N} \|\varphi(g)\|^2_{L^2(\mathcal{H}_\tau)} d(gN) \le \sum_{\kappa \in \Lambda} \sum_{\gamma \in \Gamma} \left| \langle \varphi, \pi(\kappa\gamma) f \rangle_{L^2(G/N, \mathcal{H}_\tau)} \right|^2$$

$$\le M_{\Lambda,f} \cdot \int_{G/N} \|\varphi(g)\|^2_{L^2(\mathcal{H}_\tau)} d(gN) = \|\varphi\|^2 \cdot M_{\Lambda,f}$$

and finally, $\pi(\Lambda\Gamma) f$ is a frame for $L^2\left(G/N, \mathcal{H}_\tau\right)$.

In summary, we have proved the following.

Theorem 7.6 Let G be a Lie group. Suppose that $G = NH$ such that N is a closed and normal subgroup of G and let $\pi = \text{ind}_N^{NH}(\tau)$. Suppose additionally that the following holds:

1) There exist a countable set $\Gamma \subset N$ and $f \in L^2\left(G/N, \mathcal{H}_\tau\right)$ such that

$$\sum_{\gamma \in \Gamma} \left| \langle \varphi, \pi(\gamma) f \rangle_{L^2(G/N, \mathcal{H}_\tau)} \right|^2 = \int_{G/N} \|\varphi(s)\|^2_{L^2(\mathcal{H}_\tau)} \|d(s) \cdot f(s)\|^2_{L^2(\mathcal{H}_\tau)} d(sN)$$

for all $\varphi \in L^2\left(G/N, \mathcal{H}_\tau\right)$ and some continuous function d defined on G.

2) There exists a countable subset Λ of G such that

$$\sum_{\kappa \in \Lambda} \int_{G/N} \|\varphi(g)\|^2_{L^2(\mathcal{H}_\tau)} \left\| d\left(\kappa^{-1} g\right) f\left(\kappa^{-1} g\right) \right\|^2_{L^2(\mathcal{H}_\tau)} d(gN)$$

$$= \int_{G/N} \|\varphi(g)\|^2_{L^2(\mathcal{H}_\tau)} \sum_{\kappa \in \Lambda} \left\| d\left(\kappa^{-1} g\right) f\left(\kappa^{-1} g\right) \right\|^2_{L^2(\mathcal{H}_\tau)} d(gN)$$

for every $\varphi \in L^2\left(G/N, \mathcal{H}_\tau\right)$ and additionally, there exist positive real numbers $m_{\Lambda,f} \le M_{\Lambda,f}$ such that

$$\text{esssup}_{g \in G} \left(\sum_{\kappa \in \Lambda} \left\| d\left(\kappa^{-1} g\right) f\left(\kappa^{-1} g\right) \right\|^2_{L^2(\mathcal{H}_\tau)} \right) = m_{\Lambda,f}$$

$$\text{esssup}_{g \in G} \left(\sum_{\kappa \in \Lambda} \left\| d\left(\kappa^{-1} g\right) f\left(\kappa^{-1} g\right) \right\|^2_{L^2(\mathcal{H}_\tau)} \right) = M_{\Lambda,f}.$$

Then $\pi(\Lambda\Gamma) f$ is a frame for $L^2\left(G/N, \mathcal{H}_\tau\right)$.

The first condition in Theorem 7.6 can be viewed as a generalized version of the pseudo-Parseval's equality for discrete Fourier series. This first condition would hold for instance if there exists a subset E of G/N of positive measure such that letting σ_E be the following direct integrals of representations of N

$$\sigma_E = \int_E^{\oplus} \tau_s d(sN) = \int_E^{\oplus} \tau\left(s^{-1} \cdot s\right) d(sN),$$

then $\sigma_E(\Gamma) f$ is a frame for $L^2\left(E, L^2\left(\mathcal{H}_\tau\right)\right)$.

The second condition suggests that the family of vectors

$$\left\{ g \mapsto \left\| d\left(\kappa^{-1}g\right) f\left(\kappa^{-1}g\right) \right\|_{L^2(\mathcal{H}_\tau)}^2 : \kappa \in \Lambda \right\}$$

defines a relaxed bounded partition of unity on the homogeneous space G/N. Note also that in the special case where $m_{\Lambda,f} = M_{\Lambda,f}$, the system of vectors $\pi(\Lambda\Gamma)f$ would be a Parseval frame for $L^2\left(G/N, \mathcal{H}_\tau\right)$.

References

1 Hans G. Feichtinger and Karlheinz Gröchenig. A unified approach to atomic decompositions via integrable group representations. In *Function Spaces and Applications: Proceedings of the US-Swedish Seminar held in Lund, Sweden, June 15–21, 1986*, pages 52–73. Springer, 2006.

2 Gerald B. Folland. *A course in abstract harmonic analysis*. Studies in Advanced Mathematics, pages x+276. CRC Press, Boca Raton, FL, 1995. ISBN 0-8493-8490-7.

3 Gerald B. Folland. *A course in abstract harmonic analysis*, volume 29. CRC Press, 2016.

4 Hartmut Führ. Coorbit spaces and wavelet coefficient decay over general dilation groups. *Transactions of the American Mathematical Society*, 367(10):7373–7401, 2015.

5 Gitta Kutyniok and Demetrio Labate. Introduction to shearlets. In *Shearlets*, pages 1–38. Springer, 2012.

6 Joseph Albert Wolf. *Unitary representations of maximal parabolic subgroups of the classical groups*, volume 180. American Mathematical Society, 1976.

8

Frames on Homogeneous Spaces

In this chapter, we present a pioneering approach for the construction of frames on homogeneous spaces of the form G/H, with G being a Lie group and H a closed subgroup of G. Our primary focus is on the unit sphere, a homogeneous space obtained through the quotient of a matrix group isomorphic to $SO(n)$ by a subgroup isomorphic to $SO(n-1)$. The frames we construct are defined as continuous functions, inheriting their structure from frames for $L^2(SO(n))$, and we provide rigorous and explicit frame bounds. We also demonstrate the practical utility of our method through the construction of frames on the Klein bottle.

8.1 Localized Frames on Homogeneous Spaces

Let H be a matrix Lie group endowed with a fixed Haar measure $d\mu_H$. We have seen in the previous chapter (see Theorems 7.1 7.2, and 7.3) that H can be embedded conormally into a larger matrix group, admitting a unitary representation, which can then be discretized to construct a family of localized and smooth frames for $L^2(H, d\mu_H)$. Precise bounds were obtained and in several other situations, the construction of smooth and well-localized Parseval frames for $L^2(H, d\mu_H)$ has shown to be possible. The methods developed in Theorems (see Theorems 7.1 7.2 and 7.3) are explicit and the scope of the work, broad. However, since the techniques of the previous chapter heavily rely on the assumption that H is a group, it may not be obvious how to extend such results to other types of manifolds lacking group structure.

Let us suppose that H is a Lie group with a Lie subgroup K. Let us briefly entertain the following question: How should one go about to construct well-localized frames on the homogenous space H/K? A first insight could be to initiate such a construction on H and then somehow disintegrate these frames over a smooth transversal for H/K. To enact this insight, it might be tempting to consider the adjoint action of H on its Lie algebra $\mathfrak{h}$. Indeed, the adjoint action of H determines

A Bridge Between Lie Theory and Frame Theory: Applications of Lie Theory to Harmonic Analysis, First Edition. Vignon Oussa.

a unique semi-direct product structure of the form $G = \mathfrak{h} \rtimes H$ equipped with group operation $(X, h)(Y, s) = (X + Ad(h)Y, hs)$. Let χ be a unitary character of $\mathfrak{h}$. Then, there exists a linear functional $\lambda \in \mathfrak{h}^*$ such that $\chi(X) = e^{2\pi i \langle \lambda, X \rangle}$. Letting π be a unitary representation of G obtained by inducing such a character, π acts in $L^2(H, d\mu_H)$ as:

$$\left[\pi(X, h)f\right](s) = e^{2\pi i \left\langle Ad(s^{-1})^* \lambda, X \right\rangle} f\left(h^{-1}s\right)$$

for $X \in \mathfrak{h}$, $h, s \in H$. In order to apply the methods of Theorems (7.1, 7.2, and 7.3), we require that $s \mapsto Ad\left(s^{-1}\right)^* \lambda$ is an immersion at some point on H. Note that generally, any coadjoint orbit must be of even dimension. Since any Lie group and its Lie algebra share the same dimension, we cannot expect this condition to hold in every case. For instance, if H is the three-dimensional Heisenberg Lie group, then a coadjoint orbit is of dimension at most two, and the map $s \mapsto Ad\left(s^{-1}\right)^* \lambda$ can never be locally injective. This fact extends to all Lie groups with nontrivial centers. Suppose we make the assumption that H is a linear group. Fixing a faithful representation of H inside a general linear group of sufficiently large order, we may then embed H into this general linear group. As a subgroup, H inherits the natural multiplication action of the larger group on its Lie algebra. At this point, H acts on a vector space that is generally (in terms of its dimension) larger than its Lie algebra, and one would expect that there exists some linear functional λ with respect to which, $s \mapsto Ad\left(s^{-1}\right)^* \lambda$ is an immersion.

Armed with the tools provided by Theorems 7.1, 7.2, and 7.3, we aim to turn these insights into robust methods for constructing nice frames on homogeneous spaces.

Let H be a connected Lie group with Lie algebra $\mathfrak{h}$ and let $\mathfrak{g} = \mathfrak{n} \oplus \mathfrak{h}$ such that $\mathfrak{n}$ is an exponential solvable ideal of $\mathfrak{g}$ satisfying $\dim(\mathfrak{n}) = n \geq \dim(\mathfrak{h}) = r$. Let us assume additionally that $N = \exp \mathfrak{n}$ and H are closed and connected subgroups of G. By assumption, since the exponential map defines a global diffeomorphism between $\mathfrak{n}$ and N, N is a simply connected normal (exponential) solvable subgroup of G. Moreover, we assume that $\lambda \in \mathfrak{n}^*$ such that the derived algebra $[\mathfrak{n}, \mathfrak{n}]$ is inside the kernel of λ. Consequently, $\exp X \mapsto \chi_\lambda(\exp X) = e^{2\pi i \langle \lambda, X \rangle}$ defines a unitary character of N. Next, let $\pi_\lambda = \mathrm{ind}_N^{NH}\left(\chi_\lambda\right)$ be a unitary representation obtained by inducing χ_λ. Realize π as acting in $\mathfrak{H} = L^2\left(H, d\mu_H\right)$ such that, for $x \in N$ and $z \in H$,

$$[\pi(xz)]f(h) = e^{2\pi i \left\langle Ad(h^{-1})^* \lambda, \log(x) \right\rangle} f\left(z^{-1}h\right).$$

Define $\beta_\lambda : H \to \mathfrak{n}^*$ such that $\beta_\lambda(h) = Ad\left(h^{-1}\right)^* \lambda$. Then as shown in Remark 7.4, there exists a linear functional λ such that β_λ is an immersion at the identity of the group. By Theorem 7.3, there exists a relatively compact neighborhood $\mathcal{O}$ of the identity in H such that if f is a continuous function supported on $\mathcal{O}$, then there exists a countable set $\Gamma \subset G$ such that $\pi(\Gamma)f$ is a "nice" frame for $L^2\left(H, d\mu_H\right)$.

Recall also that an H-homogeneous space is a smooth manifold endowed with a transitive smooth H-action denoted $*$. If M is a homogeneous H space, and p a point in M, then the isotropy group H_p (the collection of all elements of H fixing p) is a closed Lie subgroup of H and $hH_p \mapsto h * p$ is a diffeomorphism from H/H_p to M. Next, let K be a closed Lie subgroup of H. There is a unique way of turning $H/K = \{hK : h \in K\}$ into a smooth manifold such that the map $H \to H/K$ assigning h to hK is a smooth quotient map (see Proposition 3.37). Then H acts on H/K by the left action $(x, hK) \mapsto xhK$. This is a smooth and transitive action that turns H/K into an H-homogeneous space.

Theorem 8.1 Let H be a Lie group and let K be a closed subgroup of G. Let us assume that $\det\left(Ad_H(k)\right) = \det\left(Ad_K(k)\right)$ for all $k \in K$, so that there exists a nonzero positive invariant regular Borel measure on H/K which we denote by $d\mu_{H/K}$. Given a frame $\pi(\Gamma)f$ for $L^2\left(H, d\mu_H\right)$ such that $f \in C_c(H)$ and letting $d\mu_K$ be a Haar measure on K and E a fixed compact subset of K of positive measure in K, if

$$\psi_\gamma(x) = \int_E \left[\pi(\gamma) f\right](xk)\, d\mu_K(k)$$

for $\gamma \in \Gamma$ then the family of vectors $\left(\psi_\gamma\right)_{\gamma\in\Gamma}$ is a frame for $L^2\left(H/K, d\mu_{H/K}\right)$.

Proof: Given that $\pi(\Gamma)f$ is a frame for $L^2\left(H, d\mu_H\right)$, for arbitrary $g \in L^2\left(H, d\mu_H\right)$, there exist positive real numbers A, B such that

$$A\,\|g\|^2_{L^2(H,d\mu_H)} \le \sum_{\gamma\in\Gamma}\left|\int_H g(h)\,\overline{\left[\pi(\gamma)f\right](h)}\,d\mu_H(h)\right|^2 \le B\,\|g\|^2_{L^2(H,d\mu_H)}.$$

For a more compact notation, simple write the above compound inequality as:

$$\sum_{\gamma\in\Gamma}\left|\int_H g(h)\,\overline{\pi(\gamma)f(h)}\,d\mu_H(h)\right|^2 \asymp \|g\|^2_{L^2(H,d\mu_H)}.$$

Following Proposition 4.9, disintegrate the Haar measure on H into an H-invariant measure on H/K and a fixed left-invariant measure on K. This gives

$$\int_H g(h)\,\overline{\pi(\gamma)f(h)}\,d\mu_H(h) = \int_{H/K}\int_K g(xk)\,\overline{\left[\pi(\gamma)f\right](xk)}\,d\mu_K(k)\,d\mu_{H/K}(xK).$$

Suppose that $g = g_1 \cdot 1_E$, where E is a fixed compact subset of K and g_1 is arbitrarily fixed in $L^2\left(H/K, d\mu_{H/K}\right)$. Then the disintegration trick gives

$$\sum_{\gamma\in\Gamma}\left|\int_H g(h)\,\overline{\pi(\gamma)f(h)}\,d\mu_H(h)\right|^2$$

$$= \sum_{\gamma\in\Gamma}\left|\int_{H/K}\int_K g(xk)\,\overline{\left[\pi(\gamma)f\right](xk)}\,d\mu_K(k)\,d\mu_{H/K}(xK)\right|^2.$$

Pulling g_1 out of the inner integral,

$$\sum_{\gamma\in\Gamma}\left|\int_H g(h)\overline{\pi(\gamma)f(h)}d\mu_H(h)\right|^2$$
$$=\sum_{\gamma\in\Gamma}\left|\int_{H/K} g_1(x)\cdot\left(\int_E \overline{[\pi(\gamma)f](xk)}d\mu_K(k)\right)d\mu_{H/K}(xK)\right|^2.$$

For ease of notation, put $\psi_\gamma(x)=\int_E[\pi(\gamma)f](xk)\,d\mu_K(k)$. Since each $\pi(\gamma)f$ is continuous and compactly supported, then it is clear that each $\psi_\gamma\in C_c(H/K)$ and

$$\sum_{\gamma\in\Gamma}\left|\int_H g(h)\overline{\pi(\gamma)f(h)}d\mu_H(h)\right|^2=\sum_{\gamma\in\Gamma}\left|\int_{H/K} g_1(x)\cdot\overline{\psi_\gamma(x)}d\mu_{H/K}(xK)\right|^2.$$

Computing the norm of g with the "disintegration trick,"

$$\|g\|^2_{L^2(H,d\mu_H)}=\int_H|g(x)|^2d\mu_H(x)=\int_{H/K}\int_K|g(xk)|^2d\mu_K(k)\,d\mu_{H/K}(xK)$$

and

$$\|g\|^2_{L^2(H,d\mu_H)}=\int_{H/K}|g_1(x)|^2\cdot\left(\int_E d\mu_K(k)\right)d\mu_{H/K}(xK)$$
$$=\mu_K(E)\cdot\|g_1\|^2_{L^2(H/K,d\mu_{H/K})}.$$

Therefore,

$$\sum_{\gamma\in\Gamma}\left|\int_{H/K}\left(g_1(x)\cdot\overline{\psi_\gamma(x)}\right)d\mu_{H/K}(xK)\right|^2\asymp\|g_1\|^2_{L^2(H/K,d\mu_{H/K})}$$

and the family of vectors $(\psi_\gamma)_{\gamma\in\Gamma}$ is a frame for $L^2(H/K,d\mu_{H/K})$ with frame bounds $A\cdot\mu_K(E)\leq B\cdot\mu_K(E)$.

For the more general case where H/K does not necessarily admit an H-invariant measure, we propose the following. Suppose that $H=TK$ where T is a submanifold of H, which is a measurable transversal for H/K and additionally, $\alpha:(t,k)\mapsto tk$ defines a diffeomorphism between $T\times K$ and H. Suppose

$$d\mu_H\left(\alpha^{-1}(t,k)\right)=r(t,k)\,d\mu_T(t)\,d\mu_K(k)$$

where μ_T is a Lebesgue measure on T and μ_K is a fixed Haar measure on K. Then the following holds true:

Theorem 8.2 Given a frame $\pi(\Gamma)f$ for $L^2(H,d\mu_H)$ such that $f\in C_c(H)$ and letting E be a fixed compact subset of K of positive measure in K, if additionally,

$$\psi_\gamma(t)=\int_E\left(\left[\pi(\gamma)f\right]\left(\alpha^{-1}(tk)\right)\cdot r(t,k)\right)d\mu_K(k)$$

for $\gamma\in\Gamma$ then the family of vectors $(\psi_\gamma)_{\gamma\in\Gamma}$ is a frame for $L^2(T,d\mu_T)$.

Proof: Given that $\pi(\Gamma)f$ is a frame for $L^2(H, d\mu_H)$, for arbitrary $g \in L^2(H, d\mu_H)$, there exist positive real numbers A, B such that

$$\sum_{\gamma\in\Gamma}\left|\int_H g(h)\overline{\pi(\gamma)f(h)}d\mu_H(h)\right|^2 \asymp \|g\|^2_{L^2(H,d\mu_H)}.$$

Thus,

$$\sum_{\gamma\in\Gamma}\left|\int_H g(h)\overline{[\pi(\gamma)f](h)}d\mu_H(h)\right|^2$$

$$= \sum_{\gamma\in\Gamma}\left|\int_T\int_K g\left(\alpha^{-1}(tk)\right)\overline{[\pi(\gamma)f]\left(\alpha^{-1}(tk)\right)}\cdot r(t,k)\,d\mu_K(k)\,d\mu_T(t)\right|^2.$$

Suppose that $g\left(\alpha^{-1}(tk)\right) = g_1(t)\cdot 1_E(k)$, where g_1 is arbitrarily fixed in the Hilbert space $L^2(T, d\mu_T)$. Then

$$\sum_{\gamma\in\Gamma}\left|\int_H g(h)\overline{[\pi(\gamma)f](h)}d\mu_H(h)\right|^2$$

$$= \sum_{\gamma\in\Gamma}\left|\int_T g_1(t)\left(\int_E\overline{[\pi(\gamma)f]\left(\alpha^{-1}(tk)\right)\cdot r(t,k)\,d\mu_K(k)}\right)d\mu_T(t)\right|^2$$

and as a result, the family of vectors $\left(\int_E[\pi(\gamma)f]\left(\alpha^{-1}(tk)\right)\cdot r(t,k)\,d\mu_K(k)\right)_{\gamma\in\Gamma}$ is a frame for $L^2(T, d\mu_T)$.

Making the case that the scheme proposed in this section may be helpful for concrete applications, we will explore a few examples, in which we make explicit ingredients provided by the recipe described above.

8.2 Frames on Spheres

Let

$$H = \left\{ [0,h] = \begin{bmatrix} h & 0_n \\ 0_n & \mathrm{Id}_n \end{bmatrix} \in GL(2n,\mathbb{R}) : h \in SO(n,\mathbb{R}) \right\} \quad \text{with } n \geq 3. \tag{8.1}$$

Fix

$$p = \begin{bmatrix} e_n \\ \mathbf{0} \end{bmatrix} \in \mathbb{R}^{2n} \quad \text{where } e_n = \begin{bmatrix} 0 \\ \vdots \\ 0 \\ 1 \end{bmatrix} \in \mathbb{R}^n, \quad \mathbf{0} = \begin{bmatrix} 0 \\ \vdots \\ 0 \\ 0 \end{bmatrix} \in \mathbb{R}^n \tag{8.2}$$

and let K be the stabilizer subgroup of p. That is, $K = \{h \in H : hp = p\}$ and K is a closed subgroup of H, which is isomorphic to $SO(n-1,\mathbb{R})$. Moreover, the manifold Hp, which is diffeomorphic to H/K, is just the $(n-1)$-dimensional sphere embedded in $\mathbb{R}^{2n}$. This realization of the sphere as a submanifold of $\mathbb{R}^{2n}$ as opposed to a submanifold of $\mathbb{R}^n$ allows us to take advantage of the results for the construction of frames on Lie groups in the previous chapter. Note also that since Theorem 7.4 can be applied to the specific case, where $n=2$, since S^1 is just the one-dimensional torus (it is a Lie group), then no case is lost under the assumption that $n \geq 3$. Indeed, for the case where $n=2$, we may consider the universal covering group of the Euclidean motion group of the plane. Let $G = \mathbb{R}^2 \rtimes \mathbb{R}$ be a simply connected, connected unimodular solvable semidirect product group with multiplication law described as follows. Given $(v,s),(w,t) \in \mathbb{R}^2 \rtimes \mathbb{R}$,

$$(v,s)(w,t) = (v + s \bullet w, s+t)$$

and

$$s \bullet w = \begin{bmatrix} \cos(s) & -\sin(s) \\ \sin(s) & \cos(s) \end{bmatrix} \begin{bmatrix} w_1 \\ w_2 \end{bmatrix} = \begin{bmatrix} w_1 \cos(s) - w_2 \sin(s) \\ w_2 \cos(s) + w_1 \sin(s) \end{bmatrix}.$$

The center of G is the discrete set $\left\{ \left([0,0]^T, \ell \right) : \ell \in 2\pi\mathbb{Z} \right\}$. For a fixed nonzero linear functional

$$\lambda = \begin{bmatrix} \lambda_1 \\ \lambda_2 \end{bmatrix} \in \left(\mathbb{R}^2\right)^*,$$

we consider the monomial representation $\pi_\lambda = \mathrm{ind}_{\mathbb{R}^2}^{\mathbb{R}^2 \rtimes \mathbb{R}} \left(\chi_\lambda\right)$. Given $f \in L^2(\mathbb{R})$ and $(v,t) \in \mathbb{R}^2 \rtimes \mathbb{R}$,

$$\left[\pi_\lambda(v,t) f\right](s) = \exp\left(2\pi i \left\langle \beta_\lambda(s), \begin{bmatrix} v_1 \\ v_2 \end{bmatrix} \right\rangle\right) f(s-t)$$

and $\beta_\lambda : \mathbb{R} \to \mathbb{R}^2$

$$\beta_\lambda(s) = \begin{bmatrix} \lambda_1 \cos s - \lambda_2 \sin s \\ \lambda_2 \cos s + \lambda_1 \sin s \end{bmatrix}.$$

The Jacobian of β_λ evaluated at zero is $\begin{bmatrix} -\lambda_2 \\ \lambda_1 \end{bmatrix}$. Since λ is not the zero vector, then it is clear β_λ is an immersion at zero.

From here on, we assume that $n > 2$. Let

$$N = \left\{ [x,e] = \begin{bmatrix} \mathrm{Id}_n & x \\ 0_n & \mathrm{Id}_n \end{bmatrix} \in \mathrm{GL}(2n,\mathbb{R}) : x \in \mathfrak{gl}(n,\mathbb{R}) \right\}$$

be a simply connected abelian closed subgroup of $\mathrm{GL}(2n,\mathbb{R})$ and set $G = NH$. Then N is a normal abelian closed subgroup of G. Next, the pairing of $\left(X, \lambda_\omega\right) \in \mathfrak{n} \times \mathfrak{n}^*$ is defined as the trace of the matrix obtained by multiplying the transpose of λ_ω with X. For a linear functional λ_ω, where $\lambda_\omega = \begin{bmatrix} 0_n & \omega \\ 0_n & 0_n \end{bmatrix} \in \mathfrak{n}^*$,

$$N \ni z \mapsto \chi_{\lambda_\omega}(z) = e^{2\pi i \cdot \mathrm{Tr}\left(\lambda_\omega^T \log(z)\right)}$$

defines a unitary character of N. Next, we consider an irreducible unitary representation π of G obtained by inducing the character $\chi_{\lambda_{\mathrm{Id}_n}}$ from N to G. Then π is realized as acting on $L^2(H, d\mu_H) = L^2(SO(n,\mathbb{R}))$ as follows. For $f \in L^2(H, d\mu_H)$, let

$$[\pi(g)f](h) = f(g^{-1}h)$$

where $g \in H$ and $g = [x, e] \in N$. Then

$$[\pi(g)]f(h) = e^{2\pi i \cdot \mathrm{Tr}\left(\begin{bmatrix} 0_n & h \\ 0_n & 0_n \end{bmatrix}^T \begin{bmatrix} 0_n & x \\ 0_n & 0_n \end{bmatrix}\right)} f(h). \tag{8.3}$$

For a positive real number $\epsilon > 0$, put

$$\mathfrak{o}_\epsilon = \{A \in \mathfrak{h} : \|A\|_{\mathrm{HS}} < \epsilon\},$$

$\exp \mathfrak{o}_\epsilon = \mathcal{O}_\epsilon \subset H$ and define a linear projection projof rank $\frac{n^2-n}{2} = \dim H$ on the vector space $\mathfrak{gl}(n,\mathbb{R})$ as follows. The function proj maps a square matrix of order n to a matrix obtained by replacing all entries on and below the diagonal by zeros. Moreover, all entries strictly above the diagonal remain unchanged under the action of proj.

Define $\Theta_\lambda : \mathfrak{o}_\epsilon \longrightarrow \mathfrak{n}^*$ such that for $a = (a_1, \ldots, a_{\frac{n^2-n}{n}}) \in \mathbb{R}^{\frac{n^2-n}{n}}$,

$$\Theta_\lambda(E(a)) := \Theta_\lambda(a) = \begin{bmatrix} 0_n & \mathrm{proj}(\exp(E(a))) \\ 0_n & 0_n \end{bmatrix}$$

and

$$E(a) = \begin{bmatrix} 0 & -a_1 & \cdots & -a_{n-2} & -a_{n-1} \\ a_1 & 0 & -a_n & \cdots & -a_{2n-3} \\ \vdots & a_n & 0 & \cdots & \vdots \\ a_{n-2} & \cdots & \cdots & \ddots & -a_{\frac{n^2-n}{2}} \\ a_{n-1} & a_{2n-3} & \cdots & a_{\frac{n^2-n}{2}} & 0 \end{bmatrix}.$$

For a sufficiently small positive ϵ, Θ_λ maps $\mathfrak{o}_\epsilon$ diffeomorphically into its image. Fixing such ϵ, let $W_\lambda : \Theta_\lambda(\log(\mathcal{O}_\epsilon)) \to (0,\infty)$ such that

$$W_\lambda(\xi) = \frac{d\mu_H\left(\exp\left(\Theta_\lambda^{-1}(\xi)\right)\right)}{d\xi}.$$

Note that the function W_λ is just the Radon–Nikodym derivative of the push-forward of the Haar measure on H via the diffeomorphism $\Theta_\lambda \circ \log$ with respect to the Lebesgue measure on the range of Θ_λ. By pulling W_λ back through $\Theta_\lambda \circ \log$, we obtain $w : \mathcal{O}_\epsilon \to (0,\infty)$ such that $w(h) = W_\lambda(\Theta_\lambda(\log(h)))$. Next, w is a positive and smooth function on $\mathcal{O}_\epsilon$ (see (8.23), which plays a central role in the construction of localized and structured frames on the Orthogonal Lie groups (see Theorem 7.4). An additional essential ingredient for our scheme is a discrete set

$$\Lambda \subset \left\{ \begin{bmatrix} \mathrm{Id}_n & \mathrm{proj}\,(x) \\ 0_n & \mathrm{Id}_n \end{bmatrix} : x \in \mathfrak{gl}\,(n, \mathbb{R}) \right\} \subset N$$

together with a compact set

$$C \subset \mathfrak{p}^* = \left\{ \begin{bmatrix} 0_n & \mathrm{proj}\,(\omega) \\ 0_n & 0_n \end{bmatrix} : \omega \in \mathfrak{gl}\,(n, \mathbb{R}) \right\} \subset \mathfrak{n}^*$$

such that

$$\left\{ \xi \mapsto \frac{e^{2\pi i \langle \xi, Y \rangle}}{|C|^{1/2}} : Y \in \log(\Lambda) \right\}$$

is an orthonormal basis for $L^2(C, d\xi)$. The existence of the pair (Λ, C) is a direct consequence of our construction. For any function f, which is continuous (or smooth) and supported on a connected neighborhood of the identity $\mathcal{O}_\epsilon$, there is a finite set $\Gamma_H \subset H$ such that $\left\{ \pi(\gamma) f : \gamma \in \Gamma_H^{-1} \Lambda \right\}$ is a frame for $L^2(SO(n, \mathbb{R}))$ with frame bounds

$$\begin{aligned} 0 &< \operatorname{essinf}_{h \in H} \left(\sum_{\ell \in \Gamma_H} \left| f(\ell h) \sqrt{w(\ell h)} \right|^2 \right) \cdot |C| \\ &\leq \operatorname{esssup}_{h \in H} \left(\sum_{\ell \in \Gamma_H} \left| f(\ell h) \sqrt{w(\ell h)} \right|^2 \right) \cdot |C| < \infty. \end{aligned}$$

Let

$$Z_1 = \begin{bmatrix} 0 & & -1 & & & \\ & 0 & & & & \\ & & \ddots & & & \\ 1 & & 0 & & & \\ & & & 0 & \cdots & 0 \\ & & & \vdots & \ddots & \vdots \\ & & & 0 & \cdots & 0 \end{bmatrix}, \dots, Z_{n-1} = \begin{bmatrix} 0 & & & & & \\ & \ddots & & & & \\ & & 0 & -1 & & \\ & & 1 & 0 & & \\ & & & & 0 & \cdots & 0 \\ & & & & \vdots & \ddots & \vdots \\ & & & & 0 & \cdots & 0 \end{bmatrix} \tag{8.4}$$

where $\in \mathfrak{h} \subset \mathfrak{gl}\,(2n, \mathbb{R})$. To integrate functions on the unit sphere Hp, let $d\sigma$ be the standard surface measure on the sphere S^{n-1}, which is explicitly given by the following formula. For

$$\mathbf{I} = \underbrace{\left[-\frac{\pi}{2}, \frac{\pi}{2}\right) \times \cdots \times \left[-\frac{\pi}{2}, \frac{\pi}{2}\right)}_{n-2 \text{ times}} \times [-\pi, \pi)$$

and a suitable subset B of S^{n-1},

$$\sigma(B) = \int_{\mathbf{I}} \left(\mathbf{1}_B\left(\exp\left(t_{n-1}Z_1\right)\cdots\exp\left(t_1 Z_1\right)p\right)\right.$$
$$\left.\cdot\cos^{n-2}\left(t_1\right)\cos^{n-3}\left(t_2\right)\cdots\cos\left(t_{n-2}\right)\right)dt. \tag{8.5}$$

Then σ is an H-invariant measure on the sphere and such a measure is uniquely determined up to multiplication by a constant (see Proposition 4.9). Put

$$s(t) = \exp\left(t_{n-1}Z_{n-1}\right)\cdots\exp\left(t_1 Z_1\right) \in H.$$

For $\Gamma \subset G, f \in C(H)$, let

$$\mathcal{F}(f,\Gamma) = \left\{ s(t)p \mapsto \int_K \left[\pi(\gamma)f\right](s(t)k)\,d\mu_K(k) : \gamma \in \Gamma \right\}$$

be a collection of functions defined on the sphere $Hp = \{s(t)p : t \in \mathbf{I}\}$. It is worth highlighting that if $\{\pi(\gamma)f : \gamma \in \Gamma\}$ is a countable family of functions in $L^2(SO(n,\mathbb{R}))$ then the corresponding elements of $\mathcal{F}(f,\Gamma)$ are obtained by integrating $K \ni k \mapsto \left[\pi(\gamma)f\right](s(t)k)$ with respect to the Haar measure on $K \simeq SO(n-1,\mathbb{R})$.

The following shows that the procedure described above produces frames (consisting of continuous functions) on the unit sphere Hp with good localization property.

Theorem 8.3 Let $f \in C(H)$ be a nonnegative function which is properly supported on $\mathcal{O}_\epsilon \subset H$. There is a finite subset Γ_H of H such that the collection $\mathcal{F}\left(f, \Gamma_H^{-1}\Lambda\right)$ is a frame for $L^2\left(S^{n-1}, d\sigma\right)$ with frame bounds

$$0 < \|f \cdot w\|_{L^\infty(H)} \cdot |C| \le \|fw\|_{L^\infty(H)} \cdot \sharp\left(\Gamma_H\right) \cdot |C| < \infty.$$

Moreover, $\mathcal{F}\left(f, \Gamma_H^{-1}\Lambda\right)$ consists of continuous functions defined on the unit sphere and the $L^2\left(S^{n-1}, d\sigma\right)$-norm of every element of $\mathcal{F}\left(f, \Gamma_H^{-1}\Lambda\right)$ is bounded above by $\|f\|_{L^\infty(H)}$.

Proof of Theorem 8.3

Our main aim here is to use Theorem 7.4 to give an explicit construction of frames for $L^2\left(S^{n-1}\right)$. To this end, we proceed in stages.

Step 1: We first give an explicit formula for β_λ. Given $\left[0,h\right] \in H, h \in SO(n,\mathbb{R})$, since h is an orthogonal matrix, its transpose and inverse coincide and as such

$$\beta_\lambda\left(\left[0,h\right]\right) = \begin{bmatrix} 0_n & \left(h^{-1}\right)^T \\ 0_n & 0_n \end{bmatrix} = \begin{bmatrix} 0_n & h \\ 0_n & 0_n \end{bmatrix}. \tag{8.6}$$

Step 2: Secondly, we compute Θ_λ. To this end, let $A_1, \dots, A_{\frac{n^2-n}{2}} \in \mathfrak{so}(n, \mathbb{R})$ such that

$$\sum_{k=1}^{\frac{n^2-n}{2}} a_k A_k = \begin{bmatrix} 0 & -a_1 & \cdots & -a_{n-2} & -a_{n-1} \\ a_1 & 0 & -a_n & \cdots & -a_{2n-3} \\ \vdots & a_n & 0 & \cdots & \vdots \\ a_{n-2} & \cdots & \cdots & \ddots & -a_{\frac{n^2-n}{2}} \\ a_{n-1} & a_{2n-3} & \cdots & a_{\frac{n^2-n}{2}} & 0 \end{bmatrix} = E(a) \tag{8.7}$$

where $a = \left(a_1, a_2, \dots, a_{\frac{n^2-n}{2}}\right) \in \mathbb{R}^{\frac{n^2-n}{2}}$. Then the collection of matrices

$$\left\{ \begin{bmatrix} A_k & 0_n \\ 0_n & 0_n \end{bmatrix} : 1 \leq k \leq \frac{n^2-n}{2} \right\}$$

forms a basis for the Lie algebra $\mathfrak{h}$. Since the exponential map is a local diffeomorphism at the identity of H, there is a positive real number ϵ such that for

$$\mathfrak{o} = \mathfrak{o}_\epsilon = \left\{ A \in \mathfrak{h} : \|A\|_{HS} = \sqrt{\sum_{k=1}^{n} \sum_{j=1}^{n} A_{j,k}^2} < \epsilon \right\} \quad \text{and} \quad \exp \mathfrak{o} = \mathcal{O} \subset H, \tag{8.8}$$

the exponential map defines an analytic diffeomorphism between $\mathfrak{o}$ and $\mathcal{O}$. Its inverse is the logarithmic function, which is an analytic function defined as:

$$\log(A) = \sum_{k=1}^{\infty} (-1)^{k+1} \frac{(A - \mathrm{Id})^k}{k}$$

for $A \in \mathfrak{o}$. Thus, for any matrix $[0, h] \in \mathcal{O}$, there is a unique matrix

$$F(a) = \begin{bmatrix} E(a) & 0_n \\ 0_n & 0_n \end{bmatrix} \in \mathfrak{o} \subset \mathfrak{h} \text{ such that } [0, h] = \begin{bmatrix} \exp(E(a)) & 0_n \\ 0_n & \mathrm{Id}_n \end{bmatrix}. \tag{8.9}$$

Proposition 8.1 The exponential map defines an analytic diffeomorphism between $\mathfrak{o}_\pi = \{A \in \mathfrak{h} : \|A\|_{\mathrm{HS}} < \pi\}$ and $\exp(\mathfrak{o}_\pi) = \mathcal{O}_\pi$.

Proof: By the spectral theorem, given $a \in \mathbb{R}^{\frac{n^2-n}{2}}$, since $E(a)$ is skew-symmetric, there is an orthogonal matrix $Q(a)$ such that

$$\exp(E(a)) = Q(a) \exp(B(a)) Q(a)^{-1}$$

and

$$\exp(B(a)) = \begin{cases} \mathrm{Diag}\left(R(\tau_1(a)), \ldots, R\left(\tau_{\frac{n}{2}}(a)\right)\right) & \text{if } n \text{ is even} \\ \mathrm{Diag}\left(R(\tau_1(a)), \ldots, R\left(\tau_{\frac{n-1}{2}}(a)\right), 1\right) & \text{if } n \text{ is odd} \end{cases} \tag{8.10}$$

where

$$R(\theta) = \begin{bmatrix} \cos\theta & -\sin\theta \\ \sin\theta & \cos\theta \end{bmatrix}, \qquad \theta \in \mathbb{R}$$

is a rotation matrix and each τ_j is a real-valued multivariable function of $\mathbb{R}^{\frac{n^2-n}{2}}$. Regarding the injectivity of the exponential map, suppose that $\exp(tA(a)) = \exp(sA(b))$ for real numbers s, t. There are two cases to consider. First, if $a, b \in \mathbb{R}^{\frac{n^2-n}{2}}$ are distinct, then it is clear that $A(a), A(b)$ generate two distinct one-parameter groups and as such,

$$\exp(tA(a)) = \exp(sA(b)) \Leftrightarrow t = s = 0.$$

Otherwise if $a = b$, then

$$Q(a)\exp(tB(a))\,Q(a)^{-1} = Q(a)\exp(sB(a))\,Q(a)^{-1},$$

and as such, it is necessarily the case that

$$\text{either (even } n) \begin{cases} R(t\tau_1(a)) = R(s\tau_1(a)) \\ \vdots \\ R\left(t\tau_{\frac{n}{2}}(a)\right) = R\left(s\tau_{\frac{n}{2}}(a)\right) \end{cases} \quad \text{or}$$

$$(\text{odd } n) \begin{cases} R(t\tau_1(a)) = R(s\tau_1(a)) \\ \vdots \\ R\left(t\tau_{\frac{n-1}{2}}(a)\right) = R\left(s\tau_{\frac{n-1}{2}}(a)\right) \end{cases}.$$

In light of these observations, in order to find an open set around the identity on which the exponential map is injective, fix

$$\mathfrak{o} = \begin{cases} \left\{F(a) \in \mathfrak{h} : \left|\tau_j(a)\right| < \pi, 1 \le j \le \frac{n}{2}\right\} & \text{if } n \text{ is even} \\ \left\{F(a) \in \mathfrak{h} : \left|\tau_j(a)\right| < \pi, 1 \le j \le \frac{n-1}{2}\right\} & \text{if } n \text{ is odd} \end{cases}. \tag{8.11}$$

Note that, in a generic sense, it is extremely hard to compute the eigenvalues of a skew-symmetric matrix. However, we can easily compute tractable bounds for the eigenvalues of $E(a)$. To this end, we remark that the singular values of the skew-symmetric matrix $E(a)$ coincide with the absolute values of the eigenvalues of $E(a)$. Letting, $|||A|||_2$ be the (spectral) norm obtained by taking the largest singular value of A, it is then clear that

$$|||E(a)|||_2 = \begin{cases} \max\left\{|\tau_1(a)|, \cdots, \left|\tau_{\frac{n}{2}}(a)\right|\right\} & \text{if } n \text{ is even} \\ \max\left\{|\tau_1(a)|, \cdots, \left|\tau_{\frac{n}{2}-1}(a)\right|, 1\right\} & \text{if } n \text{ is odd} \end{cases}.$$

Furthermore, [3, Chapter 5] although $|||E(a)|||_2$ may be hard to calculate, it is known that the spectral norm of any matrix is bounded above by its Hilbert Schmidt norm, which is much easier to obtain. In other words, for $a \in \mathbb{R}^{\frac{n^2-n}{2}}$,

$$\|E(a)\|_2 \leq \|E(a)\|_{HS} = \sqrt{\sum_{k=1}^{n}\sum_{j=1}^{n}\left(E(a)_{j,k}\right)^2}$$

and the desired result follows from this inequality together with (8.11).

Define a linear projection proj : $\mathfrak{gl}(n, \mathbb{R}) \to \mathfrak{gl}(n, \mathbb{R})$ of rank $\frac{n^2-n}{2}$ such that

$$\text{proj}\left(\begin{bmatrix} \omega_1 & \omega_2 & \cdots & \omega_{n-1} & \omega_n \\ \omega_{n+1} & \omega_{n+2} & \cdots & \omega_{2n-1} & \omega_{2n} \\ \vdots & \vdots & \ddots & \vdots & \vdots \\ \omega_{n^2-2n+1} & \cdots & \cdots & \omega_{n^2-n-1} & \omega_{n^2-n} \\ \omega_{n^2-n+1} & \omega_{n^2-n+2} & \cdots & \omega_{n^2-1} & \omega_{n^2} \end{bmatrix}\right) = \begin{bmatrix} 0 & \omega_2 & \cdots & \omega_{n-1} & \omega_n \\ 0 & 0 & \cdots & \omega_{2n-1} & \omega_{2n} \\ \vdots & \vdots & \ddots & \vdots & \vdots \\ 0 & \cdots & \cdots & 0 & \omega_{n^2-n} \\ 0 & 0 & \cdots & 0 & 0 \end{bmatrix}. \tag{8.12}$$

Slightly abusing notation by naturally identifying the Lie algebra of $\mathfrak{h}$ with the vector space $\mathbb{R}^{\frac{n^2-n}{2}}$, define $\Theta_\lambda : \mathfrak{o} \to \mathfrak{n}^*$ such that

$$\Theta_\lambda(a) = \begin{bmatrix} 0_n & \text{proj}(\exp(E(a))) \\ 0_n & 0_n \end{bmatrix}.$$

Let

$$B_\mathfrak{n} = \left\{ \begin{bmatrix} 0_n & X^{(j,k)} \\ 0_n & 0_n \end{bmatrix} : 1 \leq j \leq n, \quad 1 \leq k \leq n \right\} \tag{8.13}$$

be a basis for $\mathfrak{n}$ where

$$\left(X^{(j,k)}\right)_{(j',k')} = \begin{cases} 1 & \text{if } (j,k) = (j',k') \\ 0 & \text{if } (j,k) \neq (j',k') \end{cases}$$

and let

$$B_\mathfrak{h} = \left\{ \begin{bmatrix} A_k & 0_n \\ 0_n & 0_n \end{bmatrix} : 1 \leq k \leq \frac{n^2-n}{2} \right\} \tag{8.14}$$

be a basis for $\mathfrak{h}$ where A_k is as defined in (8.7). The result below shows that Θ_λ is a local diffeomorphism at the zero matrix in $\mathfrak{h}$.

Lemma 8.1 $\det J_{\Theta_\lambda}(0) = (-1)^{\frac{n^2-n}{2}} \neq 0.$

Proof: Let $I = \{(j,k) : 1 \le j,k \le n \text{ and } j < k\}$ Then it is not hard to verify that $\sharp(I) = \frac{n^2-n}{2}$. Next, let $M = \left[\left\langle \lambda, \left[X^\kappa, A_\ell\right]\right\rangle\right]_{\kappa \in I, 1 \le \ell \le \frac{n^2-n}{2}}$ be a matrix of order $\frac{n^2-n}{2}$. Since $\left\langle \lambda, \left[X^\kappa, A_\ell\right]\right\rangle = -\mathrm{Tr}\left(A_\ell X^\kappa\right)$, there is an ordering for the elements of I such that

$$M = \left[-\mathrm{Tr}\left(A_\ell X^\kappa\right)\right]_{\kappa \in I, 1 \le \ell \le \frac{n^2-n}{2}} = -\mathrm{Id}_{\frac{n^2-n}{2}}.$$

M is a submatrix of D that is a matrix representing the differential of the map Θ_λ at the zero element in $\mathfrak{h}$. We conclude that

$$\det\left(J_{\Theta_\lambda}(0)\right) = \det M = (-1)^{\frac{n^2-n}{2}} \neq 0$$

as desired.

Step 3: We shall now compute W_λ. Since Θ_λ is a local diffeomorphism around $0 \in \mathbb{R}^{\frac{n^2-n}{2}}$, from here on, fix (appealing to the inverse function theorem) $\mathfrak{o} = \mathfrak{o}_\epsilon$ for some positive real number ϵ such that

$$\Theta_\lambda(a) = \begin{bmatrix} 0_n & \mathrm{proj}\left(\exp\left(E(a)\right)\right) \\ 0_n & 0_n \end{bmatrix}$$

defines a diffeomorphism between $\mathfrak{o}$ and its image $\Theta_\lambda(\mathfrak{o})$.

Proposition 8.2 $W_\lambda : \Theta_\lambda(\mathfrak{o}) \to (0,\infty)$ such that

$$W_\lambda(\xi) = \frac{d\mu_H\left(\exp\left(\Theta_\lambda^{-1}(\xi)\right)\right)}{d\xi}.$$

Then

$$W_\lambda(\xi) = \prod_{\tau\left(\Theta_\lambda^{-1}(\xi)\right) \in \mathbf{Eigenvalues}\left(-ad\left(E\left(\Theta_\lambda^{-1}(\xi)\right)\right)\right)} \left|\frac{e^{\tau\left(\Theta_\lambda^{-1}(\xi)\right)} - 1}{\tau\left(\Theta_\lambda^{-1}(\xi)\right)}\right| > 0 \tag{8.15}$$

for $\xi \in \Theta_\lambda(\mathfrak{o}) \subseteq \mathfrak{n}^*$. Moreover, $e^{-n\epsilon} \le W_\lambda(\xi) \le e^{n\epsilon}$ for $\xi \in \Theta_\lambda(\mathfrak{o})$.

Proof: Let f be a positive smooth function defined on $\mathcal{O}$. Then

$$\int_{\mathcal{O}} f(h)\, d\mu_H(h) = \int_{\mathfrak{o}} f\left(\begin{bmatrix} \exp(E(a)) & 0_n \\ 0_n & \mathrm{Id}_n \end{bmatrix}\right) \cdot \left|\det\left(\frac{\mathrm{Id} - \exp(-adE(a))}{adE(a)}\right)\right| da.$$

Next, [1, Page 24] the eigenvalues of the endomorphism $\frac{\exp(-adE(a))-\mathrm{Id}}{-adE(a)}$ are precisely

$$\left\{ \frac{e^{\tau(a)}-1}{\tau(a)} : \tau(a) \text{ is an eigenvalue for } -adE(a) \right\}.$$

In other words,

$$\left| \det \left(\frac{\exp(-adE(a)) - \mathrm{Id}}{-adE(a)} \right) \right| = \prod_{\tau(a) \in \mathbf{Eigenvalues}(-adE(a))} \left| \frac{e^{\tau(a)}-1}{\tau(a)} \right|.$$

Computing the inverse of Θ_λ by solving the equation

$$\xi = \Theta_\lambda(a) = \begin{bmatrix} 0_n & P^* \exp(E(a)) \\ 0_n & 0_n \end{bmatrix}$$

for unknown variable $a \in \mathbb{R}^{\frac{n^2-n}{2}}$, we obtain (8.15). Note that given a complex number z,

$$e^{-|z|} \le \left| \frac{e^z - 1}{z} \right| \le e^{|z|}$$

and as such, for an eigenvalue $\tau\left(\Theta_\lambda^{-1}(\xi)\right)$ of $\left(-ad\left(E\left(\Theta_\lambda^{-1}(\xi)\right)\right)\right)$, we have

$$e^{-\left|\tau\left(\Theta_\lambda^{-1}(\xi)\right)\right|} \le \left| \frac{e^{\tau\left(\Theta_\lambda^{-1}(\xi)\right)} - 1}{\tau\left(\Theta_\lambda^{-1}(\xi)\right)} \right| \le e^{\left|\tau\left(\Theta_\lambda^{-1}(\xi)\right)\right|}.$$

On the one hand,

$$\left|\tau\left(\Theta_\lambda^{-1}(\xi)\right)\right| \le \left\| ad\left(E\left(\Theta_\lambda^{-1}(\xi)\right)\right)\right\|_2 \le \left\| ad\left(E\left(\Theta_\lambda^{-1}(\xi)\right)\right)\right\|_{HS}.$$

On the other,

$$-\left\| ad\left(E\left(\Theta_\lambda^{-1}(\xi)\right)\right)\right\|_{HS} \le -\left\| ad\left(E\left(\Theta_\lambda^{-1}(\xi)\right)\right)\right\|_2 \le -\left|\tau\left(\Theta_\lambda^{-1}(\xi)\right)\right|.$$

Therefore,

$$e^{-\left\|ad\left(E\left(\Theta_\lambda^{-1}(\xi)\right)\right)\right\|_{HS}} \le \left| \frac{e^{\tau\left(\Theta_\lambda^{-1}(\xi)\right)} - 1}{\tau\left(\Theta_\lambda^{-1}(\xi)\right)} \right| \le e^{\left\|ad\left(E\left(\Theta_\lambda^{-1}(\xi)\right)\right)\right\|_{HS}}.$$

Letting, $\xi = \Theta_\lambda(a)$ for $a \in \mathbb{R}^{\frac{n^2-n}{2}}$, we clearly have

$$e^{-\|ad(E(a))\|_{HS}} \le \left| \frac{e^{\tau\left(\Theta_\lambda^{-1}(\xi)\right)} - 1}{\tau\left(\Theta_\lambda^{-1}(\xi)\right)} \right| \le e^{\|ad(E(a))\|_{HS}}.$$

Finally, the second claim follows from the fact that $ad(E(a)) = E(a)$ together with the assumption that $\|ad(E(a))\|_{HS} \le \epsilon$.

Step 4: We present an explicit method for integration on the unit sphere S^{n-1}. Let

$$\iota : H/K \to S^{n-1} \subset \mathbb{R}^{2n}$$

such that $\iota(hK) = hp$ is a diffeomorphism and $c : H/K \to T \subset H$ is a cross-section mapping such that $c \circ \iota^{-1}$ maps the sphere into the transversal T.

Lemma 8.2 Let r be a continuous function defined on H and let $d\sigma = d\mu_{H/K} \circ \iota^{-1}$ be a measure on the unit sphere S^{n-1}. Then

$$\int_H r(h)\, d\mu_H(h) = \int_{S^{n-1}} \int_K r\left(c\left(\iota^{-1}(s)\right) k\right) d\mu_K(k)\, d\sigma(s). \tag{8.16}$$

Proof: Let r be a continuous function defined on H. Since H and K are compact Lie groups, $|\det Ad_H(h)| = |\det Ad_K(h)| = 1$ for all h in H, we obtain

$$\int_{H/K} \int_K r(zk)\, d\mu_K(k)\, d\mu_{H/K}(zK) = \int_H r(h)\, d\mu_H(h).$$

Given $zK \in H/K$, there exists a unique point s on the sphere such that $zK = \iota^{-1}(s)$ and $c\left(\iota^{-1}(s)\right) K = z$. As a result,

$$\int_H r(h)\, d\mu_H(h) = \int_{S^{n-1}} \int_K r\left(c\left(\iota^{-1}(s)\right) k\right) d\mu_K(k)\, d\mu_{H/K}\left(\iota^{-1}(s)\right). \tag{8.17}$$

Letting $d\sigma = d\mu_{H/K} \circ \iota^{-1}$ we obtain (8.16) as expected.

For concrete cases, we need a systematic and explicit method for computing the integral described in (8.16). The result below provides such a method. In fact, we will see that the measure $d\sigma = d\mu_{H/K} \circ \iota^{-1}$ is just the standard surface measure on the sphere. Recall that

$$\mathcal{B}_{\mathfrak{h}} = \left\{ \begin{bmatrix} A_k & 0_n \\ 0_n & 0_n \end{bmatrix} : 1 \le k \le \frac{n^2 - n}{2} \right\}$$

is a basis for $\mathfrak{h}$ (see (8.7). Next, define $\mathcal{B}_{\mathfrak{so}(n-1,\mathbb{R})}, \mathcal{B}'_{\mathfrak{so}(n-1,\mathbb{R})} \subset \mathcal{B}_{\mathfrak{h}}$ such that

$$\begin{aligned} \mathcal{B}_{\mathfrak{so}(n-1,\mathbb{R})} &= \left\{ A \in \mathcal{B}_{\mathfrak{h}} : \mathrm{col}_n(A) = 0 \right\} \text{ and} \\ \mathcal{B}'_{\mathfrak{so}(n-1,\mathbb{R})} &= \left\{ A \in \mathcal{B}_{\mathfrak{h}} : \|\mathrm{col}_n(A)\| = 1 \right\}. \end{aligned} \tag{8.18}$$

Then $\sharp\left(\mathcal{B}_{\mathfrak{so}(n-1,\mathbb{R})}\right) = \dim\left(\mathfrak{so}(n-1,\mathbb{R})\right)$ and $\sharp\left(\mathcal{B}'_{\mathfrak{so}(n,\mathbb{R})}\right) = \dim\left(S^{n-1}\right) = n - 1$. Note that

$$\sharp\left(\mathcal{B}_{\mathfrak{so}(n-1,\mathbb{R})}\right) + \sharp\left(\mathcal{B}'_{\mathfrak{so}(n-1,\mathbb{R})}\right) = \dim\left(\mathfrak{so}(n,\mathbb{R})\right).$$

In fact, $\left\{ \mathcal{B}_{\mathfrak{so}(n-1,\mathbb{R})}, \mathcal{B}'_{\mathfrak{so}(n-1,\mathbb{R})} \right\}$ is a partition of $\mathcal{B}_{\mathfrak{h}}$ and $\mathcal{B}_{\mathfrak{so}(n-1,\mathbb{R})}$ is a basis for the Lie algebra for K. Let p be as defined in (8.2) and define $\mathcal{B}'_{\mathfrak{so}(n-1,\mathbb{R})} = \left\{ Z_1, \dots, Z_{n-1} \right\}$. Next, let

$$\mathbf{I} = \underbrace{\left[-\frac{\pi}{2}, \frac{\pi}{2}\right) \times \cdots \times \left[-\frac{\pi}{2}, \frac{\pi}{2}\right)}_{n-2} \times [-\pi, \pi). \tag{8.19}$$

The following proposition gives a more transparent method for integrating functions defined on the unit sphere Hp with respect to the surface measure $d\sigma$.

Proposition 8.3 Let $n \in \mathbb{N}$, $n \geq 3$. For a suitable function s defined on S^{n-1},

$$\int_{Hp} s(y)\, d\sigma(y) = \int_{\mathbf{I}} \left(s\left(\exp\left(t_{n-1}Z_1\right)\cdots\exp\left(t_1 Z_1\right)p\right)\right) \cdot \cos^{n-2}\left(t_1\right)\cos^{n-3}\left(t_2\right)\cdots\cos\left(t_{n-2}\right)\right) dt \tag{8.20}$$

where $\mathbf{I}$ is as defined in (8.19).

Proof: Put

$$\exp\left(t_{n-1}Z_1\right)\cdots\exp\left(t_1 Z_1\right)(rp) = \begin{bmatrix} f(t,r) \\ 0 \\ \vdots \\ 0 \end{bmatrix} \quad \text{(spherical coordinates in } \mathbb{R}^n \text{).}$$

For fixed $r > 0$, it is clear that $\left\{\exp\left(t_{n-1}Z_{n-1}\right)\cdots\exp\left(t_1 Z_1\right)(rp) : t \in \mathbf{I}\right\}$ is a sphere of dimension $n-1$ with radius r. For

$$f(t,r) = \begin{bmatrix} f_1(s,r) \\ \vdots \\ f_{n-1}(s,r) \\ f_n(s,r) \end{bmatrix},$$

let $\vartheta_n(t)$ be the function obtained by evaluating

$$\mathbb{R}^{n-1} \times \mathbb{R} \ni (s,r) \mapsto \left|\det\begin{bmatrix} \frac{\partial f_1(s,r)}{\partial s_1} & \cdots & \frac{\partial f_1(s,r)}{\partial s_{n-1}} & \frac{\partial f_1(s,r)}{\partial r} \\ \vdots & \ddots & \vdots & \vdots \\ \frac{\partial f_{n-1}(s,r)}{\partial s_1} & \cdots & \frac{\partial f_{n-1}(s,r)}{\partial s_{n-1}} & \frac{\partial f_{n-1}(s,r)}{\partial r} \\ \frac{\partial f_n(s,r)}{\partial s_1} & \cdots & \frac{\partial f_n(s,r)}{\partial s_{n-1}} & \frac{\partial f_n(s,r)}{\partial r} \end{bmatrix}\right| \tag{8.21}$$

at the point $(t,1) \in \mathbb{R}^{n-1} \times \mathbb{R}$. Observe that

$$T = \left\{\exp\left(t_{n-1}Z_1\right)\cdots\exp\left(t_1 Z_1\right) : t \in \mathbf{I}\right\} \tag{8.22}$$

is a subset of H which is a Borel cross section (or Borel transversal) for H/K such that the mapping $d : S^{n-1} \to T$ given by

$$d\left(\exp\left(t_{n-1}Z_{n-1}\right)\cdots\exp\left(t_1 Z_1\right)p\right) = \exp\left(t_{n-1}Z_{n-1}\right)\cdots\exp\left(t_1 Z_1\right)$$

is a Borel measurable bijection. Appealing to (8.16), we have the following integration formula

$$\int_H r(h)\,d\mu_H(h) = \int_{S^{n-1}} \int_K r\left(c\left(\iota^{-1}(s)\right)k\right) d\mu_K(k)\,d\sigma(s)$$

for any suitable function r defined on H. For $s \in S^{n-1}$,

$$\iota^{-1}(s) = \iota^{-1}\left(\exp\left(t_{n-1}Z_1\right)\cdots\exp\left(t_1Z_1\right)p\right) = \exp\left(t_{n-1}Z_1\right)\cdots\exp\left(t_1Z_1\right)K.$$

Letting $c : H/K \to T$ be the cross-section mapping, it follows that

$$c\left(\iota^{-1}(s)\right) = \exp\left(t_{n-1}Z_1\right)\cdots\exp\left(t_1Z_1\right)$$

and

$$\int_H r(h)\,d\mu_H(h) = \int_{\mathrm{I}} \vartheta_n(t)\cdot\left(\int_K r\left(\exp\left(t_{n-1}Z_1\right)\cdots\exp\left(t_1Z_1\right)k\right)d\mu_K(k)\right)dt.$$

As such, for a suitable function s defined on S^{n-1},

$$\int_{S^{n-1}} s(y)\,d\sigma(y) = \int_{\mathrm{I}} s\left(\exp\left(t_{n-1}Z_1\right)\cdots\exp\left(t_1Z_1\right)p\right)\cdot\vartheta_n(t)\,dt.$$

We shall now prove that

$$\vartheta_n(t) = \cos^{n-2}\left(t_1\right)\cos^{n-3}\left(t_2\right)\cdots\cos\left(t_{n-2}\right).$$

Next, let J_n be the Jacobian of the map

$$(t,r) \mapsto \exp\left(t_{n-1}Z_{n-1}\right)\cdots\exp\left(t_1Z_1\right)(rp).$$

We will prove by induction on the variable n that

$$\det\left(J_n\right) = \pm r^{n-1}\cos^{n-2}\left(t_1\right)\cos^{n-3}\left(t_2\right)\cdots\cos\left(t_{n-2}\right).$$

For the base case ($n = 3$), it is easy to verify that

$$\begin{aligned}\det\left(J_3\left(t_1,t_2\right)\right) &= \det\begin{bmatrix} -r\sin t_1\sin t_2 & r\cos t_1\cos t_2 & \cos t_1\sin t_2 \\ r\cos t_1 & 0 & \sin t_1 \\ -r\cos t_2\sin t_1 & -r\cos t_1\sin t_2 & \cos t_1\cos t_2 \end{bmatrix} \\ &= -r^2\cos\left(t_1\right)\end{aligned}$$

and as such,

$$\vartheta_3\left(t_1,t_2\right) = \cos\left(t_1\right) \geq 0 \text{ for } t_1 \in \left[-\frac{\pi}{2},\frac{\pi}{2}\right).$$

Suppose next that

$$\det\left(J_{n-1}\right) = \pm r^{n-2}\cos^{n-3}\left(t_1\right)\cdots\cos\left(t_{n-3}\right)$$

for $n-1 \geq 2$. With straightforward calculations (we multiply the left-hand side of J_n by some elementary matrices), it is not hard to verify that given

$$B\left(t_{n-1}\right) = \begin{bmatrix} 1 & & & & \\ & 1 & & & \\ & & \ddots & & \\ & & & 1 & \\ -1 & & & & 1 \end{bmatrix} \begin{bmatrix} \sin t_{n-1} & & & & \\ & 1 & & & \\ & & \ddots & & \\ & & & 1 & \\ & & & & \cos t_{n-1} \end{bmatrix},$$

and

$$Q(t) = \left[\begin{array}{cccc|c} * & * & \cdots & * & -r\cos t_1 \cdots \cos t_{n-1} \sin t_{n-1} \\ & & & & 0 \\ & J_{n-1} & & & \vdots \\ & & & & 0 \\ & & & & 0 \end{array}\right],$$

we have $B\left(t_{n-1}\right) J_n = Q(t) \Leftrightarrow J_n = B\left(t_{n-1}\right)^{-1} Q(t)$. Appealing to the inductive hypothesis,

$$\begin{aligned} \det\left(J_n\right) &= \det\left(B\left(t_{n-1}\right)^{-1}\right) \cdot \det\left(Q(t)\right) \\ &= (-1)^{1+n} \cdot \det\left(B\left(t_{n-1}\right)^{-1}\right) \cdot \left(-r\cos t_1 \cdots \cos t_{n-1} \sin t_{n-1}\right) \\ &\quad \cdot \det\left(J_{n-1}\right) \\ &= \frac{\pm r \cdot \cos\left(t_1\right) \cdots \cos\left(t_{n-2}\right) \cos\left(t_{n-1}\right) \cdot \sin\left(t_{n-1}\right) \cdot \det\left(J_{n-1}\right)}{\sin\left(t_{n-1}\right) \cdot \cos\left(t_{n-1}\right)} \\ &= \pm r^{n-1} \cdot \cos^{n-2}\left(t_1\right) \cdots \cos^2\left(t_{n-3}\right) \cos\left(t_{n-2}\right). \end{aligned}$$

Evaluating the function above at $(t, 1)$ gives

$$\begin{aligned} \vartheta_n\left(t_1, \cdots, t_{n-1}\right) = \cos^{n-2}\left(t_1\right) \cos^{n-3}\left(t_2\right) \cdots \cos\left(t_{n-2}\right) \geq 0 \\ \left(t_1, \cdots, t_{n-1}\right) \in \mathbf{I} \end{aligned}$$

as desired. Finally, for a suitable function s defined on S^{n-1},

$$\begin{aligned} \int_{S^{n-1}} s(y)\, d\sigma(y) = \int_{\mathbf{I}} & s\left(\exp\left(t_{n-1} Z_1\right) \cdots \exp\left(t_1 Z_1\right) p\right) \\ & \cdot \cos^{n-2}\left(t_1\right) \cos^{n-3}\left(t_2\right) \cdots \cos\left(t_{n-2}\right) dt. \end{aligned}$$

Step 5: We are now ready to prove Theorem 8.3. Referring to (8.15), let

$$w(z) = \prod_{\tau(\log(z)) \in \mathrm{Eigenvalues}(-ad(E(\log(z))))} \left| \frac{e^{\tau(\log(z))} - 1}{\tau\left(\log(z)\right)} \right| > 0. \tag{8.23}$$

Moreover, define $P^* : \mathfrak{n}^* \to \mathfrak{n}^*$ such that

$$P^*\left(\begin{bmatrix} 0_n & \omega \\ 0_n & 0_n \end{bmatrix}\right) = \begin{bmatrix} 0_n & \text{proj}(\omega) \\ 0_n & 0_n \end{bmatrix}.$$

Clearly P^* is a linear projection satisfying

$$\Theta_\lambda(a) = P^*\beta_\lambda\left(\begin{bmatrix} \exp E(a) & 0_n \\ 0_n & \text{Id}_n \end{bmatrix}\right)$$

and the adjoint of P^* is given by

$$P(X) = \begin{bmatrix} 0_n & \text{proj}(x) \\ 0_n & 0_n \end{bmatrix} \quad \text{for } X = \begin{bmatrix} 0_n & x \\ 0_n & 0_n \end{bmatrix} \in \mathfrak{n}.$$

Put

$$\mathfrak{p}^* = \left\{\begin{bmatrix} 0_n & \text{proj}(\omega) \\ 0_n & 0_n \end{bmatrix} : \begin{bmatrix} 0_n & \omega \\ 0_n & 0_n \end{bmatrix} \in \mathfrak{n}^*\right\}$$

and let

$$\Lambda \subset \exp \mathfrak{p} \text{ with } \mathfrak{p} = \left\{\begin{bmatrix} 0_n & \text{proj}(x) \\ 0_n & 0_n \end{bmatrix} : \begin{bmatrix} 0_n & x \\ 0_n & 0_n \end{bmatrix} \in \mathfrak{n}\right\}$$

such that

$$\mathcal{E}(C, \Lambda) = \left\{\frac{e^{2\pi i\langle\xi, Y\rangle}}{|C|^{1/2}} : Y \in \log(\Lambda)\right\}$$

is an orthonormal basis for $L^2(C, d\xi)$, where C is a compact subset of $\mathfrak{p}^*$.

The following results gives a construction of frames on $\mathfrak{H}$.

Proposition 8.4 Let $f \in C^\infty(H)$ be a nonnegative function defined on H, which is properly supported on $\mathcal{O}$. Then, there exists a finite subset Γ_H of H such that $S\left(f, \Gamma_H^{-1}\Lambda\right)$ is a frame for $\mathfrak{H}$ with frame bounds

$$0 < \left\|f\sqrt{w}\right\|_{L^\infty(H)} \cdot |C| \leq \left\|f\sqrt{w}\right\|_{L^\infty(H)} \cdot \sharp\left(\Gamma_H\right) \cdot |C|.$$

Proof: Given a nonnegative function $f \in C^\infty(H)$ such that the support of f is properly contained in $\mathcal{O}$, suppose additionally that U is an open subset around the identity in H such that supp$(f) \subset U \subset \mathcal{O}$. It is clear that $\left\{h^{-1}U : h \in H\right\}$ is an open cover of H. By compactness of H, $\left\{h^{-1}U : h \in H\right\}$ admits a finite subcover which we denote by $\left\{h^{-1}U : h \in \Gamma_H\right\}$ for some finite subset Γ_H of H. Next, $f \cdot \sqrt{w}$ is also smooth and its support coincides with that of f and

$$\left\|f\sqrt{w}\right\|_{L^\infty(H)} \leq \sum_{\ell \in \Gamma_H}\left|f(\ell h)\sqrt{w(\ell h)}\right|^2 \leq \left\|f\sqrt{w}\right\|_{L^\infty(H)} \cdot \sharp\left(\Gamma_H\right).$$

Finally, the fact that $S\left(f,\Gamma_H^{-1}\Lambda\right)$ is a frame for $\mathfrak{H}$ with frame bounds

$$0<\left\|f\sqrt{w}\right\|_{L^\infty(H)}\cdot|C|\leq\left\|f\sqrt{w}\right\|_{L^\infty(H)}\cdot\sharp\left(\Gamma_H\right)\cdot|C|$$

is a direct application of Theorem 7.4.

Remark 8.1 In order to construct a Parseval frame $S\left(f,\Gamma_H^{-1}\Lambda\right)$ for $\mathfrak{H}$, it suffices to find a function $f\in C(H)$ and a corresponding finite set Γ_H such that

$$\sum_{\ell\in\Gamma_H}\left|f(\ell h)\sqrt{w(\ell h)}\right|^2\equiv 1. \tag{8.24}$$

To the best of our knowledge, it is an open question whether for an arbitrarily fixed positive real number ϵ, it is possible to find $f\in C(H)$ and a finite set $\Gamma_H(\epsilon)\subset H$ such that (a) f is properly supported on $\exp\left(\mathfrak{o}_\epsilon\right)\subset H$ and (b) (8.24) holds. If in addition to (a) and (b), f is also a unit vector in $\mathfrak{H}$, then the Parseval frame $S\left(f,\Gamma_H^{-1}\Lambda\right)$ would be an orthonormal basis for $\mathfrak{H}$ consisting of continuous functions with good localization property on $SO(n,\mathbb{R})$.

Let C be a compact subset of $\mathfrak{p}^*$ with $\Theta_\lambda\left(\log(\mathcal{O})\right)\subseteq C$ such that the trigonometric system $\mathcal{E}(C,\Lambda)$ is an orthonormal basis for $L^2(C,d\xi)$. Put $\Gamma_H^{-1}\Lambda$ and let

$$\mathcal{F}\left(f,\Gamma_H^{-1}\Lambda\right)=\left\{s\mapsto\int_K[\pi(\gamma)f]\left(c\left(\iota^{-1}(s)\right)k\right)d\mu_K(k):\gamma\in\Gamma_H^{-1}\Lambda\right\}$$

be a collection of functions defined over the unit sphere S^{n-1}. More precisely, recall that for a point s on the unit sphere, there exists $t=\left(t_1,\ldots,t_{n-1}\right)\in\mathbf{I}$ (see (8.19)) such that

$$c\left(\iota^{-1}(t)\right)=\exp\left(t_{n-1}Z_{n-1}\right)\cdots\exp\left(t_1Z_1\right)\in H.$$

We may rewrite $\mathcal{F}(f,\Gamma)$ as

$$\left\{\begin{array}{l}c\left(\iota^{-1}(t)\right)p\mapsto\displaystyle\int_{SO(n-1,\mathbb{R})}\begin{array}{l}e^{2\pi i\left\langle Ad_n\left(\left(\ell\exp(t_{n-1}Z_1)\cdots\exp(t_1Z_1)k\right)^{-1}\right)^*\lambda,X\right\rangle}\\ \times f\left(\exp\left(t_{n-1}Z_1\right)\cdots\exp\left(t_1Z_1\right)k\right)d\mu_{SO(n-1,\mathbb{R})}(k)\end{array}\right|\\ \qquad\qquad\qquad\qquad\qquad\qquad\ell\in\Gamma_H,X\in\log(\Lambda)\end{array}\right\}.$$

Proposition 8.5 Let $f\in C^\infty(H)$ be a nonnegative function defined on H, which is properly supported on $\mathcal{O}$ (as defined in Proposition 8.4.) Then $\mathcal{F}\left(f,\Gamma_H^{-1}\Lambda\right)$ is a frame for $L^2\left(S^{n-1},d\sigma\right)$ with frame bounds

$$0<\left\|f\sqrt{w}\right\|_{L^\infty(H)}\cdot|C|\leq\left\|f\sqrt{w}\right\|_{L^\infty(H)}\cdot\sharp\left(\Gamma_H\right)\cdot|C|<\infty.$$

Proof: Let us assume that the Haar measure on K is normalized such that $\mu_K(K) = 1$. We aim to establish that

$$\begin{aligned}\left\|f \cdot \sqrt{w}\right\|_{L^\infty(H)} \cdot |C| \cdot \|z\|^2_{L^2(S^{n-1},d\sigma)} &\leq \sum_{\gamma \in \Gamma_H^{-1}\Lambda} \left| \int_{S^{n-1}} z(s) \cdot \overline{\int_K \left[\pi(\gamma) f\right](c(\iota^{-1}(s))k)\, d\mu_K(k) d\sigma(s)} \right|^2 \\ &\leq \left\|f\sqrt{w}\right\|_{L^\infty(H)} \cdot |C| \cdot \|z\|^2_{L^2(S^{n-1},d\sigma)}\end{aligned}$$

for every $z \in L^2\left(S^{n-1}, d\sigma\right)$. To this end, fix $z \in L^2\left(S^{n-1}, d\sigma\right)$ and define a related function g on H such that

$$g(h) = g\left(c\left(\iota^{-1}(s)\right)k\right) = z(s), \quad s \in S^{n-1}, \quad k \in K.$$

Note that g is well defined since every element $h \in H$ admits a unique factorization of the type $c\left(\iota^{-1}(s)\right)k$. Indeed, given $h \in H$, let qK be the unique coset containing h as a point such that q belongs to the transversal $T \subset H$ (see (8.22).) The point h can be uniquely factored as qk for some $k \in K$. Letting s be the unique point on the sphere corresponding to q, it follows that $h = qk = c\left(\iota^{-1}(s)\right)k$. According to (8.16), we have

$$\begin{aligned}\int_H |g(h)|^2 d\mu_H(h) &= \int_{S^{n-1}} \left(\int_K \left| g\left(c\left(\iota^{-1}(s)\right)k\right)\right|^2 d\mu_K(k) \right) d\sigma(s) \\ &= \int_K \int_{S^{n-1}} \left| g\left(c\left(\iota^{-1}(s)\right)k\right)\right|^2 d\sigma(s)\, d\mu_K(k).\end{aligned}$$

Since $z(s) = g\left(c\left(\iota^{-1}(s)\right)k\right)$,

$$\begin{aligned}\int_H |g(h)|^2 d\mu_H(h) &= \int_K \int_{S^{n-1}} |z(s)|^2 d\sigma(s)\, d\mu_K(k) \\ &= \|z\|^2_{L^2(S^{n-1},d\sigma)} \cdot \mu_K(K) = \|z\|^2_{L^2(S^{n-1},d\sigma)}\end{aligned}$$

and we conclude that

$$\|g\|_{\mathfrak{H}} - \|z\|_{L^2(S^{n-1},d\sigma)}. \tag{8.25}$$

Thus, g is square-integrable with respect to the Haar measure on H. Next, fixing $\gamma \in \Gamma_H^{-1}\Lambda$, according to (8.17),

$$\langle g, \pi(\gamma) f\rangle_{\mathfrak{H}} = \int_{S^{n-1}} \left(\int_K g\left(c\left(\iota^{-1}(s)\right)k\right) \overline{\left[\pi(\gamma) f\right](c(\iota^{-1}(s))k)} d\mu_K(k) \right) d\sigma(s).$$

By assumption, since $g\left(c\left(\iota^{-1}(s)\right)k\right) = z(s)$,

$$\langle g, \pi(\gamma) f\rangle_{\mathfrak{H}} = \int_{S^{n-1}} z(s) \cdot \left(\int_K \overline{\left[\pi(\gamma) f\right](c(\iota^{-1}(s))k)} d\mu_K(k) \right) d\sigma(s).$$

For fixed $\gamma \in \Gamma_H^{-1}\Lambda$, put

$$v_\gamma(s) = \int_K [\pi(\gamma) f] \left(c\left(\iota^{-1}(s)\right) k\right) d\mu_K(k). \tag{8.26}$$

Thus far, we have proved that

$$\sum_{\gamma \in \Gamma_H^{-1}\Lambda} \left| \langle g, \pi(\gamma) f \rangle_{\mathfrak{H}} \right|^2 = \sum_{\gamma \in \Gamma_H^{-1}\Lambda} \left| \int_{S^{n-1}} z(h) \cdot \overline{v_\gamma(s)} d\sigma(s) \right|^2. \tag{8.27}$$

Since $\{\pi(\gamma) f : \gamma \in \Gamma_H^{-1}\Lambda\}$ is a frame for $\mathfrak{H}$ with frame bounds

$$\left\| f\sqrt{w} \right\|_{L^\infty(H)} \cdot |C|, \ \left\| f\sqrt{w} \right\|_{L^\infty(H)} \cdot \sharp\left(\Gamma_H\right) \cdot |C|,$$

we have

$$\begin{aligned} \left\| f\sqrt{w} \right\|_{L^\infty(H)} \cdot |C| \cdot \|g\|_{\mathfrak{H}}^2 &\le \sum_{\gamma \in \Gamma} \left| \langle g, \pi(\gamma) f \rangle_{\mathfrak{H}} \right|^2 \\ &\le \left\| f\sqrt{w} \right\|_{L^\infty(H)} \cdot \sharp\left(\Gamma_H\right) \cdot |C| \cdot \|g\|_{\mathfrak{H}}^2. \end{aligned}$$

Finally, (8.25) together with (8.27) gives

$$\begin{aligned} \left\| f\sqrt{w} \right\|_{L^\infty(H)} \cdot |C| \cdot \|z\|_{L^2(S^{n-1},d\sigma)}^2 &\le \sum_{\gamma \in \Gamma} \left| \int_{S^{n-1}} z(h) \cdot \overline{v_\gamma(s)} d\sigma(s) \right|^2 \\ &\le \left\| f\sqrt{w} \right\|_{L^\infty(H)} \cdot \sharp\left(\Gamma_H\right) \cdot |C| \cdot \|z\|_{L^2(S^{n-1},d\sigma)}^2. \end{aligned}$$

Proposition 8.6 Let $f \in C^\infty(H)$ be a nonnegative function defined on H, which is properly supported on $\mathcal{O}$ (as defined in Proposition 8.4.) The frame $\mathcal{F}\left(f, \Gamma_H^{-1}\Lambda\right)$ consists of continuous functions with good localization property and $\|f_1\|_{L^2(S^{n-1},d\sigma)}^2 \le \|f\|_{L^\infty(H)}^2$ for every f_1 in $\mathcal{F}\left(f, \Gamma_H^{-1}\Lambda\right)$.

Proof: We will show next that $\left\| v_\gamma \right\|_{L^2(S^{n-1},d\sigma)} \le \|f\|_{L^\infty(H)}$ for each $\gamma \in \Gamma$. Note that for fixed $\gamma \in \Gamma$,

$$\begin{aligned} \left\| v_\gamma \right\|_{L^2(S^{n-1},d\sigma)}^2 &= \int_{S^{n-1}} \left| v_\gamma(s) \right|^2 d\sigma(s) \\ &\le \left\| v_\gamma \right\|_{L^\infty(S^{n-1})}^2 \cdot \sigma\left(S^{n-1}\right) \\ &= \left\| v_\gamma \right\|_{L^\infty(S^{n-1})}^2 \end{aligned}$$

and

$$\begin{aligned} \left| v_\gamma(s) \right| &= \left| v_{\gamma(\ell,X)}(s) \right| \\ &\le \int_K \left| e^{-2\pi i \left\langle Ad_n\left(\left(\ell^{-1} c(\iota^{-1}(s)) k\right)^{-1}\right)^* \lambda, X \right\rangle} \cdot \overline{f\left(\ell^{-1} c\left(\iota^{-1}(s)\right) k\right)} \right| d\mu_K(k). \end{aligned}$$

Thus,

$$\left|v_\gamma(s)\right| = \int_K \left|f\left(\ell^{-1}c\left(\iota^{-1}(s)\right)k\right)\right| d\mu_K(k) \le \int_K \|f\|_{L^\infty(H)} d\mu_K(k) \le \|f\|_{L^\infty(H)}$$

and for each $\gamma \in \Gamma$, $\left\|v_\gamma\right\|^2_{L^2(S^{n-1},d\sigma)} \le \|f\|^2_{L^\infty(H)} < \infty$. Finally, letting $\gamma = \gamma(\ell, X) = \ell \exp(X)$ for $\ell \in \Gamma_H^{-1}$, $\exp X \in \Lambda$, it is clear from (7.25) for trivial ℓ,

$$v_{\gamma(\mathrm{Id}_{2n},X)}(s) = \int_K e^{2\pi i\left\langle Ad_n\left(\left(c\left(\iota^{-1}(s)\right)k\right)^{-1}\right)^* \lambda, X\right\rangle} f\left(c\left(\iota^{-1}(s)\right)k\right) d\mu_K(k)$$

and the support of the corresponding vector $v_{\gamma(\mathrm{Id}_{2n},X)}$ can be made as small as desired since we have complete control over the support of f. Next, since each, $h \mapsto [\pi(\gamma)f](h)$ is a continuous function, it follows from [2, Lemma 1.10] that

$$\begin{aligned}&\exp\left(t_{n-1}Z_1\right)\cdots\exp\left(t_1Z_1\right)p\\ &\mapsto \int_K [\pi(\gamma)f]\left(\exp\left(t_{n-1}Z_1\right)\cdots\exp\left(t_1Z_1\right)k\right)d\mu_K(k)\end{aligned}$$

defines continuous function on the unit sphere as well.

8.3 Frames on the Klein Bottle

Let

$$H = \left\{\begin{bmatrix} e^{2\pi it} & z & 0 & 0 \\ 0 & 1 & 0 & 0 \\ 0 & 0 & 1 & 0 \\ 0 & 0 & 0 & 1 \end{bmatrix} \in GL(4,\mathbb{C}) : z \in \mathbb{C}, \quad t \in \mathbb{R}\right\} \tag{8.28}$$

and let

$$N = \left\{\begin{bmatrix} 1 & 0 & x_1 & x_2 \\ 0 & 1 & x_3 & x_4 \\ 0 & 0 & 1 & 0 \\ 0 & 0 & 0 & 1 \end{bmatrix} \in GL(4,\mathbb{C}) : x_1, x_2, x_3, x_4 \in \mathbb{C}\right\}$$

be a simply connected Lie group which is Lie isomorphic to $\mathbb{C}^4$. Let $G = NH$ and let

$$\lambda_{\mathrm{Id}_4} = \begin{bmatrix} 0 & 0 & 1 & 0 \\ 0 & 0 & 0 & 1 \\ 0 & 0 & 0 & 0 \\ 0 & 0 & 0 & 0 \end{bmatrix} \in \mathfrak{n}^*$$

such that

$$N \ni z \mapsto \chi_{\lambda_{\omega}}(z) = e^{2\pi i \cdot \mathrm{Re}\left(\mathrm{Tr}\left(\lambda_{\mathrm{Id}_4}^T \log(z)\right)\right)}$$

defines a unitary character of N and $\pi = \mathrm{ind}_N^{NH}\left(\chi_{\lambda_{\mathrm{Id}_n}}\right)$ is realized as acting on $L^2(H, d\mu_H)$. Then

$$\beta_{\lambda_{\mathrm{Id}_n}}\left(\begin{bmatrix} e^{2\pi it} & z & 0 & 0 \\ 0 & 1 & 0 & 0 \\ 0 & 0 & 1 & 0 \\ 0 & 0 & 0 & 1 \end{bmatrix}\right) = \begin{bmatrix} 0 & 0 & e^{-2i\pi t} & 0 \\ 0 & 0 & -ze^{-2i\pi t} & 1 \\ 0 & 0 & 0 & 0 \\ 0 & 0 & 0 & 0 \end{bmatrix}$$

and this shows that $\beta_{\lambda_{\mathrm{Id}_n}}$ is an immersion at the identity element of H. Thus, there exists an open set U around the identity element in H and a countable subset Γ of H such that if f is a continuous function supported on U then $\pi(\Gamma)f$ is a frame for $L^2(H, d\mu_H)$ (see Theorem 7.1).

Let

$$K = \left\{\begin{bmatrix} \pm 1 & n+iy & 0 & 0 \\ 0 & 1 & 0 & 0 \\ 0 & 0 & 1 & 0 \\ 0 & 0 & 0 & 1 \end{bmatrix} \in GL(4, \mathbb{C}) : n \in \mathbb{Z} \quad \text{and} \quad y \in \mathbb{R}\right\}$$

be a closed subgroup of H. The homogeneous space H/K is called the Klein bottle. Letting

$$T = \left\{\begin{bmatrix} e^{2\pi it} & x & 0 & 0 \\ 0 & 1 & 0 & 0 \\ 0 & 0 & 1 & 0 \\ 0 & 0 & 0 & 1 \end{bmatrix} \in GL(4, \mathbb{C}) : t \in \left[0, \frac{1}{2}\right) \quad \text{and} \quad x \in [0, 1)\right\}$$

then T is a measurable transversal for the set of cosets H/K. To simplify our presentation, naturally identity every element

$$\begin{bmatrix} e^{2\pi it} & x_1 + ix_2 & 0 & 0 \\ 0 & 1 & 0 & 0 \\ 0 & 0 & 1 & 0 \\ 0 & 0 & 0 & 1 \end{bmatrix}$$

of H with the ordered triple $(x_1, x_2, t) \in \mathbb{R} \times \mathbb{R} \times [0, 1)$. Given that $\pi(\Gamma)f$ is assumed to be a frame for $L^2(H, d\mu_H)$, for arbitrary $g \in L^2(H, d\mu_H) := L^2(\mathbb{R} \times \mathbb{R} \times [0, 1), dx_1 dx_2 dt)$ there exist positive real numbers A, B such that

$$A \|g\|^2_{L^2(\mathbb{C}\times\mathbb{R})} \leq \sum_{\gamma \in \Gamma} \left| \int_{\mathbb{R}} \int_{\mathbb{R}} \int_{[0,1)} g(x_1, x_2, t) \overline{[\pi(\gamma)f](z, t)} dt dx_1 dx_2 \right|^2$$
$$\leq B \|g\|^2_{L^2(\mathbb{C}\times\mathbb{R})}.$$

Next,

$$\int_{\mathbb{C}}\int_{\mathbb{R}} g(z,t)\,\overline{[\pi(\gamma)f](z,t)}dtdz$$

$$:= \sum_{j\in\left\{0,\frac{1}{2}\right\}} \int_{\mathbb{R}}\int_{\mathbb{R}}\int_{\left[0,\frac{1}{2}\right)} g(x_1,x_2,t+j)\,\overline{[\pi(\gamma)f](x_1,x_2,t+j)}dtdx_1dx_2$$

$$= \int_{[0,1)}\int_{\left[0,\frac{1}{2}\right)} \sum_{j\in\left\{0,\frac{1}{2}\right\}} \int_{\mathbb{R}} \sum_{k\in\mathbb{Z}} g(x_1+k,x_2,t+j)$$

$$\times\,\overline{[\pi(\gamma)f](x_1+k,x_2,t+j)}dx_2dx_1dt.$$

Put

$$g(x_1+k,x_2,t+j) = \begin{cases} g_1(x_1,t)\cdot 1_{[0,1)\times(0,\infty)}(x_1,t)\cdot 1_{[0,1]}(x_2) & \text{if } k=0 \\ 0 \quad \text{if } k\neq 0 \end{cases}$$

where g_1 is arbitrarily fixed in the Hilbert space $L^2\left([0,1)\times\left[0,\frac{1}{2}\right)\right)$. Then

$$\int_{\mathbb{C}}\int_{\mathbb{R}} g(z,t)\,\overline{[\pi(\gamma)f](z,t)}dtdz$$

$$:= \int_{[0,1)}\int_{\left[0,\frac{1}{2}\right)} g_1(x_1,t)\left(\int_{\mathbb{R}} \sum_{j\in\left\{0,\frac{1}{2}\right\}} \overline{[\pi(\gamma)f](x_1,x_2,t+j)}dx_2\right)dx_1dt.$$

Finally, letting

$$\psi_\gamma(x_1,t) = \int_{\mathbb{R}} \sum_{j\in\left\{0,\frac{1}{2}\right\}} \overline{[\pi(\gamma)f](x_1,x_2,t+j)}dx_2$$

it follows that $(\psi_\gamma)_{\gamma\in\Gamma}$ is a frame for $L^2(T) := L^2(H/K)$.

References

1 Johannes Jisse Duistermaat and Johan A. C. Kolk. *Lie groups*. Springer Science & Business Media, 2012.

2 S. Helgason. *Groups and geometric analysis*. Academic Press, London, 1984.

3 Roger A. Horn, Roger A. Horn, and Charles R. Johnson. *Matrix analysis*. Cambridge University Press, 1990.

9

Groups with Frames of Translates

The insights discussed in this chapter are rooted in a critical observation made initially by Hartmut Führ. This observation asserts that the left regular representation of the affine group can be deconstructed into a direct sum of two nonequivalent unitary irreducible representations, each appearing with infinite multiplicity. Suppose one can demonstrate that this infinite multiplicity does not impede the construction of frames through the discretization of the representation. In that case, it implies that the left regular representation of the affine group contains a frame within at least one of its orbits. Next, assume a Lie group contains a closed subgroup, which is Lie isomorphic to the affine group. Under this condition, the restriction of the left regular representation of the group to this closed subgroup can be considered unitarily equivalent to a direct sum of the left regular representation of the affine group, with each component recurring infinitely. This observation can be utilized to establish the existence of frames within an orbit of the left regular representation of the larger group. Many of the ideas in this chapter are inspired by collaborative work with Fuhr and Oussa [1].

Let G be a second countable locally compact group with Haar measure μ_G. Recall that G acts on itself by left translations and this action induces the left regular representation of G on $L^2(G)$ acting as:

$$L(x)f(y) = f\left(x^{-1}y\right) \quad x, y \in G, \quad f \in L^2(G).$$

Let Γ be a countable subset of G. Recall that Γ is (left) relatively separated, if

$$\sup_{x \in G} \sharp(\Gamma \cap xU) < \infty$$

for some, and hence all relatively compact neighborhoods U of the identity. In addition, Γ is V-separated if for some relatively compact neighborhood V of the identity, the family $(xV)_{x\in\Gamma}$ consists of pairwise disjoint sets. Γ is called separated if it is V-separated for some suitable V. Recall also that a system of the

A Bridge Between Lie Theory and Frame Theory: Applications of Lie Theory to Harmonic Analysis, First Edition. Vignon Oussa.

type $L(\Gamma)\varphi$ is called a Bessel sequence if there exists a positive real number B such that $\sum_{\gamma\in I} |\langle f, L(\gamma)\varphi\rangle|^2 \leq B \cdot \|f\|^2_{L^2(G)}$ for every vector f in $L^2(G)$.

In this chapter, we address the Discretization Problem for the regular representation of connected Lie groups. Since the left and right regular representations are unitarily equivalent to each other, any solution for one can be easily modified to suit the other. That being said, we will only focus on the left regular representation, and a vector $f \in L^2(G, d\mu_G)$ is called a frame if there exist positive real numbers A, B such that

$$A\|g\|^2 \leq \sum_{\gamma\in I} |\langle f, L(\gamma)\varphi\rangle|^2 \leq B\|g\|^2 \quad \forall g \in L^2(G).$$

When the precise values of the frame bounds are not of interest, we as previously done, we simply write

$$\sum_{\gamma\in I} |\langle f, L(\gamma)\varphi\rangle|^2 \asymp \|g\|^2 \quad \forall g \in L^2(G)$$

and we ask, if is possible to embed $L^2(G, d\mu_G)$ into $l^2(\Gamma)$ with a continuously invertible operator $f \mapsto (\langle f, L(\gamma)\varphi\rangle)_{\gamma\in\Gamma}$.

Perhaps it pays to view the problem at hand through the lens of a concrete example. For instance, should the Lie group G be the real line $\mathbb{R}$, then one is searching for a frame of the type $L(\Gamma)f$ where Γ is a countable subset of real numbers. That is, a vector f in $L^2(\mathbb{R})$ such that

$$\sum_{\gamma\in\Gamma} \left| \int_{\mathbb{R}} g(x) \cdot \overline{f(x-\gamma)} dx \right|^2 \asymp \|g\|^2 \quad \forall g \in L^2(\mathbb{R}).$$

If G is the three-dimensional Heisenberg Lie group, then the question is, is there a vector f in $L^2(\mathbb{R}^3)$ such that

$$\sum_{(c,a,b)\in\Gamma} \left| \int_{\mathbb{R}} g(z,y,x) \cdot \overline{f(z-c+ab-ay, y-b, x-a)} dx \right|^2 \asymp \|g\|^2 \quad \forall g \in L^2(\mathbb{R}^3).$$

Observe that for the Heisenberg group, and for all connected commutative Lie groups, multiplication is a polynomial function. However, there exist groups for which does not hold true. For example, if G is the $ax+b$ Lie group, the group operation is $(x, a)(y, b) \mapsto (x + e^a y, a + b)$. Moreover, existence of a frame of translates is equivalent to the existence of a vector $f \in L^2(\mathbb{R}^2, e^{-a} dxda)$ satisfying

$$\sum_{(y,b)\in\Gamma} \left| \int_{\mathbb{R}} \int_{\mathbb{R}} \left(g(x,a) \cdot \overline{f(e^{-b}(x-y), a-b)} e^{-a} \right) dxda \right|^2 \asymp \|g\|^2$$

$$\forall g \in L^2(\mathbb{R}^2, e^{-a} dxda).$$

In each of the cases described above, one might also be interested in the following questions. Could such a frame (if it even exists) be regular? Could it have compact support? Is there a frame of translates that is generated by a rapidly decaying window? How do even measure such properties, and for frames of translates of the form $L(\Gamma)f$, as a subset of G, which topological features of Γ are critical to the frame structure itself?

As a first result of consequential outcome, Lemma 9.1 suggests the following. A sampling set that is too "dense" cannot produce a frame. In other words, the upper bound of a countable system of translates increases jointly with the density of the set parametrizing it.

Lemma 9.1 The following holds true.

1. If $L(\Gamma)f$ is a Bessel sequence for $L^2(G)$, then Γ must be a relatively separated subset of G.
2. If $\Gamma \subset G$ is a relatively separated set of a locally compact group G then Γ must be a finite union of separated sets.

Proof: To prove the first part, suppose there exists a Bessel sequence $L(\Gamma)f$ for $L^2(G)$ and Γ is not a relatively separated subset of G. Since $x \mapsto \langle \varphi, L(x)\varphi\rangle_{L^2(G)}$ is continuous on G, there exists an open set V around the neutral element in G such that

$$\inf\left\{\left|\langle \varphi, L(x)\varphi\rangle_{L^2(G)}\right| : x \in V\right\} = \ell > 0.$$

Let n be a fixed natural number. Since Γ is not relatively separated, there exists an element $y \in G$ such that yV contains at least n elements from Γ. Letting $\Gamma_n = yV \cap \Gamma$,

$$\begin{aligned}\sum_{\gamma\in\Gamma}\left|\langle L(y)\varphi, L(\gamma)\varphi\rangle_{L^2(G)}\right|^2 &\geq \sum_{\gamma\in\Gamma_n}\left|\langle L(y)\varphi, L(\gamma)\varphi\rangle_{L^2(G)}\right|^2\\ &= \sum_{\gamma\in\Gamma_n}\left|\left\langle \varphi, L\left(y^{-1}\gamma\right)\varphi\right\rangle_{L^2(G)}\right|^2\end{aligned}$$

and as a result,

$$\sum_{\gamma\in\Gamma}\left|\langle L(y)\varphi, L(\gamma)\varphi\rangle_{L^2(G)}\right|^2 \geq \sum_{y\alpha\in\Gamma_n}\left|\left\langle \varphi, L\left(y^{-1}y\alpha\right)\varphi\right\rangle_{L^2(G)}\right|^2.$$

Therefore

$$\sum_{\gamma\in\Gamma}\left|\langle L(y)\varphi, L(\gamma)\varphi\rangle_{L^2(G)}\right|^2 \geq \sum_{\alpha\in y^{-1}\Gamma_n}\left|\langle \varphi, L(\alpha)\varphi\rangle_{L^2(G)}\right|^2 \geq \sharp\left(y^{-1}\Gamma_n\right)\cdot \ell^2$$

and

$$\sum_{\gamma \in \Gamma} \left| \langle L(y)\varphi, L(\gamma)\varphi \rangle_{L^2(G)} \right|^2 \geq n \cdot \ell^2 = \left(\frac{n \cdot \ell^2}{\|\varphi\|^2_{L^2(G)}} \right) \|L(y)\varphi\|^2_{L^2(G)}.$$

Since n is arbitrary, $L(\Gamma)f$ cannot be a Bessel sequence.

To prove the second part, let us suppose that V is a relatively compact and symmetric open subset of G containing in its interior, the identity element of the group. Suppose that Γ is relatively separated. Then

$$\sup_{\gamma \in \Gamma} \sharp\{\gamma' \in \Gamma : \gamma' V \cap \gamma V \neq \emptyset\} \leq \sup_{x \in G} \sharp\{\gamma' \in \Gamma : \gamma' \in xV^2\} = m < \infty.$$

Using Zorn's Lemma, choose a subset $\Gamma_1 \subset \Gamma$ that is V-discrete and maximal with respect to inclusion. If $\Gamma_1 = \Gamma$, then Γ is V-discrete. Iterate this procedure of choosing a maximal V-discrete set $\Gamma_{s+1} \subset \Gamma \backslash \bigcup_{j \leq s} \Gamma_j$, whenever the complement is nonempty. This process can be completed in at most $m + 1$ steps. Assuming that $\gamma_0 \in \Gamma \backslash \bigcup_{j \leq m+1} \Gamma_j$, we find for every fixed $1 \leq j \leq m + 1$ that $\gamma_0 \in \Gamma \backslash \bigcup_{i<j} \Gamma_i$, and since Γ_j is assumed to be maximal, $\Gamma_j \cup \{\gamma_0\}$ cannot be V-discrete. Hence, there exists $\gamma_j \in \Gamma_j$ such that $\gamma_0 V \cap \gamma_j V \neq \emptyset$. Since elements of $\left\{\Gamma_j : 1 \leq j \leq m + 1\right\}$ are pairwise disjoint, the number of elements in the set $\{\gamma' \in \Gamma : \gamma' V \cap \gamma_0 V \neq \emptyset\}$ must be at least equal to $m + 1$.

Fix a compact neighborhood U of the identity. Given a measurable function f on G, define $f_U^\sharp(x) = \operatorname{ess\,sup}_{y \in xU} |f(y)|$. Next, given $1 \leq p \leq \infty$, define the Wiener amalgam spaces $W(L^\infty(G), L^p(G))$ as the space of Borel functions f for which

$$\|f\|_{W(L^\infty(G), L^p(G))} = \|x \mapsto f_U^\sharp(x)\|_{L^p(G)}$$

is finite. Recall also that the convolution of two functions f, φ on G is defined as the integral: $(f * \varphi)(x) = \int_G f(y)\varphi\left(y^{-1}x\right) dy$ and $\varphi^*(x) = \overline{\varphi(x^{-1})}$. Moreover, it is easy to verify that $f * \varphi^*(x) = \langle f, L(x)\varphi \rangle_{L^p(G)}$, for all $f, \varphi \in L^2(G)$.

Proposition 9.1 The following holds true.

1. Let G be nondiscrete Lie group. Let $\varphi \in L^2(G)$ be such that $\varphi^* \in W(L^\infty, L^2)$. Then there does not exist a discrete set $\Gamma \subset G$ such that $L(\Gamma)\varphi$ is a frame of $L^2(G)$.
2. If G is a nondiscrete Lie group and if f is compactly supported, then it is not possible to find a countable subset of $\Gamma \subset G$ such that $L(\Gamma)\varphi$ is a frame of $L^2(G)$.
3. If G is compact and nonfinite, then it is not possible to find a countable subset of $\Gamma \subset G$ such that $L(\Gamma)\varphi$ is a frame of $L^2(G)$.
4. Let G be nondiscrete such that the inverse of any subset of G, which is relatively separated is also relatively separated. Then it is not possible to find a countable subset of $\Gamma \subset G$ such that $L(\Gamma)\varphi$ is a frame of $L^2(G)$.

Proof: For Part (1), suppose by contradiction that $L(\Gamma)\varphi$ is a frame with a lower frame bound A. For any fixed $f \in L^2(G)$, one has:

$$\|f\|_2^2 \leq A^{-1} \sum_{x\in\Gamma} |f * \varphi^*(x)|^2. \tag{9.1}$$

By Lemma 9.1, write $\Gamma = \bigcup_{i=1}^{n} \Gamma_i$ disjointly, such that each Γ_i is U-discrete, for some neighborhood U of the identity satisfying $U^{-1} = U$. Summing $f * \varphi^*(\gamma)$ over Γ gives

$$\sum_{\gamma\in\Gamma} |f * \varphi^*(\gamma)|^2 = \sum_{i=1}^{n} \sum_{\gamma\in\Gamma_i} |f * \varphi^*(\gamma)|^2.$$

Writing convolution as an integral and using the fact that each Γ_i is U-discrete,

$$\sum_{\gamma\in\Gamma} |f * \varphi^*(\gamma)|^2 = \sum_{i=1}^{n} \sum_{\gamma\in\Gamma_i} \frac{1}{|U|} \int_{\gamma U} |f * \varphi^*(\gamma)|^2 dy.$$

Since, the quantity $|f * \varphi^*(\gamma)|$ is above by $\left|(f * \varphi^*)_U^\sharp(y)\right|$,

$$\sum_{\gamma\in\Gamma} |f * \varphi^*(\gamma)|^2 \leq \sum_{i=1}^{n} \sum_{\gamma\in\Gamma_i} \frac{1}{|U|} \int_{\gamma U} \left|(f * \varphi^*)_U^\sharp(y)\right|^2 dy.$$

Since the integral above is just the L^2-norm of $(f * \varphi^*)_U^\sharp$,

$$\sum_{\gamma\in\Gamma} |f * \varphi^*(\gamma)|^2 \leq\leq \frac{n}{|U|} \| (f * \varphi^*)_U^\sharp \|_{L^2(G)}^2.$$

Next,

$$(f * g)_U^\sharp(x) = \text{ess sup}_{y\in xU} \left| \int_G f(z) g(z^{-1}y)\, dz \right| \leq (|f| * g_U^\sharp)(y),$$

for arbitrary measurable functions f, g, and it is safe to assert that

$$\| (f * \varphi^*)_U^\sharp \|_{L^2(G)} \leq \left\| |f| * (\varphi^*)_U^\sharp \right\|_{L^2(G)}.$$

Using the estimate given in (9.1), conclude that

$$\|f\|_2 \leq \left(\frac{n}{A|U|}\right)^{1/2} \left\| |f| * (\varphi^*)_U^\sharp \right\|_{L^2(G)}.$$

By Young's inequality, we also have

$$\left\| |f| * (\varphi^*)_U^\sharp \right\|_{L^2(G)} \leq \|f\|_1 \cdot \|(\varphi^*)_U^\sharp\|_{L^2(G)} = \|f\|_{L^1(G)} \cdot \|\varphi^*\|_{W(L^\infty,L^2)},$$

and the Wiener amalgam norm is finite by assumption. In summary, we have shown

$$\|f\|_{L^2(G)} \leq \left(\frac{n}{A|U|}\right)^{1/2} \|\varphi^*\|_{W(L^\infty,L^2)} \cdot \|f\|_{L^1(G)}, \tag{9.2}$$

for all $f \in L^1(G) \cap L^2(G)$. Replacing f with the indicator function of E_n, for a sequence of nonempty open sets E_n with $|E_n| \to 0$, we reach the desired contradiction.

For Part (2), let f be a compactly supported function. Then

$$\|\varphi^*\|_{W(L^\infty,L^p)} = \|x \mapsto \varphi_U^{*\sharp}(x)\|_p < \infty,$$

$\varphi^* \in W(L^\infty, L^2)$ and the result follows from Part (1).

Part (3) is a direct consequence of Lemma 9.1 together with the observation that relatively separated subsets of compact groups are necessarily finite.

Finally, to prove (4), suppose there exists a nondiscrete group G with the property that the inverse of any subset of G, which is relatively separated, is also relatively separated. Suppose additionally that there exist $\varphi \in L^2(G)$ and a countable set $\Gamma \subset G$ such that $L(\Gamma)\varphi$ is a frame for $L^2(G)$. Since $L(\Gamma)\varphi$ must also be a Bessel sequence, according to Lemma 9.1, Γ must necessarily be a relatively separated subset of G. By assumption, Γ^{-1} is also relatively separated and by Lemma 9.1, Γ^{-1} can be written as a disjoint union of separated sets. Let $\Gamma^{-1} = \bigcup_{k=1}^{s} \Psi_k$ such that each collection $\{xE : x \in \Psi_k\}$ consists of essentially disjoint subsets of G. Then

$$\sum_{\gamma\in\Gamma} \left|\langle 1_E, L(\gamma)\varphi\rangle_{L^2(G)}\right|^2 = \sum_{\gamma\in\Gamma} \left|\langle 1_E, 1_E \cdot L(\gamma)\varphi\rangle_{L^2(G)}\right|^2$$

and

$$\sum_{\gamma\in\Gamma} \left|\langle 1_E, L(\gamma)\varphi\rangle_{L^2(G)}\right|^2 \leq \|1_E\|_{L^2(G)}^2 \sum_{k=1}^{s} \sum_{\gamma\in\Psi_k} \|1_E \cdot L(\gamma)\varphi\|_{L^2(G)}^2.$$

Since

$$\sum_{k=1}^{s} \sum_{\gamma\in\Psi_k} \|1_E \cdot L(\gamma)\varphi\|_{L^2(G)}^2 = \sum_{k=1}^{s} \sum_{\gamma\in\Psi_k} \int_E \left|\varphi(\gamma^{-1}x)\right|^2 dx$$

we obtain

$$\sum_{\gamma\in\Gamma} \left|\langle 1_E, L(\gamma)\varphi\rangle_{L^2(G)}\right|^2 \leq \|1_E\|_{L^2(G)}^2 \sum_{k=1}^{s} \sum_{\gamma\in\Psi_k} \int_E \left|\varphi(\gamma^{-1}x)\right|^2 dx$$

and

$$\sum_{\gamma\in\Gamma} \left|\langle 1_E, L(\gamma)\varphi\rangle_{L^2(G)}\right|^2 \leq \|1_E\|^2 \sum_{k=1}^{s} \int_{\bigcup_{\gamma\in\Psi_k}\gamma^{-1}E} |\varphi(x)|^2 dx.$$

Since φ is square-integrable, it is also clear that

$$\int_{\bigcup_{\gamma\in\Psi_k}\gamma^{-1}E} |\varphi(x)|^2 dx \leq \int_G |\varphi(x)|^2 dx = \|\varphi\|^2 < \infty.$$

By Lebesgue's dominated convergence theorem, taking a nested family of relatively compact and open sets converging to the singleton containing the identity in G,

$$\lim_{E\to\{e\}} \int_{\cup_{\gamma\in\Psi_k}\gamma^{-1}E} |\varphi(x)|^2 dx = 0.$$

Therefore, for any $\epsilon > 0$, there exists a sufficiently small open set E around the identity such that $\sum_{\gamma\in\Gamma} |\langle 1_E, \lambda(\gamma)\varphi\rangle|^2 \leq \epsilon \|1_E\|^2$. This means that the lower frame bound of this system must coincide with zero.

Let G be an [IN]-group (a group with a compact neighborhood of the identity which is invariant under all inner automorphisms). Observe that every relatively separated subset of G has a relatively separated inverse. By definition, there exists a compact neighborhood $O \subset G$ of the identity, which is conjugation-invariant. Next, let Γ be a subset of G which is relatively separated. By assumption, $\sup_{x\in G} \sharp(\Gamma \cap xO) < \infty$. On the other hand, for any $x \in G$,

$$\Gamma \cap xO = \Gamma \cap \left(xOx^{-1}\right)x = \Gamma \cap Ox,$$

and as a result, $(\Gamma \cap xO)^{-1} = \Gamma^{-1} \cap x^{-1}O^{-1}$. Since $x \mapsto x^{-1}$ is a bijection on G,

$$\sup_{y\in G} \sharp\left(\Gamma^{-1} \cap yO^{-1}\right) = \sup_{x\in G} \sharp\left(\Gamma^{-1} \cap x^{-1}O^{-1}\right) = \sup_{x\in G} \sharp(\Gamma \cap xO) < \infty$$

and this proves that Γ^{-1} is relatively separated. In light of these observations, nondiscrete [IN]-groups do not admit frames of translates. This class of groups contain, e.g., nonabelian discrete groups and the connected Heisenberg group with a compact center.

9.1 Frames and Bases of Translates on the *ax+b* Lie Group

Recall that a Riesz basis for a Hilbert space $\mathcal{H}$ is a family of the form $\{Ue_k : k \in \mathbb{N}\}$ where $\{e_k : k \in \mathbb{N}\}$ is an orthonormal basis $\mathcal{H}$ and $U : \mathcal{H} \to \mathcal{H}$ is a bounded invertible operator. In fact, the upper and lower frame bounds are given by $\|U\|^2$ and $\|U^{-1}\|^{-2}$, respectively. Note also that if $\{L(\gamma)f : \gamma \in \Gamma\}$ is a Riesz basis for $L^2(G)$ then there exists a unique sequence $\{g_\gamma : \gamma \in \Gamma\}$ in $L^2(G)$ such that

$$\varphi = \sum_{\gamma\in\Gamma} \langle \varphi, g_\gamma \rangle L(\gamma)f \tag{9.3}$$

for all $\varphi \in L^2(G)$. In fact, $\{g_\gamma : \gamma \in \Gamma\}$ is a Riesz basis also and the series (9.3) converges unconditionally for all $\varphi \in L^2(G)$.

Let

$$N = \left\{ \begin{bmatrix} 1 & x \\ 0 & 1 \end{bmatrix} : x \in \mathbb{R} \right\}, \quad H = \left\{ \begin{bmatrix} e^t & 0 \\ 0 & 1 \end{bmatrix} : t \in \mathbb{R} \right\}$$

be matrix Lie subgroups of the general linear group $GL(2, \mathbb{R})$. Put

$$G = NH = \left\{ \begin{bmatrix} e^t & x \\ 0 & 1 \end{bmatrix} : t, x \in \mathbb{R} \right\}.$$

Then G is the $ax+b$ Lie group with Lie algebra $\mathfrak{g}$ spanned by

$$X = \begin{bmatrix} 0 & 1 \\ 0 & 0 \end{bmatrix}, \quad A = \begin{bmatrix} 1 & 0 \\ 0 & 0 \end{bmatrix};$$

and nontrivial Lie brackets given by $[A, X] = X$. In contrast to the class of groups discussed in the preceding section, we will see that in this case, $L^2(G)$ admits Riesz bases (and thus frames of translates.) However, as guaranteed by Proposition 9.1 and Proposition 9.1, any generator of such a frame must in a sense, be pathological.

Theorem 9.1 There exist a countable subset Γ of G and a vector f in $L^2(G)$ such that $L(\Gamma)f$ is an orthonormal basis for $L^2(G)$.

We associate to every real number λ, a unitary character χ_λ of the normal abelian subgroup N defined as follows $\chi_\lambda(\exp X) = e^{2\pi i \lambda x}$. For each unitary character χ_λ, there is an associated induced unitary representation

$$\pi_\lambda = \operatorname{ind}_N^{NH}(\chi_\lambda) \tag{9.4}$$

of G which is realized as acting in $L^2(\mathbb{R})$ as follows. Given real numbers a, x and $f \in L^2(\mathbb{R})$,

$$\pi_\lambda(\exp(xX))f(t) = e^{2\pi i \lambda e^{-t} x} f(t) \text{ and } \pi_\lambda(\exp(aA))f(t) = f(t-a).$$

Referring back to [2], for nonzero λ, each representation π_λ is irreducible and in fact, given two nonzero real numbers λ, λ', π_λ, and $\pi_{\lambda'}$ are unitarily equivalent if and only if λ, λ', have the same sign. Thus, up to unitary equivalence, G admits two infinite-dimensional irreducible representations: π_1 and π_{-1}. These two representations play an important role in the analysis of the left regular representation of this group as shown in Lemma 9.2.

The main strategy in proving Theorem 9.1 is to realize the left regular representation as an infinite direct sum of $\pi_1 \oplus \pi_{-1}$. Let $\mathfrak{K} = \mathfrak{Z} \oplus \mathfrak{Z}$, where

$$\mathfrak{Z} = \bigoplus_{\ell=1}^{\infty} L^2(\mathbb{R})$$

and define $\tau = \tau_1 \oplus \tau_{-1}$ such that

$$\tau_1 = \bigoplus_{\ell=1}^{\infty} \pi_1 = \bigoplus_{\ell=1}^{\infty} \operatorname{ind}_N^{NH}(\chi_1)$$

and

$$\tau_{-1} = \bigoplus_{\ell=1}^{\infty} \pi_{-1} = \bigoplus_{\ell=1}^{\infty} \operatorname{ind}_N^{NH} \left(\chi_{-1}\right).$$

Theorem 9.1 follows from an application of Lemma 9.4, which itself is derived from Lemmas 9.2 and 9.3.

Lemma 9.2 Fixing a left-invariant measure μ_G on G, there is a linear isometry $P : L^2\left(G, d\mu_G\right) \to \mathfrak{K}$ intertwining the left regular representation of G with the direct sum $\tau = \tau_1 \oplus \tau_{-1}$.

Lemma 9.3 Let $f = \left(\left(f_\ell^+\right)_{\ell=1}^{\infty}, \left(f_\ell^-\right)_{\ell=1}^{\infty}\right)$ such that for each $\ell \in \mathbb{N}, f_\ell^\pm (t) = e^{-\frac{t}{2}} \cdot 1_{E_\ell}(t)$ with $E_\ell = \ln(2) \cdot [\ell, \ell+1)$. Then f is a unit vector in $\mathfrak{K}$.

Let

$$\Gamma = \exp(\ln(2) \cdot \mathbb{Z}A) \exp(\mathbb{Z}X) \subset G.$$

Next, put

$$f = \left(\left(t \mapsto e^{-\frac{t}{2}} \cdot 1_{\ln 2\cdot[\ell,\ell+1)}(t)\right)_{\ell=1}^{\infty}, \left(t \mapsto e^{-\frac{t}{2}} \cdot 1_{\ln 2\cdot[\ell,\ell+1)}(t)\right)_{\ell=1}^{\infty}\right),$$

and let $\mathfrak{B} = \{\tau(\gamma) f : \gamma \in \Gamma\}$. Then

Lemma 9.4 If $f = \left(\left(f_\ell^+\right)_{\ell=1}^{\infty}, \left(f_\ell^-\right)_{\ell=1}^{\infty}\right)$ such that for each $\ell \in \mathbb{N}, f_\ell^\pm (t) = e^{-\frac{t}{2}} \cdot 1_{E_\ell}(t)$ with $E_\ell = \ln(2) \cdot [\ell, \ell+1)$. Then $\mathfrak{B}$ is an orthonormal basis for $\mathfrak{K}$. Since P is a unitary operator, then $L(\Gamma) P^{-1}(f)$ is an orthonormal basis for $L^2\left(G, d\mu_G\right)$.

Theorem 9.1 together with Theorem 9.2 open the doorway for a large class of groups for which, the Discretization Problem for the regular representation admits a positive answer (see Theorem 9.3).

Theorem 9.2 ***(The down-up inheritance property)*** Let G be a second countable locally compact group. Suppose that G admits a closed subgroup H such that the left regular representation of H has infinite multiplicity. If $L^2(H)$ admits a Riesz basis (frame) of translates then $L^2(G)$ admits a Riesz basis (frame) of translates.

Theorem 9.3 Let G be a Lie group and suppose that G admits a Lie closed subgroup H which is isomorphic to

$$H = \left\{ \begin{bmatrix} e^t & x \\ 0 & 1 \end{bmatrix} : t, x \in \mathbb{R} \right\}.$$

Then $L^2(G)$ admits an orthonormal basis of translates.

Example 9.1 If G is either a noncommutative general linear group or a noncommutative Special Linear Group then $L^2(G)$ has an orthonormal basis of translates.

Proof of Lemma 9.2: Define an unbounded and self-adjoint operator D densely on $L^2(\mathbb{R})$ as

$$[Du](t) = e^{\delta(t)} \cdot u(t) \quad \left(u \in L^2(\mathbb{R})\right),$$

where δ is a smooth function to be determined. Suppose that $f \in L^1(G, d\mu_G) \cap L^2(G, d\mu_G)$. Define a bounded operator $\mathcal{X} = \hat{f}(\pi_\lambda) \circ D$ acting on $L^2(\mathbb{R})$ as follows. For every $u \in L^2(\mathbb{R})$ in the domain of D, and for $\lambda \in \{1, -1\}$,

$$\left[\mathcal{X}_\lambda u\right](t) = \int_{\mathbb{R}}\int_{\mathbb{R}} f(\exp xX \exp aA)\, \pi_\lambda(\exp xX \exp aA)[Du](t)\, e^{-a} da dx.$$

Then

$$\begin{aligned}\pi_\lambda(\exp xX \exp aA)[Du](t) &= e^{2\pi i\lambda e^{-t}x}[Du](t-a)\\ &= e^{2\pi i\lambda e^{-t}x} \cdot e^{\delta(t-a)-a} \cdot u(t-a)\end{aligned}$$

and substituting the above in the explicit definition of $\mathcal{X}u$,

$$\left[\mathcal{X}_\lambda u\right](t) = \int_{\mathbb{R}}\int_{\mathbb{R}} f(\exp xX \exp aA)\, e^{2\pi i\lambda e^{-t}x} e^{\delta(t-a)-a} u(t-a)\, da dx.$$

The change of variable $r = t - a$ gives $a = t - r$ and

$$\left[\mathcal{X}_\lambda u\right](t) = \int_{\mathbb{R}} u(r) \cdot \left(\int_{\mathbb{R}} f(\exp xX \exp((t-r)A))\, e^{2\pi i\lambda e^{-t}x} e^{\delta(r)+r-t} dx\right) dr. \tag{9.5}$$

By isolating the inner integral in (9.5), define

$$K_f(t, r) = \int_{\mathbb{R}} f(\exp xX \exp((t-r)A))\, e^{2\pi i\lambda e^{-t}x} e^{\delta(r)+r-t} dx$$

so that $\left[\mathcal{X}_\lambda u\right](t) = \int_{\mathbb{R}} K_f(t, r)\, u(r)\, dr$. This shows that $\mathcal{X}_\lambda$ is merely an integral operator on $L^2(\mathbb{R})$ with kernel $K_f(t, r)$ and the Hilbert–Schmidt norm of $\mathcal{X}_\lambda$ is related to the L^2-norm of the kernel as follows:

$$\|\mathcal{X}_\lambda\|_{\mathrm{HS}}^2 = \int_{\mathbb{R}}\int_{\mathbb{R}} \left|K_f(t, r)\right|^2 dt dr.$$

To simplify notation, write $f(x, t)$ for $f(\exp xX \exp tA)$ and denote by $\mathcal{F}$, the partial Euclidean Fourier transform acting on $L^2(\mathbb{R} \times \mathbb{R})$ in the first input argument of f. Then, the kernel K_f is a multiple of $\mathcal{F}f$ partially evaluated at λe^{-t}. Symbolically, we may write

$$K_f(t, r) = e^{\delta(r)} \cdot \int_{\mathbb{R}} f(x, t-r)\, e^{2\pi i\lambda e^{-t}x} dx = e^{\delta(r)+r-t} \cdot \mathcal{F}f\left(\lambda e^{-t}, t-r\right).$$

Going back to (9.5),

$$[\mathcal{X}_\lambda u](t) = \int_{\mathbb{R}} u(r) \cdot \left(\int_{\mathbb{R}} f(\exp xX \exp((t-r)A)) e^{2\pi i \lambda e^{-t} x} e^{\delta(r)} dx \right) dr$$
$$= \int_{\mathbb{R}} u(r) \cdot e^{\delta(r)+r-t} \left(\mathcal{F}f\left(\lambda e^{-t}, t-r\right)\right) dr,$$

the change of variable $s = t - r$ gives $r = t - s$ and additionally,

$$\|\mathcal{X}_\lambda\|_{\mathrm{HS}}^2 = \int_{\mathbb{R}} \int_{\mathbb{R}} e^{2\delta(t-s)-2s} \left|\mathcal{F}f\left(\lambda e^{-t}, t-r\right)\right|^2 dtdr.$$

Thus on the one hand,

$$\|\mathcal{X}_1\|_{\mathrm{HS}}^2 + \|\mathcal{X}_{-1}\|_{\mathrm{HS}}^2 = \int_{\mathbb{R}} \int_{\mathbb{R}} e^{2\delta(t-s)-2s} \left(\left|\mathcal{F}f\left(e^{-t}, s\right)\right|^2 + \left|\mathcal{F}f\left(-e^{-t}, s\right)\right|^2 \right) dtds.$$

On the other hand, by restricting f onto the positive and negative reals, a change of variable in the first argument also gives

$$\int_{\mathbb{R}} \int_{\mathbb{R}} |f(x,s)|^2 e^{-s} dxds = \int_{\mathbb{R}} \int_{\mathbb{R}} \left(\left|\mathcal{F}f\left(e^{-t}, s\right)\right|^2 + \left|\mathcal{F}f\left(-e^{-t}, s\right)\right|^2 \right) e^{-(s+t)} dtds.$$

Letting $\delta(t) = -\frac{t}{2}$,

$$2\delta(t-s) - 2s = 2\left(-\frac{1}{2}\right)(t-s) - 2s = -s - t = -(s+t).$$

Thus,

$$\|\mathcal{X}_1\|_{\mathrm{HS}}^2 + \|\mathcal{X}_{-1}\|_{\mathrm{HS}}^2 = \int_{\mathbb{R}} \int_{\mathbb{R}} e^{-(s+t)} \left(\left|\mathcal{F}f\left(e^{-t}, s\right)\right|^2 + \left|\mathcal{F}f\left(-e^{-t}, s\right)\right|^2 \right) dtds$$

showing that the square of the norm of f coincides with the sum $\|\mathcal{X}_1\|_{\mathrm{HS}}^2 + \|\mathcal{X}_{-1}\|_{\mathrm{HS}}^2$ and this effectively implies that

$$\|f\|_{L^2(G)}^2 = \|\mathcal{X}_1\|_{\mathrm{HS}}^2 + \|\mathcal{X}_{-1}\|_{\mathrm{HS}}^2.$$

In other words, the map that takes f to the pair of Hilbert Schmidt operators $\left(\mathcal{X}_1, \mathcal{X}_{-1}\right)$ defines an isometry between $L^1(G) \cap L^2(G)$ and

$$\mathfrak{K}' = \left(L^2(\mathbb{R}) \otimes L^2(\mathbb{R})\right) \oplus \left(L^2(\mathbb{R}) \otimes L^2(\mathbb{R})\right).$$

Since $L^1(G) \cap L^2(G)$ is dense in $L^2(G)$, $f \mapsto \left(\mathcal{X}_1, \mathcal{X}_{-1}\right)$ extends to an isometry between $L^2(G)$ and the Hilbert space $\mathfrak{K}'$. Moreover, for arbitrary $x \in G$, given $u, v \in L^2(\mathbb{R})$ in the domain of D, we verify that

$$\left\langle \widehat{L(x)f}\left(\pi_{\pm 1}\right) Du, v \right\rangle_{L^2(\mathbb{R})} = \int_G f\left(x^{-1}y\right) \left\langle \pi_{\pm 1}(y) Du, v \right\rangle_{L^2(\mathbb{R})} d\mu_G(y)$$
$$= \int_G f(y) \left\langle \pi_{\pm 1}(xy) Du, v \right\rangle_{L^2(\mathbb{R})} d\mu_G(y).$$

Thus,

$$\begin{aligned}\left\langle \widehat{L(x)f}\left(\pi_{\pm 1}\right) Du, v\right\rangle_{L^2(\mathbb{R})} &= \int_G f(y)\left\langle \pi_{\pm 1}(y) Du, \pi_{\pi_{\pm 1}}(x)^* v\right\rangle_{L^2(\mathbb{R})} d\mu_G(y) \\ &= \left\langle \widehat{f}\left(\pi_{\pm 1}\right) Du, \pi_{\pm 1}(x)^* v\right\rangle_{L^2(\mathbb{R})} \\ &= \left\langle \pi_{\pm 1}(x) \widehat{f}\left(\pi_{\pm 1}\right) Du, v\right\rangle_{L^2(\mathbb{R})},\end{aligned}$$

and as a result,

$$\left(\widehat{L(x)f}\left(\pi_1\right) \circ D, \widehat{L(x)f}\left(\pi_{-1}\right) \circ D\right) = \left(\pi_1(x) \widehat{f}\left(\pi_1\right) \circ D, \pi_{-1}(x) \widehat{f}\left(\pi_{-1}\right) \circ D\right).$$

The stated result is thus obtained by naturally identifying $L^2(\mathbb{R}) \otimes L^2(\mathbb{R})$ with an infinite direct sum of copies of $L^2(\mathbb{R})$.

Proof of Lemma 9.3: The fact that

$$f = \left(\left(t \mapsto e^{-\frac{t}{2}} \cdot 1_{E_\ell}(t)\right)_{\ell=1}^{\infty}, \left(t \mapsto e^{-\frac{t}{2}} \cdot 1_{E_\ell}(t)\right)_{\ell=1}^{\infty}\right)$$

is a unit vector is a consequence of the following calculations. First

$$\|f\|_{\mathfrak{K}} = \left(\sum_{\ell=1}^{\infty} \int_{\mathbb{R}} \left|f_\ell^+(t)\right|^2 dt + \sum_{\ell=1}^{\infty} \int_{\mathbb{R}} \left|f_\ell^-(t)\right|^2 dt\right)^{1/2}$$

and

$$\|f\|_{\mathfrak{K}} = \left(\sum_{\ell=1}^{\infty} \left(\int_{\ln(2)\cdot[\ell,\ell+1)} e^{-t} dt\right) + \sum_{\ell=1}^{\infty} \left(\int_{\ln(2)\cdot[\ell,\ell+1)} e^{-t} dt\right)\right)^{1/2}.$$

Secondly,

$$\int_{\ell \ln(2)}^{(\ell+1)(\ln(2))} e^{-t} dt = 2^{-\ell-1} \text{ and } \sum_{\ell=1}^{\infty} 2^{-\ell-1} = \frac{1}{2}$$

and as a result, $\|f\|_{\mathfrak{K}} = \left(2 \cdot \left(\sum_{\ell=1}^{\infty} 2^{-\ell-1}\right)\right)^{1/2}$. But since, $\sum_{\ell=1}^{\infty} 2^{-\ell-1} = \frac{1}{2}$, it is clear that $\|f\|_{\mathfrak{K}} = \left(2 \cdot \left(\frac{1}{2}\right)\right)^{1/2} = 1$. Finally, appealing to Lemma 9.2, the following holds true: $P \circ L(g) P^{-1} = \tau(g)$ for all $g \in G$ and for $h = \left(\left(h_\ell^+\right)_{\ell=1}^{\infty}, \left(h_\ell^-\right)_{\ell=1}^{\infty}\right) \in \mathfrak{K}$,

$$\tau(g) h = \left(\left(\pi_1(g) h_\ell^+\right)_{\ell=1}^{\infty}, \left(\pi_{-1}(g) h_\ell^-\right)_{\ell=1}^{\infty}\right) \quad (g \in G).$$

Proof of Theorems 9.2 and 9.3: For the first part, let L_G be the left regular representation of G. Let $Q \subseteq G$ be a Borel transversal for G/H and define $c : H \times Q \to G$ such that $c(h, q) = hq$. Then c is a measurable bijection. Next, let ω be the push-forward of the Haar measure under the inverse of c. Define $\Phi : L^2\left(G, \mu_G\right) \to L^2(H \times Q, \omega)$ such that $\Phi f(h, q) = f(hq) = f(c(h, q))$. Since Φ is surjective and

$$\int_{H\times Q} |\Phi f(h, q)|^2 d\omega(h, q) = \int_G \left|\Phi f\left(c^{-1}(g)\right)\right|^2 d\mu_G(g) = \int_G |f(g)|^2 d\mu_G(g),$$

it follows that Φ is a unitary map. Observe that for $z \in H$,

$$\Phi\left(L_G(z)f\right)(h,q) = f\left(z^{-1}hq\right) = f\left(c\left(z^{-1}h,q\right)\right) = \Phi f\left(z^{-1}h,q\right).$$

Let L_H be the left regular representation of H. By identifying $L^2(H \times Q, \omega)$ with $L^2(H) \otimes L^2(Q)$, it follows that the restriction of the left regular representation of G to H is unitarily equivalent to an infinite direct sum of L_H. However, an infinite direct sum of the left regular representation of H is unitarily equivalent to L_H. Thus, there is unitary map between $L^2\left(G, \mu_G\right)$ and $L^2\left(H, \mu_H\right)$, intertwining the restriction of L to H with L_H. Since unitary maps take Riesz bases (frames) to Riesz bases (frames), we conclude that If $L^2(H)$ admits a Riesz basis (frame) of translates then $L^2(G)$ admits a Riesz basis (frame) of translates as well.

Let L_H be the left regular representation of H. According to Lemma 9.2–9.4, we know that L_H is unitarily equivalent to the following direct sum of irreducible representations $\bigoplus_{\ell=1}^{\infty} \pi_1 \oplus \bigoplus_{\ell=1}^{\infty} \pi_{-1}$. This shows that the regular representations H has infinite multiplicities. Finally, Theorem 9.3 follows from a straightforward application of the down-up inheritance property established by Theorem 9.2.

References

1 Hartmut Fuhr and Vignon Oussa. Groups with frames of translates. *arXiv preprint arXiv:1802.02658*, 2018.

2 Eberhard Kaniuth and Keith F. Taylor. *Induced representations of locally compact groups*, volume 197. Cambridge University Press, 2013.

10

Sampling and Interpolation on Unimodular Lie Groups

It is a well-known fact that a function defined on the real line, whose Fourier transform is restricted to a unit interval, can be reconstructed using the Whittaker–Kotel'nikov–Shannon sampling series, given its values at points on the integer lattice. A more recent area of inquiry in the field of harmonic analysis has been the search for analogs of the Whittaker–Kotel'nikov–Shannon sampling series for noncommutative groups. This chapter explores this question within the context of the discretization problem. To this end, we shall first revisit the notion of admissible representations. Subsequently, we shall delve into a methodology based on oscillation estimates, owing to Gröchenig and Führ, for discretizing an admissible representation to construct frames. Additionally, we shall expand the concept of bandlimitation on the real line, as exemplified by Paley–Wiener-type spaces, to a broader class of noncommutative Lie groups and provide sufficient conditions under which the discretization problem admits a favorable resolution.

10.1 Admissible Representations

The concept of admissibility has proved to be an influential bridge between representation and generalized wavelets theories. Our modest objective for this section is to convince the reader that admissibility is a theory that not only is worthy of independent interest but also provides us with powerful tools to detect relevant features of representations, allowing the existence of a class of functions called continuous frames or generalized wavelets. We refer the reader to the influential monograph of Hartmut Führ [1] for a thorough treatment of admissibility. In contrast to the work of Führ, our exposition of topics related to admissibility will be rather brief, as we will only emphasize aspects of the theory for which, we judge that new differential geometric insights might be useful.

A Bridge Between Lie Theory and Frame Theory: Applications of Lie Theory to Harmonic Analysis, First Edition. Vignon Oussa.

Let π be a strongly continuous unitary representation of a Lie group G acting in some Hilbert space $\mathcal{H}_\pi$. We equip G with a fixed left-invariant measure denoted by $d\mu_G$.

Definition 10.1 We say π is an admissible representation if there exists a vector $f \in \mathcal{H}_\pi$ such that $V_f : h \mapsto V_f h$ is a continuous function defined on G as $\left[V_f h\right](x) = \langle h, \pi(x) f\rangle_{\mathcal{H}_\pi}$, then V_f defines an isometry between Hilbert spaces $\mathcal{H}_\pi$ and $L^2\left(G, d\mu_G\right)$. When such a vector f exists, we say that f is an admissible vector for π or a continuous Parseval frame or a generalized wavelet.

For a fixed vector $f \in \mathcal{H}_\pi$ called the analyzing generalized wavelet, the map V_f : $\mathcal{H}_\pi \to L^2\left(G, d\mu_G\right)$ establishes a correspondence between an arbitrary element φ in the space $\mathcal{H}_\pi$ and the coefficient matrix $x \mapsto \langle \varphi, \pi(x) f\rangle_{\mathcal{H}_\pi}$ which can then be viewed as a continuous function defined on G. The admissibility of π suggests that $\varphi \mapsto V_f\varphi$ is norm-preserving, and one wonders, under which conditions is it possible to fully reconstruct some unknown vector $\varphi \in \mathcal{H}_\pi$ if we only have information about $V_f\varphi$?

As a starter, Proposition 10.1 provides useful answers.

Proposition 10.1 The following holds true:

1) If f is an admissible vector for π then V_f is a closed operator.
2) If π is an admissible representation with admissible vector f, then for any vector $F \in V_f\left(\mathcal{H}_\pi\right) \subset L^2\left(G, d\mu_G\right), F = F * V_f f$.
3) Let $\left(b_k\right)_{k\in I}$ be an orthonormal basis for $\mathcal{H}_\pi$. For a fixed vector $f \in \mathcal{H}_\pi$ such that V_f is a bounded operator between $\mathcal{H}_\pi$ and $V_f\left(\mathcal{H}_\pi\right)$, letting $F \in V_f\left(\mathcal{H}_\pi\right)$,

$$V_f^* F = \sum_{k\in I} \left(\int_G \left\langle F(x)\, \pi(x) f, b_k \right\rangle_{\mathcal{H}_\pi} d\mu_G(x) \right) b_k.$$

Proof: **Proof of Part (1)**: Let $\left(h_n\right)_{n=1}^{\infty}$ be a sequence of vectors in $\mathcal{H}_\pi$ converging to h. Then

$$\left|\left\langle h_n, \pi(x) f\right\rangle_{\mathcal{H}_\pi} - \langle h, \pi(x) f\rangle_{\mathcal{H}_\pi}\right| = \left|\left\langle h_n - h, \pi(x) f\right\rangle_{\mathcal{H}_\pi}\right| \le \left\|h_n - h\right\|_{\mathcal{H}_\pi} \cdot \|f\|_{\mathcal{H}_\pi}$$

and

$$\lim_{n\to\infty} \left|\left\langle h_n, \pi(x) f\right\rangle_{\mathcal{H}_\pi} - \langle h, \pi(x) f\rangle_{\mathcal{H}_\pi}\right| \le \lim_{n\to\infty} \left\|h_n - h\right\|_{\mathcal{H}_\pi} \cdot \|f\|_{\mathcal{H}_\pi} = 0.$$

Suppose $\lim_{n\to\infty} V_f h_n = F$ in $L^2(G)$. Then $V_f h = F$ and F belongs to the range of V_f

Proof of Part (2): Observe that $V_f V_f^*$ is a projection on the range of V_f. Indeed, for every $x \in G$,

$$\begin{aligned} F(x) = V_f V_f^* F(x) &= \left\langle V_f^* F, \pi(x) f \right\rangle_{\mathcal{H}_\pi} \\ &= \left\langle F, V_f \pi(x) f \right\rangle_{\mathcal{H}_\pi} \\ &= \left\langle F, L(x) V_f f \right\rangle_{\mathcal{H}_\pi} \\ &= F * \left(V_f f\right)^*(x). \end{aligned}$$

Secondly,

$$\left(V_f f\right)^*(x) = \left\langle \pi\left(x^{-1}\right) f, f \right\rangle_{\mathcal{H}_\pi} = \langle f, \pi(x) f \rangle_{\mathcal{H}_\pi} = V_f f.$$

This shows that $F = V_f V_f^* F = F * V_f f$ as desired.

Proof of Part (3): Given $F \in V_f\left(\mathcal{H}_\pi\right)$ and $\varphi \in \mathcal{H}_\pi$,

$$\begin{aligned} \left\langle V_f \varphi, F \right\rangle_{L^2(G, d\mu_G)} &= \int_G V_f \varphi(x) \overline{F(x)} d\mu_G(x) \\ &= \int_G \langle \varphi, \pi(x) f \rangle_{\mathcal{H}_\pi} \overline{F(x)} d\mu_G(x) \\ &= \int_G \langle \varphi, F(x) \pi(x) f \rangle_{\mathcal{H}_\pi} d\mu_G(x) \end{aligned}$$

and

$$\left\langle \varphi, V_f^* F \right\rangle_{L^2(G, d\mu_G)} = \int_G \langle \varphi, F(x) \pi(x) f \rangle_{\mathcal{H}_\pi} d\mu_G(x).$$

Letting $\left(b_k\right)_{k \in I}$ be an orthonormal basis for $\mathcal{H}_\pi$, since $V_f^* F = \sum_{k \in I} \left\langle V_f^* F, b_k \right\rangle b_k$,

$$V_f^* F = \sum_{k \in I} \left(\int_G \left\langle F(x) \pi(x) f, b_k \right\rangle_{\mathcal{H}_\pi} d\mu_G(x) \right) b_k.$$

Thus, if f is an admissible vector for π, any vector $\varphi \in \mathcal{H}_\pi$ can be reconstructed as follows:

$$\varphi = \sum_{k \in I} \left(\int_G \left\langle V_f \varphi(x) \pi(x) f, b_k \right\rangle_{\mathcal{H}_\pi} d\mu_G(x) \right) b_k$$

with convergence in the norm of $\mathcal{H}_\pi$.

We will show next, that the infinite-dimensional irreducible representations of the Heisenberg and the $ax+b$ groups are admissible.

Let G be the three-dimensional Heisenberg–Lie group with Lie algebra $\mathfrak{g}$ spanned by vectors X, Y, Z satisfying the nontrivial Lie brackets $[X, Y] = Z$. Put

$$\pi_{\lambda Z^*+\upsilon Y^*+\omega X^*} = \mathrm{ind}_{\exp \mathbb{R}Z+\mathbb{R}Y}^{G} \left(\chi_{\lambda Z^*+\upsilon Y^*+\omega X^*}\right),$$

where $\ell = \lambda Z^* + \upsilon Y^* + \omega X^*$ is a linear functional in $\mathfrak{g}^*$ and λ is a nonzero real number. Then each $\pi_{\lambda Z^*+\upsilon Y^*+\omega X^*}$ may be realized as acting in $L^2(\mathbb{R})$ as follows:

$$\pi_\ell (\exp zZ \exp yY \exp xX) f(t) = e^{2\pi i \left\langle \begin{bmatrix} \lambda \\ \upsilon - t\lambda \end{bmatrix}, \begin{bmatrix} z \\ y \end{bmatrix} \right\rangle} f(t-x).$$

Since the $\left|\pi_\ell (\exp zZ) f(t)\right| = |f(t-x)|$ for every $z, t, x \in \mathbb{R}$, it is clear that π_ℓ cannot be admissible. This situation can, however, be remedied as follows. Let $N_\ell = G/\ker \pi_\ell$ and define a representation τ_ℓ of N_ℓ such that

$$\tau_\ell \left((\exp zZ \exp yY \exp xX) \ker \pi_\ell\right) = \pi_\ell (\exp zZ \exp yY \exp xX).$$

We will prove that τ_ℓ is an admissible representation of N_ℓ. Indeed, observe that N_ℓ is a connected Lie group with a compact center and

$$\begin{aligned} \left\|V_f h\right\|_{L^2(N_\ell)}^2 &= \int_{\mathbb{R}}\int_{\mathbb{R}} \left|\left\langle h, \pi_\ell (\exp yY \exp xX) f\right\rangle_{L^2(\mathbb{R})}\right|^2 dydx \\ &= \int_{\mathbb{R}}\int_{\mathbb{R}} \left|\int_{\mathbb{R}} h(t)\overline{f(t-x)e^{2\pi i y t\lambda}}dt\right|^2 dydx. \end{aligned}$$

The change of variable $s = t\lambda$ in the integral $\int_{\mathbb{R}} h(t)\overline{f(t-x)e^{2\pi i y t\lambda}}dt$ gives

$$\begin{aligned} \left\|V_f h\right\|_{L^2(N_\ell)}^2 &= \frac{1}{|\lambda|^2}\int_{\mathbb{R}}\int_{\mathbb{R}} \left|\int_{\mathbb{R}} h\left(\frac{s}{\lambda}\right) \cdot \overline{f\left(\frac{s}{\lambda}-x\right)e^{2\pi i y s}}ds\right|^2 dydx \\ &= \frac{1}{|\lambda|^2}\int_{\mathbb{R}}\int_{\mathbb{R}} \left|\mathcal{F}\left(s \mapsto h\left(\frac{s}{\lambda}\right)\overline{f\left(\frac{s}{\lambda}-x\right)}\right)(y)\right|^2 dydx \end{aligned}$$

and as a result,

$$\begin{aligned} \left\|V_f h\right\|_{L^2(N_\ell)}^2 &= \frac{1}{|\lambda|^2}\int_{\mathbb{R}}\int_{\mathbb{R}} \left|h\left(\frac{s}{\lambda}\right)\overline{f\left(\frac{s}{\lambda}-x\right)}\right|^2 dsdx \\ &= \frac{1}{|\lambda|}\int_{\mathbb{R}} |h(s)|^2 \cdot \left(\int_{\mathbb{R}} |f(s-x)|^2 dx\right) ds \\ &= \int_{\mathbb{R}} |h(s)|^2 \cdot \left(\frac{1}{|\lambda|}\int_{\mathbb{R}} |f(x)|^2 dx\right) ds. \end{aligned}$$

Assuming finally that $\frac{1}{|\lambda|}\int_{\mathbb{R}} |f(x)|^2 dx = 1$, it follows that f is an admissible vector for τ_ℓ. Fix a vector $\eta \in L^2(\mathbb{R})$ such that $\frac{1}{|\lambda|}\int_{\mathbb{R}} |\eta(x)|^2 dx = 1$. Next, let $V_\eta : L^2(\mathbb{R}) \to$

$L^2(\exp[0,1)Z\exp\mathbb{R}Y\exp\mathbb{R}X)$ such that

$$\begin{aligned} V_\eta\psi(\exp zZ\exp yY\exp xX) &= \langle\psi, \pi(\exp zZ\exp yY\exp xX)\eta\rangle_{L^2(\mathbb{R})} \\ &= e^{-2\pi i\lambda z}\cdot\int_{\mathbb{R}}\psi(t)\overline{\eta(t-x)e^{2\pi iyt\lambda}}dt. \end{aligned}$$

Let L be the left regular representation of $\exp[0,1)Z\exp\mathbb{R}Y\exp\mathbb{R}X$. Then the range of V_η is a closed L-invariant subspace of $L^2(\exp[0,1)Z\exp\mathbb{R}Y\exp\mathbb{R}X)$ consisting of continuous functions. Additionally, for any $F\in V_\eta\left(L^2(\mathbb{R})\right)$, $F=F*V_\eta\eta$.

Let $G=\mathbb{R}\times\mathbb{R}$ be the semidirect product group equipped with group operation $(x,t)(y,s)=\left(x+e^ty,t+s\right)$. For a fixed $\xi\neq 0$, π_ξ is an irreducible representation of G realized as acting in $L^2(\mathbb{R})$ as follows:

$$\left[\pi_\xi(x,t)f\right](s)=e^{2\pi ie^{-s}x\xi}f(s-t)$$

for $(x,t)\in\mathbb{R}^2$. We will prove that each π_ξ is admissible. Indeed,

$$\begin{aligned} \left\|V_fh\right\|^2_{L^2(N)} &= \int_{\mathbb{R}}\int_{\mathbb{R}}\left|\left\langle h,\pi_\xi(x,t)f\right\rangle_{L^2(\mathbb{R})}\right|^2e^{-t}dxdt \\ &= \int_{\mathbb{R}}\int_{\mathbb{R}}\left|\int_{\mathbb{R}}h(r)\overline{f(r-t)e^{2\pi ixe^{-r}\xi}}dr\right|^2e^{-t}dxdt. \end{aligned}$$

The change of variable $z=e^{-r}\xi$ yields

$$\begin{aligned} \left\|V_fh\right\|^2_{L^2(N)} &= \int_{\mathbb{R}}\int_{\mathbb{R}}\left|\int_{e^{\mathbb{R}}\xi}h\left(\ln\left(\frac{\xi}{z}\right)\right)\frac{\overline{f\left(\ln\left(\frac{\xi}{z}\right)-t\right)e^{2\pi ixz}}}{|z|}dz\right|^2e^{-t}dxdt \\ &= \int_{\mathbb{R}}\int_{e^{\mathbb{R}}\xi}\left|h\left(\ln\left(\frac{\xi}{z}\right)\right)|z|^{-\frac{1}{2}}f\left(\ln\left(\frac{\xi}{z}\right)-t\right)\right|^2|z|^{-1}dze^{-t}dt \end{aligned}$$

and

$$\begin{aligned} \left\|V_fh\right\|^2_{L^2(N)} &= \int_{\mathbb{R}}\int_{\mathbb{R}}|h(r)f(r-t)|^2\frac{e^{r-t}}{|\xi|}drdt \\ &= \int_{\mathbb{R}}|h(r)|^2\int_{\mathbb{R}}|f(r-t)|^2\frac{e^{r-t}}{|\xi|}dtdr \\ &= \int_{\mathbb{R}}|h(r)|^2\left(\int_{\mathbb{R}}|f(t)|^2\frac{e^t}{|\xi|}dt\right)dr. \end{aligned}$$

Assuming that $\int_{\mathbb{R}}|f(t)|^2\frac{e^t}{|\xi|}dt=1$, it follows that f is an admissible vector for π_ξ.

Proposition 10.2 generalizes the previous examples to a class of semi-direct product groups. Let $G=NH$ be a connected Lie group where N is a closed and simply connected abelian subgroup of G, and H is a closed subgroup of G which normalizes N. Suppose additionally that N admits a unitary character χ such that

$\chi(\exp X) = e^{2\pi i\langle\lambda,X\rangle}, X \in \mathfrak{n}$ where λ is some fixed linear functional in $\mathfrak{n}^*$. Next, let $\pi = \mathrm{ind}_N^G(\chi)$ be a representation of G obtained by inducing χ. Then there exists an invariant measure $d\mu_H$ on H such that π acts in $L^2(H, d\mu_H)$ as follows. For $s, h \in H, n \in H$ and $f \in L^2(H, d\mu_H)$,

$$[\pi(nh)f](s) = \chi(s^{-1}ns) \cdot f(h^{-1}s).$$

For fixed $h \in H$, let $\delta(h) = \frac{d(Ad(h^{-1})X)}{dX}$. Let $\beta_\lambda(s) = Ad(s^{-1})^*\lambda$. Suppose that $\beta_\lambda : O \to \mathfrak{n}^*$ is a diffeomorphism around the neutral element $e \in H$, where O is a relatively compact neighborhood of the neutral element. In other words, β_λ is a local diffeomorphism around e. Put

$$d\mu_H(\beta_\lambda^{-1}(\xi)) = W(\xi)\,d\xi.$$

Proposition 10.2 If φ is a fixed vector in $L^2(H, d\mu_H)$ supported on O such that

$$\int_H \left|\varphi(h^{-1}s)\right|^2 \cdot W(Ad(s^{-1})^*\lambda) \cdot \delta(h)\,d\mu_H(h) = 1$$

for almost every $s \in H$ then π is an admissible representation with admissible vector φ.

Proof: Fixing $\varphi, \psi \in L^2(H, d\mu_H)$,

$$V_\varphi\psi(nh) = \langle\psi, \pi(nh)\varphi\rangle_{L^2(H,d\mu_H)} = \int_H \left(\psi(s) \cdot \overline{f(h^{-1}s)}\right) \cdot \chi(s^{-1}n^{-1}s)\,d\mu_H(s).$$

Let $n = \exp X$ and put $\beta_\lambda(s) = Ad(s^{-1})^*\lambda$. Then β_λ is a smooth map taking H into $\mathfrak{n}^*$ and for any $nh \in NH$,

$$V_\varphi\psi(nh) = \int_H \left(\psi(s) \cdot \overline{f(h^{-1}s)}\right) e^{-2\pi i\langle\beta_\lambda(s),X\rangle}\,d\mu_H(s).$$

Suppose that β_λ defines a diffeomorphism around the identity element e in H. Then there exists an open subset O of H around the element e in the set H such that the restriction of β_λ onto O is a diffeomorphism. Suppose that f is a continuous function supported in O. Letting $\Phi_h(s) = \psi(s) \cdot \overline{f(h^{-1}s)}$, the change of variable ξ gives

$$V_\varphi\psi(\exp(X)h) = \int_{\beta_\lambda^{-1}(O)} \Phi_h(\beta_\lambda^{-1}(\xi))\,e^{-2\pi i\langle\xi,X\rangle}\,d\mu_H(\beta_\lambda^{-1}(\xi)).$$

Setting $d\mu_H(\beta_\lambda^{-1}(\xi)) = W(\xi)\,d\xi$

$$V_\varphi\psi(\exp(X)h) = \int_{\beta_\lambda^{-1}(O)} \left(\Phi_h(\beta_\lambda^{-1}(\xi)) \cdot W(\xi)\right) e^{-2\pi i\langle\xi,X\rangle}\,d\xi.$$

For fixed X, the function $h \mapsto V_\varphi \psi(\exp(X)h)$ is the Fourier transform of $\xi \mapsto \Phi_s\left(\beta_\lambda^{-1}(\xi)\right)\cdot W(\xi)$ evaluated at $X \in \mathfrak{n}$. Write

$$V_\varphi \psi(\exp(X)h) = \left(\xi \mapsto \Phi_h\left(\beta_\lambda^{-1}\cdot(\xi)\right)\cdot W(\xi)\right)^\wedge(X).$$

Applying Plancherel's equality,

$$\int_{\mathfrak{n}} |V_\varphi \psi(\exp(X)h)|^2 dX = \int_{\mathfrak{n}} |(\xi \mapsto \Phi_h\left(\beta_\lambda^{-1}\cdot(\xi)\right)\cdot W(\xi))(X)|^2 d\xi$$

$$= \int_{\mathfrak{n}^*} |\Phi_h\left(\beta_\lambda^{-1}(\xi)\right)\cdot W(\xi)|^2 d\xi$$

and

$$\int_{\mathfrak{n}} \left|V_\varphi \psi(\exp(X)h)\right|^2 dX = \int_H \left|\psi(s)\cdot\overline{\varphi(h^{-1}s)}\cdot\sqrt{W(\beta_\lambda(s))}\right|^2 d\mu_H(s)$$

$$= \int_H \left|\psi(s)\cdot\varphi\left(h^{-1}s\right)\right|^2 W\left(\beta_\lambda(s)\right)d\mu_H(s).$$

For fixed $h \in H$, let $\delta(h) = \frac{d(Ad(h^{-1})X)}{dX}$. Then $\mu_G(\exp(X)h) = \delta(h)\,dXd\mu_H(h)$ defines a left-invariant measure on G and

$$\int_H \int_{\mathfrak{n}} \left|V_\varphi \psi(\exp(X)h)\right|^2 \delta(h)\,dXd\mu_H(h)$$

$$= \int_H |\psi(s)|^2 \cdot \left(\int_H \left|\varphi\left(h^{-1}s\right)\right|^2 \cdot W\left(\beta_\lambda(s)\right)\cdot\delta(h)\,d\mu_H(h)\right) d\mu_H(s).$$

If additionally,

$$\int_H \left|\varphi\left(h^{-1}s\right)\right|^2 \cdot W\left(\beta_\lambda(s)\right)\cdot\delta(h)\,d\mu_H(h) = 1$$

for almost every $s \in H$, then

$$\int_H \int_{\mathfrak{n}} \left|V_\varphi \psi(\exp(X)h)\right|^2 \delta(h)\,dXd\mu_H(h) = \int_H |\psi(s)|^2 d\mu_H(s)$$

and under such assumptions φ would be an admissible vector for π. Fix a vector $\eta \in L^2\left(H, d\mu_H\right)$ such that η is admissible for π. Next, let $V_\eta : L^2\left(H, d\mu_H\right) \to L^2(NH)$ such that

$$V_\eta \psi(nh) = \langle \psi, \pi(nh)\eta\rangle_{L^2(H,d\mu_H)} = \int_H \psi(s)\,\chi\left(s^{-1}n^{-1}s\right)\cdot\overline{\eta(h^{-1}s)}d\mu_H(h).$$

Then $V_\eta\left(L^2\left(H, d\mu_H\right)\right)$ is a left-invariant subspace of $L^2(NH)$ consisting of continuous functions. Next, for any $F \in V_\eta\left(L^2\left(H, d\mu_H\right)\right)$, letting

$$V_\eta \eta(nh) = \int_H \eta(s)\,\chi\left(s^{-1}n^{-1}s\right)\cdot\overline{\eta(h^{-1}s)}d\mu_H(h)$$

we have $F = F * V_\eta \eta$.

10.2 Gröchenig–Führ's Method of Oscillations

In this section, we will discuss a technique based on oscillation estimates (10.1) due to Führ and Gröchenig [2] for discretizing an admissible representation to construct frames. Although the scope of the method extends to all locally compact groups, our focus will remain on Lie groups.

Let G be a Lie group equipped with a fixed left-invariant measure μ_G and let $\mathcal{H}$ be a left-invariant and closed subspace of $L^2(G)$ consisting of continuous functions. Let Γ be a countable subset of G. We say that $(\mathcal{H}, \Gamma)$ is a sampling pair for G if there exist positive constants $A \leq B$ such that for any $f \in \mathcal{H}$,

$$A \cdot \|f\|_{\mathcal{H}}^2 \leq \sum_{\gamma \in \Gamma} |f(\gamma)|^2 \leq B \cdot \|f\|_{\mathcal{H}}^2 .$$

In other words, $(\mathcal{H}, \Gamma)$ is a sampling pair for G, if the l^2-norm of any sequence obtained by restricting a vector f in $\mathcal{H}$ to some fixed countable set Γ is equivalent to the $\mathcal{H}$-norm of f. In a sense, if $(\mathcal{H}, \Gamma)$ is a sampling pair for G, then compression of information induced by the restriction $f \mapsto f|_\Gamma$ (this is analogous to the compression from analog to digital data), is a stable and reversible process. For suitable functions f, g defined on G, recall that the operations of convolution and involution are defined as:

$$f * g(x) = \int f(y) g\left(y^{-1}x\right) dy \text{ and } f^*(x) = \overline{f(x^{-1})}$$

respectively. A function $s \in \mathcal{H}$ is said to be self-adjoint convolution idempotent if $s * s^* = s = s^*$. In this section, we will assume the following. First, there exists a trivial continuous function $s \in \mathcal{H}$ such that s is self-adjoint convolution idempotent and, secondly, for any $f \in \mathcal{H}, f * s = f$. In other words, for arbitrary $x \in G$, $f \in \mathcal{H}$,

$$f(x) = \int_G f(y) s\left(y^{-1}x\right) d\mu(y).$$

Our first result draws a straight line between the concept of a sampling pair and frame theory.

Proposition 10.3 $(\mathcal{H}, \Gamma)$ is a sampling pair for G if and only if there exist positive real numbers A, B such that

$$A \cdot \|f\|_{L^2(G)}^2 \leq \sum_{\gamma \in \Gamma} \left| \langle f, L(\gamma) s \rangle_{L^2(G)} \right|^2 \leq B \cdot \|f\|_{L^2(G)}^2 .$$

for every $f \in \mathcal{H}$. In other words, $(\mathcal{H}, \Gamma)$ is a sampling pair for G if and only if the system of vectors $L(\Gamma) s$ is a frame for the Hilbert space $\mathcal{H}$.

Proof: Given $f \in \mathcal{H}$, since by assumption, $f * s = f$,

$$\sum_{\gamma \in \Gamma} |f(\gamma)|^2 = \sum_{\gamma \in \Gamma} \left| [f * s^*](\gamma) \right|^2 \cdot \sum_{\gamma \in \Gamma} \left| \langle f, L(\gamma) s \rangle_{L^2(G)} \right|^2.$$

As a result,

$$A \cdot \|f\|^2_{L^2(G)} \leq \sum_{\gamma \in \Gamma} |f(\gamma)|^2 \leq B \cdot \|f\|^2_{L^2(G)}$$

for some positive real numbers A, B if and only if

$$A \cdot \|f\|^2_{L^2(G)} \leq \sum_{\gamma \in \Gamma} \left| \langle f, L(\gamma) s \rangle_{L^2(G)} \right|^2 \leq B \cdot \|f\|^2_{L^2(G)}$$

for some positive real numbers A and B.

Definition 10.2 Let G be a connected Lie group. For a function f defined on G and a set $U \subset G$, the oscillation of f with respect to U is defined as:

$$\operatorname{osc}_U(f)(x) = \sup_{y \in U} \left| f(x) - f\left(xy^{-1}\right) \right|. \tag{10.1}$$

Moreover, for a relatively compact and open subset U of G containing the identity, (U, Γ) is a tiling pair if $\{\gamma U : \gamma \in \Gamma\}$ is a partition of G (see Theorem 3.8).

For a continuous function f, the oscillation with respect to some open subset U evaluated at x measures the extent to which point evaluation at varying inputs in U deviates from the value obtained by evaluating f at x. Thus, a function with bounded oscillation at some point x, cannot grow or decay too quickly when evaluated near x.

Example 10.1 For a concrete case, suppose that G is the group of real numbers and let $U = \left(-\frac{1}{2}, \frac{1}{2}\right)$ be an open interval of unit length around zero. Next, let $g(t) = e^{-t^2}$ be a Gaussian. Then the oscillation of g with respect to $\left(-\frac{1}{2}, \frac{1}{2}\right)$ and evaluated at zero is just

$$\operatorname{osc}_U(g)(0) = \sup_{y \in \left(-\frac{1}{2}, \frac{1}{2}\right)} \left| 1 - e^{-y^2} \right| = 1 - e^{-\frac{1}{4}} \approx 0.221\,20.$$

Generally, to compute $\operatorname{osc}_U(g)(x)$, consider the following function $(x, y) \Leftarrow \left| e^{-x^2} - e^{-(x-y)^2} \right|$. For each fixed x, we only need to compute a maximum for each map $y \mapsto \left| e^{-x^2} - e^{-(x-y)^2} \right|$ on the closed interval $\left[-\frac{1}{2}, \frac{1}{2}\right]$. Since the Gaussian g is either always strictly increasing or strictly decreasing, the value of its oscillation

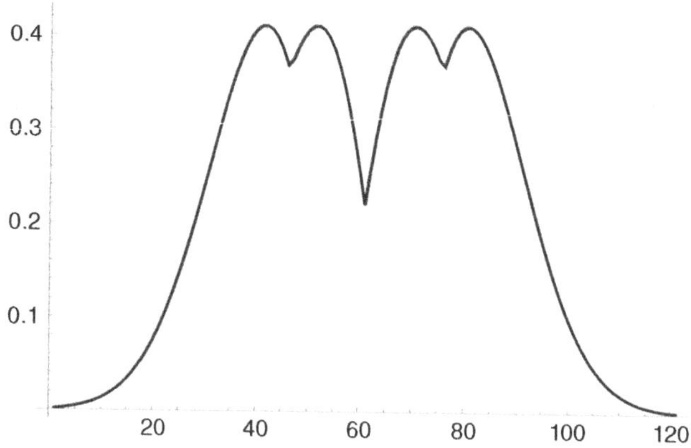

Figure 10.1 Graph of the oscillation of a Gaussian.

is obtained by evaluation for fixed x, the function $y \mapsto \left| e^{-x^2} - e^{-(x-y)^2} \right|$ at one of the boundary points of $\left[-\frac{1}{2}, \frac{1}{2}\right]$. In other words,

$$\mathrm{osc}_U(g)(x) = \max \left\{ \left| e^{-x^2} - e^{-\left(x-\frac{1}{2}\right)^2} \right|, \left| e^{-x^2} - e^{-\left(x+\frac{1}{2}\right)^2} \right| \right\}.$$

The reader is invited to verify that the following illustration is an accurate depiction of the graph of $\mathrm{osc}_U(g)$ (Figure 10.1).

Theorem 10.1 Suppose (V, Γ) is a tiling pair for G (see Theorem 3.8) and suppose additionally that there exists some positive real number $\epsilon < 1$ such that $\|\mathrm{osc}_U(f)\|_{L^2(G)} \leq \epsilon \|f\|_{L^2(G)}$ for all $f \in \mathcal{H}$. Then letting L be the left regular representation of G

$$\frac{(1-\epsilon)^2}{\mu_G(V)} \|f\|^2_{L^2(G)} \leq \sum_{\gamma \in \Gamma} |f(\gamma)|^2 \leq \frac{(1+\epsilon)^2}{\mu_G(V)} \|f\|^2_{L^2(G)}$$

and $L(\Gamma)s$ is a frame for $\mathcal{H}$ with frame bounds $\frac{(1-\epsilon)^2}{\mu_G(U)}$ and $\frac{(1+\epsilon)^2}{\mu_G(U)}$. Furthermore, if there exists an open subset U around the identity such that $\|\mathrm{osc}_V(s)\|_{L^1(G)} < 1$, then $\mathcal{H}$ is sampling set.

Proof: To prove the first part of the result, define (formally for now) $Q : l^2(\Gamma) \to Q(l^2(\Gamma)) \subseteq L^2(G)$ such that

$$Q\left(\left(c_\gamma\right)_{\gamma \in \Gamma}\right) = \sum_{\gamma \in \Gamma} c_\gamma L(\gamma) 1_V.$$

Q is just an operator that outputs a vector obtained by taking a linear combination of elements in $\{L(\gamma) 1_V : \gamma \in \Gamma\}$ associated with the complex numbers in terms of the input sequence. For any input sequence $(c_\gamma)_{\gamma\in\Gamma}$ of unit $l^2(\Gamma)$-norm, one verifies that the square of the norm of this linear combination satisfies the following string of equalities:

$$\left\|\sum_{\gamma\in\Gamma} c_\gamma 1_{\gamma V}\right\|^2_{L^2(G)} = \sum_{\gamma\in\Gamma}\int_G \left|c_\gamma 1_V\left(\gamma^{-1}x\right)\right|^2 d\mu_G(x) = \left(\sum_{\gamma\in\Gamma}\left|c_\gamma\right|^2\right)\cdot \mu_G(V) = \mu_G(V).$$

It is perhaps worth pointing out to the reader that the second equality above is merely a consequence of the left-invariance property of the Haar measure. The equality above, also shows that Q is in fact bounded, and its norm is $\|Q\|_{l^2(\Gamma)\to Q(l^2(\Gamma))} = \mu_G(V)^{1/2}$. Furthermore, any function in the range of Q is of the form $\sum_{\gamma\in\Gamma} c_\gamma L(\gamma) 1_V$ for some sequence $(c_\gamma)_{\gamma\in\Gamma_r^{-1}} \in l^2(\Gamma)$, and it has for norm

$$\left\|\sum_{\gamma\in\Gamma} c_\gamma L(\gamma) 1_V\right\|_{L^2(G)} = \mu_G(V)\cdot\sum_{\gamma\in\Gamma}\left|c_\gamma\right|^2.$$

Suppose next that $\sum_{\gamma\in\Gamma} c_\gamma L(\gamma) 1_V = \sum_{\gamma\in\Gamma} c_\gamma 1_{\gamma V}$ is a unit vector in $L^2(G)$. Then $\sum_{\gamma\in\Gamma}\left|c_\gamma\right|^2 = \mu_G(V)^{-1}$ and the inverse of Q is also a bounded operator, with norm $\|Q^{-1}\|_{Q(l^2(\Gamma))\to l^2(\Gamma)} = \mu_G(V)^{-1}$. Moving forward, let $R : f \mapsto (f(\gamma))_{\gamma\in\Gamma}$ be the restriction map of $f \in L^2(G)$ onto Γ. Since $f - QRf = f - \sum_{\gamma\in\Gamma} f(\gamma) L(\gamma) 1_V$, for each $x \in G$,

$$\begin{aligned}
\left[f - QRf\right](x) &= f(x) - \sum_{\gamma\in\Gamma} f(\gamma) L(\gamma) 1_V(x)\\
&= f(x) - \sum_{\gamma\in\Gamma} f(\gamma) 1_V\left(\gamma^{-1}x\right)\\
&= f(x) - \sum_{\gamma\in\Gamma} f(\gamma) 1_{\gamma V}(x).
\end{aligned}$$

Using the assumption that $\{\gamma V : \gamma\in\Gamma\}$ is a partition of G,

$$\begin{aligned}
\|f - QRf\|^2_{L^2(G)} &= \int_G \left|f(x) - \sum_{\gamma\in\Gamma} f(\gamma) 1_{\gamma V}(x)\right|^2 d\mu_G(x)\\
&= \sum_{\gamma'\in\Gamma}\int_{\gamma' V}\left|f(x) - \sum_{\gamma\in\Gamma} f(\gamma) 1_{\gamma V}(x)\right|^2 d\mu_G(x)\\
&= \sum_{\gamma\in\Gamma}\int_{\gamma V} |f(x) - f(\gamma)|^2 dx.
\end{aligned}$$

Since $|f(x) - f(\gamma)| \le \left|\mathrm{osc}_U(f)(x)\right|$, the square of the norm of $f - QRf$ is estimated as follows:

$$\|f - QRf\|^2_{L^2(G)} = \sum_{\gamma\in\Gamma}\int_{\gamma V} |f(x) - f(\gamma)|^2 dx \le \sum_{\gamma\in\Gamma}\int_{\gamma U}\left|\mathrm{osc}_U(f)(x)\right|^2 dx.$$

By assumption, $\left\|\mathrm{osc}_U(f)\right\|_{L^2(G)} \le \epsilon \cdot \|f\|_{L^2(G)}$ and

$$\|f - QRf\|_{L^2(G)}^2 = \left\|\mathrm{osc}_U(f)\right\|_{L^2(G)}^2 \le \epsilon^2 \cdot \|f\|_{L^2(G)}^2 .$$

Moreover, a factorization of the identity into the composition $Q^{-1}Q$, yields

$$\|Rf\|_{l^2(\Gamma)} = \left\|Q^{-1}QRf\right\|_{l^2(\Gamma)} \le \left\|Q^{-1}\right\|_{Q(l^2(\Gamma))\to l^2(\Gamma)} \cdot \|QRf\|_{L^2(G)}.$$

This implies that

$$\|Rf\|_{l^2(\Gamma)} \le \left\|Q^{-1}\right\|_{Q(l^2(\Gamma))\to l^2(\Gamma)} \cdot \|QRf - f + f\|_{L^2(G)}.$$

Applying triangle inequality to the second norm on the right side of the inequality above, we derive

$$\begin{aligned}
\|Rf\|_{l^2(\Gamma)} &\le \left\|Q^{-1}\right\|_{Q(l^2(\Gamma))\to l^2(\Gamma)} \cdot \left(\|QRf - f\|_{L^2(G)} + \|f\|_{L^2(G)}\right) \\
&\le \frac{1}{\left|\mu_G(V)\right|}\left(\epsilon\|f\|_{L^2(G)} + \|f\|_{L^2(G)}\right) \\
&= \left(\frac{1+\epsilon}{\left|\mu_G(V)\right|}\right) \cdot \|f\|_{L^2(G)}.
\end{aligned}$$

Finally,

$$\|QRf\|_{L^2(G)} \le \|Q\|_{l^2(\Gamma)\to Q(l^2(\Gamma))} \cdot \|Rf\|_{l^2(\Gamma)} \tag{10.2}$$

and since

$$\|f\|_{L^2(G)} = \|f - QRf + QRf\|_{L^2(G)} \le \|f - QRf\|_{L^2(G)} + \|QRf\|_{L^2(G)}$$

then $\|f\|_{L^2(G)} - \|f - QRf\|_{L^2(G)} \le \|QRf\|_{L^2(G)}$. Solving the inequality (10.2) for $\|Rf\|_{l^2(\Gamma)}$ gives

$$\begin{aligned}
\|Rf\|_{l^2(\Gamma)} &\ge \frac{1}{\|Q\|_{l^2(\Gamma)\to Q(l^2(\Gamma))}}\|QRf\|_{L^2(G)} \\
&\ge \frac{1}{\|Q\|_{l^2(\Gamma)\to Q(l^2(\Gamma))}}\left(\|f\|_{L^2(G)} - \|f - QRf\|_{L^2(G)}\right) \\
&\ge \frac{1-\epsilon}{\|Q\|_{l^2(\Gamma)\to Q(l^2(\Gamma))}}\|f\|_{L^2(G)} \\
&= \frac{1-\epsilon}{\left|\mu_G(V)\right|}\|f\|_{L^2(G)}.
\end{aligned}$$

Thus far, we have shown that

$$\left(\frac{1-\epsilon}{\left|\mu_G(V)\right|}\right) \cdot \|f\|_{L^2(G)} \le \|Rf\|_{l^2(\Gamma)} \le \left(\frac{1+\epsilon}{\left|\mu_G(V)\right|}\right) \cdot \|f\|_{L^2(G)}.$$

To prove the second part, fix $f \in \mathcal{H}$. By definition

$$\mathrm{osc}_V(f)(x) = \mathrm{osc}_V(f * s)(x) = \sup_{y\in V}\left|(f * s)(x) - (f * s)\left(xy^{-1}\right)\right|$$

and additionally,

$$\mathrm{osc}_V(f)(x) = \sup_{y\in V} \left|\langle f, L(x)s\rangle_{L^2(G)} - \left\langle f, L\left(xy^{-1}\right)s\right\rangle_{L^2(G)}\right|.$$

It is perhaps helpful to rewrite the above formula explicitly as:

$$\mathrm{osc}_V(f)(x) = \sup_{y\in V} \left|\int_G f(z)\overline{s(x^{-1}z)}d\mu_G(z) - \int_G f(z)\overline{s(yx^{-1}z)}d\mu_G(z)\right|.$$

Combining the two integrals above into a single one,

$$\mathrm{osc}_V(f)(x) = \sup_{y\in V} \left|\int_G f(z)\left(\overline{s(x^{-1}z)} - \overline{s(yx^{-1}z)}\right)d\mu_G(z)\right|.$$

Since s is invariant under involution,

$$\mathrm{osc}_V(f)(x) = \sup_{y\in V} \left|\int_G f(z)\left(s\left(z^{-1}x\right) - s\left(z^{-1}xy^{-1}\right)\right)d\mu_G(z)\right|$$

and

$$\mathrm{osc}_V(f) \le \int_G |f(z)| \cdot \sup_{y\in V} \left|s\left(z^{-1}x\right) - s\left(z^{-1}xy^{-1}\right)\right| d\mu_G(z).$$

Next

$$\left[|f| * \mathrm{osc}_V(s)\right](x) = \int_G |f(z)| \cdot \left[\mathrm{osc}_V(s)\right]\left(z^{-1}x\right)d\mu_G(z)$$

and consequently, $\mathrm{osc}_V(f) \le |f| * \mathrm{osc}_V(s)$. Computing the L^2-norm of the oscillation of f, we obtain

$$\|\mathrm{osc}_V(f)\|^2_{L^2(G)} = \int_G \left|\int_G |f(z)| \cdot \left[\mathrm{osc}_V(s)\right]\left(z^{-1}x\right)d\mu_G(x)\right|^2 d\mu_G(z).$$

Changing variable $x \mapsto zx$, and appealing to the left invariance of μ_G,

$$\|\mathrm{osc}_V(f)\|^2_{L^2(G)} = \int_G \left|\int_G |f(z)| \cdot \left[\mathrm{osc}_V(s)\right](x)\,d\mu_G(x)\right|^2 d\mu_G(z).$$

Moving $|f(z)|$ out of the first integral above, we obtain:

$$\|\mathrm{osc}_V(f)\|^2_{L^2(G)} = \int_G |f(z)|^2 \cdot \left|\int_G \left[\mathrm{osc}_V(s)\right](x)\,d\mu_G(x)\right|^2 d\mu_G(z).$$

Next, by triangle inequality,

$$\left|\int_G \left[\mathrm{osc}_V(s)\right](x)\,d\mu_G(x)\right|^2 \le \left(\int_G \left|\left[\mathrm{osc}_V(s)\right](x)\right| d\mu_G(x)\right)^2 = \|\mathrm{osc}_U(s)\|^2_{L^1(G)}$$

and as such,

$$\|\mathrm{osc}_V(f)\|^2_{L^2(G)} \le \int_G |f(z)|^2 \cdot \|\mathrm{osc}_V(s)\|^2_{L^1(G)}\,d\mu_G(z).$$

Moving the constant $\|\mathrm{osc}_U(s)\|^2_{L^1(G)}$ out of the integrand above,

$$\|\mathrm{osc}_V(f)\|^2_{L^2(G)} \leq \|\mathrm{osc}_V(s)\|^2_{L^1(G)} \cdot \int_G |f(z)|^2 d\mu_G(z).$$

Finally, applying square-root to each side of the inequality above,

$$\|\mathrm{osc}_V(f)\|_{L^2(G)} \leq \|\mathrm{osc}_V(s)\|_{L^1(G)} \cdot \|f\|_{L^2(G)}.$$

and

$$\epsilon = \|\mathrm{osc}_V(s)\|_{L^1(G)} \Rightarrow \|\mathrm{osc}_V(f)\|_{L^2(G)} \leq \epsilon \cdot \|f\|_{L^2(G)}.$$

The second part of this proposition is at this point, an immediate application of the first statement.

Example 10.2 Let G be the three-dimensional Heisenberg Lie group with Lie algebra $\mathfrak{g}$ spanned by vectors X, Y, Z satisfying the nontrivial Lie brackets $[X, Y] = Z$. Put $\pi_{Z^*} = \mathrm{ind}^G_{\exp \mathbb{R}Z+\mathbb{R}Y}(\chi_{Z^*})$ and define τ_{Z^*} such that

$$\tau_{Z^*}((\exp zZ \exp yY \exp xX) \ker(\pi_\ell)) = \pi_{Z^*}(\exp zZ \exp yY \exp xX).$$

Next, let $f : t \mapsto f(t) = \left(\frac{2}{\pi}\right)^{1/4} e^{-t^2}$ be a Gaussian suitably normalized such that $\|f\|_{L^2(\mathbb{R})} = 1$. Since f is a unit vector, we know that f is admissible for τ_{Z^*}. Define $V_f : L^2(\mathbb{R}) \to \mathcal{K} = V_f(L^2(\mathbb{R}))$ such that $V_f h(g) = \langle f, \pi_{Z^*}(g) f\rangle_{L^2(\mathbb{R})}$. Then $V_f \circ \tau_{Z^*} \circ V_f$ is a subrepresentation of the regular representation of $N = G/\ker \pi_{Z^*}$ acting in $\mathcal{K}$. Furthermore, $\mathcal{K}$ is a closed left-invariant subspace of $L^2(N)$ and an admissible vector $V_f(f)$ for the representation $V_f \circ \tau_{Z^*} \circ V_f$ is explicitly calculated as follows:

$$\begin{aligned}
V_f f(z, y, x) &:= V_f f((\exp zZ \exp yY \exp xX) \ker \pi_{Z^*}) \\
&= \langle f, \pi_{Z^*}(\exp zZ \exp yY \exp xX) f\rangle_{L^2(\mathbb{R})} \\
&= e^{-2\pi i z}\langle f, \pi_{Z^*}(\exp yY \exp xX) f\rangle_{L^2(\mathbb{R})} \\
&= e^{-2\pi i z}\left(\frac{2}{\pi}\right)^{1/2} \int_{\mathbb{R}} f(t) f(t - x) e^{2\pi i t y} dt \\
&= e^{-2\pi i z}\left(\frac{2}{\pi}\right)^{1/2} e^{i\pi xy} e^{-\frac{1}{2}(x^2+\pi^2 y^2)}.
\end{aligned}$$

Identifying the Lie group N with the transversal $\exp\left(\left(-\frac{1}{2}, \frac{1}{2}\right) Z\right) \exp(\mathbb{R}Y) \exp(\mathbb{R}X) \subset G$ for fixed $\epsilon > 0$, let $\Gamma_\epsilon = \exp(\epsilon \mathbb{Z} Y) \exp(\epsilon \mathbb{Z} X)$ and let

$$U_\epsilon = \exp\left(\left(-\frac{1}{2}, \frac{1}{2}\right) Z\right) \exp\left(\left(-\frac{\epsilon}{2}, \frac{\epsilon}{2}\right) Y\right) \exp\left(\left(-\frac{\epsilon}{2}, \frac{\epsilon}{2}\right) X\right)$$

an open subset around the identity in the group. Then it is not hard to verify that $(\Gamma_\epsilon^{-1}, U_\epsilon)$ is a tiling pair for N. Furthermore, there exists a sufficiently positive real number ϵ such that $\left\|\operatorname{osc}_{U_\epsilon}\left(V_f f\right)\right\|_{L^1(G)} < 1$. The method of Gröchenig and Führ guarantees then, that the system of vectors $L\left(\Gamma_\epsilon^{-1}\right) f$ is a frame for $\mathcal{H}$ with frame bounds $\frac{(1-\epsilon)^2}{\epsilon^2}$ and $\frac{(1+\epsilon)^2}{\epsilon^2}$.

10.3 Sampling on Locally Compact Groups

The following is a fundamental question in information theory. How often can we sample a signal without losing the ability to reconstruct it fully? Consider a signal modeled as a continuous square-integrable function f on the real line. Can we fully recover f if we only knew its values on the integers? Stated in this general form, this question is undoubtedly ill-posed, since it is not difficult to construct two distinct square-integrable functions that agree on the integers. For instance, define continuous functions β, υ as follows:

$$\beta(t) = \left[1_{\left[-\frac{1}{2},\frac{1}{2}\right]} * 1_{\left[-\frac{1}{2},\frac{1}{2}\right]}\right](2t) \text{ and } \upsilon(t) = \left[1_{\left[-\frac{1}{2},\frac{1}{2}\right]} * 1_{\left[-\frac{1}{2},\frac{1}{2}\right]}\right](3t).$$

Next, let

$$b(t) = \sum_{k \in r\mathbb{Z}} \frac{\beta(t-2k)}{1+k^2} \text{ and } u(t) = \sum_{k \in r\mathbb{Z}} \frac{\upsilon(t-2k)}{1+k^2}.$$

Then $u, v \in L^2(\mathbb{R})$, $\|u - v\| > 0$ and $u|_{\mathbb{Z}} = v|_{\mathbb{Z}}$ as illustrated in Figure 10.2.

Under certain sufficient conditions, this is possible. According to Shannon Nyquist sampling theorem, a function containing no frequency higher than ωHz is entirely determined by sampling at $2\omega Hz$. That is, to be able to resolve all frequencies in such a signal, it must be sampled at a rate that is at least twice the highest frequency present. Since the Fourier transform tells us precisely those frequencies that are active in a function, bandlimited vectors are a natural class of vectors to consider for this problem. A bandlimited vector is a vector whose Fourier transform is supported in a fixed compact subset Ω of the real line. The set of such functions is often referred to in the literature as Ω-bandlimited space.

Suppose we knew that the Euclidean Fourier transform of f was supported on some bounded interval of the type $[-\delta, \delta]$ for some fixed positive real number δ. Suppose additionally that h is a continuous function whose Fourier transform is equal to 1 on the interval $[-\delta, \delta]$ and let C_h be a convolution operator formally defined as follows: $C_h : f \mapsto f * h$. Applying (at least formally) the Fourier transform to a vector $f * h$ in the range of C_h, one obtains the product $\hat{f} \cdot \hat{h}$. Since $\hat{h} = 1$ on $[-\delta, \delta]$, it is clear that $\hat{h} \cdot \hat{f} = \hat{f}$. Next, let

$$PW^{(\delta)} := \left\{f \in L^2(\mathbb{R}) \cap C(\mathbb{R}) : \operatorname{supp}\hat{f} \subseteq [-\delta, \delta]\right\}.$$

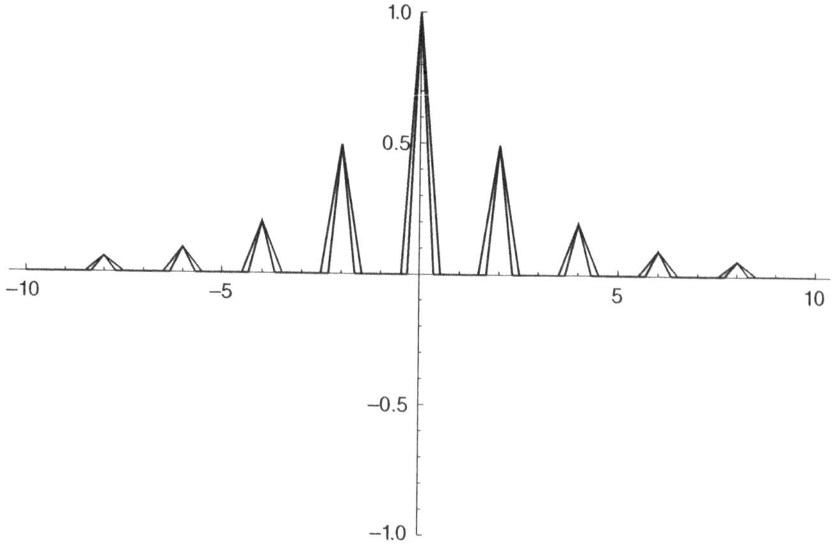

Figure 10.2 Graphs of u and v.

Then $PW^{(\delta)}$ is a left-invariant subspace of $L^2(\mathbb{R})$ consisting of continuous functions. Additionally, it is easy to verify that the operator C_h is a projection from $L^2(\mathbb{R})$ onto its range, $PW^{(\delta)}$. Coming back to the problem at hand, let L be the left regular representation of $\mathbb{R}$. We wish to construct an invertible endomorphism Ψ_r acting on $PW^{(\delta)}$ such that

$$\Psi_r(f) = \sum_{\kappa \in r\mathbb{Z}} f(\kappa) \cdot L_\kappa h \text{ for a fixed } r > 0.$$

This allows viewing $PW^{(\delta)}$ as continuously embedded into $l^2(r\mathbb{Z})$ via a map the restriction map $f \mapsto (f(\gamma))_{\gamma \in \Gamma}$. Moreover, since

$$h(\gamma) = h * h(\gamma) = \langle h, L(\gamma) h \rangle = \int_{[-\delta,\delta]} e^{2\pi i \tau \gamma} d\tau,$$

to ensure the existence of the endomorphism Ψ_r, the following will suffice. First, we define h such that the restriction of h onto $r\mathbb{Z}$ is the indicator function of a singleton. Secondly, we must also verify that $f \mapsto (f(\gamma))_{\gamma \in \Gamma}$ is isometric. Although sampling is certainly possible for bandlimited spaces $PW^{(\delta)}$, $\delta > 0$, the specific case where $\delta = 1$ is easier to present. The vector space

$$PW^{(1)} = \left\{ f \in L^2(\mathbb{R}) : \operatorname{supp} \hat{f} \subseteq \left[-\frac{1}{2}, \frac{1}{2}\right] \right\}$$

is classically called a Paley–Weiner space. This space serves as a prototype for the sampling problem, as it reveals with great simplicity that the sampling phenomenon is just an instantiation of a much more general principle.

As observed earlier, bandlimitation imposes a stringent condition on the oscillation behavior of vectors in $PW^{(1)}$. In principle, it seems reasonable to trust that once one has control over the oscillations of all vectors in a space, one should be able to establish a sampling rate. To substantiate these intuitions, let

$$\operatorname{sinc}(x) = \begin{cases} \frac{\sin(\pi x)}{\pi x} \text{ if } x \neq 0 \\ 1 \text{ if } x = 0 \end{cases}$$

be the sinc function. Then, sinc is a continuous function, explicitly given by the Fourier inverse of the indicator function of $\left[-\frac{1}{2}, \frac{1}{2}\right]$. Moreover, since the system of trigonometric polynomials

$$\left\{x \mapsto e^{2\pi i x \kappa} \cdot 1_{\left[-\frac{1}{2}, \frac{1}{2}\right]}(x) : \kappa \in \mathbb{Z}\right\}$$

is just an orthonormal basis for $L^2\left(\left[-\frac{1}{2}, \frac{1}{2}\right]\right)$ and since the Fourier transform is a unitary operator intertwining left translations with multiplication by complex exponentials, the set of all integer translates of the sinc function must be an orthonormal basis for $PW^{(1)}$. A solution to the sampling problem of the Paley–Wiener is classically referred to in the literature as Shannon's sampling theorem (see Theorem 10.2 and its proof). For an in-depth exposure and treatment, the interested reader is referred to the following [4].

Theorem 10.2 If $f \in L^2(\mathbb{R})$ is a continuous function with Fourier transform supported on the unit interval $\left[-\frac{1}{2}, \frac{1}{2}\right]$ then $\|f\|^2 = \sum_{k\in\mathbb{Z}} |f(k)|^2$. In other words, the l^2-norm of the sequence obtained by restricting f to the integers is equal to the L^2-norm of the given function. Furthermore, $f(t) = \sum_{k\in\mathbb{Z}} f(k)\operatorname{sinc}(t-k)$ with convergence in $L^2(\mathbb{R})$.

Proof: Let $\mathcal{F}$ be the Fourier transform defined on the real line. Since

$$\left\{t \mapsto e^{2\pi i k(t)} 1_{(-1/2,1/2)}(t)\right\}_{k\in\mathbb{Z}}$$

is an orthonormal basis for $L^2(-1/2, 1/2)$, it is clear that

$$\mathcal{F}\left(t \mapsto e^{2\pi i k(t)} \cdot \chi_{(-1/2,1/2)}(t)\right)(\gamma) = \int_{-1/2}^{1/2} e^{2\pi i k x} e^{-2\pi i x \gamma} dx = \operatorname{sinc}(\gamma - k).$$

Next, since $\mathcal{F}$ is unitary, it takes orthonormal bases to orthonormal bases. As a result, the set of translates of the sinc function by the integers: $\{\operatorname{sinc}(x-k) : k \in \mathbb{Z}\}$ forms an orthonormal basis for $PW^{(1)}$. Fix $f \in PW^{(1)} \cap C(\mathbb{R})$. Then $\hat{f}(\gamma) = \sum_{k\in\mathbb{Z}} c_k e^{-2\pi i k \gamma}$ and c_k is a Fourier coefficient of the form $c_k = \int_{-1/2}^{1/2} \hat{f}(\gamma) e^{-2\pi i k \gamma} d\gamma$. Since the partial sums of the Fourier series converge in the norm of $L^2(-1/2, 1/2)$,

$$\lim_{N\to\infty} \int_{-1/2}^{1/2} \left| \hat{f}(\gamma) - \sum_{n=-N}^{N} c_k e^{-2\pi i k \gamma} \right|^2 d\gamma = 0. \tag{10.3}$$

In fact $c_k = \int_{-1/2}^{1/2} \hat{f}(\gamma) e^{2\pi i(-k)\gamma} d\gamma = f(-k)$. For any fixed $x \in \mathbb{R}$, appealing to (10.3)

$$f(x) = \int_{\mathbb{R}} \hat{f}(\gamma) e^{2\pi i x\gamma} d\gamma = \int_{-1/2}^{1/2} \left(\sum_{k\in\mathbb{Z}} f(-k) e^{2\pi i k\gamma} \right) e^{2\pi i x\gamma} d\gamma$$

and

$$f(x) = \sum_{k\in\mathbb{Z}} f(-k) \left(\int_{-1/2}^{1/2} e^{2\pi i\gamma(x+k)} d\gamma \right).$$

Finally,

$$f(x) = \sum_{k\in\mathbb{Z}} f(-k)\ \mathrm{sinc}(x+k) = \sum_{k\in\mathbb{Z}} f(k)\ \mathrm{sinc}(x-k).$$

To show that this series converges in $L^2(\mathbb{R})$, use the fact that $(f(k))_{k\in\mathbb{Z}} \in l^2(\mathbb{Z})$ and derive that

$$\lim_{N\to\infty} \left\| f - \sum_{k=-N}^{N} f(k)\ \mathrm{sinc}(\cdot - k) \right\| = \lim_{N\to\infty} \sqrt{\sum_{|k|>N} |f(k)|^2} = 0.$$

Having established a practical framework under which the sampling problem is tractable, one naturally wonders to which extent can we generalize sampling on Lie groups that may not even be commutative or even solvable.

Let G be a Lie group, and let Γ be a countable subset of G for which no group structure is required. Next, let $\mathcal{H}$ be a left-invariant closed subspace of $L^2(G)$ consisting of continuous functions.

Definition 10.3 $\mathcal{H}$ is a sampling space with respect to Γ if (1) the map $R_\Gamma : \mathcal{H} \to l^2(\Gamma), R_\Gamma f = (f(\gamma))_{\gamma\in\Gamma}$ is a perturbed isometry: there exist positive constants $A \leq B$ such that for every $f \in \mathcal{H}$, $A\|f\|_{\mathcal{H}}^2 \leq \sum_{\gamma\in\Gamma} |f(\gamma)|^2 \leq B\|f\|_{\mathcal{H}}^2$ and secondly, there exists a vector $s \in \mathcal{H}$ such that for any $f \in \mathcal{H}, f(x) = \sum_{\gamma\in\Gamma} f(\gamma) s(\gamma^{-1}x)$ with convergence in the L^2-norm of $\mathcal{H}$. The function s is called a sinc-type function. Moreover, if R_Γ is a surjective isometry then $\mathcal{H}$ is called a sampling space with the interpolation property. This means that every element of $l^2(\Gamma)$ can be interpolated by some fixed vector in $\mathcal{H}$.

As observed in the case of the real line, for any connected Lie group and for any discrete subset Γ of G, it is always possible to construct two distinct L^2-functions that agree everywhere on Γ. Thus, for the existence of a sampling space, one needs a proper definition of bandlimitation.

Proposition 10.4 is a first successful step in generalizing the concept of sampling spaces and bandlimitation on arbitrary Lie groups.

Proposition 10.4 Let π be a unitary representation of a Lie group G acting in some Hilbert space $\mathcal{H}$. If f is an admissible vector for $\mathcal{H}$ and if $\pi(\Gamma)f$ is a frame for

$\mathcal{H}$ with associated frame operator S then $V_f(\mathcal{H})$ is a sampling space with sinc-type function $V_f\left(S^{-1}f\right)$. Moreover, if $\pi(\Gamma)f$ is an orthonormal basis, then $V_f(\mathcal{H})$ has the interpolation property.

Proof: First, observe that the range of V_f consists only of continuous functions defined on G. Given $V_f(\varphi)$ in the range of V_f, assuming that $\pi(\Gamma)f$ is a frame for $\mathcal{H}$, with frame operator S, since

$$\varphi = \sum_{\gamma\in\Gamma} \langle \varphi, \pi(\gamma) f\rangle_{\mathcal{H}} \pi(\gamma) S^{-1} f,$$

the linearity of V_f gives

$$V_f\varphi = V_f\left(\sum_{\gamma\in\Gamma} \langle \varphi, \pi(\gamma) f\rangle_{\mathcal{H}} \cdot \pi(\gamma) S^{-1} f\right) = \sum_{\gamma\in\Gamma} \langle \varphi, \pi(\gamma) f\rangle_{\mathcal{H}} \cdot L(\gamma) V_f\left(S^{-1}f\right)$$

and the sum above is convergent in the norm of $L^2\left(G, d\mu_G\right)$. Moving forward, since $\langle \varphi, \pi(\gamma) f\rangle_{\mathcal{H}} = V_f(\varphi)(\gamma)$, it follows that

$$V_f\varphi = \sum_{\gamma\in\Gamma} \underbrace{\left(\langle \varphi, \pi(\gamma) f\rangle_{\mathcal{H}}\right)}_{V_f(\varphi)(\gamma)} \cdot L(\gamma) V_f\left(S^{-1}f\right) = \sum_{\gamma\in\Gamma} V_f(\varphi)(\gamma) \cdot L(\gamma) V_f\left(S^{-1}f\right).$$

This verifies the second property of Definition 10.3.

To verify the first property of Definition 10.3, we argue as follows. Since $\pi(\Gamma)f$ is a frame for $\mathcal{H}$ with frame bounds A, B, and since f is an admissible vector for the representation π, we have $\left\|V_f(\varphi)\right\|^2_{L^2(G,d\mu_G)} = \|\varphi\|^2_{\mathcal{H}}$ and as a result,

$$\begin{aligned} A\left\|V_f(\varphi)\right\|^2_{L^2(G,d\mu_G)} &= A\|\varphi\|^2_{\mathcal{H}} \le \sum_{\gamma\in\Gamma} \left|\langle \varphi, \pi(\gamma) f\rangle_{\mathcal{H}}\right|^2 \le \|\varphi\|^2_{\mathcal{H}} \\ &= A\left\|V_f(\varphi)\right\|^2_{L^2(G,d\mu_G)}. \end{aligned}$$

Suppose now that $\pi(\Gamma)f$ is an orthonormal basis. Assume additionally that there exists a sequence c, which is orthogonal to the range of R_Γ. Given $\kappa, \gamma \in \Gamma$, we observe that $V_f(\pi(\kappa)f)(\gamma) = \left\langle f, \pi\left(\kappa^1\gamma\right) f\right\rangle_{\mathcal{H}}$, and

$$\left\langle f, \pi\left(\kappa^1\gamma\right) f\right\rangle_{H} = \begin{cases} 1 & \text{if } \kappa^1\gamma = e \\ 0 & \text{otherwise} \end{cases}.$$

Therefore, for every $\kappa \in \Gamma$,

$$0 = \left\langle \left(V_f(\pi(\kappa) f)(\gamma)\right)_{\gamma\in\Gamma}, c\right\rangle_{l^2(\Gamma)} = \sum_{\gamma\in\Gamma} V_f(\pi(\kappa) f)(\gamma) \cdot \overline{c_\gamma}$$

and

$$\sum_{\gamma\in\Gamma} \left\langle f, \pi\left(\kappa^1\gamma\right) f\right\rangle_{H} \cdot \overline{c_\gamma} = \overline{c_\kappa} = 0$$

and this implies that c must be the zero sequence. Therefore, the restriction function R_Γ is indeed a surjective isometry.

We present next a few examples which illustrate the main result of this section.

Example 10.3 Let G be the three-dimensional Heisenberg Lie group, with Lie algebra spanned by X, Y, Z such that $[X, Y] = Z$. Next, let $H = G/\exp \mathbb{Z}Z$ be a connected three-dimensional Heisenberg Lie group with a compact center. Let π be an irreducible representation of H acting in $L^2(\mathbb{R})$ such that

$$\left[\pi(\exp(yY)\exp(xX)\exp\mathbb{Z}Z)f\right](t) = e^{-2\pi iyt} \cdot f(t-x).$$

Then π is an admissible representation of the quotient group H. Moreover (see the discussion subsequent to the proof of Proposition 10.1), it has been shown that any unit vector h in $L^2(\mathbb{R})$ is admissible. For instance, let

$$h = C \cdot \left(1_{\left[-\frac{1}{2},\frac{1}{2}\right]} * 1_{\left[-\frac{1}{2},\frac{1}{2}\right]}\right), \text{ where } C > 0$$

be a triangular continuous function defined on the real line. Here, C serves only as a normalizing factor so that h is an admissible vector for π. Furthermore, by Theorem 6.3, there exist constants $a, b > 0$ such that for

$$\Gamma = \exp(a\mathbb{Z}Y)\exp(b\mathbb{Z}X) \bmod \exp(\mathbb{Z}Z)$$

the system of vectors $\pi(\Gamma)f$ is a frame for $L^2(\mathbb{R})$. By definition,

$$V_h h(z, y, x) := V_h h(\exp(zZ)\exp(yY)\exp(xX)\exp(\mathbb{Z}Z))$$

and

$$V_h h(z, y, x) = e^{2\pi iz} \cdot \int_{\mathbb{R}} \left(h(t)\overline{h(t-x)}\right) e^{-2\pi iyt} dt.$$

More explicitly, h is defined pointwise (up to multiplication by a constant) as follows:

$$\begin{cases} \dfrac{e^{-2i\pi(x+1)y}\left(-\pi(x+2)y - e^{2i\pi(x+2)y}(\pi(x+2)y+i)+i\right)}{4\pi^3y^3} & -2 < x < -1 \\ \dfrac{\pi y - \frac{1}{2}\sin(2\pi y)}{\pi^3y^3} & x = 0 \\ \dfrac{e^{-i\pi xy}(\pi(x-2)y\cos(\pi(x-2)y) - \sin(\pi(x-2)y))}{2\pi^3y^3} & 1 \le x < 2 \\ \dfrac{e^{i\pi y}(\sin(\pi y) - \pi y\cos(\pi y))}{2\pi^3y^3} & x = -1 \\ \dfrac{e^{-i\pi xy}(\sin(\pi(x-2)y) + 2\sin(\pi xy) - \pi xy\cos(\pi(x-2)y) - 2\pi(x-1)y\cos(\pi xy))}{2\pi^3y^3} & 0 < x < 1 \\ \dfrac{e^{-2i\pi(2x+1)y}\left(e^{2i\pi xy}(\pi xy - i) + e^{4i\pi(x+1)y}(\pi xy + i) + 2e^{2i\pi(x+1)y}(\pi(x+1)y - i) + 2e^{2i\pi(2x+1)y}(\pi(x+1)y + i)\right)}{4\pi^3y^3} & -1 < x < 0 \end{cases}$$

In light of Proposition 10.4, it follows that $\mathcal{H} = L^2(H) * V_h h$ is a left-invariant closed subset consisting of continuous functions in $L^2(H)$, which is also a sampling set with sinc-type function $V_h h$. Note that the results of Theorem 6.3 ensure that all of the statements above remain valid even if h is replaced by some arbitrary function in $C_c(\mathbb{R})$.

Example 10.4 Let $G = \mathbb{R} \times \mathbb{R}$ be the semidirect product group equipped with group operation $(x, t)(y, s) = \left(x + e^t y, t + s\right)$. Let π be an irreducible representation of G realized as acting in $L^2(\mathbb{R})$ as follows: $\left[\pi(x, t) f\right](s) = e^{2\pi i e^{-s} x} f(s - t)$ for $(x, t) \in \mathbb{R}^2$. Next, let h be a vector in $L^2(\mathbb{R})$ satisfying the following condition: $\int_{\mathbb{R}} \left| h(t) e^{-\frac{t}{2}} \right|^2 dt = 1$. Then h is an admissible vector for π. Suppose additionally that h satisfies all conditions provided in the statement of Theorem 6.1. This ensures that $\pi(\Gamma) h$ is also a frame for $L^2(\mathbb{R})$. Next, let $\mathcal{K} = L^2\left(G, e^{-t} dxdt\right) * V_h h$ be a left-invariant Hilbert space consisting of continuous functions. Here of course, convolution is with respect to the group operation of $\mathbb{R} \times \mathbb{R}$ and $\mathcal{K}$ is a sampling space with respect to Γ.

10.4 Bandlimitation for Extensions of $\mathbb{R}^n$

Let G be a connected Lie group obtained by extending a simply connected normal and maximal abelian closed subgroup A. Since the left regular representation L of any group is equivalent to a representation obtained by inducing the trivial representation of the identity group, stage induction shows that $L \simeq \int_{\hat{A}}^{\oplus} \mathrm{ind}_A^G(\chi)\, d\chi$ and

G	A
$\mathbb{R}$	$\mathbb{R}$
Heisenberg group $G = \exp \mathbb{R}Z \exp \mathbb{R}Y \exp \mathbb{R}X$ $[X, Y] = Z$	$A = \exp \mathbb{R}Z \exp \mathbb{R}Y$
ax + b group $G = \exp \mathbb{R}Y \exp \mathbb{R}X$ $[Y, X] = X$	$A = \exp \mathbb{R}Y$
Grélaud group $G = \exp \mathbb{R}X \exp \mathbb{R}Y \exp \mathbb{R}W$ $[W, X] = X + Y, [W, Y] = X - Y$	$A = \exp \mathbb{R}X \exp \mathbb{R}Y$

One may then be tempted to define a subspace $\mathcal{K}$ of $L^2(G)$ as bandlimited, if via the unitary equivalence intertwining L with $\int_{\hat{A}}^{\oplus} \mathrm{ind}_A^G(\chi)\, d\chi$, there exists a measurable subset E of $\hat{A}$ such that $\mathcal{K}$ corresponds to $\int_E^{\oplus} L^2(G/A)\,d\chi$. Note that as observed above, when $G = \mathbb{R}^n$ then $G = A$, and our definition of bandlimitation coincides with the classical definition associated with Paley–Wiener-type spaces. However, we will see that a definition this general is pathological in a sense which we will clarify.

First, unlike the abelian case, in general the fiber representations in $\int_E^{\oplus} \mathrm{ind}_A^G(\chi)\,d\chi$ need not even be free of multiplicity. In fact, it is possible for $\mathrm{ind}_A^G(\chi)$ to be equivalent to infinite copies of some irreducible representation.[1] In addition, the multiplicity function of each $\mathrm{ind}_A^G(\chi)$ given by such a decomposition may not even be uniform on E. However, these limitations are pretty pale compared to the measure-theoretic issues inevitable in situations such as the one illustrated by the Mautner group and its relatives.

10.4.1 The Mautner Group and Its Relatives

Let $G = \mathbb{R}^4 \rtimes \mathbb{R}$ be a semi-direct product group with multiplication $(x,t)(y,s) = (x + t \bullet y, t + s)$. Here $t \bullet y$ is defined by multiplying a vector $y \in \mathbb{R}^4$ by an element of the one-parameter group

$$H = \left\{ e^{tA} = \begin{bmatrix} \cos t & -\sin t & 0 & 0 \\ \sin t & \cos t & 0 & 0 \\ 0 & 0 & \cos t\sqrt{2} & -\sin t\sqrt{2} \\ 0 & 0 & \sin t\sqrt{2} & \cos t\sqrt{2} \end{bmatrix} : t \in \mathbb{R} \right\}.$$

Note that as a subgroup of $GL(4,\mathbb{R})$, H is not even closed. In addition, the conjugation action of the one-parameter group H induces a dual action of H on the unitary dual of $\mathbb{R}^4$ defined as $t \star \chi = e^{-tA^T}\chi$. It is easily seen that the differential of $t \mapsto e^{-tA^T}\chi$ at the identity is a linear operator of rank at most one with matrix representation:

$$D_\chi = \begin{bmatrix} -\chi_2 \\ \chi_1 \\ -\sqrt{2}\chi_4 \\ \sqrt{2}\chi_3 \end{bmatrix}.$$

Moreover, for any nonzero element, the H-orbit is a curve densely defined in some unique connected and compact set containing $\chi \in \mathbb{R}^4$. Note also that whenever χ is a nonzero vector, $H \star \chi$ is not an open subset of its closure (with respect to the subspace topology). Next, let χ acts irreducibly on $\mathbb{R}^4$ as $\chi(x) = e^{2\pi i \langle \chi, x\rangle}$ and let π_χ

1 This is the case for instance when G is the $ax+b$ group.

be a unitary representation of G obtained by inducing χ. Then π_χ acts in $L^2(\mathbb{R})$ as follows:

$$\pi_\chi(x,t)f(s) = e^{2\pi i\left\langle e^{-sA^T}\chi,x\right\rangle}\cdot f(s-t).$$

We claim that π_χ is irreducible if and only if χ is a nonzero vector. To this end, suppose that h is orthogonal to the orbit $\pi_\chi(G)f$, where f is some nonzero vector. Then for every $(x,t)\in\mathbb{R}^4\rtimes\mathbb{R}$,

$$0 = \left\langle h, \pi_\chi(x,t)f\right\rangle = \int_{\mathbb{R}} e^{-2\pi i\left\langle e^{-sA^T}\chi,x\right\rangle}\overline{f(s-t)}\cdot h(s)\,ds.$$

Suppose that χ is a nonzero vector in $\mathbb{R}^4$. Then $t\mapsto e^{-tA^T}\chi$ is everywhere a local immersion. This means that there exists a projection P such that

$$0 = \left\langle h, \pi_\chi(Px,t)f\right\rangle = \int_{\mathbb{R}} e^{-2\pi i\left\langle Pe^{-sA^T}\chi,x\right\rangle}\overline{f(s-t)}\cdot h(s)\,ds.$$

Let $\beta(t) = Pe^{-sA^T}\chi$ and put $\omega(\xi)\,d\xi = d\beta^{-1}(\xi)$. Then the change of variable $\xi = Pe^{-sA^T}\chi$ gives

$$\left\langle h, \pi_\chi(Px,t)f\right\rangle = \int_{\beta(\mathbb{R})} e^{-2\pi i\langle\xi,x\rangle}\overline{f\left(\beta^{-1}(\xi)-t\right)}\cdot h\left(\beta^{-1}(\xi)\right)d\beta^{-1}(\xi).$$

and in summary, we obtain that for every $(x,t)\in\mathbb{R}^4\rtimes\mathbb{R}$,

$$\left\langle h, \pi_\chi(Px,t)f\right\rangle = \int_{\beta(\mathbb{R})} e^{-2\pi i\langle\xi,x\rangle}\left(\overline{f\left(\beta^{-1}(\xi)-t\right)}\cdot h\left(\beta^{-1}(\xi)\right)\omega(\xi)\right)d\xi = 0.$$

Therefore, the matrix coefficient $\left\langle h, \pi_\chi(Px,t)f\right\rangle$ is a Fourier transform of the product $\overline{f\left(\beta^{-1}(\xi)-t\right)}\cdot h\left(\beta^{-1}(\xi)\right)\omega(\xi)$ and for every real parameter t,

$$\overline{f\left(\beta^{-1}(\xi)-t\right)}\cdot h\left(\beta^{-1}(\xi)\right)\omega(\xi) = 0 \text{ a.e.}$$

This can only be possible if $h\left(\beta^{-1}(\xi)\right) = 0$ a.e. and therefore h must be zero. These arguments prove that $\pi_\chi = \mathrm{ind}_{\mathbb{R}^4}^G(\chi)$ is irreducible if and only if χ is a nonzero vector. In summary, the set of representations

$$\Theta = \left\{\mathrm{ind}_A^G(\chi) : \chi\in\mathbb{R}^4 \text{ and } \chi\neq 0\right\}$$

is a relatively large collection of irreducible representations. But Θ is not even a subset of the unitary dual of the group (Θ is highly redundant in terms of unitary equivalence). To verify this, let χ and ω be two distinct elements belonging to the same orbit. Then

$$\left\langle h, \pi_\chi(x,t)f\right\rangle = \int_{\mathbb{R}} e^{-2\pi i\left\langle e^{-sA^T}\chi,x\right\rangle}\overline{f(s-t)}\cdot h(s)\,ds$$

and since there exists a real number r such that $e^{-rA^T}\omega = \chi$,

$$\left\langle h, \pi_\chi(x,t)f\right\rangle = \int_{\mathbb{R}} e^{-2\pi i\left\langle e^{-sA^T}e^{-rA^T}\omega,x\right\rangle}\overline{f(s-t)}\cdot h(s)\,ds.$$

Let T_r be a unitary operator acting by left-translation on $L^2(\mathbb{R})$. Then

$$\langle h, \pi_\chi(x,t)f\rangle = \int_{\mathbb{R}} e^{-2\pi i\left\langle e^{-(s+r)A^T}\omega, x\right\rangle}\overline{f(s-t)}\cdot h(s)\,ds$$

and the change of variable $s \mapsto s-r$ gives

$$\langle h, \pi_\chi(x,t)f\rangle = \int_{\mathbb{R}} e^{-2\pi i\left\langle e^{-sA^T}\omega, x\right\rangle}\overline{f(s-r-t)}\cdot h(s-r)\,ds = \langle T_r h, \pi_\omega(x,t)\,T_r f\rangle.$$

The above shows that any two elements χ and ω belonging to the same orbit give rise to equivalent induced representations, and this begs the following question. Can we construct a smooth transversal for unitarily equivalent representations in Θ? In other words, is there a submanifold of $\mathbb{R}^4$ that meets every orbit at precisely one point? According to a known result of Glimm [3, Theorem 1] such a set does not exist. In fact, the orbit space $\mathbb{R}^4/e^{\mathbb{R}A}$ cannot be countably separated.

Groups of Mautner types: Let G be a Lie group. Suppose that $G = AH$, such that A is a simply connected and connected abelian and normal Lie subgroup of G. Moreover, H is a closed Lie subgroup of G and G acts on the unitary dual $\hat{A}$ of A as follows: for arbitrary $g \in G$ and a unitary character $\chi \in \hat{A}$,

$$[g \star \chi](a) = \chi\left(g^{-1}ag\right) \quad (g \in G, a \in A).$$

Next, let $H \star \chi$ be an orbit of H. Then $h \mapsto h \star \chi$ defines a natural bijection between H and its orbit. We equip orbit $H \star \chi$ with the manifold structure of H with the identification $h \mapsto h \star \chi$. Then naturally G acts on $L^2(H \star \chi)$ as follows:

$$\sigma_\lambda(ah)f(s) = (s \star \chi)(a)\cdot f\left(h^{-1}s\right).$$

Each such σ_λ is an irreducible representation of G. If there exists a unitary character χ such that $H \star \chi$ is not relatively open in its closure in $\hat{A}$, we say that G is a Mautner-type group.

For a concrete example of a group of Mautner type, consider the following. Let $G = \left(\mathbb{C}\times\mathbb{C}\times\mathbb{R}^3\right)\times\mathbb{R}^2$ such that $\mathbb{R}^2$ acts linearly on $\mathbb{C}\times\mathbb{C}\times\mathbb{R}^3$ as $t \bullet x = e^{t_1A_1+tA_2}x$ and

$$A_1 = \begin{bmatrix} i & & & & \\ & 0 & & & \\ & & 0 & 0 & 1 \\ & & 0 & 0 & 0 \\ & & 0 & 0 & 0 \end{bmatrix},\quad A_2 = \begin{bmatrix} 0 & & & & \\ & i & & & \\ & & 0 & 0 & 0 \\ & & 0 & 0 & 1 \\ & & 0 & 0 & 0 \end{bmatrix}.$$

The dual action of H takes the form $(t_1,t_2)\star\chi = e^{-t_1A_1^T - tA_2^T}\chi$ and

$$e^{t_1A_1+t_2A_2}\chi = \begin{bmatrix} e^{-t_1} & 0 & 0 & 0 & 0 \\ 0 & e^{-t_2} & 0 & 0 & 0 \\ 0 & 0 & 1 & 0 & 0 \\ 0 & 0 & 0 & 1 & 0 \\ 0 & 0 & -t_1 & -t_2 & 1 \end{bmatrix}\begin{bmatrix}\chi_1\\ \chi_2\\ \chi_3\\ \chi_4\\ \chi_5\end{bmatrix} = \begin{bmatrix} e^{-t_1}\chi_1 \\ e^{-t_2}\chi_2 \\ \chi_3 \\ \chi_4 \\ \chi_5 - t_1\chi_3 - t_2\chi_4 \end{bmatrix}.$$

Suppose that χ_3 and χ_4 are nonzero real numbers. Then for any integer k_1 and k_2,

$$e^{2\pi k_1 A_1 + 2\pi k_2 A_2}\chi = \begin{bmatrix} \chi_1 \\ \chi_2 \\ \chi_3 \\ \chi_4 \\ \chi_5 - 2\pi k_1 \chi_3 - 2\pi k_2 \chi_4 \end{bmatrix}$$

One can easily verify that $e^{\mathbb{R}A_1+\mathbb{R}A_2} \star \chi$ fails to be relatively open in its closure. Appealing once again to the result of Glimm [3, Theorem 1] the orbit space $\left(\mathbb{C}\times\mathbb{C}\times\mathbb{R}^3\right)/e^{\mathbb{R}A_1+\mathbb{R}A_2}$ cannot be countably separated.

10.4.2 Bandlimitation on a Class of Lie Groups

In this section, we will introduce a useful notion of bandlimitation. To avoid the types of complications we have observed earlier with Mautner-type groups, we make the following assumptions. Let G be a Lie group and let us assume that $G = AH$, such that A is a simply connected and connected abelian and normal Lie subgroup of G. Moreover, H is a closed Lie subgroup of G and G acts on the unitary dual $\hat{A}$ of A as follows. For arbitrary $g \in G$ and a unitary character $\chi \in \hat{A}$,

$$\left[g \star \chi\right](a) = \chi\left(g^{-1}ag\right) \quad (g \in G, a \in A).$$

We make the following additional assumptions

- There exists an open and G-invariant subset Ω of $\hat{A}$ such that $\hat{A}\backslash\Omega$ is a subset of measure zero in $\hat{A}$.
- There exists a subset $\Sigma \subseteq \Omega$ such that Σ is a submanifold of Ω for the G-orbits in $\hat{A}$. In other words, the map $\beta : H \times \Sigma \to \Omega$ such that $\beta(s,\chi) = s \star \chi$ is a diffeomorphism of manifolds.
- For each $\chi \in \Sigma$, letting G_χ be the stabilizer of χ in G, $G_\chi = A$.

We say that a left-invariant subspace $\mathcal{K}$ of $L^2(G)$ is a multiplicity-free bandlimited space, if there exists a relatively compact subset E of the manifold Σ and a unitary operator U taking $\mathcal{K}$ into $L^2\left(E, L^2(H), d\chi\right)$ such that U intertwines the regular representation restricted to $\mathcal{K}$ to the direct integral $\int_E^{\oplus} \mathrm{ind}_A^G(\chi)\,d\chi$. In the definition above, $d\chi$ is the usual Lebesgue measures on E.

Generally, we say that $\mathcal{K}$ is a bandlimited space of multiplicity m if $\mathcal{K}$ is a left-invariant subspace of $\mathcal{K}$ corresponding to the direct sum $\oplus_{k=1}^m L^2(E, L^2(H), d\chi)$.

Example 10.5 To justify the importance of this definition of bandlimitation, we list a few concrete cases:

1) Let $G = \mathbb{R}^d$. Then $\mathbb{R}^d$ is self-dual and in this case and $\Sigma = \mathbb{R}^d$. Letting E be Lebesgue measurable subset of $\mathbb{R}^d$ and letting $d\chi$ be the usual Lebesgue measure on $\mathbb{R}^d$, $\pi_E = \int_E^{\oplus} \chi d\chi$.

2) Let G be the three-dimensional Heisenberg–Lie group with Lie algebra spanned by X, Y, Z such that $[X, Y] = Z$, and let $\Sigma = \mathbb{R}\setminus\{0\}\,Z^*$ and let E be a Lebesgue measurable subset of $\mathbb{R}\setminus\{0\}$. Then $\pi_E = \int_E^{\oplus} \mathrm{ind}^G_{\exp \mathbb{R}Z \exp \mathbb{R}Y}(\chi_\lambda)\, d\lambda$.
3) Let G be the ax+b Lie group with Lie algebra spanned by Y, X such that $[Y, X] = X$. Let $\Sigma = \{X^*, -X^*\}$. Of course in this example, Σ is equipped with the point-mass measure and for any subset E of $\{X^*, -X^*\}$, $\pi_E = \oplus_{\lambda\in E}\mathrm{ind}^G_{\exp \mathbb{R}Y}(\chi_\lambda)$.
4) Let G be a three-dimensional Lie group with Lie algebra $\mathfrak{g}$ spanned by

$$\sum_{k=1}^{3} c_k X_k = \begin{bmatrix} x_3 & -x_3 & x_1 \\ x_3 & x_3 & x_2 \\ 0 & 0 & 0 \end{bmatrix}.$$

Put $\Omega = \left(\mathbb{R}X_1^* + \mathbb{R}X_2^*\right)\setminus\{0\}$. Then, letting Σ be a unit circle in Ω centered at zero, it is clear that Σ is a transversal for the orbits in Ω. Next,

$$\pi_E = \int_E^{\oplus} \mathrm{ind}^G_{\exp \mathbb{R}X_1 \exp \mathbb{R}X_2}(\chi_\lambda)\, d\lambda$$

and E can be taken to be a measurable subset of the unit circle.
5) Let $G = \mathbb{R}^n \rtimes H$ where H is a closed subgroup of the General Linear group and additionally, there exists an open and G-invariant subset Ω of $\hat{\mathbb{R}^n}$ such that $\hat{\mathbb{R}^n}\setminus\Omega$ is a subset of measure zero in $\hat{\mathbb{R}^n}$. We also assume the existence of a subset $\Sigma \subseteq \Omega$ such that Σ is a submanifold of Ω for the G-orbits in $\hat{\mathbb{R}^n}$ and for each $\chi \in \Sigma$, $G_\chi = \mathbb{R}^n$. In this case, π_E is a representation of the form $\int_E^{\oplus} \mathrm{ind}^{\mathbb{R}^n \rtimes H}_{\mathbb{R}^n}(\chi_\lambda)\, d\mu(\lambda)$.

Let ω be a strictly positive function on the transversal Σ. We first prove following:

Proposition 10.5 **(Multiplicity-free case)** Let $\pi = \int_E^{\oplus} \mathrm{ind}^G_A(\chi)\,\omega(\chi)\,d\chi$ such that E has positive measure in $\hat{A}$ and is relatively compact. Then there exists a vector $f \in L^2\left(E, L^2(H), \omega(\chi)\,d\chi\right)$ and a countable set $\Gamma \subset G$ such that $\pi(\Gamma)f$ is a frame for

$$L^2\left(E, L^2(H), \omega(\chi)\,d\chi\right).$$

Proposition 10.6 **(Non-multiplicity-free case)** Let m be a fixed natural number. Then there exists a vector $f \in \oplus_{k=1}^m L^2\left(E, L^2(H)\right)$ and a countable set $\Gamma \subset G$ such that $\oplus_{k=1}^m \pi(\Gamma)f$ is a frame for

$$\oplus_{k=1}^m L^2\left(E, L^2(H), \omega(\chi)\,d\chi\right).$$

In light of Propositions 10.5 and 10.6, to construct sampling spaces associated with the bandlimited spaces of interest, we simply need to appeal to Proposition 10.4.

Proof of Proposition 10.5 For fixed $u, v \in L^2(E, L^2(H))$, it is clear that

$$\langle u, \pi(h)\pi(a)v\rangle_{L^2(E,L^2(H))} = \int_E \left\langle \left[\pi\left(h^{-1}\right)u\right](\chi), \chi\left(s^{-1}as\right)v(\chi)\right\rangle_{L^2(H)}\omega(\chi)\,d\chi.$$

By assumption, for every $\chi \in \Sigma$, $G_\chi = A$, $G_\chi \cap H$ is trivial and β is a diffeomorphism between the manifolds $H \times \Sigma$ and Ω. Since $\chi\left(s^{-1}as\right) = \beta(s,\chi)(a)$,

$$\langle u, \pi(h)\pi(a)v\rangle_{L^2(E,L^2(H))} = \int_E \left\langle \pi\left(h^{-1}\right)u(\chi), \left[\beta(s,\chi)(a)\right]\omega(\chi)v(\chi)\right\rangle_{L^2(H)}d\chi.$$

Suppose further that (a) $(\chi, h) \mapsto u(\chi)(h)$ and (b) $(\chi, h) \mapsto v(\chi)(h)$ are compactly supported on $E \times K$. Defining $d(h,\chi) = d\mu_H(s)\,d\chi$, $U_h(s,\chi) = \pi\left(h^{-1}\right)u(\chi)(s)$ and $V(s,\chi) = \omega(\chi)\overline{v(\chi)(s)}$,

$$\begin{aligned}&\langle u, \pi(h)\pi(a)v\rangle_{L^2(E,L^2(H))}\\ &= \int_E\int_K \left(\pi\left(h^{-1}\right)u(\chi)(s)\cdot\left[\beta(s,\chi)\left(a^{-1}\right)\right]\omega(\chi)\overline{v(\chi)(s)}\right)d\mu_H(s)\,d\chi.\end{aligned}$$

Setting $d(s,\chi) = d\mu_H(s)\,d\chi$,

$$\langle u, \pi(h)\pi(a)v\rangle_{L^2(E,L^2(H))} = \int_{K\times E} U_h(s,\chi)\left[\beta(s,\chi)\left(a^{-1}\right)\right]V(s,\chi)\,d(s,\chi).$$

The change of variable $\gamma = \beta(s,\chi)$ yields

$$\langle u, \pi(h)\pi(a)v\rangle_{L^2(E,L^2(H))} = \int_{\beta(K\times E)} U_h\left(\beta^{-1}(\gamma)\right)\gamma\left(a^{-1}\right)V\left(\beta^{-1}(\gamma)\right)d\left(\beta^{-1}(\gamma)\right).$$

Letting

$$d\left(\beta^{-1}(\gamma)\right) = w(\gamma)\,d\gamma, \tag{10.4}$$

we obtain

$$\langle u, \pi(h)\pi(a)v\rangle_{L^2(E,L^2(H))} = \int_{\beta(K\times E)} \left[U_h\left(\beta^{-1}(\gamma)\right)V\left(\beta^{-1}(\gamma)\right)w(\gamma)\right]\cdot\gamma\left(a^{-1}\right)d\gamma.$$

Now, since $\beta(K \times E)$ is a compact subset of the unitary dual of A, there exist a countable subset Γ_A of A, $C \subset \hat{A}$ such that $\beta(K \times E) \subseteq C$ and $\left\{\gamma \mapsto \frac{\gamma(a^{-1})}{|C|^{1/2}} : a \in \Gamma_A\right\}$ is an orthonormal basis for $L^2(C)$. Consequently,

$$\begin{aligned}&\sum_{a\in\Gamma_A}\left|\langle u, \pi(h)\pi(a)v\rangle_{L^2(E,L^2(H))}\right|^2\\ &= |C|\sum_{a\in\Gamma_A}\left|\int_{\beta(K\times E)} U_h\left(\beta^{-1}(\gamma)\right)V\left(\beta^{-1}(\gamma)\right)w(\gamma)\frac{\gamma\left(a^{-1}\right)}{|C|^{1/2}}d\gamma\right|^2\end{aligned}$$

and

$$\begin{aligned}&\sum_{a\in\Gamma_A}\left|\langle u, \pi(h)\pi(a)v\rangle_{L^2(E,L^2(H))}\right|^2\\ &= |C|\int_{\beta(K\times E)}\left|U_h\left(\beta^{-1}(\gamma)\right)V\left(\beta^{-1}(\gamma)\right)w(\gamma)^{1/2}\right|^2 w(\gamma)\,d\gamma.\end{aligned}$$

Letting $(s,\chi)=\beta^{-1}(\gamma)$

$$\sum_{a\in\Gamma_A}\left|\langle u,\pi(h)\pi(a)v\rangle_{L^2(E,L^2(H))}\right|^2$$
$$=|C|\int_{K\times E}\left|U_h(s,\chi)V(s,\chi)w(\beta(s,\chi))^{1/2}\right|^2 d(s,\chi)$$
$$=|C|\int_{K\times E}\left|\pi\left(h^{-1}\right)u(\chi)(s)V(s,\chi)w(s\star\chi)^{1/2}\right|^2 d(s,\chi)$$

and

$$\sum_{a\in\Gamma_A}\left|\langle u,\pi(h)\pi(a)v\rangle_{L^2(E,L^2(H))}\right|^2$$
$$=|C|\int_{K\times E}\left|u(\chi)(s)V\left(h^{-1}s,\chi\right)w\left(h^{-1}s\star\chi\right)^{1/2}\right|^2 d(h,\chi)$$
$$=\int_{H\times E}|u(\chi)(s)|^2\left(|C|\cdot\left|V\left(h^{-1}s,\chi\right)w\left(h^{-1}s\star\chi\right)^{1/2}\right|^2\right)d(h,\chi).$$

Now, letting Γ_H be a countable subset of H,

$$\sum_{h\in\Gamma_H}\sum_{a\in\Gamma_A}\left|\langle u,\pi(h)\pi(a)v\rangle_{L^2(E,L^2(H))}\right|^2$$
$$=\int_{H\times E}|u(\chi)(s)|^2\left(|C|\cdot\sum_{h\in\Gamma_H}\left|V\left(h^{-1}s,\chi\right)w\left(h^{-1}s\star\chi\right)^{1/2}\right|^2\right)d(h,\chi).$$

Since $V\left(h^{-1}s,\chi\right)=\omega(\chi)\overline{v(\chi)(h^{-1}s)}$,

$$\sum_{h\in\Gamma_H}\sum_{a\in\Gamma_A}\left|\langle u,\pi(h)\pi(a)v\rangle_{L^2(E,L^2(H))}\right|^2$$
$$=\int_{H\times E}|u(\chi)(s)|^2\left(|C|\cdot\sum_{h\in\Gamma_H}\left|v(\chi)\left(h^{-1}s\right)\omega(\chi)^{1/2}w\left(h^{-1}s\star\chi\right)^{1/2}\right|^2\right)$$
$$\times\,\omega(\chi)\,d(h,\chi).$$

Choosing a pair $\left(\Gamma_H,v\right)$ such that

$$0<\operatorname{essinf}\left((s,\chi)\mapsto|C|\cdot\sum_{h\in\Gamma_H}\left|v(\chi)\left(h^{-1}s\right)\omega(\chi)^{1/2}w\left(h^{-1}s\star\chi\right)^{1/2}\right|^2\right)=A_v$$
$$\leq\operatorname{esssup}\left((s,\chi)\mapsto|C|\cdot\sum_{h\in\Gamma_H}\left|v(\chi)\left(h^{-1}s\right)\omega(\chi)^{1/2}w\left(h^{-1}s\star\chi\right)^{1/2}\right|^2\right)$$
$$<B_v<\infty,$$

it follows that

$$A_v\|u\|^2_{L^2(E,L^2(H))}\leq\sum_{h\in\Gamma_H}\sum_{a\in\Gamma_A}\left|\langle u,\pi(h)\pi(a)v\rangle_{L^2(E,L^2(H))}\right|^2\leq B_v\|u\|^2_{L^2(E,L^2(H))}.$$

Since the set of functions $(s,h)\mapsto u(\chi)(h)$, which are compactly supported on $E\times H$ is dense in $L^2(E,L^2(H))$, it follows that $\pi(\Gamma_H\Gamma_A)v$ is a frame for $L^2(E,L^2(H))$ with frame bounds $0<A_v\leq B_v<\infty$.

Proof of Proposition 10.6 Let m be a natural number. Next, let $v_1,\ldots,v_m\in L^2(E,L^2(H))$ such that $v=\sum_{k=1}^m v_k\in\bigoplus_{k=1}^m L^2(E,L^2(H))$ and each $(\chi,h)\mapsto v_k(\chi)(h)$ is compactly supported. Next, letting $u=\sum_{k=1}^m u_k$ be a vector in the direct sum of Hilbert spaces: $\bigoplus_{k=1}^m L^2(E,L^2(H))$, for any countable subset Γ_A of A,

$$\sum_{a\in\Gamma_A}\left|\langle u,\pi(h)\pi(a)v\rangle_{\bigoplus_{k=1}^m L^2(E,L^2(H))}\right|^2=\sum_{a\in\Gamma_A}\left|\sum_{k=1}^m\left\langle\pi(h)u_k,\pi(a)v_k\right\rangle_{L^2(E,L^2(H))}\right|^2.$$

Letting $d(h,\chi)=d\mu_H(s)\,d\chi$, $U_h^{(k)}(s,\chi)=\pi\left(h^{-1}\right)u_k(\chi)(h)$ and $V^{(k)}(s,\chi)=\overline{\omega(\chi)v_k(\chi)(h)}$, the inner summand of the preceding equation takes the form

$$\begin{aligned}&\sum_{k=1}^m\left\langle\pi\left(h^{-1}\right)u_k,\pi(a)v_k\right\rangle_{L^2(E,L^2(H))}\\&\quad=\sum_{k=1}^m\int_{K\times E}U_h^{(k)}(s,\chi)\left[\beta(s,\chi)\left(a^{-1}\right)\right]V^{(k)}(s,\chi)\,d(s,\chi)\end{aligned}$$

for some compact subset K of H. The change of variable $\gamma=\beta(s,\chi)$ gives

$$\begin{aligned}&\sum_{k=1}^m\left\langle\pi\left(h^{-1}\right)u_k,\pi(a)v_k\right\rangle_{L^2(E,L^2(H))}\\&\quad=\sum_{k=1}^m\int_{\beta(K\times E)}\left[U_h^{(k)}\left(\beta^{-1}(\gamma)\right)V^{(k)}\left(\beta^{-1}(\gamma)\right)w(\gamma)\right]\gamma\left(a^{-1}\right)d\gamma.\end{aligned}$$

Suppose now that each vector $\gamma\mapsto V^{(k)}\left(\beta^{-1}(\gamma)\right)$ is supported on some compact subset $S_k\subset\hat{A}$. Then

$$\begin{aligned}&\sum_{k=1}^m\left\langle\pi\left(h^{-1}\right)u_k,\pi(a)v_k\right\rangle_{L^2(E,L^2(H))}\\&\quad=\sum_{k=1}^m\int_{\beta(S_k\times E)}\left[U_h^{(k)}\left(\beta^{-1}(\gamma)\right)V^{(k)}\left(\beta^{-1}(\gamma)\right)w(\gamma)\right]\gamma\left(a^{-1}\right)d\gamma.\end{aligned}$$

Let us now assume that v is fixed in $\bigoplus_{k=1}^m L^2(E,L^2(H))$ such that the collection of sets $\left\{\beta\left(S_k\times E\right):1\leq k\leq m\right\}$ are pairwise disjoint and every $\beta\left(S_k\times E\right)$ is relatively compact. Since m is assumed to be a natural number, we let C be a relatively compact subset of the unitary dual of A such that

$$\bigcup_{k=1}^m\beta\left(S_k\times E\right)\subseteq C. \tag{10.5}$$

Then there exists a countable subset Γ_A of A, such that the system of vectors $\left\{\gamma \mapsto \frac{\gamma(a^{-1})}{|C|^{1/2}} : a \in \Gamma_A\right\}$ is an orthonormal basis for $L^2(C)$. Consequently,

$$\begin{aligned}
&\sum_{a\in\Gamma_A}\left|\sum_{k=1}^{m}\left\langle \pi\left(h^{-1}\right)u_k, \pi(a)v_k\right\rangle_{L^2(E,L^2(H))}\right|^2 \\
&\quad = |C|\int_C\left|\sum_{k=1}^{m}U_h^{(k)}\left(\beta^{-1}(\gamma)\right)\sum_{k=1}^{m}V^{(k)}\left(\beta^{-1}(\gamma)\right)w(\gamma)^{1/2}1_{\beta(S_k\times E)}(\gamma)\right|^2 w(\gamma)\,d\gamma \\
&\quad = \int_{H\times E}\left|\sum_{k=1}^{m}u^{(k)}(\chi)(s)\right|^2 |C|\left|\sum_{k=1}^{m}V^{(k)}\left(h^{-1}s,\chi\right)w\left(\beta\left(h^{-1}s,\chi\right)\right)^{1/2}\right|^2 d(s,\chi).
\end{aligned}$$

Now, letting Γ_H be a fixed countable subset of H, and summing the above over Γ_H, we obtain

$$\begin{aligned}
&\sum_{h\in\Gamma_H}\sum_{a\in\Gamma_A}\left|\sum_{k=1}^{m}\left\langle \pi\left(h^{-1}\right)u_k, \pi(a)v_k\right\rangle_{L^2(E,L^2(H))}\right|^2 \\
&\quad = \int_{H\times E}\left|\sum_{k=1}^{m}u^{(k)}(\chi)(s)\right|^2 \\
&\quad \times\left(\sum_{h\in\Gamma_H}|C|\cdot\left|\sum_{k=1}^{m}\overline{v_k(\chi)(h^{-1}s)}\omega(\chi)^{1/2}w\left(\beta\left(h^{-1}s,\chi\right)\right)^{1/2}\right|^2\right)\omega(\chi)\,d(s,\chi).
\end{aligned}$$

Choosing a pair (Γ_H, v) such that

$$\begin{aligned}
0 &< \operatorname{essinf}\left((s,\chi)\mapsto\sum_{h\in\Gamma_H}|C|\cdot\left|\sum_{k=1}^{m}\overline{v_k(\chi)(h^{-1}s)}\omega(\chi)^{1/2}w\left(\beta\left(h^{-1}s,\chi\right)\right)^{1/2}\right|^2\right) = A_v \\
&\leq \operatorname{esssup}\left((s,\chi)\mapsto\sum_{h\in\Gamma_H}|C|\cdot\left|\sum_{k=1}^{m}\overline{v_k(\chi)(h^{-1}s)}\omega(\chi)^{1/2}w\left(\beta\left(h^{-1}s,\chi\right)\right)^{1/2}\right|^2\right) \\
&\times < B_v < \infty,
\end{aligned}$$

it follows that

$$\begin{aligned}
A_v\|u\|^2_{\oplus_{k=1}^m L^2(E,L^2(H))} &\leq \sum_{h\in\Gamma_H}\sum_{a\in\Gamma_A}\left|\sum_{k=1}^{m}\left\langle \pi\left(h^{-1}\right)u_k, \pi(a)v_k\right\rangle_{L^2(E,L^2(H))}\right|^2 \\
&\leq B_v\|u\|^2_{\oplus_{k=1}^m L^2(E,L^2(H))}.
\end{aligned}$$

Since the set of vectors $(s,h)\mapsto\sum_{k=1}^{m}u^{(k)}(\chi)(h)$, which are compactly supported on $E\times H$ is dense in $\oplus_{k=1}^{m}L^2\left(E, L^2(H), \omega(\chi)\right)$, it follows that $\pi\left(\Gamma_H\Gamma_A\right)v$ is a frame for $\oplus_{k=1}^{m}L^2\left(E, L^2(H), \omega(\chi)\right)$ with frame bounds $0 < A_v \leq B_v < \infty$.

Remark 10.1 The following is worth noting.

1) Proposition 10.5 gives a systematic method for constructing a frame of the type $\left[\int_E^{\oplus} \text{ind}_A^G(\chi)\,\omega(\chi)\,d\chi\right](\Gamma)\,v$, where ω is a strictly positive function on the transversal Σ, which we summarize as follows. First, choose $v \in L^2(E, L^2(H))$ such that $v = (\varphi)_{\chi \in E}$ and φ is a smooth function that is compactly supported on a compact subset K of H. Secondly, letting $\beta : H \times \Sigma \to \Omega$ such that $\beta(s, \chi) = s \star \chi$, since $\beta(K \times E)$ is a compact subset of $\hat{A}$, there exist a countable subset Γ_A of A, $C \subset \hat{A}$ such that $\beta(K \times E) \subseteq C$ and $\left\{\gamma \mapsto \frac{\gamma(a^{-1})}{|C|^{1/2}} : a \in \Gamma_A\right\}$ is an orthonormal basis for $L^2(C)$. Thirdly, for a countable set $\Gamma_H \subset H$, let

$$m_{(\Gamma_H, v, E, C)} = \text{essinf}\left((s, \chi) \mapsto |C| \cdot \sum_{h \in \Gamma_H} \left|\varphi\left(h^{-1}s\right)\omega(\chi)^{1/2} w\left(h^{-1}s \star \chi\right)^{1/2}\right|^2\right)$$

$$M_{(\Gamma_H, v, E, C)} = \text{esssup}\left((s, \chi) \mapsto |C| \cdot \sum_{h \in \Gamma_H} \left|\varphi\left(h^{-1}s\right)\omega(\chi)^{1/2} w\left(h^{-1}s \star \chi\right)^{1/2}\right|^2\right)$$

where w satisfies $d\beta^{-1}(\gamma) = w(\gamma)\,d\gamma$. Finally, $\left[\int_E^{\oplus} \text{ind}_A^G(\chi)\,\omega(\chi)\,d\chi\right]\left(\Gamma_H \Gamma_A\right) v$ is a frame for $L^2(E, L^2(H))$ with frame bounds $m_{(\Gamma_H, v, E, C)} \leq M_{(\Gamma_H, v, E, C)}$.

2) It is worth noting that the proof of Proposition 10.5 establishes the following. If $m_{(\Gamma_H, v, E, C)} = M_{(\Gamma_H, v, E, C)} = 1$, then there exists a vector $v \in \oplus_{k=1}^m L^2(E, L^2(H))$ and $\Gamma_H \Gamma_A \subset G$ such that $\pi\left(\Gamma_H \Gamma_A\right) v$ is a Parseval frame for $\oplus_{k=1}^m L^2(E, L^2(H))$. If additionally, v can be selected such that

$$\|v\|^2_{\oplus_{k=1}^m L^2(E, L^2(H))} = \int_E \sum_{k=1}^m \|v_k(\chi)\|^2_{L^2(H)}\,\omega(\chi)\,d\chi = 1,$$

then $\pi\left(\Gamma_H \Gamma_A\right) v$ would be an orthonormal basis for $\oplus_{k=1}^m L^2(E, L^2(H))$.

10.4.2.1 Spectral Analysis of Induced Representations

Let G be a Lie group and let H be a closed subgroup of G. Next, let $\tau = \text{ind}_H^G(\upsilon)$ be a unitary representation of G where υ is a unitary representation of H. The following question is of general interest in representation theory. How do we decompose τ into unitary irreducible representations of G?

If H is trivial, then τ is the left regular representation of the group. In this case, the tools of abstract Plancherel theorems give satisfactory answers for a class of groups such as compact, nilpotent, exponentially solvable Lie groups, for example. On the other hand, when H is not trivial, we are not aware of a general theory. In addition, for a useful decomposition, the following ingredients are needed: (1) the spectrum of the decomposition, (2) the multiplicity function, and (3) a spectral measure. The spectrum consists of all irreducible representations occurring in the decomposition. The multiplicity function returns the number of times each irreducible representation of the group occurs. Finally, a spectral measure provides a central foundation for describing a Hilbert space on which the decomposition of τ is realized. Thus, in general, we seek to resolve the equation

$$\mathrm{ind}_H^G(\upsilon) \simeq \int_{\Sigma}^{\oplus} \oplus_{k=1}^{n(\pi)} \pi \, d\mu(z)$$

for the following variables: Σ (the spectrum), $\pi \mapsto n(\pi)$ (the multiplicity function), and $d\mu$ (the spectral measure.)

Generally, in the case where H is not normal, the action of τ is extremely difficult to describe explicitly. The decomposition problem of an induced representation is of independent interest and has been extensively studied. We refer the reader to a perusal of Lipsman, Fujiwara, and Currey's body of work for this specific program which is beyond the scope of this book. To illustrate the difficulty involved in decomposing $\mathrm{ind}_H^G(\upsilon)$, let $\mathcal{K}_\upsilon$ be the set of continuous $\mathcal{H}_\upsilon$-valued functions f defined over G satisfying the following conditions. First, the image of the support of f under the quotient map $\pi : G \to G/H$ is compact, and secondly,

$$f(gh) = \left[\upsilon(h)^{-1} f\right](g) \tag{10.6}$$

for $g \in G$ and $h \in H$. Next, let $\mu_{G/H}$ be a quasi-invariant measure on the quotient space G/H associated with a continuous rho-function ρ. Then

$$(f, f') \mapsto \langle f, f' \rangle_{\mathcal{K}_\upsilon} = \int_{G/H} \langle f(x), f'(x) \rangle_{\mathcal{H}_\upsilon} d\mu_{G/H}(xH)$$

determines an inner product on $\mathcal{K}_\upsilon$. Let $\mathcal{H}_{(G/H,\upsilon)}$ be the Hilbert completion of $\mathcal{K}_\upsilon$ with respect to this inner product space. The action of τ is as follows. For arbitrary $x \in G$ and $f \in \mathcal{H}_{(G/H,\upsilon)}$,

$$\left[\tau(x) f\right](y) = \left[\frac{d\mu_{G/H}(L_x)}{d\mu_{G/H}}\right]^{-1/2} (yH) f\left(x^{-1} y\right).$$

If H is normal such that $G = HM$, where M is a smooth transversal for the cosets of H, letting $x = x_h x_m$ and $y = y_h y_m$ such that $x_h, y_h \in H$ and $x_m, y_m \in M$,

$$x^{-1} y = \left(x_h x_m\right)^{-1} \left(y_h y_m\right) = x_m^{-1} x_h^{-1} \left(y_h y_m\right) = x_m^{-1} y_m \left(y_m^{-1} \left(x_h^{-1} y_h\right) y_m\right).$$

Since H is assumed to be normal, $y_m^{-1}\left(x_h^{-1} y_h\right) y_m$ is in H, but $x_m^{-1} y_m$ need not be in M since it is not assumed that M has a group structure. We can however, write

$$x_m^{-1} y_m = z_m\left(x_m, y_m\right) w_h\left(x_m, y_m\right)$$

for some $z_m \in M$ and w_h in H and

$$x^{-1} y = x_m^{-1} y_m \left(y_m^{-1} \left(x_h^{-1} y_h\right) y_m\right) = z_m\left(x_m, y_m\right) w_h\left(x_m, y_m\right) \left(y_m^{-1} \left(x_h^{-1} y_h\right) y_m\right).$$

In this case,

$$[\tau(x)f](y) = \left[\frac{d\mu_{G/H}(L_x)}{d\mu_{G/H}}\right]^{-1/2}(yH)$$
$$\cdot\left[\upsilon\left(w_h(x_m, y_m)\left(y_m^{-1}\left(x_h^{-1}y_h\right)y_m\right)\right)^{-1}f\right]\left(z_m(x_m, y_m)\right).$$

Observe however that this procedure breaks down without the additional assumption that H is normal.

For the class of groups $G = AH$ of interest in this section, the notion of bandlimitation coincides with a direct integral decomposition of the form

$$\oplus_{k=1}^{L}\int_{E}^{\oplus}\mathrm{ind}_A^G(\chi)\,d\chi \text{ where } E \subset \Sigma$$

and Proposition 10.6 gives a satisfactory answer to the Discretization Problem of

$$\oplus_{k=1}^{L}\int_{\Sigma}^{\oplus}\mathrm{ind}_A^G(\chi)\,d\chi$$

If it is also assumed that H is a closed subgroup of G then the quasiregular representation $\tau = \mathrm{ind}_H^G(1)$ acts in $L^2(A)$ as follows. For $n, m \in A$, $\tau(n)f(m) = f\left(n^{-1}m\right)$ and for $h \in H$, $\tau(h)f(m) = f\left(h^{-1}mh\right)$. Moreover, it is a useful exercise for the reader to verify that the Fourier transform on A intertwines τ with a multiplicity-free direct integral of the form

$$\int_{\Sigma}^{\oplus}\mathrm{ind}_A^G(\chi)\,d\chi.$$

Thus, the results of Proposition 10.6 apply to a large class of quasiregular representations which naturally occurs in the study of wavelets, shearlets, and their various generalizations.

References

1 Hartmut Führ. *Abstract harmonic analysis of continuous wavelet transforms*, volume 1863 of *Lecture Notes in Mathematics*. Springer-Verlag, Berlin, 2005.

2 H. Führ and K. Gröchenig. Sampling theorems on locally compact groups from oscillation estimates. *Mathematische Zeitschrift*, 255(1):177–194, 2007.

3 James Glimm. Locally compact transformation groups. *Transactions of the American Mathematical Society*, 101:124–138, 1961.

4 Robert J. Marks II. *Introduction to Shannon sampling and interpolation theory*. Springer Science & Business Media, 2012.

11

Finite Frames Maximally Robust to Erasures

How should one go about, to construct a finite frame for $\mathbb{R}^n$? As a starter, let us restrict our discussion to the case where $n = 2$. Since Differential Geometry is one of the main interests of this book, a reasonable suggestion would be to sample, the graph

$$\left\{ \begin{bmatrix} x \\ f(x) \end{bmatrix} : x \in \mathbb{R} \right\}$$

of a real-valued function f defined on $\mathbb{R}$. Sampling could then be either done regularly (say along some discrete subgroup), or irregularly by random sampling. In fact, one might even aim to sample along

$$\mathcal{S}(f, \Gamma) = \left\{ \begin{bmatrix} \gamma \\ f(\gamma) \end{bmatrix} : \gamma \in \Gamma \right\}$$

where Γ is a nondiscrete subgroup of $\mathbb{R}$. Regardless of the sampling strategy, the length $\sqrt{\gamma^2 + f(\gamma)^2}$, of each sample is not consequential to $\mathcal{S}(f, \Gamma)$ being a frame. Thus, from here on, it is safe to replace $\mathcal{S}(f, \Gamma)$ with the collection of unit-norm vectors:

$$\left\{ \frac{1}{\sqrt{\gamma^2 + f(\gamma)^2}} \begin{bmatrix} \gamma \\ f(\gamma) \end{bmatrix} : \gamma \in \Gamma \right\}.$$

The sampling problem, as described above is now equivalent to a sampling problem on the unit circle. This is, in fact a compact manifold, which is also a Lie group. Indeed, let

$$X = \begin{bmatrix} 0 & -1 \\ 1 & 0 \end{bmatrix}$$

A Bridge Between Lie Theory and Frame Theory: Applications of Lie Theory to Harmonic Analysis, First Edition. Vignon Oussa.

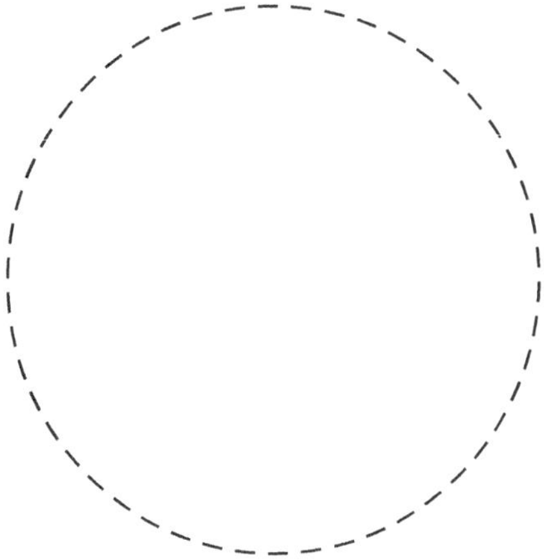

Figure 11.1 Sampling along a unit circle.

be a skew-symmetric real matrix of order 2, and let $\mathfrak{g}$ be the real Lie algebra spanned by X. Its Lie group, is the unit circle of matrices of the form

$$\exp(tX) = \begin{bmatrix} \cos t & -\sin t \\ \sin t & \cos t \end{bmatrix} \quad \text{where } t \in \mathbb{R}.$$

A sampling along the integers for example produces a selection of the type

$$\exp(\mathbb{Z}X)e_1 = \exp(\mathbb{Z}X)\begin{bmatrix} 1 \\ 0 \end{bmatrix} = \left\{ \begin{bmatrix} \cos k \\ \sin k \end{bmatrix} : k \in \mathbb{Z} \right\},$$

which is of course, a frame for $\mathbb{R}^2$ as illustrated in Figure 11.1.

In fact, this set is much more than a frame. It is also, a full-spark frame. That is, every collection of two vectors in $\exp(\mathbb{Z}X)e_1$ is a basis for $\mathbb{R}^2$ since

$$\det \begin{bmatrix} \cos k_1 & \cos k_2 \\ \sin k_1 & \sin k_2 \end{bmatrix} = -\sin(k_1 - k_2) \neq 0 \quad \text{for any } k_1 \neq k_2 \in \mathbb{Z}.$$

Note also that $\exp(\mathbb{Z}X)e_1$ is a dense subset of the unit circle. In fact, rotation by $\exp X$ defines an ergodic action on the unit circle. Moreover, Ergodic theory provides us with the right tools needed to translate a discrete problem into a continuous one. Indeed, appealing to the Ergodic theorem, one obtains:

$$\begin{aligned} \lim_{n\to 0} \frac{1}{n} \sum_{k=0}^{n-1} \left|\langle v, \exp(kX)e_1\rangle\right|^2 &= \int_0^1 \left|\langle v, \exp(tX)e_1\rangle\right|^2 dt \\ &= \frac{1}{4}\left(4v_1v_2 \sin(1)^2 - v_2^2(\sin 2 - 2) + v_1^2(2 + \sin 2)\right). \end{aligned}$$

Moreover, the value of the integral above is zero if and only if v is the zero vector. This demonstrates that the matrix coefficients $t \mapsto \langle v, \exp(tX)e_1\rangle$ is not the zero

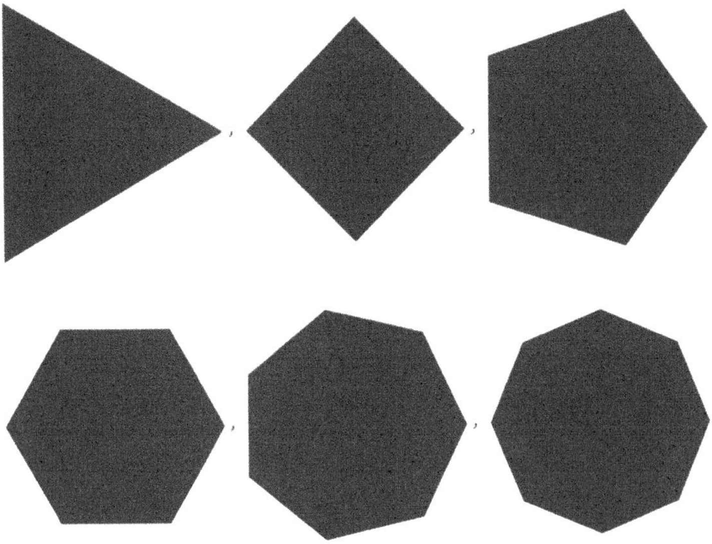

Figure 11.2 Sampling over the finite subgroups of the unit circle, $\exp\left(\frac{2\pi k}{3}X\right)$, where $k = 0, \ldots, n-1$ amounts to sampling along the corners of the regular polygons.

function whenever, v is nonzero. The linear map which takes v to the continuous function $t \mapsto \langle v, \exp(tX)\, e_1 \rangle$ is the analysis operator associated with the exponential function $t \mapsto \exp(tX)$. One verifies that the analysis operator intertwines the action of $\exp(tX)$ with a translation action on $\mathbb{R}$.

Sampling over the finite subgroups of the unit circle, $\exp\left(\frac{2\pi k}{3}X\right)$, where $k = 0, \ldots, n-1$ amounts to sampling along the corners of the regular polygons (Figure 11.2).

Such frames can be also studied as orbits of a unitary action of the cyclic group $\mathbb{Z}_n$.

For the special case where $n = 3$, the unit sphere may be parametrized as:

$$\exp\begin{bmatrix} 0 & 0 & -\alpha \\ 0 & 0 & 0 \\ \alpha & 0 & 0 \end{bmatrix} \exp\begin{bmatrix} 0 & -\tau & 0 \\ \tau & 0 & 0 \\ 0 & 0 & 0 \end{bmatrix} \begin{bmatrix} 1 \\ 0 \\ 0 \end{bmatrix}, \alpha, \tau \in [0, 2\pi)$$

and the problem is thus reduced to sampling on the unit sphere with an action of $SO(3, \mathbb{R})$. Up to isomorphism, $SO(3, \mathbb{R})$ only has a few finite subgroups. The rotational symmetry group of an n-pyramid (cyclic group of order n), the rotational symmetry group of an n-prism (the Dihedral groups), the rotational symmetry group A_4 of a regular tetrahedron, the rotational symmetry group

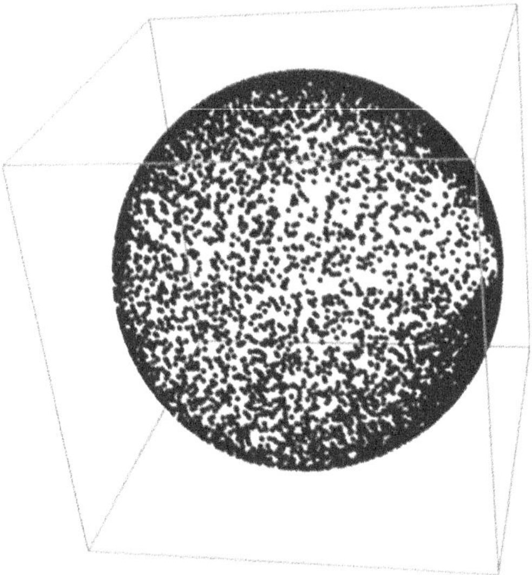

Figure 11.3 A dense subset of S^2.

S_4 of a cube (or a regular octahedron), and the rotational symmetry group A_5 of a regular dodecahedron (a regular icosahedron). Thus, the study of frames generated by finite cyclic groups is a tractable endeavor. Generally, however, an orbit of the type $\exp(\mathbb{Z}X)v$ for a skew-symmetric matrix, will generate a dense subset of a unit circle, and this action can be studied via the tools of Fourier series on the integers. In addition,

$$\Gamma e_1 = \left\{ \exp \begin{bmatrix} 0 & k & j \\ -k & 0 & l \\ -j & -l & 0 \end{bmatrix} \begin{bmatrix} 1 \\ 0 \\ 0 \end{bmatrix} : k, j, l \in \mathbb{Z}^3 \right\}$$

is a dense countable subset of the unit sphere (Figure 11.3).

Consequently, any orthonormal basis of $\mathbb{R}^3$ can be approximated by a triple in Γe_1 and the error made by such an approximation can be made as small as desired. Generalizing the constructions described here to $\mathbb{R}^n$ is a worthwhile exercise, left to the reader, since $SO(n)$ (the Lie group of orthogonal and unit determinant matrices), acts smoothly on the unit sphere. Under such assumptions, the problem becomes now to find ways to sample along the sphere $SO(n)\,b \simeq SO(n)/SO(n-1)$, where b is a fixed element of a standard basis.

Since the theory of finite frames, often focuses its study on complex Hilbert spaces, we will turn our main focus to the action of the unitary group of matrices on $\mathbb{C}^n$. The reader will find in the diagram below a map that might be useful in navigating around the stream of ideas

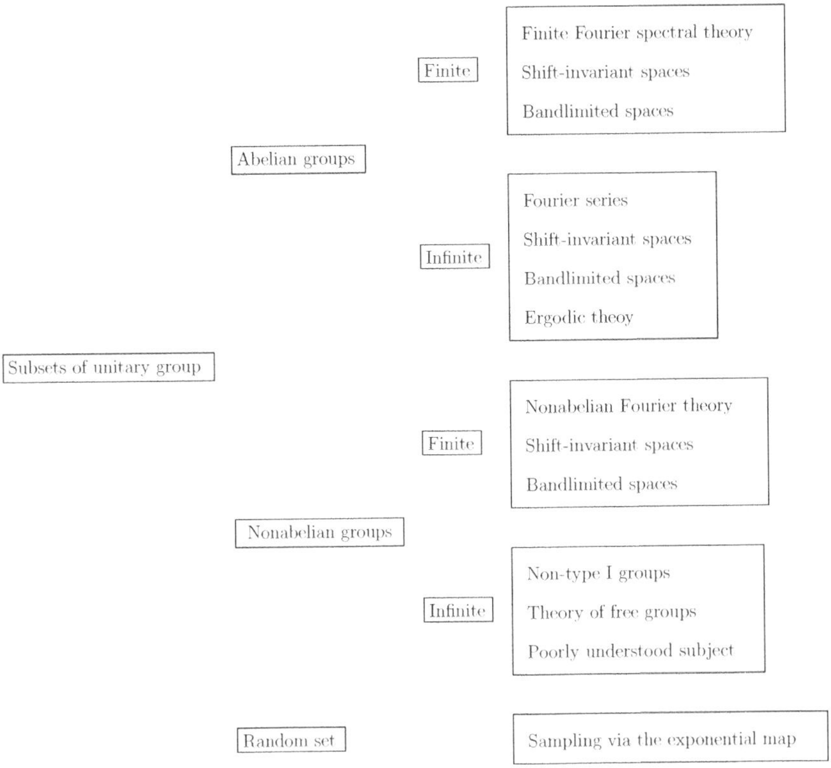

Let $M(n,\mathbb{C})$ be the Lie algebra of complex matrices of order n and let $B : M(n,\mathbb{C}) \to M(n,\mathbb{C})$ be an invertible endomorphism sending every matrix to its adjoint. Viewed as an operator acting on vector space $M(n,\mathbb{C})$, B^2 is the identity operator and consequently, $1, -1$ are the only possible eigenvalues of B. Let $\mathfrak{g}$ be the eigenspace associated with -1. Then a matrix X belongs to the space $\mathfrak{g}$ if and only if $X^* = -X$ (X is a skew-Hermitian matrix) and it is not difficult to verify that for $X, Y \in \mathfrak{g}$, $[X, Y] \in \mathfrak{g}$. Thus $\mathfrak{g}$ is a Lie subalgebra of $M(n,\mathbb{C})$ and it is our chief objective in this chapter to convince the reader that Lie theory provides useful insights into the world of finite frames. Moreover, although frame theory is generally not regarded as adjacent to Lie theory, the reader will find in this chapter a bridge, designed to allow navigation from one subject to the other.

Recall that an endomorphism P is a projection if it satisfies the equation $P^2 = P$. In other words, P leaves its range invariant. Moreover, a projection is called orthogonal if not only it is true that $P^2 = P$, but additionally, one has that P and its adjoint are the same operator. Any projection that is not orthogonal is called oblique.

Let $U(n)$ be the group of unitary matrices of order n. Write

$$U(n) = \{M \in M(n, \mathbb{C}) : M^*M = \mathrm{Id}_n\}.$$

For a real parameter t, $\exp(tX)$ is unitary if and only if $\exp(tX^*) = \exp(-tX)$. Computing the differential of the curve $t \mapsto \exp(tX^*) - \exp(-tX)$ at zero, one verifies that X is a skew-Hermitian matrix. Therefore, $\mathfrak{g}$ is the Lie algebra of the unitary group $U(n)$. Put

$$G = U(n) \quad \text{and} \quad \mathfrak{g} = \mathfrak{u}(n) \quad \text{and write } \mathfrak{u}(n) = \mathrm{Lie}(U(n)).$$

Lemma 11.1 $U(n)$ is a connected and closed subgroup of the general linear group.

Proof: First, recall that any unitary matrix can be diagonalized by conjugating it by another unitary matrix. Since every eigenvalue of a unitary matrix must be a complex number of modulus one, for a fixed matrix $M \in U(n)$, there exists a unitary map P such that

$$PMP^{-1} = \begin{bmatrix} e^{i\lambda_1} & & \\ & \ddots & \\ & & e^{i\lambda_n} \end{bmatrix} = \exp(D).$$

This means that $M = \exp\left(P^{-1}DP\right)$ and $t \mapsto \exp\left(t \cdot \left(P^{-1}DP\right)\right)$ is a curve passing through the identity in $U(n)$ and M. Therefore, $U(n)$ is a connected.

Lemma 11.2 The exponential map is a smooth surjection on the unitary group $U(n)$.

Proof: For a fixed $X \in \mathfrak{g}$, viewed as a linear operator acting on $\mathbb{C}^n$, if a complex number λ is an eigenvalue of X, the assumption that $X^* = -X$ automatically implies that $|\lambda| = 1$. Therefore, the range of the exponential map restricted to $\mathfrak{g}$ is compact. Since the range of the exponential map coincides with the connected component of the identity, and since the unitary group is connected group, it follows that the exponential map defines a surjection of the set of skew-Hermitian complex matrices onto the group of unitary operators on $\mathbb{C}^n$.

11.1 Inductive Construction of All Complex *n*-Frames

By n-frame, we mean, a collection of n-vectors forming a frame for some Hilbert space $\mathbb{C}^m$. Our main intent for this section, is to discuss a number of ways in which, Lie theory gives frame theory a language, useful in organizing and categorizing "smoothly" all orthonormal bases, Riesz bases, n-Parseval frames and n-frames. Indeed, every orthonormal basis in $\mathbb{C}^n$ corresponds to the column vectors of a fixed unitary matrix in $U(n)$, and vice versa. Therefore, there is a one-to-one correspondence between the set of all orthonormal bases of $\mathbb{C}^n$ and the set of all unitary matrices in $U(n)$. Similarly, every Riesz basis is the image of some fixed orthonormal basis, viewed through an invertible operator. In other words, the set of Riesz bases in $\mathbb{C}^n$ can be parametrized by the following fibered set:

$$\mathcal{R}(n) = \bigcup_{X \in U(n)} \{PX : \det P \neq 0\}.$$

Next, since every n-Parseval frame is obtained by viewing an orthonormal basis through a fixed orthogonal projection, Parseval frames are naturally, in one-to-one correspondence with the set

$$\mathcal{P}(n) = \bigcup_{X \in U(n)} \left\{PX : P^2 = P \quad \text{and} \quad P = P^*\right\}.$$

Moreover, every n-frame is parametrized by

$$\mathcal{F}(n) = \bigcup_{X \in U(n)} \left\{PX : P^2 = P\right\}.$$

Note that, each set $\mathcal{R}(n), \mathcal{P}(n)$, and $\mathcal{F}(n)$ has for base space the unitary group $U(n)$. Thus, to construct $\mathcal{R}(n), \mathcal{P}(n)$, and $\mathcal{F}(n)$, it helps to have at one's disposal a smooth parametrization of the unitary group. That is, a diffeomorphism between a manifold $\mathcal{M}$ and the group of unitary operators $U(n)$. For example, any n-Parseval frame can be identified with the columns of a matrix of the type

$$P \exp\left(X_1\right) \cdots \exp\left(X_m\right)$$

where $X_1, \ldots, X_m \in \mathfrak{u}(n)$, $P^2 = P$, and $P = P^*$.

Example 11.1 Let

$$P = \begin{bmatrix} 1 & 0 & 0 \\ 0 & 1 & 0 \\ 0 & 0 & 0 \end{bmatrix} \quad \text{and} \quad X = \begin{bmatrix} 0 & it & 0 \\ it & 0 & it \\ 0 & it & 0 \end{bmatrix}.$$

Since P is a rank-2 projection, computing $P \exp tX$, for every real parameter t,

$$\begin{bmatrix} \cos^2\left(\frac{t}{\sqrt{2}}\right) \\ \frac{i\sin\left(\sqrt{2}t\right)}{\sqrt{2}} \end{bmatrix}, \begin{bmatrix} \frac{i\sin\left(\sqrt{2}t\right)}{\sqrt{2}} \\ \cos\left(\sqrt{2}t\right) \end{bmatrix}, \begin{bmatrix} -\sin^2\left(\frac{t}{\sqrt{2}}\right) \\ \frac{i\sin\left(\sqrt{2}t\right)}{\sqrt{2}} \end{bmatrix}$$

is a 3-Parseval frame for $\mathbb{C}^2$.

Generally however for a frame of the form $P\exp(X_1)\cdots\exp(X_m)$ very little can be said about m and how $X_1,\ldots,X_m$ are systematically selected and ordered. Moreover, since the exponential map fails to define a diffeormophism between the Lie algebra of skew-Hermitian matrices and the unitary group, exp is not an ideal parametrizing function for the unitary group. What would be helpful for us, is a step-by-step procedure that gives a canonical (or algorithmic) construction of all unitary matrices.

For the first case where $n=1$, the problem is easy. The adjoint of every matrix M is a scalar given by complex conjugation, and

$$|M|^2 = 1 \Leftrightarrow M = e^{2\pi i\theta} \quad \text{for some } \theta \in [0,1).$$

When n is assumed to be two, the problem becomes quite involved already. Indeed, $U(2)$ is the set of complex matrices of order 2

$$\begin{bmatrix} x+iy & u+iv \\ r+it & z+iw \end{bmatrix} \quad \text{such that} \begin{cases} r^2+t^2+x^2+y^2=1 \\ u^2+v^2-w^2-z^2=1 \\ irw+tw+rz+ux-itz-iuy+ivx+vy=0 \end{cases}. \tag{11.1}$$

To parametrize every element in $U(2)$, one might think, that we should only need to solve a system of three (polynomial) equations and four complex variables. Although, solving the system of equations in (11.1) is certainly possible, it would not be a wise use of time. Plus, eventually for large n, this method is unsustainable.

As an alternative, we will put to use a natural action of the unitary group on the unit S^{2n-1} in $\mathbb{C}^n$. The simplest interesting case is for $n=2$. Then, the one-parameter group

$$U(2)_v = \left\{ \begin{bmatrix} e^{i\phi_{(1,1)}} & 0 \\ 0 & 1 \end{bmatrix} : \phi_{(1,1)} \in [0,2\pi) \right\} = \mathbb{T}$$

fixes

$$v = \begin{bmatrix} 0 \\ 1 \end{bmatrix}.$$

But since $U(2)$ acts transitively on S^3, the homogeneous space $U(2)/U(2)_v$ is just S^3. Parametrizing the unit sphere in $\mathbb{C}^2$ with complex spherical coordinates, the set of matrices

$$\left\{ \begin{bmatrix} e^{i\phi_{(1,2)}}\cos\theta_{(1,2)} & e^{i\phi_{(1,2)}}\sin\theta_{(1,2)} \\ -e^{i\phi_{(1,2)}}\sin\theta_{(1,2)} & e^{i\phi_{(2,2)}}\cos\theta_{(1,2)} \end{bmatrix} : \phi_{(1,2)},\phi_{(2,2)} \in [0,2\pi) \text{ and } \theta_{(1,2)} \in \left[0,\frac{\pi}{2}\right] \right\}$$

is a transversal for the cosets of $U(2)_v$. This shows that

$$\left(\phi_{(1,1)}, \phi_{(1,2)}, \phi_{(2,2)}\right) \mapsto \begin{bmatrix} e^{i\phi_{(1,2)}} \cos\theta_{(1,2)} & e^{i\phi_{(1,2)}} \sin\theta_{(1,2)} \\ -e^{i\phi_{(1,2)}} \sin\theta_{(1,2)} & e^{i\phi_{(2,2)}} \cos\theta_{(1,2)} \end{bmatrix} \begin{bmatrix} e^{i\phi_{(1,1)}} & 0 \\ 0 & 1 \end{bmatrix}$$

is a smooth parametrization of the unitary group. To generalize this procedure, fix

$$v = \begin{bmatrix} 0 \\ \vdots \\ 0 \\ 1 \end{bmatrix} \in \mathbb{C}^n.$$

Then $U(n)v = S^{2n-1}$ and to parametrize the homogeneous space $U(n)/U(n)_v$, where $U(n)_v$ is the stabilizer of v, some familiarity with coordinates of complex hyperspheres is helpful. To this end, define inductively,

$$\Theta_1\left(\phi_1\right) = e^{i\phi_1}, \quad \Theta_2\left(\phi_1, \phi_2, \theta_1\right) = \begin{bmatrix} e^{i\phi_1} \sin\theta_1 \\ e^{i\phi_2} \cos\theta_1 \end{bmatrix},$$

and generally,

$$\Theta_n^{(k)}\left(\phi_1, \dots, \phi_n, \theta_1, \dots, \theta_{n-1}\right) = \begin{cases} e^{i\phi_1} \sin\theta_1 \cdots \sin\theta_{n-1} & \text{if } k = 1 \\ e^{i\phi_k} \cos\theta_{k-1} \sin\theta_k \cdots \sin\theta_{n-1} & \text{if } 2 \leq k \leq 1 \end{cases}$$

such that $\phi_k \in [0, 2\pi)$ and $\theta_j \in \left[0, \frac{\pi}{2}\right]$. Then

$$\left(\phi_1, \dots, \phi_n, \theta_1, \dots, \theta_{n-1}\right) \mapsto \Theta_n^{(k)}\left(\phi_1, \dots, \phi_n, \theta_1, \dots, \theta_{n-1}\right)$$

maps into the complex sphere S^{2n-1}, which serves a coordinate system for $U(n)/U(n)_v$.

To complete our description, we also need a useful parametrization of $U(n)_v$. But such a set is necessarily, diffeomorphic to $U(n-1)$ and we proceed inductively for a full parametrization of $U(n)$. Note also that this method for parametrizing $U(n)$ can easily be modified to adapt to the orthogonal group $O(n)$ (see Chapter 8) for the construction of all n frames defined on a real-vector space. This is left as an exercise for the interested reader.

The matrix–vector action of $U(n)$ on $\mathbb{C}^n$ is an important one. For instance, every unitary irreducible representation of a compact Lie group is finite-dimensional [2] and as such, its range is a subgroup of some unitary group $U(\ell)$, $\ell \geq 0$. Thus, the matrix coefficients of the irreducible representation of a compact Lie group can be viewed as the matrix coefficients of the matrix action of $U(n)$ on $\mathbb{C}^n$. This provides us with a universal set for which, the Discretization Problem is very relevant.

Problem 11.1 ***(The Discretization Problem for unitary groups)*** Is there a vector $v \in \mathbb{C}^n$ such that $U(n)v$ can be sampled to construct a frame (with prescribed properties)?

Without any specific properties prescribed for the sought out frames, the problem is easily solved. Specific choices of sampling sets lead to interesting paths.

If one opts to sample the orthogonal group along some abelian group, two main roads are up to exploration. In the case of finite abelian subgroups, the sampling problem is easily answered through finite Fourier analysis in terms of the spectral description of the left regular representation and its associated invariant spaces. For instance, the regular representation of the cyclic group $\mathbb{Z}_n$

$$k \mapsto \begin{bmatrix} & & 1 \\ 1 & & \\ & \ddots & \\ & & 1 \end{bmatrix}^k$$

generates the canonical orthonormal basis for $\mathbb{C}^n$. The spectral analysis of this action is best studied through the lenses of the discrete Fourier transform. Let T be a generator of the range of the regular representation of $\mathbb{Z}_n$. Then T is diagonalizable. As it turns out, up to multiplication there is only one operator that diagonalizes T. When normalized to be unitary, we call such a matrix with its familiar name: the discrete Fourier transform. By virtue of the fact that T is diagonalizable, its eigenvalues are the nth-roots of unity. The spectral theory of the Fourier transform makes the study of spaces which are invariant under the action of T possible. Moreover, the theory naturally generalizes to the class of all finitely generated abelian groups.

Note that any discrete subgroup Γ of $U(n)$ is finite since $U(n)$ is compact. Thus, sampling regularly, (along a finite subgroup) is easily articulated in the language of admissibility. Here the problem is the following: is the representation obtained by restricting the action of the unitary group to Γ admissible? It is worth mentioning that the body of work of Hartmut Führ [3] on this problem makes the answer accessible to an audience fluent in the language of admissibility.

Proposition 11.1 If the restriction to Γ is equivalent to a subrepresentation of the regular representation of the group then the answer is yes. Otherwise, no.

Proof: The idea in proving this is rather simple. The existence of an admissible vector, implies the existence of an isometry intertwining the action of Γ with the regular representation. Schur's lemma guarantees that any such map is unique up to constant multiple. Thus, should the action of Γ, induce an admissible representation, then such a representation is necessarily a subrepresentation of the regular representation. For the converse of this statement, it suffices to fix the indicator function of the identity element as a vector generating $l^2(\Gamma)$ under Γ-shift. Next, fixing the preimage a projection this indicator function onto the range some associated analysis operator gives an admissible vector associated with the action of Γ on $\mathbb{C}^n$.

Thus, finite subgroups of the unitary groups provide a general framework for discretizing the unitary group along some structured sets. This provides a wealth of advantages many of which have been rather well documented in the literature [1]. For example, the value of the following result is self-evident.

Proposition 11.2 Sampling along a finite group acting irreducibly automatically produces Parseval frames.

Proof: To prove this result, let us denote a representation of Γ acting on $\mathbb{C}^n$ by π. Then the analysis operator (or voice transform) associated with some nonzero vector $v \in \mathbb{C}^n$ is a unitary map taking $\mathbb{C}^n$ into $l^2(\Gamma)$ defined as follows:

$$\left[W_v(w)\right](x) = \langle w, \pi(x)v\rangle_{\mathbb{C}^n}.$$

Moreover, the synthesis operator an adjoint of W_v acting as follows:

$$W_v^*(c) = \sum_{\gamma\in\Gamma} c_\gamma \pi(\gamma).$$

One easily verifies that the frame operator $W_v^* \circ W_v$ belongs to the commutant of π and Schur's lemma informs us that such a map is necessarily a scaled version of the identity map. Therefore, if Γ acts irreducibly as a subgroup of the unitary group, and if additionally Γ is finite then for any nonzero vector v, Γv is a Parseval frame for $\mathbb{C}^n$.

As illustrated in the examples below, Proposition 11.2 is quite handy for the construction of finite Parseval frames.

Example 11.2 ***(The discrete Heisenberg groups)*** Fix

$$T = \begin{bmatrix} & & 1 \\ 1 & & \\ & \ddots & \\ & & 1 \end{bmatrix} \quad \text{and} \quad M = \begin{bmatrix} 1 & & & \\ & e^{\frac{2\pi i}{n}} & & \\ & & \ddots & \\ & & & e^{\frac{2\pi i(n-1)}{n}} \end{bmatrix}$$

and let Γ be the finite group of $U(n)$ generated by T and M. This is a nonabelian group and in fact, the commutator subgroup of Γ is generated by

$$Z = \begin{bmatrix} e^{\frac{2\pi i}{n}} & & & \\ & e^{\frac{2\pi i}{n}} & & \\ & & \ddots & \\ & & & e^{\frac{2\pi i}{n}} \end{bmatrix}.$$

This shows that the center of Γ is the cyclic group generated by Z. Γ is a finite nilpotent group often regarded as a discrete version of the Heisenberg group.

Moreover, it is not difficult to verify that this representation is in fact obtained by inducing a unitary character χ (such that $\chi(T) = e^{\frac{2\pi i}{n}}$) of the cyclic subgroup generated by T. The stabilizer of the action of Γ being trivial, by the theory of Mackey, this representation is irreducible.

Let us now collect the low-hanging fruits of our labor. For any nonzero vector $v \in \mathbb{C}^n$, Γv is a Parseval frame. Next, let L be the left regular representation of the discrete Heisenberg group. This representation decomposes itself into a direct integral of irreducible representations, each occurring with multiplicity, coinciding with its associated dimension. Moreover, any representation of the Heisenberg group that generates a frame, must be a subrepresentation of L. For instance if n is prime, then with the help of Mackey's analysis, it can be shown that the regular representation is unitarily equivalent to a direct sum

$$\left(\bigoplus_{k=0}^{n-1}\bigoplus_{k=0}^{n-1}\chi_{(k,j)}\right) \oplus (\pi \otimes \pi)$$

acting in

$$\left(\bigoplus_{k=0}^{n-1}\bigoplus_{k=0}^{n-1}\mathbb{C}\right) \oplus \left(\mathbb{C}^n \otimes \mathbb{C}^n\right).$$

Minus important details to be revealed at the granular and technical level, of how each irreducible representation is realized (this is an exercise left to the interested reader) it is safe to rest the case of admissible representations of Γ at this point.

Example 11.3 ***(Diheldral groups)*** Fix

$$T = \begin{bmatrix} & & 1 \\ 1 & & \\ & \ddots & \\ & 1 & \end{bmatrix} \quad \text{and} \quad D = \begin{bmatrix} 1 & & \\ & & 1 \\ & \cdots & \\ 1 & & \end{bmatrix}$$

and let Γ be the finite group of $U(n)$ generated by T and D. Then Γ is a subgroup of the unitary group, which is isomorphic to the semi-direct product group $\mathbb{Z}_n \rtimes \mathbb{Z}_2$. This is not an irreducible representation of the Dihedral group. To see this, it suffices to diagonalize T with the discrete Fourier matrix. Since D commutes with the Fourier matrix, nontrivial invariant spaces exist. In any case, this action of Γ defines a quasiregular representation of the Dihedral group, obtained by conormally inducing the trivial representations. This representation can be shown to decompose into multiplicity-free spaces, and as such, it is admissible.

Example 11.4 ***(Extension of lattices)*** Let $G = A \rtimes H$ where H is some finite subgroup acting of some finitely-generated abelian group A. Next, let χ be a unitary character of A and let π be a representation of G obtained by inducing χ.

Realize π as acting in $l^2(H)$. Via this, $\mathbb{Z}^n \rtimes H$ acts on vectors in $l^2(H)$ as follows:

$$\left[\pi(\kappa, j) f\right](s) = \chi\left(s^{-1}\kappa s\right) \cdot f(s - j).$$

Therefore, on the one hand, H acts by left-translations and $\mathbb{Z}^n$ acts diagonally. As one would suspect, the range of π is a subset of the unitary group, and this example can be viewed as a discrete version of the class of groups of the type $G = NH$ as explored in preceding chapters. Moreover all results in Chapter 7 are flexible templates that can be easily modified to suit our need. This actually makes the case for us (not to be redundant) and it would be a worthwhile exercise for the reader to work the details out independently.

The following generally result is also quite useful. Let Γ be a finite subgroup of the unitary group. Next, let $C(\Gamma)$ be the commutant of Γ. That is precisely, a subspace of complex matrices of order n, each commuting which every element of Γ. Then the following is immediately seen.

Proposition 11.3 If v is a fixed vector in $\mathbb{C}^n$ such that Γv is a frame for $\mathbb{C}^n$ and if A is a linear operator in $C(\Gamma)$ of rank r then $\Gamma(Av)$ is a frame for $\mathbb{C}^r$. If additionally, A is a unit element in the algebra $C(\Gamma)$ then $\Gamma(Av)$ is a frame for $\mathbb{C}^n$.

The structure of the commutant algebra $C(\Gamma)$ of Γ is well understood via the theory of noncommutative Fourier analysis, which is a rather mature subject. For the existence of a frame, the action of Γ must induce a representation which is equivalent to a subrepresentation of the regular representation, which is given by

$$L = \bigoplus_{\pi \in \widehat{\Gamma}} \left(\oplus_{k=1}^{\dim \pi}(\pi)\right).$$

Thus a complete understanding of non-commutative finite Fourier analysis requires a full description of the unitary dual of Γ.

11.2 Infinite Singly Generated Subgroups of $U(n)$

The unitary group $U(n)$ admits a large number of singly generated groups, which are not discrete. Indeed, every matrix A in $U(n)$ is diagonalizable, and the group generated by A is finite if and only if the vector space generated by the eigenvalues of A is a one-dimensional rational vector space. For example, in $U(1)$, the group generated by $e^{2\pi i\alpha}$ is infinite if and only if α is irrational. When α is irrational, the sequence $\left(e^{2\pi ik\alpha}\right)_{k\in\mathbb{Z}}$ of complex numbers is a dense subset of the one-dimensional torus.

In fact, the map $z \mapsto e^{2\pi i\alpha}z$ defines an ergodic action on $U(1)$. It is also interesting to observe that the vector given by the real part and imaginary part, respectively of terms of $\left(e^{2\pi ik\alpha}\right)_{k\in\mathbb{Z}}$ gives rise to a special type of system called a full-spark frame.

Definition 11.1 A finite frame in an n-dimensional space is called full spark, if its size is at least equal to n, and additionally, any n-sample of the frame is a basis for its underlying space.

For example, one easily checks that every subset of cardinality two of the sequence $\left(e^{2\pi ik\alpha}\right)_{k\in\mathbb{Z}}$ gives rise to a real basis for $\mathbb{R}^2$ (as a vector space, of course over the reals).

Full spark frames are potentially useful in the transmission of signals for various reasons. To transmit information coded in the form of some vector $v \in \mathbb{K}^n$ where $\mathbb{K} = \mathbb{R}$ or $\mathbb{C}$, a sender may wish to to represent v, redundantly as a sequence of scalars given by the components of Cv where C is a matrix representation of the analysis operator of some full spark frame $\mathfrak{P}$ consisting of m elements. In the unfortunate event that at most $m - n$ components of Cv are deleted during transmission, the fact that any n-subset of $\mathfrak{P}$ is a basis can be exploited to prevent loss of information. For instance, let z be the vector obtained by removing the damaged part of the vector Cv and let R be the matrix obtained by deleting from C the rows associated to the information that has been damaged. Since $\mathfrak{P}$ is full spark, then it is clear that R is injective. As a result, although some of the information transmitted may be destroyed beyond repair, the receiver still has the ability to fully recover the original vector from the equation $v = \left(R^T R\right)^{-1} R^T z$. For this reason, full spark frames are also called frames that are maximally robust to erasures.

For the special case where Γ is an infinite subgroup of the unitary group generated by some element X, without loss of generality, we may assume that $A = \exp X$ for a skew-Hermitian matrix X, and every element of Γ is therefore of the form $\exp kX$ for some integer k. For the next stage of our discussion, we introduce some useful terminology.

Definition 11.2 The rational rank of A is the rational dimension of the rational vector space spanned by the set containing the eigenvalues of A and $2\pi i$.

Example 11.5 Let

$$A = \begin{bmatrix} \frac{2\pi i}{3} & & & \\ & \frac{2\pi i\sqrt{2}}{3} & & \\ & & 2\pi i\sqrt{2} & \\ & & & 0 \end{bmatrix}.$$

Then the rational rank of A is equal to two. Since $\left\{2\pi i\sqrt{2}, 2\pi i\right\}$ serves a basis for the rational space spanned by the eigenvalues of A.

When the rank of A is full (this means equal to n), then $\exp X$ is a dense subset of a maximal torus in $U(n)$. Otherwise, $\exp X$ is just a dense in a torus of positive, but of lower dimension. In any case, one would need to sample the surface of a torus of positive dimension to construct a frame. Next, let G be the topological closure of the torus $\exp \mathbb{Z}X$ and let $d\mu_G$ be a Haar measure on G. Left multiplication by $\exp X$ on each such a torus, defines an ergodic action on G which transfers naturally to an ergodic action on a torus in $\mathbb{C}^n$. Indeed, for each nonzero vector $v \in \mathbb{C}^n$, $\exp(\mathbb{Z}X)v$ is a torus in $\mathbb{C}^n$. Thus, the action of G partitions $\mathbb{C}^n$ into a family of tori, and the tools of Differential Geometry lend themselves well to the study of the parametrized family of manifolds arising from this action.

Next, we will highlight an interesting connection between frame theory and a standard formulation of the Ergodic theorem, and we will explore what facts that can be easily deduced with this connection. Since for every natural number n, the sum $\frac{1}{n}\sum_{k=0}^{n-1}|\langle w, \exp(kX)v\rangle|^2$ can be viewed as a partial Riemann sum on the surface of the torus, it is clear that the Cesàro sum of a sequence $(\langle w, \exp(kX)v\rangle)_{k\in\mathbb{Z}}$ is convergent, and additionally, one obtains that

$$\lim_{n\to\infty}\frac{1}{n}\sum_{k=0}^{n-1}|\langle w, \exp(kX)v\rangle|^2 = \int_G |\langle w, gv\rangle|^2 d\mu_G(g). \tag{11.2}$$

Since the action of G is smooth, the continuous function $g \mapsto \langle w, gv\rangle$ is necessarily smooth on G. Also, since G is compact, (11.2) converges. Therefore, should the Haar measure on G be absolutely continuous with respect to the Haar measure of the maximal torus in $U(n)$, the integral (11.2) would have to be positive. Otherwise, (11.2) is zero and this implies that no orbit of v can be discretized to construct a frame.

Relevant to us, next, is the unique torus containing the fixed vector v in $\mathbb{C}^n$ such that $v_k = 1$ for all k. For arbitrary $w \in \mathbb{C}^n$, the scalar $\langle w, \exp(kX)v\rangle_{\mathbb{C}^n}$ is the coefficient associated with a projection on $\exp(kX)v$. The analysis operator W_v associated to v takes w to the sequence $(\exp(kX)v)_{k\in\mathbb{Z}}$. Thus, W_v maps $\mathbb{C}^n$ linearly into $l^2(\mathbb{Z})$ and its range is a left-invariant space of $l^2(\mathbb{Z})$. Computing the Fourier series of this coefficient (assuming for the moment this makes sense), we obtain

$$\widehat{W_v(w)}(\chi) = \int_G \langle w, gv\rangle_{\mathbb{C}^n} \cdot \overline{\chi(g)} d\mu_G(g).$$

The unitary dual of G is parametrized by the lattice $\mathbb{Z}^n$ and the linear map $w \mapsto \widehat{W_v(w)}$ diagonalizes the action of G. Indeed, for $h \in G$,

$$\widehat{W_v(hw)}(\xi) = \overline{\chi(h)} \cdot \left(\int_G \langle w, gv\rangle_{\mathbb{C}^n} \cdot \overline{\chi(g)} d\mu_G(g)\right) = \overline{\chi(h)} \cdot \widehat{W_v(w)}(\xi).$$

For instance,

$$\widehat{W_v(v)}(\chi) = \int_G \langle v, gv\rangle_{\mathbb{C}^n} \cdot \overline{\chi(g)} d\mu_G(g) = \int_G \overline{\chi(g)} d\mu_G(g) = \widehat{1_G}(\chi).$$

Thus, the linear combination $\sum_{k=1}^{n} c_k g_k v = 0$ holds if and only if

$$\left(\sum_{k=1}^{n} c_k \overline{\chi(g_k)}\right) \cdot \widehat{W_v(v)}(\chi) = 0.$$

Since $\widehat{W_v(v)}$ is nonzero on the unitary dual of G, it follows that

$$\sum_{k=1}^{n} c_k \cdot \overline{\chi(g_k)} = 0 \quad \text{for every } \chi \in \widehat{G}.$$

This implies that all the coefficients c_k must be equal to zero and as such, $\exp(\mathbb{Z}X)v$ is a full spark frame for $\mathbb{C}^n$.

Not every finitely generated subgroup of the unitary group is of course abelian. For example, the unitary group admits several pairs of matrices, each generating a free group. Since the harmonic analysis of such groups is generally quite poorly understood, it is best to replace any countable subgroup of the unitary group with its topological closure, which is of course a Lie group.

11.3 Random Sampling

Let $\mathcal{O}$ be a connected neighborhood of the identity of the Lie algebra of the Unitary group, which is being mapped diffeomorphically to a neighborhood of the identity element in $U(n)$ via the exponential map. Then for any random set (sampled randomly with respect to the Lebesgue measure) Λ contained in the open set $\mathcal{O}$, its image

$$\Gamma = \left\{ \sum_{k=0}^{n-4} \frac{X^k}{k!} : X \in \Lambda \right\}$$

is random sample of a neighborhood of the identity with a distribution function absolutely continuous with respect to the weight of the Haar measure on $U(n)$. Thus, the pair consisting of a fixed neighborhood of the identity and the exponential map, provides a general framework for studying random sample of unitary operators acting on $\mathbb{C}^n$. This road, we expect might lead to interesting connections with the theory of finite frames. For a more complete treatment of the concept of finite frames, we invite the reader refer to the following textbook by Casazza and Kutyniok [1].

References

1 Peter G. Casazza and Gitta Kutyniok. *Finite frames: Theory and applications*. Springer Science & Business Media, 2012.

2 Gerald B. Folland. *A course in abstract harmonic analysis*. Studies in Advanced Mathematics, pages x+276. CRC Press, Boca Raton, FL, 1995. ISBN 0-8493-8490-7.

3 Hartmut Führ. *Abstract harmonic analysis of continuous wavelet transforms*, volume 1863 of *Lecture Notes in Mathematics*. Springer-Verlag, Berlin, 2005.

Index

A Bridge Between Lie Theory and Frame Theory: Applications of Lie Theory to Harmonic Analysis, First Edition. Vignon Oussa.

Printed and bound by CPI Group (UK) Ltd, Croydon, CR0 4YY

14/04/2025

14657191-0004